T0190328

Communications in Computer and Information Science 1819

Rationale

The CCIS series is devoted to the publication of proceedings of computer science conferences. Its aim is to efficiently disseminate original research results in informatics in printed and electronic form. While the focus is on publication of peer-reviewed full papers presenting mature work, inclusion of reviewed short papers reporting on work in progress is welcome, too. Besides globally relevant meetings with internationally representative program committees guaranteeing a strict peer-reviewing and paper selection process, conferences run by societies or of high regional or national relevance are also considered for publication.

Topics

The topical scope of CCIS spans the entire spectrum of informatics ranging from foundational topics in the theory of computing to information and communications science and technology and a broad variety of interdisciplinary application fields.

Information for Volume Editors and Authors

Publication in CCIS is free of charge. No royalties are paid, however, we offer registered conference participants temporary free access to the online version of the conference proceedings on SpringerLink (http://link.springer.com) by means of an http referrer from the conference website and/or a number of complimentary printed copies, as specified in the official acceptance email of the event.

CCIS proceedings can be published in time for distribution at conferences or as post-proceedings, and delivered in the form of printed books and/or electronically as USBs and/or e-content licenses for accessing proceedings at SpringerLink. Furthermore, CCIS proceedings are included in the CCIS electronic book series hosted in the SpringerLink digital library at http://link.springer.com/bookseries/7899. Conferences publishing in CCIS are allowed to use Online Conference Service (OCS) for managing the whole proceedings lifecycle (from submission and reviewing to preparing for publication) free of charge.

Publication process

The language of publication is exclusively English. Authors publishing in CCIS have to sign the Springer CCIS copyright transfer form, however, they are free to use their material published in CCIS for substantially changed, more elaborate subsequent publications elsewhere. For the preparation of the camera-ready papers/files, authors have to strictly adhere to the Springer CCIS Authors' Instructions and are strongly encouraged to use the CCIS LaTeX style files or templates.

Abstracting/Indexing

CCIS is abstracted/indexed in DBLP, Google Scholar, EI-Compendex, Mathematical Reviews, SCImago, Scopus. CCIS volumes are also submitted for the inclusion in ISI Proceedings.

How to start

To start the evaluation of your proposal for inclusion in the CCIS series, please send an e-mail to ccis@springer.com.

Michela Turrin · Charalampos Andriotis ·
Azarakhsh Rafiee
Editors

Computer-Aided Architectural Design

INTERCONNECTIONS: Co-computing Beyond Boundaries

20th International Conference, CAAD Futures 2023
Delft, The Netherlands, July 5–7, 2023
Selected Papers

Springer

Editors
Michela Turrin ⓘ
Delft University of Technology
Delft, The Netherlands

Charalampos Andriotis ⓘ
Delft University of Technology
Delft, The Netherlands

Azarakhsh Rafiee ⓘ
Delft University of Technology
Delft, The Netherlands

ISSN 1865-0929   ISSN 1865-0937 (electronic)
Communications in Computer and Information Science
ISBN 978-3-031-37188-2   ISBN 978-3-031-37189-9 (eBook)
https://doi.org/10.1007/978-3-031-37189-9

This Springer imprint is published by the registered company Springer Nature Switzerland AG
The registered company address is: Gewerbestrasse 11, 6330 Cham, Switzerland

Gabriela Celani and Mine Özkar, for their invaluable and always prompt support during the preparation process. Without their initial propositions, the Co-creation track would have not been ideated and their fruitful creativity when sharing thoughts has been crucial during its entire development. We express our gratitude to all members of the Scientific Committee for their insightful reviews and availability throughout the review process, which guaranteed the highest academic standards of the outcome. Finally, we gratefully acknowledge the support received from the TU Delft Faculty of Architecture and the Built Environment, and especially from the Department of Architectural Engineering and Technology.

June 2023

Michela Turrin
Charalampos Andriotis
Azarakhsh Rafiee

Organization

CAAD Futures Board

Tom Kvan	University of Melbourne, Australia
Gabriela Celani	University of Campinas, Brazil
Mine Özkar	Istanbul Technical University, Turkey

Organizing Committee

Michela Turrin (General Chair)	Delft University of Technology, The Netherlands
Charalampos Andriotis (Scientific Chair)	Delft University of Technology, The Netherlands
Azarakhsh Rafiee (Scientific Chair)	Delft University of Technology, The Netherlands
Serdar Asut (Program Chair)	Delft University of Technology, The Netherlands
Mariana Popescu (Co-creation Chair)	Delft University of Technology, The Netherlands
Marija Mateljan	Delft University of Technology, The Netherlands
Fatemeh Mostafavi	Delft University of Technology, The Netherlands
Eftychia Kalogianni	Delft University of Technology, The Netherlands

Scientific Committee

Abdullah Kara	Delft University of Technology, The Netherlands
Abel Groenewolt	Katholieke Universiteit Leuven, Belgium
Ahmad Rafi	Multimedia University, Malaysia
Alessandra Luna-Navarro	Delft University of Technology, The Netherlands
Andrei Nejur	University of Montreal, Canada
Andressa Martinez	Judson University, USA
Andrew Vande Moere	Katholieke Universiteit Leuven, Belgium
Angelos Chronis	Austrian Institute of Technology, Austria
Anja Pratschke	University of São Paulo, Brazil
António Leitão	University of Lisbon, Portugal
Ardavan Bidgoli	Carnegie Mellon University, USA
Armando Trento	Sapienza University of Rome, Italy
Arzu Gönenç Sorguç	Middle East Technical University, Turkey

Asterios Agkathidis	University of Liverpool, UK
Aurelie de Boissieu	University of Liège, Belgium
Axel Kilian	Massachusetts Institute of Technology, USA
Bastian Wibranek	University of Texas, USA
Benachir Medjdoub	Nottingham Trent University, UK
Benay Gursoy	Pennsylvania State University, USA
Berk Ekici	Izmir Institute of Technology, Turkey
Bertug Ozarisoy	University of East London, UK
Bob Martens	Vienna University of Technology, Austria
Brandon Haworth	University of Victoria, Canada
Bruno Figueiredo	University of Minho, Portugal
Burak Pak	Katholieke Universiteit Leuven, Belgium
Cemre Çubukçuoğlu	Yaşar University, Turkey
Cheney Chen	Perkins + Will, Canada
Christoph Waibel	ETH Zürich, Switzerland
Christopher Borg Costanzi	University of Southern Denmark, Denmark
Corneel Cannaerts	Katholieke Universiteit Leuven, Belgium
Dan Baciu	Delft University of Technology, The Netherlands
Danilo Di Mascio	University of Huddersfield, UK
Davide Schaumann	Technion - Israel Institute of Technology, Israel
Davide Simeone	University of Brescia, Italy
Denis Bourgeois	rd2.ca, Canada
Deok-Oh Woo	Lawrence Technological University, USA
Ding Wen 'Nic' Bao	RMIT University, Australia
Diogo Henriques	Aarhus University, Denmark
Edward Verbree	Delft University of Technology, The Netherlands
Eleonora Brembilla	Delft University of Technology, The Netherlands
Elif Erdine	Architectural Association School of Architecture, UK
Evangelos Pantazis	University of Southern California, USA
Fernando Lima	Belmont University, USA
Francesco De Luca	Tallinn University of Technology, Estonia
Frank Petzold	Technical University of Munich, Germany
Frederico Braida	Federal University of Juiz de Fora, Brazil
Gabriel Wainer	Carleton University, Canada
Heather Ligler	Pennsylvania State University, USA
Henriette Bier	Delft University of Technology, The Netherlands
Hyoung-June Park	University of Hawaii, USA
Iestyn Jowers	Open University, UK
Igor Lacroix	University of Porto, Portugal
Ioanna Symeonidou	University of Thessaly, Greece
Irene Gallou	Foster and Partners, UK

James Park	Montana State University, USA
Jeremy Ham	Independent Researcher (PhD), Australia
Johannes Braumann	University of Art and Design Linz, Austria
John Gero	University of North Carolina at Charlotte, USA
Jose Duarte	Pennsylvania State University, USA
Jose Pedro Sousa	University of Porto, Portugal
Joshua Vermillion	University of Nevada-Las Vegas, USA
Kean Walmsley	Autodesk Research, Switzerland
Konstantinos Oungrinis	Technical University of Crete, Greece
Leman Figen Gul	Istanbul Technical University, Turkey
Mariana Popescu	Delft University of Technology, The Netherlands
Marirena Kladeftira	ETH Zürich, Switzerland
Markus Hudert	Aarhus University, Denmark
Martin Bechthold	Harvard University, USA
Martin Tamke	Royal Danish Academy, Denmark
Matan Mayer	IE University, Spain
Mathew Schwartz	New Jersey Institute of Technology, USA
Maycon Sedrez	Deakin University, Australia
Mine Özkar	Istanbul Technical University, Turkey
Mohammad Bolhassani	City College of New York, USA
Nada Tarkhan	Massachusetts Institute of Technology, USA
Nathaniel Jones	Arup, USA
Nimish Biloria	University of Technology Sydney, Australia
Nuri Cihan Kayaçetin	Middle East Technical University, Turkey
Nuria Alvarez-Lombardero	Architectural Association School of Architecture, UK
Panagiotis Parthenios	Technical University of Crete, Greece
Pedro Azambuja Varela	University of Porto, Portugal
Peter Buš	Tsinghua Shenzhen International Graduate School, China
Peter Russell	Tsinghua Shenzhen International Graduate School, China
Peter von Buelow	University of Michigan, USA
RaEd QaQish	Canadian University Dubai, United Arab Emirates
Rahman Azari	Pennsylvania State University, USA
Ramesh Krishnamurti	Carnegie Mellon University, USA
Ramon Weber	Massachusetts Institute of Technology, USA
Renate Weissenböck	Frankfurt University of Applied Sciences, Germany
Rhys Goldstein	Autodesk Research, Switzerland
Robert Klinc	University of Ljubljana, Slovenia
Rudi Stouffs	National University of Singapore, Singapore

Samuel Sohn	Rutgers University, USA
Sara Eloy	University Institute of Lisbon, Portugal
Sarah Mokhtar	Massachusetts Institute of Technology, USA
Saurabh Mhatre	Harvard Center for Green Buildings and Cities, USA
Sevil Yazici	Istanbul Technical University, Turkey
Seyran Khademi	Delft University of Technology, The Netherlands
Shady Attia	University of Liège, Belgium
Sheng-Fen Chien	National Cheng Kung University, Taiwan ROC
Spyridon Ampanavos	Perkins&Will, USA
Stijn Brancart	Delft University of Technology, The Netherlands
Stylianos Dritsas	Singapore University of Technology and Design, Singapore
Sule Tasli Pektas	OSTIM Technical University, Turkey
Sven Schneider	Bauhaus University Weimar, Germany
Taysheng Jeng	National Cheng Kung University, Taiwan ROC
Thomas Dissaux	University of Liège, Belgium
Tomohiro Fukuda	Osaka University, Japan
Tong Wang	Delft University of Technology, The Netherlands
Tuba Kocaturk	Deakin University, Australia
Tugrul Yazar	Istanbul Bilgi University, Turkey
Vesna Stojakovic	University of Novi Sad, Serbia
Vinu Subashini Rajus	Carleton University, Canada
Wilson Florio	University of Campinas, Brazil
Yun Kyu Yi	University of Illinois, USA

Contents

Urban Models and Analysis

Urban Design

Digital Design, Materials and Fabrication

Spatial Information, Data and Semantics

Building Data Analysis, Visualisation, Interaction

Building Massing and Layouts

Algorithmic Architectural Design

The Marching Shape 3D

Extensions of the Ice Ray Shape Grammar

Alexandros Tsamis and Constantina Varsami[✉]

Rensselaer Polytechnic Institute, Troy, NY, USA
constantinavarsami@gmail.com

Abstract. In recent years, voxel-based approaches have been explored to address data distributions related to space and form in diverse design disciplines. While originally used for the digital reconstruction of the human body, they are now widely used in design to represent intensive spatial properties. However, a voxel-based design process can limit design outcomes due to a predetermined voxel size and space resolution. Although variable voxel density subdivision algebras exist, a generalized algebra for volume subdivision of any size and shape as well as computing between individual subdivided parts has not been implemented. This paper presents a generalized volumetric space dividing algebra for closed, convex Polyhedra of any shape, which extends the Ice-Ray Shape Grammar and demonstrates weighted 3D boundary subdivision. Specifically, we present an algebra for subdividing any closed convex polyhedron, we demonstrate a boundary-weight algebra that interrelates properties (weights) with a variably subdivided space and allows for their direct manipulation in space as the design process takes place. We also present rules for extracting isocurves and reconstructing iso-surfaces in a variably subdivided, weighted 3D space. Finally, a use case of The Marching Shape 3D is demonstrated in a renewable energy based design process. Algebras and schemas are shown for a general shape grammar as well as for a digital computing platform.

Keywords: Marching Shape · Voxel Space · Shape Grammars

1 Introduction

In recent years, voxel-based approaches for space subdivision have gained popularity as a viable method for addressing the distribution of information related to space and form across diverse design disciplines. Initially, voxels were used to digitally reconstruct two or three-dimensional geometries, such as parts of the human body. (Lengsfeld et al. 1998). However, today they are extensively used in design fields to represent and make accessible to the designer different kinds of intensive spatial properties. For instance, voxels are used as placeholders of spatial data that represent material properties (Michalatos and Payne, 2013), (Tsamis, 2010), (Oxman, 2010), or energy accessibility in buildings and urban configurations (Kadhim PhD, 2015) (Jyoti, 2016) (Kosicki et al, 2020).

© The Author(s), under exclusive license to Springer Nature Switzerland AG 2023
M. Turrin et al. (Eds.): CAAD Futures 2023, CCIS 1819, pp. 3–20, 2023.
https://doi.org/10.1007/978-3-031-37189-9_1

In this paper we extend the Ice Ray Shape Grammar (Stiny, 1977) and the Marching Shape (Tsamis, 2017) in three dimensions, incorporating schemas that relate boundaries and weights to formalize a general space dividing algebra for closed convex polyhedra. Algebras and schemas are shown as a general shape grammar as well as for a digital computing platform. Similar to how The Marching Shape (Tsamis, 2017) demonstrates the applicability of the weighted space dividing algebra on material distributions, we are focusing on energy distributions as morphological drivers for urban design.

Knowles' solar envelopes (Knowles, 1978) are an early example of how energy availability and space subdivision can be used as a form-finding driver for buildings and neighborhoods. In urban design several more cases have employed volumetric units to determine energy performance characteristics of the built environment. For example, in "High Rise Morphologies" (Jyoti, 2016) voxels are associated with energy related parameters like solar irradiation and direct solar exposure. Energy values for each voxel are compared to specified thresholds which in turn dictate whether each voxel will remain in place or be eliminated. An equivalent approach is implemented in "HYDRA" (Kosicki et al. 2020) where voxels are used to optimize city configurations (Fig. 1).

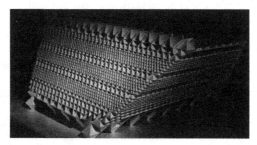

Fig. 1. Designing with variable units. Knowles, Ralph., Energy and Form, Fig. 7.14 (p. 146), © 1975 Massachusetts Institute of Technology, by permission of The MIT Press.

Integrating a voxel approach often leads to energy-efficient configurations, but the uniformity in shape and size of voxel units restrict the resulting design outcomes. From a designer's standpoint, the design process becomes a pre-determined, **countable**, combinatorial process that limits the design outcome to the initial voxel space resolution. Essentially, the number of possible "solutions" for design or even optimization is constrained by the number of configurations that a fixed voxel space can provide for any given project. The volumetric equivalent of variable grid resolution is well established today and known as octrees (Meagher, 1982). Octrees allow for more flexibility but only in terms of resolution, not shape. This means that the potential design iterations might not be countable in the same way they are with a fixed resolution, however, they are still bound by the shape of an infinitesimal small cubic unit. This paper formalizes an algebra for variable volumetric space dividing that addresses the finite nature of voxel and octree like approaches.

The general concept of *where needed and when needed* space subdivision and unit variability was demonstrated by Stiny in his Ice Ray shape grammar (Stiny, 1977).

Although this grammar was originally derived from Chinese lattices, it has been subsequently reinterpreted as a grammar for local subdivision of space and has been applied to the derivation of a painting (Stiny, 2006), an architectural plan of a building, or even a city (Duarte, 2011). The ice ray provides a way to formally describe a generalized space dividing grammar in 2D for polygons of three, four, and five vertices. It was recently extended by Tsamis in the Marching Shape (Tsamis, 2017), to account for all types of closed, convex polygons and demonstrated how relationships between weights and boundaries can be developed. The Marching Shape is, in essence, a boundary-weight algebra that variably subdivides space as the design process evolves, while allowing for properties to be referenced for each subdivision and be directly manipulated in space without the limitations of a predefined fixed grid structure.

Although the Marching Shape (Fig. 2) has been discussed as a generalized space dividing algebra of weighted units, its formalism has only been demonstrated in 2D. In this paper we extend this formalism to the third dimension and discuss affordances and limitations relative to the original two-dimensional work.

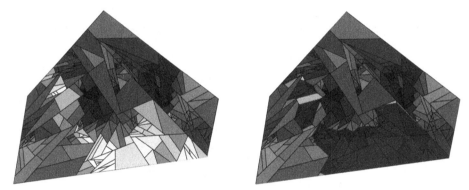

Fig. 2. Alexandros Tsamis, The Marching Shape (2017). Left: extended weighted Ice Ray. Right: Marching Shape, plotted iso-value on weighted Ice Ray.

2   Marching Shape 3D

The Marching Shape 3D (MS3D) extends the Ice Ray Grammar by following its overall formalism and developing a boundary-weight algebra in three-dimensional space. Unlike voxel-based modeling, MS3D allows space to be divided into irregular polyhedral units of varying resolution. The volume unit properties, including number, size, shape, location, and weights, are not fixed and can be dynamically defined by the designer before or during the design process. The MS3D grammar is not constrained to regular grids, but a 3D grid (a voxel space) can still be addressed as a special case of fixed-size cubic units.

2.1   The Marching Shape 3D - Subdivision Rules and Schemas

The general rules and schemas developed in the original Marching Shape (Tsamis, 2017) describe convex polygon subdivisions from the recursive application of which the two-dimensional design space can be variably (non-uniformly and not only orthogonally) subdivided. The Marching Shape rules and schemas are extended here to describe closed, convex polyhedra subdivisions enabling the three-dimensional space to be variably (non-uniformly and not only orthogonally) subdivided.

Figure 3 enumerates subdivision rules for Polyhedra PH(4), PH(5), PH(6) and PH(7). The pyramid PH(4), for example, can produce 2 possible rules that subdivide it either into a tetrahedron and a pentahedron or into two pentahedra. Similarly, a pentahedron can produce 4 possible rules that subdivide it either into a tetrahedron and a hexahedron, or into two pentahedra, or into a pentahedron and a hexahedron, or into two hexahedra.

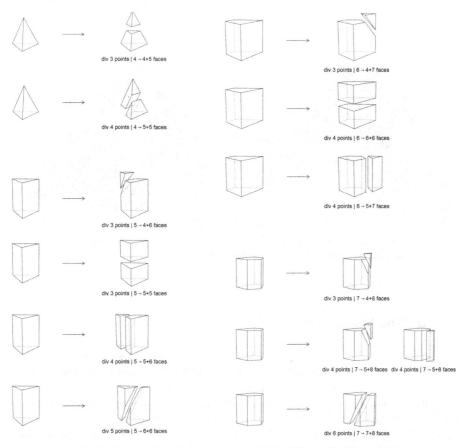

Fig. 3. Ice Ray 3D. Subdivision rules on PH(4), PH(5), PH(6), and PH(7).

As it is the case in the extended Ice Ray described in the original Marching Shape, a subdivision of a polyhedron PH(N) will produce at least one polyhedron of PH (N +

1). For this reason, an extended (un-bound) MS3D grammar is described with the use of a schema instead of an explicit list of volumetric shape rules. For any convex, closed polyhedron **PH** and a cutting plane C_p, the general MS3D grammar schema is:

$$\mathbf{PH(X_f) \rightarrow PH(Y_f) + PH(Z_f)} \text{ where:} \tag{1}$$

a. **PH** is a closed, convex polyhedron.

b. $\mathbf{X_f, Y_f, Z_f}$ are positive integers greater or equal to 4 representing the number of faces of the polyhedron and

c. $\mathbf{X_f = (Y_f + Z_f) - (p + 2)}$ where:

p is an integer defining the number of points resulting from the intersection of the polyhedron **PH** with the cutting plane C_p that subdivides it. Constraining **p** ($3 \leq p \leq$ **Xf**) ensures that the cutting plane will lie inside the polyhedron. For example, for the case of a polyhedron PH(4) with a cutting plane C_p ($3 \leq p \leq 4$) we get:

$$\text{for } p = 3, C_3 : PH(4) \rightarrow PH(4) + PH(5)$$

$$\text{for } p = 4, C4 : PH(4) \rightarrow PH(5) + PH(5)$$

Early work of Lionel March (1977) came across matters described by this schema. Particularly, while subdividing polyhedra it observed that similar shapes are produced by different sequences of cuts. Aside from this observation, polyhedra subdivision characteristics were not further explored by March. The general schema proposed by the MS3D grammar specifically addresses the interrelation between original and resulting polyhedra, given the number of intersection vertices of a polyhedron with the cutting plane.

By consecutively applying the MS3D rules, an original volume can be recursively subdivided while the design process is taking place (Fig. 4).

Fig. 4. Ice Ray 3D. Volumetric recursive subdivision.

For a **computer application** the MS3D starts with a single convex polyhedron. Similarly to the two-dimensional algorithm, all subsequently produced polyhedra are distinguished by their unique identity number (01, 02, 03) and are computationally specified as jagged arrays (Arr[01], Arr[02], Arr[03],…). The members of each jagged array are as many as the number of the polyhedron's faces; each one constitutes a collection of vertices which corresponds to the respective face of the polyhedron it represents. The vertices are stored in each member array in an ordered way (clockwise or counterclockwise). Since any subdivision performed on any polyhedron substitutes the original volume unit with two new ones, it similarly substitutes the original jagged array with two new ones recursively (Fig. 5):

$$\text{Arr}[01] \rightarrow \text{Arr}[02] + \text{Arr}[03] \tag{2}$$

$$\text{Arr}[02] + \text{Arr}[03] \rightarrow \text{Arr}[02] + \text{Arr}[04] + \text{Arr}[05] \tag{3}$$

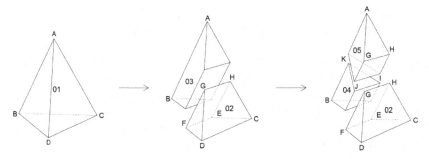

Fig. 5. Ice Ray 3D. Computer application. Ordered vertices stored in member arrays.

An important condition for preserving the topology of the subdivided unit volume is for both new arrays to maintain the vertex arrays for each of the faces of the polyhedron ordered (clockwise or counterclockwise). For example shape 01 is represented with the following array structure:

$$\text{Arr}[01] = [\text{Arr}(A, C, D), \text{Arr}(C, B, D), \text{Arr}(A, D, B), \text{Arr}(A, B, C)]$$

Arr[01] → Arr[02] + Arr[03] where :

$$\text{Arr}[02] = [\text{Arr}(G, H, C, D), \text{Arr}(C, E, F, D), \text{Arr}(G, D, F), \text{Arr}(G, F, E, H), \text{Arr}(H, E, C)]$$

$$\text{Arr}[03] = [\text{Arr}(A, H, G), \text{Arr}(G, H, E, F), \text{Arr}(E, B, F), \text{Arr}(A, G, F, B), \text{Arr}(A, B, E, H)]$$

In the case where a polyhedron does not get subdivided, but a new subdivision vertex lies on one of its edges, that point is inserted into the corresponding member array of the polyhedron and becomes part of its description. (Fig. 5 right)

Arr[02] + Arr[03] → Arr[02] + Arr[04] + Arr[05] where :

Right Arr[02] = [Arr(G, H, C, D), Arr(C, E, F, D), Arr(G, D, F, J), **Arr(G, J, F, E, I, H)**, Arr(H, I, E, C)]

Arr[04] = [Arr(J, I, E, F), Arr(E, B, F), Arr(K, J, F, B), Arr(K, B, E, I), Arr(K, I, J)]

Arr[05] = [Arr(A, H, G), Arr(G, H, I, J), Arr(I, K, J), Arr(A, G, J, K), Arr(A, K, I, H)]

Unlike voxels, which all have the same size and shape along with a predetermined relationship, units generated through the application of the rules and schemas of the MS3D allow the designer to derive a variably subdivided space of variable density, during the design - optimization process.

2.2   The Marching Shape 3D: Weights/Properties

Precedent examples of energy based urban design use voxel subdivisions as placeholders for energy quantities. Specifically, Pratt and Bosworth (2011) map on each voxel simulated yearly building energy use metrics by means of assigning properties in the form of colors, integrating energy data and projecting building performance insight on local topology. The MS3D uses the same principles of weight to boundary relations, without the fixed resolution of a voxel space or the fixed shape of a voxel.

Similar to the original Marching Shape (Fig. 6), for MS3D volumetric shape rules only concern their subdivision, therefore, any weight rule associated with them can be considered as a special case of a general color grammar (Knight, 1989) (Fig. 7).

Fig. 6. Weighted Ice Ray. Marching Shape 2017

Figure 6 enumerates weighted subdivision rules for Polyhedra PH(4), PH(5), PH(6) and PH(7) according to the Marching Shape 3D. A general schema would be:

$$PH(X_f), W_A \rightarrow PH(Y_f), W_B + PH(Z_f), W_C \text{ where:} \qquad (4)$$

a. W_A, W_B and W_C are weights that can be represented as numerical values or visualized as colors.

For a computer application, there are two schemas of interaction between weights and shapes. Color weights are defined as R, G, B, A vectors assigned to each vertex. According to the first schema, every time a polyhedron gets subdivided the weight assigned to it gets associated to all its vertices. Since those vertices are also inserted

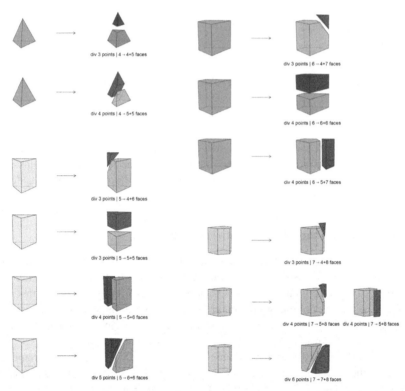

Fig. 7. Examples of weighted Ice Rays 3D for PH(4), PH(5), PH(6), PH(7).

into the descriptions of neighboring polyhedra, in the resulting configuration, weights associated with that polyhedra are re-computed to account for the weights of the new vertices. For example, in Fig. 8 (right) vertices I, J, K get the weight that was assigned to polyhedron 04 and transfer it to adjacent polyhedra affecting their weights as well.

According to the second schema, every time a polyhedron gets subdivided the weights assigned to the vertices that result from the intersection of the shape with the cutting plane are determined by the weight (R, G, B, A) of the cutting plane itself. By employing a non-uniformly colored cutting plane different colors can be assigned to each new vertex. The weight associated with each of the resulting polyhedra is, in turn, the average weight of the polyhedron's vertices. Since the new vertices are also inserted into the descriptions of neighboring polyhedra, weights associated with that polyhedra are re-computed to account for the weights of the new vertices (Fig. 9).

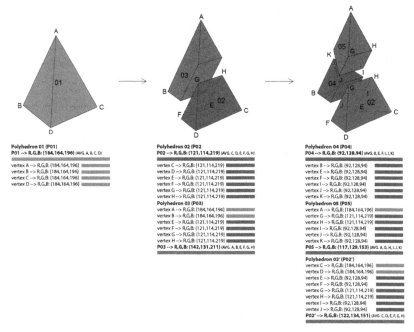

Fig. 8. Weighted Ice Ray 3D. Computer application. First schema of weight-shape interaction.

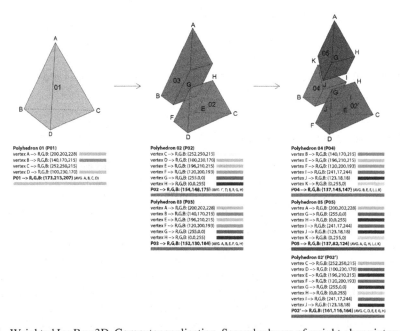

Fig. 9. Weighted Ice Ray 3D. Computer application. Second schema of weight-shape interaction.

A special case of the second schema is when the weight of the cutting plane is "transparent". In this case weights are not assigned to the new vertices by the cutting plane; they are instead computed proportionally to the distance of subdivision. This special case can be used to average weights between polyhedra as more subdivisions are added (Fig. 10).

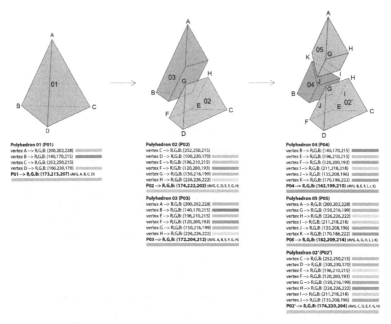

Fig. 10. Weighted Ice Ray 3D. Computer application. Second schema (a special case).

2.3   The Marching Shape 3D: Case Study

The special case of the second schema is implemented on a case study (Fig. 11); a cluster of building blocks located in Manhattan. Specifically, annual daylight exposure is initially simulated using Energy + on the original polyhedron (Fig. 11.a.) and the results are associated in the form of weights with its vertices. The polyhedron is then recursively subdivided by a transparent cutting plane (Fig. 11.b.) and the encapsulated energy gets averaged between the multiple polyhedra. The recursive nature of subdivisions allows for an increase in resolution and a unique definition of unit shape during the design process. The designer can decide where to increase the density of unit information placeholders, depending on their preference. This design decision may be influenced by the distribution of weights in the design space. For example, the designer may choose to increase resolution in areas where more energy is required or desired. Polyhedra can also be removed. In this case the dayligt analysis will be recalculated to account for the new configuration (Fig. 11.c.). The subdivision process can then resume (Fig. 11.d., Fig. 11.e.) distributing the newly simulated energy potential to the remaining polyhedra.

The sequence of operations can be repeated until a configuration that satisfies all design requirements is acquired (Fig. 11.f.). The result of the design process is a volumetric configuration for which the designer understands the energy needs (in the form of heat gains) as the design process evolves.

Fig. 11. Recursive weighted Ice Ray 3D.

3   Marching Shape 3D: Deriving Iso-Surfaces

Voxel modeling provides a unique functionality - the ability to reconstruct 3D iso-surfaces from Voxel weight data. The Marching Cubes algorithm (Lorensen, Cline, 1987) allows designers to query the weights assigned to a 3D voxel space and generate a boundary (an iso-surface) that answers the question: where in space is weight W equal to a given iso-value? The algorithm selects from a predetermined set of 256 possible configurations and determines the cases based on the number of vertices of a single cube that are greater than or equal to the desired iso-value. Through symmetry and rotation, the cases can be reduced to 15 (Fig. 12).

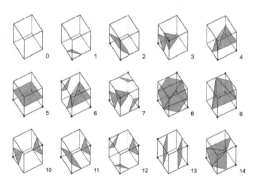

Fig. 12. Original Marching Cubes Cases. Lorensen and Cline, 1987

The Marching Cubes algorithm has been employed in a recent study (Kadhim PhD, 2015) to extract building envelopes from a weighted voxel design space and interrelate urban form with energy. The MS3D also enables the derivation of iso-value computations to derive iso-surfaces in space, but it eliminates the predetermined shape of a voxel and the restrictions of a fixed resolution. The number of MS3D cases of a polyhedron with X number of vertices can be calculated by either one of the following formulas:

$$\sum_{con} = 2^X \text{ where:} \tag{5}$$

Σ_{con} is the sum of all the possible configurations for a specific polyhedron.

X is a positive integer greater or equal to 4, representing the number of the polyhedron's vertices and

$$\sum_{k=0}^{n} \binom{n}{k} = \sum_{k=0}^{n} (n!/[k!(n-k)!]) \text{where:} \tag{6}$$

n is equal to **X**, the positive integer greater or equal to 4, representing the number of the polyhedron's vertices

k is a subset of **n** (k ≤ n); it represents the number of the polyhedron's vertices whose value is greater or equal than the currently desired iso-value $\binom{n}{k} = {}_nC_k$ (n choose k) equals the number of ways to choose the unordered subset k from the fixed set of n elements.

To demonstrate the process of deriving iso surface boundaries from closed convex polyhedra a triangular pyramid, a rectangular pyramid, a pentahedral prism, and a heptahedron are used as examples. Figure 13 indicates the full set of iso-surfaces that can be derived by applying the MS3D on the first three shapes and a few of the iso-surfaces that can be derived by applying it on the heptahedron. The total number of cases for the triangular pyramid with 4 vertices are $2^4 = 16$, for the rectangular pyramid with 5 vertices are $2^5 = 32$, for the prism with 6 vertices are $2^6 = 64$, and for the heptahedron with 10 vertices are $2^{10} = 1024$. Cases that are identical because of symmetry are not shown.

Since the polyhedra that are subdivided by the MS3D are not characterized by a fixed number of vertices, and since new vertices can be embedded along their edges as the subdivision process progresses, MS3D generalizes with a rule that is applicable to all polyhedra and helps calculate boundaries given specified iso-values. MS3D considers each polygonal face of each polyhedron as an "unfolded" polyline, determines if a point of that polyline is equal to the desired iso-value, and uses one of the two original Marching Shape rules (Fig. 14) to derive boundaries. The boundaries defined for a specified iso-value compose the exterior boundaries of the desired iso-surfaces.

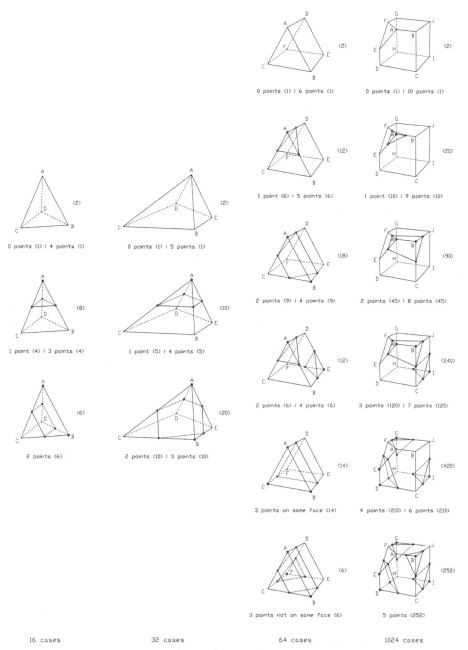

16 cases   32 cases   64 cases   1024 cases

Fig. 13. Marching Shape 3D. Full enumeration of 16 cases on a triangular pyramid with 4 vertices, full enumeration of 32 cases on a rectangular pyramid with 5 vertices, full enumeration of 64 cases on a prism with 6 vertices, partial enumeration of 1024 cases on a heptahedron with 10 vertices.

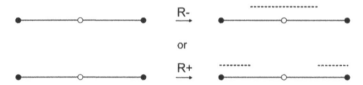

Fig. 14. Marching Shape Rules.

Each of the Marching Shape rules can be selected independently and be applied regardless of the number of vertices in the definition of the specific polygon. The rule can be applied repeatedly (without breaking the overall topology) as the designer keeps evaluating and subdividing further, addressing the ambiguity of connecting identified points or calculating boundaries in some polygons. It should be noted here that the MS3D iso-surface rule is identical to the original Marching Shape rule from 2017, only applied on all faces of any convex polyhedron. A similar approach has been demonstrated in the Cubical Marching Squares (Ho et al. 2005) but solely to extract iso-surfaces from cubical volume units.

The iso-surface rule for the MS3D can be articulated as: **pick or find any point on an edge of each face of a polyhedron (including start and end points) and connect it to any other point on any other edge of the same face.** If we analytically go through all possible results, we get to a list of possible shape grammar rules that showcase how we can derive iso-curves on each polyhedron's face (Fig. 15). The rules shown here for a pentagon are adopted from the Marching Shape and show how from a single face of a polygon one can derive iso-surface boundaries (point - lines - planes).

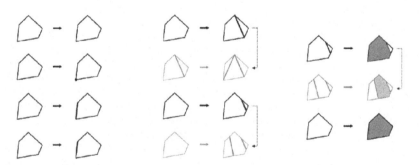

Fig. 15. Points, lines, and planes in the Marching Shape for a pentagon.

3.1 Iso-surface Reconstruction from Iso-Curve Values

In order to reconstruct iso-surfaces from the iso-curve boundaries derived from the MS3D, the resulting iso-curves of each of the polyhedron's faces are joined to form the exterior boundary of the desired iso-surfaces. If the number of boundary curves is greater than three and they are not coplanar, a triangulation method is employed

  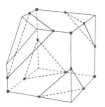 

Fig. 16. The Marching Shape 3D. Calculating iso-surfaces (red) based on assigned weights (green) on tetrahedra and heptahedra.

to appropriately define iso-surfaces. This ensures an uninterrupted poly-surface whose boundaries intersect with the faces of the volume unit. (Fig. 16).

The MS3D is applicable to all closed and convex polyhedra. The overarching goal is to derive iso-surfaces (boundaries) based on a distribution of properties (weights) within a MS3D canvas. For a **computer application** the aforementioned rule is also applicable. The algorithm first determines if a polyhedron has at least one vertex whose weight value is equal or bigger to the desired iso-value. Then, it examines each face of the polyhedron going through all of its ordered points (clockwise or anti-clockwise), finds the appropriate locations for the desired iso-values and connects between them.

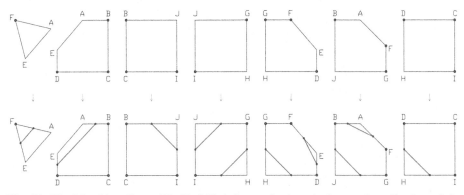

Fig. 17. The Marching Shape 3D. Unfolded faces of a heptahedron and application of the Marching Shape on each face.

To determine whether to apply a triangulation algorithm or if the iso-surface is planar, the number of points defined by the specified iso-value is counted on all faces of the polyhedron (see Fig. 17 and 18). The algorithm can be continuously applied as the designer subdivides volumes. Numerous algorithms exist for optimal triangulation of points in 3D space, including the Delaunay Refinement approach (Shewchuk, 2014) (Foteinos et al. 2011) which generates a mesh that satisfies specific constraints, or the triangulation algorithm used in the Marching Cube (Dey et al. 2008) or the Bowyer-Watson algorithm (Rebay, 1993) that apply Delaunay triangulation principles to a finite set of points irrespective of their dimensionality. The derived iso-surfaces from the MS3D are not necessarily planar, so the three-dimensional canvas and the iso-surfaces cannot collapse into a single entity, unlike the original Marching Shape. The hierarchy between iso-surface and canvas must be maintained, so MS3D iso-surfaces cannot be used to further cut the canvas. However, the user can continue to subdivide the space during the design process (as demonstrated in Fig. 11) and iso-surfaces are recalculated each time the canvas changes. Using our previous case, Fig. 18 isolates a single building volume of the final configuration of Fig. 11 and demonstrates how the energy content of the volume is initially aggregated at its vertices. The figure also demonstrates an iso-surface that subdivides the volume indicatively at 5.300.000 Watts. Such a study may help the designer deliberately subdivide the polyhedron further depending on whether the energy content exceeds or not the designated threshold value. Alternatively, it may assist designers customize the architectural program of the building based on the distribution of the energy contained in each distinct building envelope.

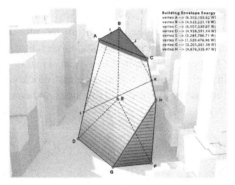

Fig. 18. The Marching Shape 3D. Calculating iso-surfaces (red) based on assigned weights (green) on a hexahedron.

4   Conclusions

The use of voxel-based modeling has proven to be an effective way to incorporate intensive spatial properties (like energy content) into the design process. This has led to the development of computer applications that can represent and manipulate both form and property concurrently. However, we believe that the voxel structure imposes limitations on the design process by making it combinatorial. To address this, the MS3D was introduced as a new type of variable shape spatial subdivision algebra that extends the Ice Ray Shape grammar to 3D space. This paper presented a formal algebra for subdividing any closed convex polyhedron, a boundary-weight algebra that interrelates weights with a variably subdivided space, and rules for extracting isocurves and reconstructing iso-surfaces from them in a variably subdivided weighted 3D space. Rules were provided for a general shape grammar as well as a digital computing platform. Although energy was used as the base case in this paper, the MS3D can be applied to incorporate other intensive spatial characteristics. It offers an alternative to volume-based modeling that maintains the weight to boundary relationships and other operations involved in voxel-based software without limiting designers to predetermined resolutions and "zero-dimensional calculations" (Stiny, 2006).

References

Dey, T.K., Levine, J.A.: Delaunay meshing of isosurfaces. Visual Comp. **24**(6), 411–422 (2008)

Duarte, J.P., Beirão, J.: Towards a methodology for flexible urban design: designing with urban patterns and shape grammars. Environment and Planning B: Planning and Design (2011)

Foteinos, P., Chrisochoides, N.: Dynamic parallel 3D delaunay triangulation. In: Quadros, W.R. (eds) Proceedings of the 20th International Meshing Roundtable. Springer, Berlin, Heidelberg. (2011). https://doi.org/10.1007/978-3-642-24734-7_1

Ho, C.C., Wu, F.C., Chen, B.Y., Chuang, Y.Y., Ouhyoung, M.: Cubical marching squares: adaptive feature preserving surface extraction from volume data. Comp. Graphics Forum **24**(3), 537–545 (2005). https://doi.org/10.1111/j.1467-8659.2005.00879.x

Jyoti, A.: High Rise Morphologies: Architectural Form Finding in a Performative Design Search Space of Dense Urban Contexts (2016)

Kadhim, N.M.M., Mourshed, M., Bray, M.: Automatic extraction of urban structures based on shadow information from satellite imagery. In: 14th International Conference of IBPSA - Building Simulation 2015, BS 2015, Conference Proceedings, December, pp. 2607–2614 (2015)

Knight, T.W.: Color grammars: designing with lines and colors. Environ. Plann. B. Plann. Des. **16**(4), 417–449 (1989). https://doi.org/10.1068/b160417

Knowles, R.L.: Energy and Form: An Ecological Approach to Urban Growth. MIT Press, Cambridge, Massachusetts (1978)

Kosicki, M., Tsiliakos, M., Tsigkari, M.: HYDRA distributed multi-objective optimization for designers. In: Gengnagel, C., Baverel, O., Burry, J., Ramsgaard Thomsen, M., Weinzierl, S. (eds.) DMSB 2019, pp. 106–118. Springer, Cham (2020). https://doi.org/10.1007/978-3-030-29829-6_9

Lengsfeld, M., Schmitt, J., Alter, P., Kaminsky, J., Leppek, R.: Comparison of geometry-based and CT voxel-based finite element modelling and experimental validation. Med. Eng. Phys. **20**(7), 515–522 (1998). https://doi.org/10.1016/S1350-4533(98)00054-X

Lorensen, W.E., Cline, H.E.: Marching cubes: a high resolution 3D surface construction algorithm. In: Computer Graphics, **21**(4) (1987)

March, L., Earl, C.F.: On counting architectural plans. Environ. Plann. B. Plann. Des. **4**(1), 57–80 (1977). https://doi.org/10.1068/b040057

Meagher, D.: Geometric modeling using octree encoding. Comput. Graphics Image Process. **19**(2), 129–147 (1982). https://doi.org/10.1016/0146-664X(82)90104-6

Oxman, N.: Material-based Design Computation. Ph.D. thesis, MIT (2010)

Pratt, K.B., Bosworth, D.E.: A method for the design and analysis of parametric building energy models. In: Proceedings of Building Simulation 2011: 12th Conference of International Building Performance Simulation Association, Schumacher 2010, pp. 2499–2506 (2011)

Rebay, S.: Efficient unstructured mesh generation by means of delaunay triangulation and bowyer-watson algorithm. J. Comput. Phys. **106**(1), 125–138 (1993)

Shewchuk, J.R.: Delaunay refinement algorithms for triangular mesh generation. Comput. Geometry: Theory Appl. **47**(7), 741–778 (2014)

Stiny, G.: Ice-ray: a note on the generation of chinese lattice designs. Environ. Planning B **4**, 89–98 (1977)

Stiny, G.: Shape: Talking about Seeing and Doing. The MIT Press, Cambridge, MA (2006)

Tsamis, A.: Go Brown. Inner-disciplinary Conjectures in Architectural Design (AD): Ecoredux. Design Remedies for an Ailing Planet, Lydia, K. (Ed), John Wiley and Sons, London (2010)

Tsamis, A.: The marching shape. In: Çağdaş, G., Özkar, M., Gül, L.F., Gürer, E. (eds.) CAADFutures 2017. CCIS, vol. 724, pp. 366–380. Springer, Singapore (2017). https://doi.org/10.1007/978-981-10-5197-5_20

Between System and Improvisation: The Design Language of Donald Judd's "100 Untitled Works in Mill Aluminum"

Mahyar Hadighi[1(✉)] and Mehrdad Hadighi[2]

[1] Texas Tech University, El Paso, TX 79901, USA
mhadighi@ttu.edu
[2] The Pennsylvania State University, University Park, PA 16802, USA

Abstract. In this paper, shape grammar is implemented as a computational design methodology to analyze the design language of Donald Judd's "100 untitled works in mill aluminum"—permanently installed at the Chinati Foundation in Marfa, TX. Judd (1928–1994), an American artist, writer, and critic installed 100 aluminum objects—with the exact same outer dimensions ($41'' \times 51'' \times 72''$) but unique configuration, "all made from ½" mill aluminum plates—over a four-year period from 1982 through 1986 in two former artillery sheds of a de-commissioned US military base. The question we are seeking to answer is whether these 100 works belong to a system, and, if so, what that system is and what delimits the system to produce 100 objects and no more. Our hypothesis, having spent considerable time examining the works and exploring the archival materials, is that there is a system, but the system is tempered by improvisational moments at multiple junctures in the project. We are interested in deciphering the system, but also the moments of artistic improvisation by Judd. The project is significant in relation to serial art works of the 50's and 60's. We have been calling them "serial" assuming there is a serial order and a system in their repetition and variation, but this very issue has not been researched in the context of the present work.

Keywords: Donald Judd · Design System · Shape Grammars

1 Introduction

1.1 Serial Art

Mel Bochner, the renowned artist, and theorist, in his 1967 *ARTFORUM* essay, titled "The Serial Attitude," mentions Donald Judd's painted wall pieces as employing serial logic [1]. In a counter-essay, "Serial Imagery" published in 1968 in *ARTFORUM*, John Coplans did not mention Judd at all [2]. In fact, in a book/exhibition catalog of the same title that included the essay as its main feature, Coplans added a "history" section to the start of the essay and in it, he specifically noted: "Although the work of such sculptors as Donald Judd has been described as Serial, this is incorrect" [2, p. 9]. Noting that both essays reflect on Judd's earlier work (wall pieces), we seek to examine one

M. Turrin et al. (Eds.): CAAD Futures 2023, CCIS 1819, pp. 21–35, 2023.
https://doi.org/10.1007/978-3-031-37189-9_2

of his late works, permanently installed at the Chinati Foundation in Marfa, Texas, "100 untitled works in mill aluminum," (Fig. 1) in the context of serial works and to analyze and describe it in the spectrum between improvisational and systemic order-types, using shape grammar as an analytical and generative tool. Not long after Bochner's and Coplans' essays, Stiny and Gips published their "Shape Grammar and the Generative Specification of Painting and Sculpture" in 1971. Their aim, as stated, was to "use formal, generative techniques to produce good art objects" [3, p.125]. In other words, they developed a grammar to generate art. Subsequently, Knight developed a transformation grammar to analyze art objects post-facto. In her case, the grammar was used to describe "stylistic change in design" and to "analyze transformations of the De Stijl style of painting in the works of two artists" [4, p. 51]. Since then, as a computational design theory, shape grammar has been used in art and architectural analysis when a pattern of design characteristics or a stylistic repetition of shapes is evident. Following this background, we utilize shape grammar as an analytical tool to analyze Judd's work, and subsequently to generate designs within the deciphered grammar.

Fig. 1. Donald Judd, 100 untitled works in mill aluminum, 1982–1986. Permanent collection, the Chinati Foundation, Marfa, Texas. Photo by Douglas Tuck, courtesy of the Chinati Foundation. Donald Judd Art © 2023 Judd Foundation/Artists Rights Society (ARS), New York.

Judd (1928–1994), an American artist, writer, and critic installed 100 aluminum objects—with the exact same outer dimensions (41″ × 51″ × 72″) but unique configuration, all made from "½" mill aluminum plates—over a four-year period from 1982 through 1986 in two former artillery sheds of a de-commissioned US military base in Marfa, Texas. The question we are seeking to answer is whether these 100 works belong to a system, and, if so, what that system is and what delimits the system to produce 100 objects and no more, identifying whether it belongs to the "serial" works analyzed by Bochner and Coplans. Given that Judd's work may be considered a part of serial art with strong evidence of shared shapes and transformation rules, a shape-grammar-based methodology is appropriate for both understanding the system and identifying and explaining the moments of improvisation.

1.2 Shape Grammar

Shape grammars in computation are a specific class of production systems based on an initial shape, or a set of finite shapes, and transformational shape rules [3]. In other words, a shape grammar is a rule-based system for analyzing, describing, and generating visual or spatial designs. Shape grammars began as a concept, with early applications focused on decorative arts (1977), architecture (1978), design (1998), fine arts (1989), and eventually urban design (2011). There are two main shape grammar categories: analytical and synthetic. Analytical grammars are used to analyze historical styles or languages of design, whereas synthetic grammars are used to create new and original designs [5]. Traditionally, there has been a separation between the two. However, in "Customizing Mass Housing: Toward a Formalized Approach" Duarte demonstrated that the two are, in fact, linked [5]. He argued that analytical grammars can be used to decode historical precedents and grammatical transformations to encode "synthetic" grammars.

This methodology, in general, is used for one or more of three reasons: (1) to describe or analyze a design, (2) to produce a design or a series of designs, and (3) to determine the design group to which a given design belongs. However, the shape grammar concept can also be used effectively to study design principles used in the past with the purpose of re-using them in contemporary design, and likewise, to adapt earlier designs for modern purposes. Here, we are expanding the use of shape grammar to include the analysis of Donald Judd's design language in his "100 untitled works in mill aluminum," not only to understand his design language (through an analytical grammar), but also to understand and explain moments of improvisation in his work. To understand moments of improvisation, shape grammar is used as a generative tool to create designs beyond the 100 objects installed in the two artillery sheds.

The shape grammar concept has been used as a powerful tool to analyze the art and architecture of the past, both for historical reasons and to produce new designs based on the predominant design principles of earlier periods. Accordingly, in his article "More Than the Sum of Parts: The Grammar of Queen Anne Houses," Ulrich Flemming described how he used the shape grammar concept to analyze his subject and the challenges encountered and the benefits gained in doing so [6]. In terms of analyzing architectural examples, the first significant work is the shape grammar of Frank Lloyd Wright's Prairie Houses created by H. Koning and J. Elizenberg [7]. The researchers studied a corpus of Wright's houses in this genre and analyzed the spatial relationships they express. And, on this basis, they created an "additive" shape grammar for Wright's Prairie Houses capable of producing not only the houses originally designed by Wright, but also new designs based on his Prairie style. Against that background, in the present paper, shape grammar (as an analytical and generative tool) is utilized to reveal, recognize, and describe the improvisational moments of the work, in this case, Judd's 100 untitled work in mill aluminum.

1.3   Serial Order

Mel Bochner opens his essay with a quote from Josiah Royce's *Principles of Logic*: "What ordertype is universally present wherever there is any order in the world? The answer is serial order." He then distinguishes between working in series, making "different versions of a basic theme" from serially ordered works. He outlines three principles for serially ordered work:

1. The derivation of the terms... of the work is by means of a... systematically predetermined process (permutation, progression, rotation, reversal).
2. The order takes precedence over the execution.
3. The completed work is fundamentally parsimonious and systematically self-exhausting [1].

With the distinction made between working in series and serially ordered work, Bochner, along with these principles, also distinguishes modular works from serial ones. "Modular works are based on the repetition of a standard unit," he notes [1]. With these delineations, he classifies Judd among the serial artists. He mentions one of Judd's untitled galvanized iron pieces as being dimensionally proportioned based on a "progression," which is listed under principle 1, thus bringing it into the serial realm, although it is a singular work of art. With these three principles, we know that Judd's "100 untitled works in mill aluminum" oscillates between modularity and seriality. We have 100 repetitive objects, all $41'' \times 51'' \times 72''$, which may classify them as modular, yet each has unique configuration which may evoke the systematicity that is mentioned by Bochner. This, however, depends on whether the variation in their individual configuration belongs to an order-type and a meta-structure.

John Coplans' essay in *ARTFORUM* later served as the centerpiece of his book of the same title, *Serial Imagery* [2]. Given the length of the book, Coplans takes a much more detailed look at the subject. He traces seriality to Claude Monet and his seven views of *Gare Saint-Lazare* of 1877 and attributes the initiation of serial art works to major discoveries in science and philosophy. In particular, he points to the publication of Dedekind's 1872 *Stetigkeit und irrationale Zahlen* [8] where he maps the mathematical theory of the continuous independent variable and notes its influence on Monet's early serial work. Coplans cites the second half of the theory presented by Cantor in 1895: *Beiträge zur Begründung der transfiniten Mengenlehre* and includes other artists, poets, and musicians producing between 1872 and the 1960's inspired by the Dedekind- Cantor theory.

Although Coplans does not enumerate his principles of "Serial Imagery" as a list in the way that Bochner did, he provides a list nonetheless, in prose form. He opens the definition chapter of the book with two quotes from Mondrian and Reinhardt, both emphasizing the relation between pieces rather than the individual artwork. He continues the logic of relations and writes:

1. "Central to Serial Imagery is the concept of macro- structure—that which is apprehended in terms of relational order and of continuity…" [2, p.11]. In Bochner's first principle, we can surmise that process results in a relational order, but here, Coplans states it with clarity and precision. He makes a definitive link between macro-structure and relational order. Expounding the same notion in a different context, Coplans writes "Meaning is enhanced, and the artist's intentions can be more fully decoded when the individual Serial work is seen within the context of its set" [2, p.11]. In the same essay, Coplans writes "Seriality is identified by a particular inter-relationship, rigorously consistent, of structure and syntax: Serial structures are produced by a single indivisible process that links the internal structure of a work to that of other works within a differentiated whole" [2, p.11]. Coplans confirms the alignment between Bochner's first principle that focuses on process with his macro-structure.

   Clearly, there is alignment here, in principle 1, between Bochner and Coplans. Although the language may be different, the proposed principle remains constant. Coplans is much clearer about relational orders, but both adhere to a process, a macro-structure as the founding principle of serial order. Coplans continues with the following definitions:

2. "Serial Imagery is a type of repeated form or structure shared equally by each work in a group of related works made by one artist" [2, p.10]. Here again, we find alignment between the two principles cited by Bochner and Coplans, although stated differently. Bochner highlights the structure over the execution (artistry) of individual pieces, while Coplans emphasizes the shared structure, but does not mention the execution.

3. "There is no limit to the quantity of works in a series other than what is determined by the artist" [2, p. 11]. This presents a conflict with Bochner's third principle, where a self-exhausting system is mentioned. Coplans, in his exploration of this principle covers the self-exhausting enumeration as one of four possibilities, making his definition more expansive than Bochner's. We see great alignment between the definitions/principles of serial order presented by Bochner and Coplans. So, what is it that made Bochner include Judd's wall pieces within serial works, and Coplans not. In the same book, in the history section, Coplans provides a clue to his reasoning. He writes in the conclusive paragraph of the history chapter: Although the work of such sculp- tors as Donald Judd has been described as Serial, this is incorrect. Judd, for example, replicates parts by having identical units manufactured; they are then positioned to form one sculpture, one unit. Judd's images have a modular structure, and his range of similar sculptures relate more to sculptors' traditional use of editions than to true Serial form [2, p. 9].

In Coplans' analysis, Judd's "wall pieces," though each made of repeated modular units, construct a single artwork. As we have seen from the definitions, a single artwork cannot have any relational order with works within a series. It is a single painting or sculpture exhibited in a gallery or museum without the possibility of relations to other pieces. If we account for the repeated units within each wall piece, those fall into the modular definition. If we contrast one wall piece with another, exhibited in a different time and place, those fall into "editions" of the same piece, rather than a serial relationship. Whether we agree with Bochner, or Coplans regarding Judd is not the point here. We are simply trying to parse their definitions of serial works with analytical precision.

2 The Design Language of Donald Judd's "100 Untitled Works in Mill Aluminum"

With the context of principles and definitions of "Serial works," we will examine Donald Judd's "100 untitled works in mill aluminum". We will first examine principle #1 to see whether there is a macro-structure, a process that is governed by the artist from outside of individual works.

We spent several full days with the 100 objects in the artillery sheds at the Chinati Foundation in Marfa, Texas. We were able to verify through field observation two methods of structuring. We referred to them as the "compression type" and "cap type." The compression type objects held internal pieces together by compressing them by the bolted sides or ends. In essence, two sides or two ends compress all perpendicular members. The cap type objects held everything together through a full "cap" a $51''$x$72''$ full "½" plate on top. This type also works in compression, however, through gravity, through the weight of the cap compressing the other members, rather than the compression type that uses bolts and an inner plate to compress. These two types described 96 of the 100 objects. There were four that did not neatly fit in this analysis. We will need to get back to those later, in a different analysis. In addition, our initial assumption was that these two types, which appeared to be inherently structural, would govern the structure and fabrication of all pieces. This also turned out not to be true.

We noticed some imperfections and surface anomalies on the aluminum, exactly where we had imagined internal blind pins. Upon further examination, confirmed with records kept by the Chinati Foundation conservator, we recognized that although the two types held, they were more "form" types than structural types. Where the form was achievable with invisible welds, brackets, and bolts that made fabrication easier, or sometimes even possible, that was the route taken, rather than the strict structural and fabrication adherence to the types. We need to reiterate that the types are ours, and not mentioned anywhere by Judd or in any of his drawings. Nonetheless, the two types were so consistent that we held on to them.

The Judd Foundation archives holds all the shop drawings produced by Lippincott Inc. of North Haven, CT, the fabricators and installers of all 100 objects. Shop drawings, in general, are produced by fabrication shops based on an artist's or architect's drawings and specifications, showing their understanding of the drawings and how they intend to fabricate things. Artists and architects examine the shop drawings to note their compliance with their ideas as documented. The shop drawings produced for this project not only describe the objects themselves, but also how they are structured, detailed, and fabricated. We recognized language in describing the fabrication, which confirmed our field observations about the two types. Although there is no mention of "types" in the drawings, there is language that creates types, nonetheless. For example, all "cap type" objects are noted as "top over side over bottom" or "top over end over bottom," depending on whether the openings were on sides or on ends. All compression types are noted as "sides over ends" or "ends over sides" depending on whether the compressed members were on ends or sides.

The Judd Foundation archives also holds Judd's sketches for the 100 objects. Examining Judd's sketches, we recognized three "form types" which defined the groupings of objects: "open or recessed top," "open or recessed side," and "open or recessed end" (Fig. 2).

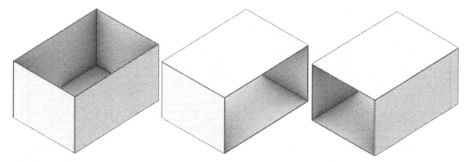

Fig. 2. Diagrams of "Open Top," "Open Side" and "Open End," Objects number 1, 13, and 12. Diagrams by authors.

We mapped every object in a chart with three columns, each related to the "form types." In addition, we added a column for anomalous objects. We then applied a color code to all the objects documenting their structural type, whether compression or cap. We discovered that all open top objects were of the compression type and all open side and open end objects were of the cap type. The strict adherence of our structural discoveries from the field observation with Judd's sketches and the fabrication shop drawings confirmed that we were on the right track and continued to pursue the types as described (Table 1).

Table 1. Chart describing all 100 objects in categories of open top, open side, and open end, including their sub-type families and construction type color codes. The 100 objects are in reference to Donald Judd's "100 untiled works in mill aluminum."

	Donald Judd's 100 untitled works in mill aluminum			
	Open or Recessed Top	Open or Recessed Side	Open or Recessed End	Anomalies
Full	1	13	12	
	10	68	27	
	16		67	
	26			
	100			
Half	2	24	25	98
	3	34	33	
	22	40	48	
	23	52	66	
	63	65		
	64	89		
	93	90		
	94	97		
		99		
Full-oblique	6	35	28	
	7	36	29	
		96	95	
Half-oblique	4	38	31	
	5			
	20			
	21			
Half/full-Oblique-offset	41	37	30	
	42	39	32	
	43	49	45	
	44	50	46	
	57	51	47	
	58	70	69	
	59	71	72	
	60	73	74	
	61	75	76	
	62	85	87	
	77	86	88	
	78			
	79			
	80			
	81			
	82			
	83			
	84			
	91			
	92			
Full-shape offset	8	18	17	11
	9	55	53	14
	15	56	54	19

Compression
Cap
Anomalie or mix

Based on these discoveries and the alignment of the results of multiple angles of research, we were convinced about the structure that governed the majority of the 100 objects. We then categorized each object under its "form type" ("open top," "open side," and "open end") and sub-categorized them in smaller families. We have noted "full" and "half" families. Each family is then tempered by other overlay rulesets of "offset," "double," and "oblique." The "full" and "half" families include objects that only have orthographically organized plates. When overlayed by the rules, they include objects that contain oblique plates, double-oblique plates, and double-oblique offset plates. Altogether, there are 37 objects of the orthographic type and 57 of the oblique type. There are two oblique objects that contain a plate with a compound angle, and four objects which remain anomalies in the deciphered systems.

Once we knew what the form families were, we extracted the grammar of each family. Different rules were developed in order to describe the way in which the outer boxes were divided, for example with different configurations of top plates, inner plates, and offset plates (Figs. 3, 4 and 5).

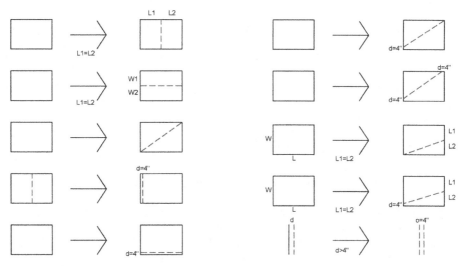

Fig. 3. Set of rules showing the way in which the outer box can be divided in plan.

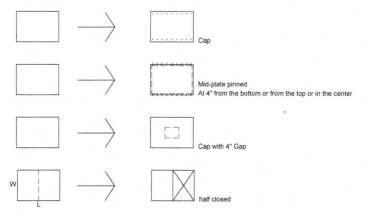

Fig. 4. Set of rules showing configurations of the top plate (cap).

Using the grammar, we mapped every object that could have been made. For example, we noted a family of objects that we called "bisectors." As the name implies, these are objects that have a bisecting plate, splitting the internal volume of the object into two equal parts. The bisectors are a sub-set of and cut across the "full" and "half" families and are tempered by the "double" and "offset" rulesets. The bisecting plate could be either perpendicular to the open side or parallel. When perpendicular, it can have two

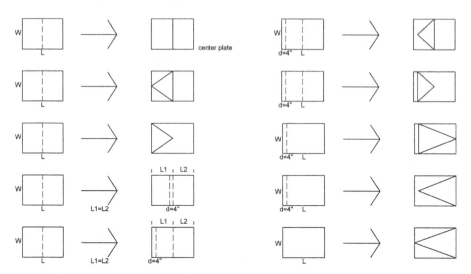

Fig. 5. Set of rules showing configurations of inner plate(s).

directions, turned ninety degrees to one another. This constitutes nine objects, three in each type, and exhausts the system completely (Fig. 6).

We noted in many of Judd's sketches the letter "D" next to the image of an object. This denoted that there would be a "double" in addition to the object. What Judd meant by "D" is that an inner plate would be doubled. In some of the early sketches the distance between the doubles is marked as 25 (cm? mm?). This dimension is fixed at 4″ in the final objects. In the case of the "bisector" family of objects, this constitutes six additional objects (Fig. 7). Where the bisecting plate is parallel to the open side, the doubling would be invisible and thus not experienced. The three objects, one in each form type (open top, side and end) were not constructed.

With a few exceptions, we have confirmed that objects dictated by the families and the rulesets were either made, or that we could easily identify why there were not, such as noted above in the double bisector family. There are, however, families of objects that were not made, or not made completely in a self-exhausting manner, pointing to an improvisational moment. For example, in the bisector family, Judd began the construction of a series with one side closed. There are two objects of this family fabricated, object numbers 89, and 90 (Fig. 8). They are like objects 40, and 52, except with one side closed. However, the same was not constructed for objects 24, 25, 33, 48, 65, and 66.

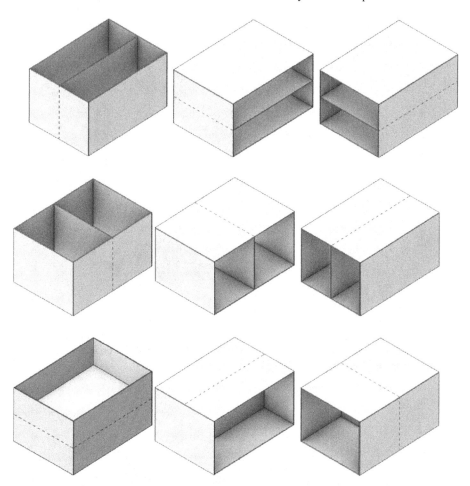

Fig. 6. Column 1: Diagrams of objects number 22, 23, and 10. Open top, bisector; Column 2: Diagrams of objects number 24, 40, and 34. Open side, bisector; Column 3: Diagrams of objects number 25, 33, and 27. Open end, bisector. Diagrams by authors.

3   Discussion

In addition to the form families described above, we have also detected "visual families" that overlay the form types. There is alignment between the two systems, but also significant differences. The visual types cut across the form types. For example, the "half" visual type describes an elevation (side, end, or top view) that is split in the middle. This split may be accomplished with a plate at the surface of the elevation, a recessed plate, an oblique plate defining the same half surface, a perpendicular to elevation plate (bisector) and so on, crossing multiple types and families.

Notwithstanding the four anomalous objects, our research confirms there is a macro-structure that governs the design of the objects. This work satisfies principle #1. The order and structure take precedence over the execution, though the execution by Lippincott is

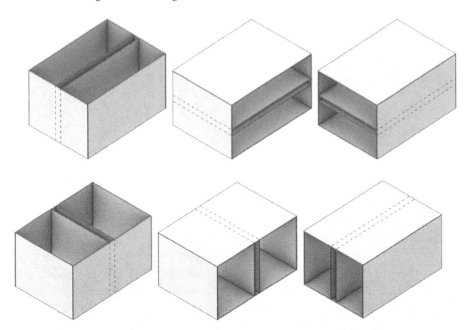

Fig. 7. Column 1: Objects number 63 and 64. Open top, double bisector; Column 2: Objects number 65 and 52. Open side, double bisector; Column 3: Objects number 66 and 48. Open end, double bisector. Diagrams by authors.

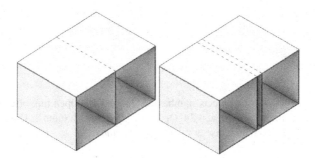

Fig. 8. Diagrams of objects number 89 and 90. Diagrams by authors.

meticulous, as dictated by Judd. This work, therefore, also satisfies principle #2. Regarding principle #3, we have already determined there is deviation between Bochner and Coplans, where Bochner insisted on a "self-exhausting" system and Coplans had a more expansive and thus inclusive principle. Judd's "100 untitled works in mill aluminum" certainly fulfills Coplans' definition. However, in relation to Bochner's more restrictive definition, it is not yet clear to us whether the 100 objects are "self-exhausting." Our initial results suggest that they are not, as was described briefly in the context of the bisectors. But we need further study to determine this point more strictly.

It is important to note that while for a thorough analysis of the objects, a 3-dimensional grammar is preferred, at this stage, only a 2-dimensinal grammar is developed. The grammar, however, has many components that describe the configurations of plates in all three dimensions. To facilitate the generation of designs and to eliminate human input in applying rules to generate objects, a computer program has been developed. The code was written in the Python scripting language for Rhino. Figure 9 shows examples of designs generated by this computer program.

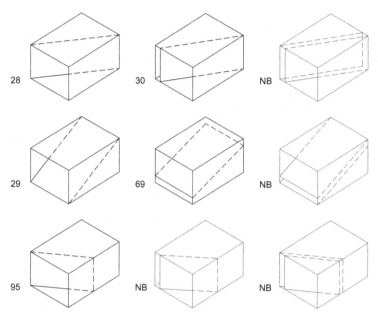

Fig. 9. Designs generated by the computer program, including designs that were not constructed as part of the 100 objects shown in red. Diagrams by authors. (Color figure online)

4   Conclusion

Donald Judd began the Marfa installation with twenty-five objects that were meant to be installed at the Wool and Mohair building. Following a previous method of work, Judd re-imagined previously fabricated objects in new dimensions and material configurations. Once the number of objects to be installed increased from twenty-five to one hundred and the location was moved to the artillery sheds, he devised a systemic strategy for making new objects. All the objects listed under "anomalous" belong to the first twenty-five. The remaining seventy-five, all but one (object 98), belong to the systemic grammar described above. In the manner of Knight, we could have devised an expanded grammar to include and explain transformation from the first 25 objects to the next 75 objects. Or we could have devised a *hybrid* grammar to identify and explain the mixture between the first 25 and the next 75 objects [9]. We could also have devised an all-encompassing, *generic* grammar, an idea defined by Benros, Hanna, and Duarte, as "a formalism that allows the design of diverse solutions," to include all one hundred objects [10]. We, however, decided that it was most important to design the tightest grammar possible to reveal and recognize the improvisational moments of the work. At some point, the artist will make decisions that do not follow the grammar of the work as initially conceived. With the grammar developed, we utilized Stiny and Gips's method to generate all possible forms to explore the exhaustive nature of the application of the grammar. Judd has stated that although his work is derived from a plan that is conceived prior to the work, he is, nonetheless, un-interested in making the plan legible didactically. He has intentionally placed objects within the group that appear to belong to the group yet are defiant of the logic that governs all objects. Object #98, in our assessment, is such object. He has also eliminated objects from the sets for which we have no explanation to date (such as the bisectors mentioned earlier). So far, our analysis has only included the study of the physical objects. As the project expands to include the installation of the 100 works and Judd's writings, we will certainly discover more of these improvisational moments. The question we sought to answer was whether these 100 works belong to a system, and, if so, what that system is and what delimits the system to produce 100 objects and no more. We have shown that there is a grammar for the design of seventy-four of the objects. The first twenty-five objects belong to grammars produced for prior sets of work. The one remaining object, #98, exists to defy all logic, as a proverbial wrench in the system. Using analytical shape grammar, we have identified that there is a system, and using generative shape grammar, we have confirmed that the system is not applied exhaustively. In the gaps between the two methods, we have identified the improvisational moments in which Judd designed objects or sets of objects that were outside the grammar of the project.

References

1. Bochner, M.: The serial attitude. ARTFORUM VI **4**, 28–33 (1967)
2. Coplans, J.: Serial Imagery. Pasadena Art Museum, Pasadena, CA (1968)
3. Stiny, G., Gips, J.: Shape grammars and the generative specification of painting and sculpture. Information Process. **71** (1971), Freiman, C.V. (ed.)
4. Knight, T.: Transformations of *De Stijl* art: the paintings of Georges Vantongerloo and Fritz Glarner. Environ. Planning B: Planning and Design, **16** (1989)
5. Duarte, J.: Customizing mass housing: toward a formalized approach. In: Mass Customization and Design Democratization, Kolarevic, B., Duarte, J. (ed.) Routledge, London and New York (2019)
6. Flemming, U.: More than the sum of parts: the grammar of queen anne houses. Environ. Planning B: Planning and Design **14** (1987)
7. Koning, H., Elizenberg, J.: The language of the Prairie: frank Lloyd Wright's prairie houses. Environ. Planning B **8**, 295–323 (1981)
8. Referenced in Coplans, J. Serial Imagery, p. 19: Huntigton, E.D.: The Continuum (and Other Types of Serial Order), second edition. Harvard University Press, Cambridge (1921)

9. Hadighi, M., Duarte, J.: Bauhaus internationalism to college town modernism: exploring bauhaus culture in Hajjar's hybrid architecture. In: Lee, J.-H. (ed.), Computer-Aided Architectural Design, "Hello Culture," 18th International Conference, CAAD Future 2019, Springer-Verlag, pp. 429–433 (2019)

10. Benrós, D., Duarte J., Hanna, S.: The inference of generic housing rules: a methodology to explain and recreate palladian villas, prairie houses and Malagueira houses. Design Comput. Cognition **14**, 401–419 (2014)

Illustrating Algorithmic Design

Renata Castelo-Branco[(✉)][iD] and António Leitão[iD]

INESC-ID/Instituto Superior Técnico, University of Lisbon, Lisbon, Portugal
{renata.castelo.branco,antonio.menezes.leitao}@tecnico.ulisboa.pt

Abstract. Architectural design is strongly based on visual and spatial reasoning, which is not easy to translate into algorithmic descriptions and, eventually, running programs, making it difficult for architects to use computational approaches, such as Algorithmic Design (AD). One of the most pressing problems is program comprehension. To overcome it, we propose an automatic illustration system for AD programs that produces annotated schemes of the program's meaning.

The illustration system focuses on a basic set of geometric elements used in most calculations to place geometry in space (points, distances, angles, vectors, etc.), and on the way they are manipulated to create more complex geometric entities. The proposed system automatically extracts the information from the AD program and the resulting illustrations can then be integrated into the AD program itself, intertwined with the instructions they intend to explain.

This article presents the implementation of this solution using an AD tool to generate the illustrations and a computational notebook to intertwine the program and the illustrations. It discusses the choices made on the system's implementation, the expected workflow for such a system, and potential future developments.

Keywords: Algorithmic Design · Illustration · Documentation · Program Comprehension

1 Introduction

In the Algorithmic Design (AD) approach, designs are represented through algorithmic descriptions that, when executed, generate the corresponding digital model [5,32]. AD's parametric nature [3] promotes design experimentation, facilitates the evaluation of multiple and often conflicting design requirements [20,33], and promotes the production of large-scale unconventional designs [7].

However, AD is less intuitive than other design methods, and architects still struggle to understand how the AD program relates to the design it represents, particularly in programs developed by others, an increasingly common scenario in collaborative work environments [35]. To help practitioners understand AD programs, we propose an automatic illustration system that can produce annotated 2D schemes explaining parameters and other relevant relations that compose the AD. The proposed system provides illustrations for a set of basic geo-

© The Author(s), under exclusive license to Springer Nature Switzerland AG 2023
M. Turrin et al. (Eds.): CAAD Futures 2023, CCIS 1819, pp. 36–50, 2023.
https://doi.org/10.1007/978-3-031-37189-9_3

metric elements, such as distances, angles, points, vectors, etc., which can then be intelligently combined to produce useful illustrations.

To support the natural evolution of the design, the planned use is for architects to piggyback the generation of illustrations on top of the AD program they are developing, making it generate not only the intended architectural model but also the illustrations explaining it. Combined with existing visual documentation techniques for AD [8, 9], these illustrations can then be intertwined with the program itself. By promoting a dialog between the algorithm and the design it represents, this proposal aims to reduce AD's comprehension-related drawbacks, improving the development, maintenance, and sharing of AD programs.

2   A Challenging Practice

Although initially met with resistance, the paper-to-digital transition brought considerable advantages to architectural design [6, 21] and the adoption of digital-based design methods rapidly increased, especially because the developed digital design tools emulated existing representation methods, only with more precision and efficiency. Currently, another big leap is taking place with the use of AD, a design approach that represents designs through algorithmic descriptions [31], i.e., computer programs with rigorous instructions for a computer to perform [3].

Even further from the hands-on nature and materiality of traditional architectural development [22], AD relies on abstractness to transcend the constraints to representation and imagination that bind prior digital design tools [22], outperforming them in terms of flexibility and expressiveness [7]. Additionally, AD can seamlessly integrate analysis and optimization in design exploration processes, allowing performance to act as a design principle [4]. This capacity is becoming increasingly critical in an era where the industry is pressed to reduce not only time and cost requirements but also its environmental impact [23].

However, with fewer analogies to traditional representation methods than previous digital design paradigms, AD has some challenges ahead. One of them is the need for programming skills. Withal, even for experienced programmers, the ability to achieve design thinking with AD remains difficult, since these are two very distinct modes of thought [15]. Design representations are meant to stimulate creativity [28] by extending architects' imagination to the physical realm and establishing a feedback loop between both types of representation (internal and external to the creator's mind) [12]. Unfortunately, with AD, the practice seems unable to intuitively allow this mutual influence.

3   Comprehension Mechanisms

The architectural design process is strongly based on visual and spatial reasoning, which is not easily translated into algorithmic descriptions. However, the challenge in converting abstract ideas to and from algorithmic representations is not a specific drawback of AD. Program comprehension is a well-known problem in computer science and several solutions have been proposed in the past.

3.1   Documentation

Donald Knuth, for instance, proposed the development of programs as literary works [17]. Other examples include the use of graphical representations to explain programs, such as flow diagrams [13] and other diagrammatic techniques [14,19], as well as animations of the program's fundamental operations [2]. Although some of these works addressed the use of images as visual explanations for textual programs, particularly those that fell under the program visualization umbrella [25,27], they were still based on a computer science outline, which completely disregarded the intrinsic visual nature of the design process itself.

Architects have learned to think and reflect upon their design through sketching, and ADs are no different. The hand-made drawings architects make during the design process represent their intentions for their AD programs, explaining the logic behind their conception and what they expect them to produce. They thus constitute fundamental information for the understanding of the architect's design idea and could be integrated with the algorithmic representation, illustrating and explaining it.

Some authors developed solutions to accommodate these assets in AD programs. Leitão et al. [18] proposed the inclusion of sketches, images, and renders in textual AD descriptions. However, their solution suffered from scalability issues. Grasshopper allows for the inclusion of imagery in the middle of AD programs, although, in the visual programming environment, it becomes hard to contain the clutter that these elements cause.

3.2   Computational Notebooks

Beyond documentation, other theories stress the importance of displaying not just the program's structure and behavior but also its evolution [11], which is crucial in a creative process such as AD. An interesting take on this problem has been forwarded by computational notebooks [29], which were designed to support computational narratives, allowing users to create and communicate their experiments in a comprehensible manner. To that end, they allow incremental development of programs with immediate feedback on their results, as well as the intertwining of code with textual and visual documentation.

The immediate feedback provided by computational notebooks touches on another comprehension aspect, which is liveliness [16]. Liveliness has been differently interpreted in different fields [26]. In the case of AD, it means that changes to the algorithmic description should have immediate repercussions in the generated design model so that users can relate the changes in the program to their respective impact on the model. Tools like Grasshopper [10] and LunaMoth [1] do this via live coding [26], whereas computation notebooks rely on other facets of liveliness, such as interactive evaluation and reactivity [9].

Relying on computational notebooks, previous works also explored the idea of integrating, in the AD program, hand-made drawings and other digital media produced during the design process (such as model screenshots and rendered images) [8,9]. However, there is a downside to these elements. Screenshots and renders typically present a global view of the digital model, rarely succeeding in explaining the relevance of particular program fragments for the generation of the intended geometry. Hand-made drawings offer finer control over which aspects the architect wishes to illustrate. The problem, however, lies in their static nature. Scanned drawings incorporated in the program rapidly get outdated as the design evolves. And whereas repurposing drawings made during design ideation has little cost for the architect, taking the time to correct outdated drawings imposes a higher penalty that not many are willing to pay.

4   Automatic Illustration

Many of the existing solutions in the field of program comprehension contemplate or derive from the computational thinking paradigm and do not necessarily comply with the visual demands of AD. There is still much to explain in the intricate dependencies that compose a parametric program and there is still a long way to go to make AD more akin to traditional architectural design processes. As a first step toward that goal, we turn our attention toward the essence of design experimentation and the most popular means used to iterate over design ideas: drawings.

Building upon the solution presented in Sect. 3.2, we propose to overcome the previously-mentioned shortcomings with (1) the automatic generation of computer-made geometric illustrations explaining relevant aspects of the algorithmic description, and (2) their subsequent integration in the AD program. The envisioned workflow is for architects to generate and integrate automatic illustrations with as little extra work as possible, as well as automatically update them given any changes to the algorithmic description. To that end, the proposed system extracts as much information as it can from the existing AD description, requiring users to write additional instructions only if and when they wish to alter the system's default behavior. Since the illustrations are automatically extracted from the algorithmic descriptions, they can be regenerated at any time, ensuring the program's visual documentation is up to date.

For evaluation purposes, we implemented the proposal on top of the Khepri AD tool [30], which uses the Julia programming language and communicates with several modeling, rendering, analysis, and optimization tools. From this set, we chose two visualization tools to generate the illustrations: TikZ, a procedural drawing tool, and AutoCAD, a 2D drafting and 3D modeling tool known to most architects. As for the integration of the generated illustrations in the AD program, we follow the workflow proposed in [9], using computational notebooks to intertwine visual and textual documentation with the AD program and, thus, create a more comprehensible programming experience. The computational notebook used to exemplify the proposed system is Pluto [24].

4.1   Geometric Elements

Khepri has a large set of pre-defined modeling operations capable of creating 2D and 3D geometry in multiple tools. The shape and positioning of that geometry typically depend on a series of calculations performed with the most basic geometric elements, such as points, vectors, distances, and angles. Using a multiplicity of coordinate systems (namely, cartesian, polar, cylindrical, and spherical) these elements can be used to form more complex geometric entities, such as arcs, circles, polygons, etc. These eventually get extruded, swept, lofted, and combined in other ways to create 3D shapes. However, more often than not, it is in these first steps that most geometric calculations are done and that most design parameters imprint their influence. These operations are thus the focus of our illustration proposal.

Figure 1 shows, on the left, a 3D model generated by an AD program. The building's profile mimics an Islamic pattern and is achieved with two superimposed squares and eight circles centered on each of the eight intersection points and tangent to the bounding octagon. In this case, the tangency was achieved by calculating the correct circle radius using trigonometry. The resulting 2D surface is replicated in height with decreasing size and then extruded to make slabs and glass panels. The parameters defining the base shape are explained in the geometric illustrations on the right, with the recursive distribution of the circles represented with increasing transparency (further explained in Sect. 4.5).

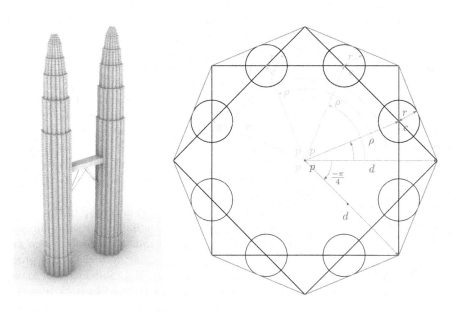

Fig. 1. Petronas Towers inspired 3D model and 2D geometric illustration generated in TikZ.

For the illustration process to piggyback the AD program itself with as little additional programming as possible, we capitalize on the parameters that users must already provide to create each element. For instance, in order to create a circle, users must specify a center and a radius (c and r, in the previous example), whereas to create a regular polygon, besides center (p) and radius (d), an initial angle ($\pi/4$) and the number of sides are also required. The automatic illustration of each element is, in most cases, simply comprised of the parameters required to create it. Naturally, beyond the automation, users are also given mechanisms to include additional information they may find relevant to the explanation (e.g., the 'invisible' octagon in the previous example).

4.2   The Evaluator

The illustration system is implemented as a specialized Julia evaluator, which recognizes specific program patterns in the evaluated program and acts upon those, generating the appropriate illustration, before sending the instructions to the standard Julia evaluator, who resumes the normal evaluation cycle producing the program's originally intended results. The resulting behavior is similar to that of code injection in selected parts of the program.

The patterns recognized by the evaluator include operations between points and vectors in the same or different coordinate systems, and the basic geometry creation operations mentioned above. For each case, the evaluator explores the expressions used as arguments in the corresponding function calls to produce the annotation labels for the illustration. For example, in the definition of the arrow shape shown in Fig. 2, the evaluator intercepts the point-vector sums and draws the elements involved in each one, using the name of the points for their labels. Polar vectors, however, are represented with two elements - a distance and an angle - whose labels the evaluator fetches from the vectors' expressions (e.g., ρ and α, or θ and $\alpha + \pi - \beta$).

The illustrations intend to show the underlying process that leads to the final result. Since the same geometric outcome may be achieved using different program patterns, each of these will yield specific illustrations. Figure 3 shows two versions of a function that places four tangent circles around a point. The first uses cartesian coordinates and the second uses polar coordinates. Accordingly, the automatic illustration of these instructions will produce different drawings, although the intended output of the functions is the same.

4.3   Programming Style

Illustration allows us to visualize the entire generation process, whereas without it we would only see the final result. If not for illustration, users would likely choose between possible implementation styles on a whim. However, knowing that the way they write their program impacts its illustration will likely lead them to a more conscious choice, perhaps even to a new programming style.

Take the two definitions in Fig. 4 for an egg shape. The first one calculates the center points for the arcs directly within the arc calls, whereas the second one defines the center points as local variables, giving them specific names. The illustration below corresponds to the definition on the left, where the egg's larger radius is annotated with the local name r_2 instead of its corresponding but less readable definition $(r_0 - r_1 * cos(\alpha))/(1 - cos(\alpha))$. The illustration of the

```
arrow(p, ρ, α, θ, β) =
    let p1 = p + vpol(ρ, α)
        p2 = p1 + vpol(θ, α + π + β)
        p3 = p1 + vpol(θ, α + π - β)
        line(p, p1, p2, p3, p1)
    end
```

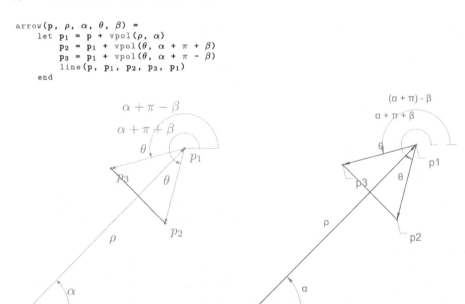

Fig. 2. Arrow shape illustration generated in TikZ (left) and AutoCAD (right).

second case would be even more intelligible since the same happens with all the lengthy expressions representing the points (an illustration corresponding to this definition can be found in Fig. 5). Despite the added trouble, having the points as separate entities from the arcs makes both the program and the illustration easier to read. As a general rule, properly naming variables over which we may want to operate is a good programming practice.

4.4 Too Much Information (TMI)

Some of the examples shown portrayed rather simple computations, with just the right amount of information to produce an intelligible illustration. However, illustrating all identified patterns, in most cases, will likely result in an excessively cluttered drawing with superimposed information.

To avoid overlapping annotations, the evaluator keeps track of all the elements generated in a single illustration, through their insertion points, lengths, angles, labels, etc. If it identifies collisions, the system distributes the colliding elements through the available space. For instance, coincident radii will be placed at different angles on the circle, coincident angles will be given different radii for their arcs (see Fig. 3 right), coincident lengths will be given different offsets from the line, and coincident labels will be placed at different angles and distances from the insertion point. Cases where both point and label coincide

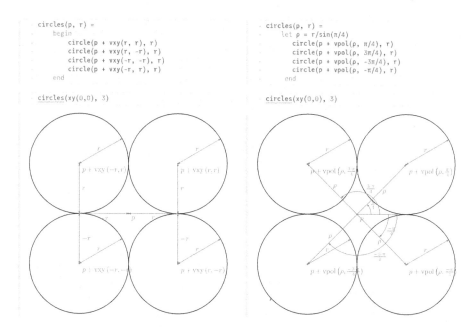

Fig. 3. Two possible definitions to place four tangent circles around a point: using cartesian (left) and polar coordinates (right). Both algorithmic description and illustration are shown in the Pluto notebook.

will simply not be generated. Naturally, not all elements possess such flexibility. Vectors, for instance, cannot change place, but their labels can.

Collision avoidance strategies work wonders at small scales, but they cannot perform miracles when there are simply too many elements to illustrate. As such, user-configurable flags are also available for users to control which elements get generated in each illustration, and if they wish to visualize the illustration step by step. In the latter case, the evaluator generates multiple illustrations, one step at a time, until the AD program fragment is completely evaluated. This results in a group of images that can then be combined in an animation. As an example, by applying this mechanism to the egg shape, we obtain a step-by-step illustration where the comprehension of each individual arc is improved (Fig. 5).

4.5   Repeated Illustrations

Loop instructions, such as `for` and `while` cycles, and, more importantly, recursive definitions are strong candidates for the generation of cluttered illustrations. If we repeat the illustration of the geometric elements for each iteration of the loop or each recursive call, we are likely to get not only repeated information but also superimposed geometry, which will be difficult to differentiate. To avoid this, the system increases the transparency of the annotations with each loop.

```
egg(p, r₀, r₁, h) =
    let α = 2*atan(r₀-r₁, h-r₀-r₁),
        r₂ = (r₀-r₁*cos(α))/(1-cos(α))
        arc(p, r₀, 0, -π)
        arc(p + vx(r₀-r₂), r₂, 0, α)
        arc(p + vx(r₂-r₀), r₂, π-α, α)
        arc(p + vy((r₂-r₁)*sin(α)),
            r₁, α, π-α-α)
    end
```

```
egg(p, r₀, r₁, h) =
    let α = 2*atan(r₀-r₁, h-r₀-r₁),
        r₂ = (r₀-r₁*cos(α))/(1-cos(α))
        p₁ = p + vx(r₀-r₂)
        p₂ = p + vx(r₂-r₀)
        p₃ = p + vy((r₂-r₁)*sin(α))
        arc(p, r₀, 0, -π)
        arc(p₁, r₂, 0, α)
        arc(p₂, r₂, π-α, α)
        arc(p₃, r₁, α, π-α-α)
    end
```

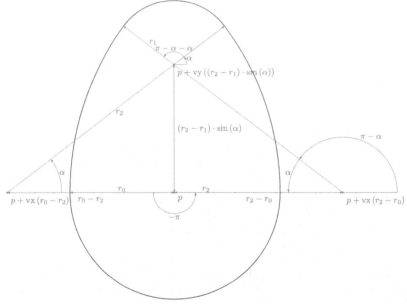

Fig. 4. Two definitions of an egg shape, either calculating the center points for the arcs directly within the arc calls (left) or defining the center points as local variables (right). The illustration below corresponds to the first definition.

If the illustration remains cluttered, users can also choose to illustrate a limited number of steps. As an example, consider the recursive definition of a spiral in Fig. 6. On the left, we see an unconstrained illustration with increasing transparency. On the right, we only illustrate the first recursive step. Other mechanisms, such as the step-by-step illustration option presented above, may be equally useful to illustrate the execution of a recursive program.

In most cases, our approach is enough to get an idea of the process and understand where the errors lie when the program is not producing the expected result. Take, for instance, the recursive diamond-shaped pattern in Fig. 7. The leftmost definition contains a bug that is easily perceived in the illustration. The one on the right is correct. Either way, illustrations might require style decisions on the user's part, as the system's default behavior is unlikely to suit all cases.

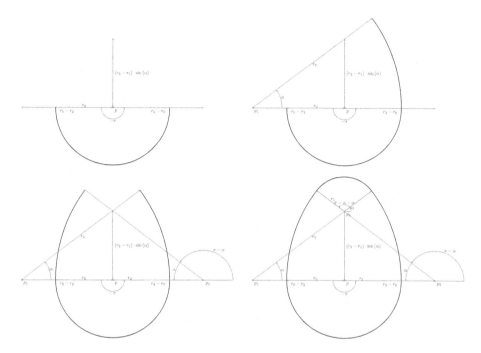

Fig. 5. Step-by-step illustration of the egg, using the second definition in Fig. 4. The arc centers are annotated with the names of the corresponding local variables.

5   Discussion and Follow-Ups

Explanations are all about simplifying or breaking complex problems down into smaller bits. In architecture, we frequently resort to simplified 2D plan or section depictions, 2D line-based schematics, etc. to explain a complex design idea in simpler terms. Capitalizing on this concept, the proposed system focused on the creation of 2D geometric illustrations explaining the behavior of AD program fragments, based on the operations and parameters used in it.

We foresee two main scenarios that strongly benefit from this proposal: (1) collaborative work endeavors involving shared AD programs, and (2) learning environments, including not only novice programmers trying to get a grip on their creations, but also teachers and learning-content creators.

Withal, the research presented here is a work in progress, with the potential to integrate more ideas. There are many fields left to explore. For instance, AD programs do not describe geometry modeling operations only. They frequently integrate descriptions of simulation and optimization routines. Graphically explaining these routines and subsequent processing of the output data is an entirely new illustration challenge. We leave below some unexplored research paths that we believe to be logical next steps in this investigation.

```
spiral_arc(p, r, α, Δα) = arc(p, r, α, Δα)

spiral(p, r, α, Δα, ω, f) =
    if ω-α < Δα
        spiral_arc(p, r, α, ω-α)
    else
        spiral_arc(p, r, α, Δα)
        spiral(p + vpol(r*(1 - f), α+Δα), r*f, α+Δα, Δα, ω, f)
    end
```

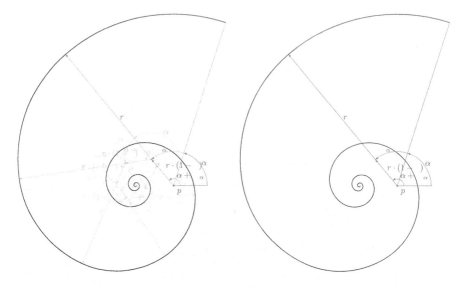

Fig. 6. Recursive definition of a spiral (top), with an unconstrained illustration of the entire process (left), and an illustration limited to the first recursive step (right).

Geometric Constraints. In the Petronas example (Fig. 1), the radius that guarantees the *tangency* between the circles and the octagon was mathematically calculated in the program without a single reference to either the octagon or the term 'tangent'. As a result, since it is not explicitly expressed in the program, this geometric relation cannot easily be inferred by the evaluator either. We made the octagon visible using the mechanisms available in the illustrator to add additional information to the images. However, it remains a far cry from an expressive illustration of geometric constraints. Research work has been done on ways to explicitly state these relations in the program in order to facilitate the associated calculations, prevent errors, and make the program more intelligible [34]. Building upon these principles, we plan to extend our illustration library to include geometric constraint concepts.

Organizing Labels. The proposed mechanism to infer if there is any information juxtaposition currently considers label insertion points only. For most geometric elements this approach will suffice. However, if the illustration contains long labels, the chances of juxtaposition increase and the system is none the wiser. We could further develop the existing label placement algorithm to consider the bounding boxes of previously generated labels as well.

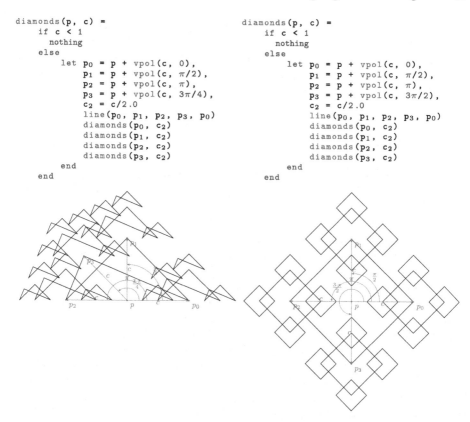

```
diamonds(p, c) =
   if c < 1
      nothing
   else
      let p0 = p + vpol(c, 0),
          p1 = p + vpol(c, π/2),
          p2 = p + vpol(c, π),
          p3 = p + vpol(c, 3π/4),
          c2 = c/2.0
          line(p0, p1, p2, p3, p0)
          diamonds(p0, c2)
          diamonds(p1, c2)
          diamonds(p2, c2)
          diamonds(p3, c2)
      end
   end
```

```
diamonds(p, c) =
   if c < 1
      nothing
   else
      let p0 = p + vpol(c, 0),
          p1 = p + vpol(c, π/2),
          p2 = p + vpol(c, π),
          p3 = p + vpol(c, 3π/2),
          c2 = c/2.0
          line(p0, p1, p2, p3, p0)
          diamonds(p0, c2)
          diamonds(p1, c2)
          diamonds(p2, c2)
          diamonds(p3, c2)
      end
   end
```

Fig. 7. Recursive placing of diamond shapes. The illustration of the program on the left shows that the angle used to calculate the position of point $p3$ is wrong. The definition on the right is correct.

Illustration Visualizer. In this implementation, we used a CAD tool (Auto-CAD) and a dedicated drawing program (TikZ) to generate the illustrations to be inserted into the Pluto computational notebook. Since this insertion is not literal, but rather a reference to a file in a folder, updating the illustration does not require a new insertion; if we re-generate the image with the same name, Pluto will fetch the updated version automatically. We could, nevertheless, explore other options for image generation, such as browser-based visualizers that produce the graphs directly in the notebook (e.g., Plotly). A downside to this approach is that users have to effectively run the program to visualize the illustrations, whereas the approach we chose to pursue keeps a version of the illustrations available for anyone to see, even without executing the program.

3D. The best explanations are often simplifications or depictions of isolated parts of the problem; hence, our initial focus on 2D geometry. However, we have already begun extending the system to 3D illustrations. We are currently studying label positioning techniques that consider the rendering viewpoint, automatic sectioning methods, and step-by-step decompositions particularized to some geometry modeling operations (for instance, in a sweep operation it may be interesting to show not only the extrusion of the section along the path step-by-step but also any scaling or rotation factor applied to the section in a separate animation).

6   Conclusion

This article proposed an automatic illustration system to produce annotated 2D geometric schemes explaining Algorithmic Design (AD) programs. The aim was to improve the comprehension, and thus, learning, development, maintenance, and sharing of AD programs among architects.

The illustrations comprise a set of basic geometric elements, such as points, distances, angles, vectors, etc., which can help users understand the geometric calculations done in the AD program, the meaning of the symbols used, and the impact they have on the overall geometry if changed. The proposed system allows architects to piggyback the illustration process on top of AD programs, making them generate not only the intended architectural models but also the illustrations explaining them. The resulting illustrations are then integrated into the AD program, intertwined with the instruction they intend to explain.

The proposal builds upon previous work regarding the inclusion of visual documentation in AD programs, now contemplating two types of documentation: hand-made drawings and computer-generated illustrations. Comparing one with the other at any stage of the process can also help users understand if the program is producing the expected results.

This research addresses and considerably simplifies the hardworking task of illustrating AD programs, frequently automating it completely. Nevertheless, this is ongoing research and several logical follow-ups to this proposal were discussed, such as extending the illustration library to include more complex concepts like geometric constraints, a more holistic approach to annotation placement that is sensitive to the space occupied by pre-existing labels, considering other visualizers for the generation of the images, and extending the system to 3D illustrations.

Acknowledgments. This work was supported by national funds of *Fundação para a Ciência e a Tecnologia* (FCT) with references UIDB/50021/2020, PTDC/ART-DAQ/31061/2017, and DFA/BD/4682/2020.

References

1. Alfaiate, P., Caetano, I., Leitão, A.: Luna Moth: supporting creativity in the cloud. In: Proceedings of the 37th Annual Conference of the Association for Computer Aided Design in Architecture (ACADIA), pp. 72–81. Cambridge (2017)
2. Brown, M.H., Najork, M.A.: Algorithm animation using 3D interactive graphics. In: Proceedings of the 6th Annual Symposium on User Interface Software and Technology (UIST 1993), pp. 93–100. ACM (1993)

3. Burry, M.: Scripting Cultures: Architectural Design and Programming. Architectural Design Primer. Wiley, Hoboken (2013). https://doi.org/10.1002/9781118670538

4. Caetano, I., Garcia, S., Pereira, I., Leitão, A.: Creativity inspired by analysis: an algorithmic design system for designing structurally feasible façades. In: Proceedings of the 25th International Conference of the Association for Computer-Aided Architectural Design Research in Asia (CAADRIA), vol. 1, pp. 599–608. Bangkok (2020)

5. Caetano, I., Santos, L., Leitão, A.: Computational design in architecture: defining parametric, generative, and algorithmic design. Front. Archit. Res. **9**(2), 287–300 (2020). https://doi.org/10.1016/j.foar.2019.12.008

6. Carpo, M.: The Alphabet and the Algorithm, 1st edn. MIT Press, Cambridge (2011)

7. Castelo-Branco, R., Caetano, I., Leitão, A.: Digital representation methods: the case of algorithmic design. Front. Archit. Res. **11**(3), 527–541 (2022). https://doi.org/10.1016/j.foar.2021.12.008

8. Castelo-Branco, R., Caetano, I., Pereira, I., Leitão, A.: Sketching algorithmic design. J. Archit. Eng. **28**(2), 04022010 (2022). https://doi.org/10.1061/(ASCE)AE.1943-5568.0000539

9. Castelo-Branco, R., Leitão, A.: Comprehending algorithmic design. In: Gerber, D., Pantazis, E., Bogosian, B., Nahmad, A., Miltiadis, C. (eds.) CAAD Futures 2021. CCIS, vol. 1465, pp. 15–35. Springer, Singapore (2022). https://doi.org/10.1007/978-981-19-1280-1_2

10. Davidson, S.: Grasshopper: algorithmic modeling for rhino (2023). https://www.grasshopper3d.com. Accessed 14 Feb 2023

11. Diehl, S.: Software Visualization: Visualizing the Structure, Behaviour, and Evolution of Software. Springer, Heidelberg (2007). https://doi.org/10.1007/978-3-540-46505-8

12. Goldschmidt, G.: Design representation: private process, public image. In: Goldschmidt, G., Porter, W.L. (eds.) Design Representation, pp. 203–217. Springer, London (2004). https://doi.org/10.1007/978-1-85233-863-3_9

13. Goldstine, H.H., Von Neumann, J.: Planning and coding of problems for an electronic computing instrument: report on the mathematical and logical aspects of an electronic computing instrument (1947)

14. Jackson, M.A.: Principles of Program Design. Academic Press Inc, USA (1975)

15. Kelly, N., Gero, J.S.: Design thinking and computational thinking: a dual process model for addressing design problems. Des. Sci. **7**(May), 1–15 (2021). https://doi.org/10.1017/dsj.2021.7

16. Kery, M.B., Myers, B.: Exploring exploratory programming. In: Symposium on Visual Languages and Human-Centric Computing (VL/HCC), pp. 25–29. IEEE (2017). https://doi.org/10.1109/VLHCC.2017.8103446

17. Knuth, D.E.: Literate programming. Comput. J. **27**(2), 97–111 (1984). https://doi.org/10.1093/comjnl/27.2.97

18. Leitão, A., Lopes, J., Santos, L.: Illustrated programming. In: Proceedings of the 34th Annual Conference of the Association for Computer Aided Design in Architecture (ACADIA), pp. 291–300. Los Angeles (2014)

19. Nassi, I., Shneiderman, B.: Flowchart techniques for structured programming. SIGPLAN Not. **8**(8), 12–26 (1973). https://doi.org/10.1145/953349.953350

20. Nguyen, A.T., Reiter, S., Rigo, P.: A review on simulation-based optimization methods applied to building performance analysis. Appl. Energy **113**, 1043–1058 (2014). https://doi.org/10.1016/j.apenergy.2013.08.061

21. Oppenheimer, N.: An enthusiastic sceptic. Archit. Des. **79**(2), 100–105 (2009). https://doi.org/10.1002/ad.862
22. Picon, A.: Architecture and the virtual: towards a new materiality? Prax. J. Philos. 114–121 (2004)
23. Picon, A.: Beyond digital Avant-gardes: The materiality of architecture and its impact. Archit. Des. **90**(5), 118–125 (2020). https://doi.org/10.1002/ad.2618
24. van der Plas, F., Bochenski, M.: Pluto.jl (2021). https://github.com/fonsp/Pluto.jl. Accessed 14 Feb 2023
25. Price, B.A., Baecker, R.M., Small, I.S.: A principled taxonomy of software visualization. J. Vis. Lang. Comput. **4**(3), 211–266 (1993). https://doi.org/10.1006/jvlc.1993.1015
26. Rein, P., Ramson, S., Lincke, J., Hirschfeld, R., Pape, T.: Exploratory and live, programming and coding: a literature study comparing perspectives on liveness. Program. J. **3**(1), 1:1–1:33 (2018). https://doi.org/10.22152/programming-journal.org/2019/3/1
27. Roman, G.C., Cox, K.C.: Program visualization: the art of mapping programs to pictures. In: International Conference on Software Engineering, pp. 412–420. IEEE (1992). https://doi.org/10.1145/143062.143157
28. Ruck, A.: Abacus and sketch. In: Kara, H., Bosia, D. (eds.) Design Engineering Refocused, chap. 5, pp. 76–87. AD Smart 03. Wiley, Hoboken (2017)
29. Rule, A., Tabard, A., Hollan, J.D.: Exploration and explanation in computational notebooks. In: Proceedings of the 2018 CHI Conference on Human Factors in Computing Systems, pp. 1–12. CHI 2018, Association for Computing Machinery, New York (2018). https://doi.org/10.1145/3173574.3173606
30. Sammer, M.J., Leitão, A., Caetano, I.: From visual input to visual output in textual programming. In: Proceedings of the 24th International Conference of the Association for Computer-Aided Architectural Design Research in Asia (CAADRIA), vol. 1, pp. 645–654. Wellington, New Zealand (2019)
31. Terzidis, K.: Expressive Form: A Conceptual Approach to Computational Design. Spon Press, London and New York (2003)
32. Terzidis, K.: Algorithmic Architecture. Architectural Press, New York (2006)
33. Turrin, M., von Buelow, P., Stouffs, R.: Design explorations of performance driven geometry in architectural design using parametric modeling and genetic algorithms. Adv. Eng. Inform. **25**(4), 656–675 (2011). https://doi.org/10.1016/j.aei.2011.07.009
34. Ventura, R.: Geometric Constraints in Algorithmic Design. Master's thesis, Instituto Superior Técnico, Universidade de Lisboa, Lisbon, Portugal (2021)
35. Wang, A.Y., Mittal, A., Brooks, C., Oney, S.: How data scientists use computational notebooks for real-time collaboration. Proc. ACM Hum. Comput. Interact. **3**, 1–30 (2019). https://doi.org/10.1145/3359141

AI-powered Architectural Ideation

Architectural Sketch to 3D Model: An Experiment on Simple-Form Houses

Hong-Bin Yang$^{(\boxtimes)}$ ⓘ, Mikhael Johanes ⓘ, Frederick Chando Kim ⓘ, Mathias Bernhard ⓘ, and Jeffrey Huang ⓘ

École Polytechnique Fédérale de Lausanne, Lausanne, Switzerland
`hong-bin.yang@epfl.ch`

Abstract. Transforming sketches into digital 3D models has been an enduring practice in the design process since the first digital turn in architecture. However, 3D modeling is time-consuming, and 3D modeling software usually has a cumbersome interface. Aiming to bridge the gap between sketch and 3D model, we propose a framework that can turn a hand-drawn 2D sketch into a 3D mesh. The user can draw a sketch on the canvas through a web-based interface, and the corresponding 3D model will automatically be generated and shown aside. The 3D model can be downloaded or synchronized into the Rhino directly through Grasshopper.

The proposed framework uses a machine learning-based approach to generate a 3D mesh from a single hand-drawn sketch by deforming a template shape. Since the generated models have an uneven surface, we apply an optimization step to refine the form, creating a more usable architectural 3D model with planar faces and sharper edges. We create the Simple House Dataset, which consists of 5000 single-volume houses, to train the neural network. We defined five categories of house typologies - distinguished by roof shape and other geometric properties - and generated 1000 models for each class with parameters chosen randomly. Each model includes a 3D mesh and 20 perspective line drawings from different angles.

Although the limitation of the generalization ability makes it unlikely to replace the conventional 3D modeling software today, the fast sketch to 3D transformation allows architects to explore the possibility of various architectural forms and may speed up the design process in the early stage. The code of this project and the Simple House dataset has been published on Github [29].

Keywords: 3D Modeling · Machine Learning · Sketch to 3D · Mesh Optimization · Architectural Form

1 Introduction

Freehand sketching is a highly effective and intuitive method for quickly conveying ideas during architectural design. Using simple strokes, individuals can

M. Turrin et al. (Eds.): CAAD Futures 2023, CCIS 1819, pp. 53–67, 2023.
https://doi.org/10.1007/978-3-031-37189-9_4

transfer 3D forms in their minds and intuitively communicate concepts in design processes [18]. On the other hand, the 3D model provides a better experience for reviewing the overall design from different perspectives. However, constructing a 3D model is often time-consuming and requires significant expertise in 3D modeling software. Furthermore, conventional 3D modeling tools usually impose design bias depending on various factors, such as the user-friendliness of particular modeling interfaces and the underlying algorithm of the software [25].

To bridge the creative explorations of hand sketching with digital 3D modeling tools, an algorithm could be developed to reconstruct 3D models from hand-drawn sketches automatically. With this approach, architects could intuitively develop designs through sketching while simultaneously documenting and evaluating the results in 3D.

Although humans can easily infer 3D structures from 2D sketches, it is challenging for machines to automate this process. The skill depends heavily on the observer's experience, who unconsciously aligns the strokes with the shape of the commonly perceived objects. Mathematical analysis of these strokes reveals their non-deterministic nature and ability to represent an infinite number of structures, making it too ambiguous to design an algorithm that can reconstruct 3D scenes correctly, as depicted in Fig. 1.

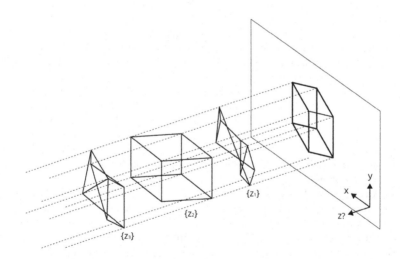

Fig. 1. The ambiguity makes developing a reliable algorithm for converting a sketch into a 3D model difficult. This image is credited to [19].

In addition to projection ambiguity, the presence of occluded objects brings an additional challenge for machines to reconstruct 3D models from sketches. Consider a chair with one leg hidden from view; humans typically infer that the chair has four legs rather than three, which may not be the case for machines. Successful reconstruction of a 3D model requires prior knowledge about the object in question. One approach to address this problem is to leverage machine learning techniques to train a model that learns the typical structure of a specific class of objects. However, this class-wise approach might lack generalization ability as it only learns the information from each category in isolation.

With the rapid evolution of deep neural networks in recent years, more and more machine learning-based methods have emerged in this field. However, their practical application in architectural design or 3D modeling processes remains unknown. To make this 2D to 3D generators accessible to users, an easy-to-use system needs to be developed. The main focus of this paper is to introduce a system that integrates a machine learning-based 2D sketch to 3D model generator into conventional 3D modeling software. The system consists of three main components:

1. A web-based interface that provides a canvas for users to sketch and view the generated 3D model.
2. A machine learning-based 3D model generator converts the 2D sketch into a 3D model. This is followed by a refinement approach, which flattens the uneven surfaces of the generated 3D model, resulting in a more accurate and realistic representation of the design.
3. A Rhino Grasshopper program that can connect to the web interface and automatically imports the generated 3D model into the scene, facilitating further design iterations and modifications.

The system presented in this paper offers users a rapid and intuitive method for generating 3D models from 2D sketches. Architects can benefit from this system during schematics design phases by inspecting the 3D while forming the concept. For clients or those with limited technical knowledge of 3D modeling software, this tool can also be utilized to be more involved in design conversations. While this is an experimental project with some limitations and areas for improvement, we believe that with further development, such a system can be updated gradually to bring new possibilities to the architectural design workflow. It may revolutionize how architects and designers conceptualize and communicate their ideas, leading to more effective and inclusive design processes.

2   Related Works

Nowadays, constructing a 3D model from 2D images has become more and more realistic due to the improvements in hardware and software capabilities. Either way, the 3D model can be built and modified instinctively. Although no commercial CAD software has included this feature, several academic research has been proposed. Here, we separate them into two categories and discuss them respectively.

2.1   Sketch-Based Modeling

Sketch-based modeling is a more direct and interactive way to build a 3D model by sketching. Most of them have a specific interface to capture the path of every single stroke. For example, some identify each stroke as a single object's silhouette and generate a corresponding 3D model by "inflating" techniques. [12] is one of the earliest research using this method. They demonstrated a stand-alone software with an interactive interface that allows the user to draw, erase, and edit the 3D model. Later research [13] further takes occlusion into account by analyzing contours containing T-junctions and cusps. [24] asks users to draw the silhouettes of each component from the front and side view, and the algorithm will generate the corresponding geometry by intersecting the two silhouettes. With a similar idea, [16] further introduces the "bending line," allowing the user to draw auxiliary strokes to indicate the degree of the bending surface. A recent research [6] presents a framework that automatically joins different parts and can be animated smoothly by user-specified control points and motion trajectory.

The above-mentioned sketch-based modeling typically requires users to draw the projection outline of an object onto a plane in the 3D space, with the system subsequently transforming the sketched line into a 3D component and merging them to form the final model. In contrast, we focus more on converting a 2D sketch with perspective information into a 3D model directly. It is more intuitive and less time-consuming since architects and designers frequently initiate their design process with a 2D sketch.

2.2   3D Model Reconstruction from 2D Image

In contrast to sketch-based modeling, we can take the entire 2D image as input and reconstruct the 3D geometry by analyzing the object(s) present in the scene. Consequently, training an end-to-end machine learning model is the common approach in this category, especially with the presence of large-scale dataset [3] and differential rendering [14,17]. These learning-based methods are springing up today [10]; some take a single image as input, and some require multiple photos from different views. The model can be generated as voxel [4,15], polygon mesh [28], and point cloud [7,9,23]. We consider polygon mesh the most suitable format for integration with most 3D modeling tools.

Nevertheless, many methods mentioned here are hampered by low general-ization ability, which is an intrinsic issue related to their machine learning model. Since the overall structure of the generated models is limited, these methods typ-ically train class-specific models for objects within the same category. As a result, they can only generate models within the same class with similar structures. To tackle this problem, [26,31] suggest using a 2-stage process by estimating the normal map, depth map, and silhouette as intermediate results through the first network, using them as the input to the 2nd network to generate the final 3D model.

Recently, multiple machine-learning models were proposed to reconstruct 3D models from hand-drawn sketches. [27] trained a network that generates a point cloud from a 2D sketch. [1] used a 2-stage structure, where the network first converts the sketch into multiple 2.5D normal maps and then takes it as input to generate the 3D mesh in the next stage. In this paper, we trained a Sketch2Model

[30] network with the dataset we created for the initial model generation, which will be discussed in detail in the following section.

3   Method

3.1   Synthetic Data Generation

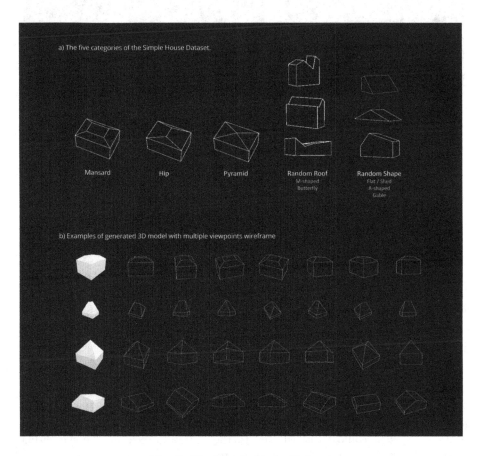

Fig. 2. The Simple House Dataset.

In the beginning, we combined the models generated by Synthetic Building Dataset Generator [8] to form the training dataset for our experiment. However, we quickly realized the algorithm used in the generator is focused on complex urban-block building typologies, which does not fit our purpose. In this initial experiment, we would like to train the model with various but simple architectural formal typologies, which led to the creation of our own dataset.

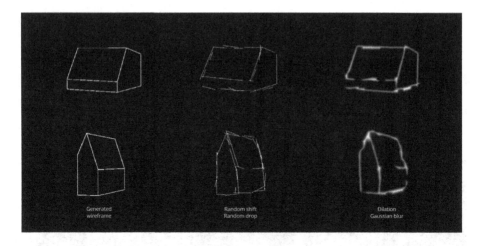

Fig. 3. Examples of the hand-drawn sketch style simulation.

We created the Simple House Dataset that contains single-volume houses with 5-different types of roofs, including flat, gable, mansard, hip, pyramid, M-shape, butterfly, and gambrel as shown in Fig. 2a. For each generated house model, 20 images from the randomly chosen viewpoint are rendered in a wire-frame style. Examples are shown in Fig. 2b. The viewpoint is described by azimuth and elevation, where the sphere center aligns with the center of the model's bounding box. The distance from the center to the camera is fixed. For each type of roof, 1000 pairs of data are generated, resulting in 5000 samples. Among them, 4000 samples are used for training, and the rest 1000 are preserved for evaluation and testing.

3.2   Hand-Drawn Sketch Style Simulation

To increase the model's performance on the hand-drawn sketch, we first tried to train a Contrastive Unpaired Translation (CUT) [22] network, seeing if it can convert the hand-drawn sketch into a refined wire-frame image. However, the generated image does not preserve the key structure of the object. The failure of the conversion could be attributed to the lack of context in the sketch-like illustrations to infer the conversion in comparison to photographic images.

As an alternative, we designed a pre-processing pipeline for data augmentation, where the rendered wire-frame images would go through during training. First, Hough Line Detection [5] is used to extract straight lines in the picture, and some of them are randomly dropped. For each preserved line, the two end-points are randomly shifted by some pixels, creating an imperfect structure as a hand-drawn sketch. Then, dilation and Gaussian blur are applied to union the scattered lines into a single stroke. The example is shown in Fig. 3.

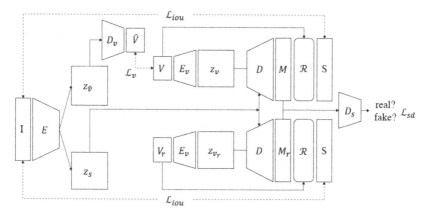

Fig. 4. Network Architecture of Sketch2Model [30]. The purple and blue blocks represent inputs and trainable parameters, respectively. The image is taken from the original paper. (Color figure online)

3.3 3D Model Generation

To reconstruct the 3D model from a 2D sketch, we trained a Sketch2Model [30] network as our generator. It takes a single 2D image and optional viewpoint parameters as input to generate a 3D mesh by deforming a template shape. In the original paper, the model was trained and tested with the synthetic data they created with ShapeNet dataset [3], including 13 different classes, such as airplane, car, chair, or table. In this experiment, we keep the model architecture and optimization target unchanged and train it with the Simple House Dataset we created. Instead of training the model class-wise, we trained it with the five architectural categories to see its performance in learning various architectural forms within the classes.

Model Structure. The model structure is shown in Fig. 4. It aims to generate a 3D mesh M from the input sketch I drawn from viewpoint V. The input sketch first goes through an encoder E, extracting a compact feature vector z, further mapped to a latent view space z_V and a latent shape space z_S. The decoder D takes a set of latent view codes z_V and a shape code z_S as input and generates a 3D mesh M. After generating mesh M, a differentiable rendering module R renders a 2D image S, allowing the network to be trained end-to-end by comparing the rendered sketch S and input sketch I. A view auto-encoder (E_v, D_v) is trained together for view estimation between the latent code and the decoder.

To ensure that the decoder D takes z_v into consideration, instead of only considering z_s, a random viewpoint V_r is fed into another branch to generate another mesh M_r. While adding a constraint to ensure the consistency of the projected silhouette of M_r under V_r and the input silhouette, the network would learn how to re-project one sketch from its original viewpoint to another. However, in this way, the network may generate oddly shaped meshes that only satisfy the constraint but are not meaningful to humans. To prevent this side effect, a discriminator D_S is introduced to ensure the quality of the generated model. In the implementation, ResNet-18 [11] and SoftRas [17] is used for the encoder and differentiable rendering modules, respectively.

Optimization Objectives. The loss function consists of 4 parts:

1. Silhouette loss L_s: the combination of the IoU loss L_{iou} between the silhouettes of the input sketch I and the generated one S.
2. Geometry regularization loss L_r: flatten loss and Laplacian loss as regularization for mesh structures, same as [14, 17], which enhance the visual quality of generated meshes.
3. Viewpoint prediction loss L_v: the Euler angle of the estimated viewpoint and the ground truth, where the distance is fixed.
4. Shape discriminator and domain discriminator L_{sd} and L_{dd}: the shape discriminator is expected to output 1 for M and 0 for M_r, and for the domain discriminator, 1 for synthetic sketches and 0 for hand-drawn sketches. L_{sd} guarantee the consistency between the mesh, viewpoint, and rendered image, and L_{dd} helps improve the robustness of the hand-drawn sketch by classifying features from the two domain.

The model is trained from scratch for 2000 epochs using Adam optimizer with learning rate $= 1e - 4$, and the hyper-parameters are set to $\lambda_r = \lambda_{sd} = \lambda_{dd} = 0.1, \lambda_v = \lambda_{vr} = 10$.

3.4   Mesh Optimization

The generated model comes with uneven surfaces and blurred edges, which is not the desired artifact in most cases. Thus, we use PolyFit [21] to refine the mesh, turning it into a simpler mesh with flattened surfaces and sharper edges. Since the algorithm is working on point cloud data, the generated mesh is first converted into a point cloud by uniform sampling, and then, the point clouds are clustered by plane-fitting based on their normal and distance. Finally, an optimization process is applied to trim the plane, forming a manifold and watertight mesh. An example of this procedure is shown in Fig. 5, and the results of the refined meshes are shown in the third row of Fig. 6.

3.5   User Interface

A user interface for generating 3D models is developed as a web-based application. This platform lets users sketch directly on a canvas within their browser while simultaneously interacting with the resulting 3D model. The most significant benefit of the web application is the accessibility across various devices, such as smartphones and tablets, so the user can draw naturally with a stylus pen or finger rather than being constrained by a mouse.

Generated Mesh   Point Cloud Sampling +   Clustering +   Plane Segmentation   Optimized Mesh
Normal Estimation   Plane Estimation

Fig. 5. The process of the mesh optimization. The procedure takes a mesh with uneven surfaces as input and generates an optimized mesh with flat surfaces and sharper edges.

The system comprises a web application, a back-end server, and a Grasshopper script with a synchronization tool. The drawing component of the web application is powered by p5.js [20], an open-source JavaScript library that simplifies the process of creating visual and interactive applications in the web browser. Three.js [2], a library specializing in computer graphic rendering on the web, is utilized to display the 3D model.

The back-end server is developed based on Flask, one of the most popular web frameworks. The 2D-to-3D model is integrated into the back end to reduce the computational load on the front end. The conversion of a 2D sketch to a 3D model is designed to be easily modifiable, allowing for seamless upgrades to newer and better models in the future without affecting the rest of the system.

To synchronize the 3D model with Rhino, a Grasshopper script is created to automate the process. However, due to the low scalability of IronPython (the Python scripting engine of Grasshopper), it is hard to develop a web-based application on top of it. As a compromise, an auxiliary synchronization tool is developed, using Python to interact with the server and automatically download the 3D model to a temporary folder for the Grasshopper script to import.

When a user first accesses the web application, a new session is created with a unique ID. This ID identifies the session and enables sharing of 3D models between the web application and Grasshopper. Once a user has sketched their design, the image is sent to the back-end server for conversion into a 3D model. After conversion, the resulting model is sent back to the user's browser and the synchronization tool that shares the same session ID. Finally, by triggering the Grasshopper script with a button, the model will be imported into Rhino.

4   Result and Discussion

4.1   Generated Results

The testing results on the synthetic dataset and the hand-drawn sketch are shown in Fig. 6. The model can process these sketches and generate satisfactory results from those examples. However, as mentioned above, this model still lacks generalization ability. First, the model is sensitive to defects, where the model would generate unexpected results (Fig. 7, right). The model cannot produce a matching result for cases further away from examples in the dataset (Fig. 7,

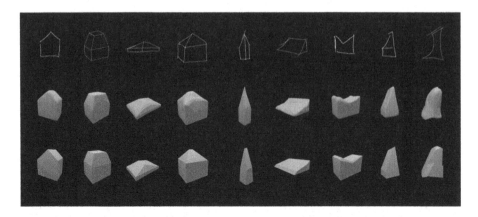

Fig. 6. Results of the 3D models generated from 2D hand-drawn sketches. From the top to bottom row is the input 2D sketch, the raw 3D mesh generated by the machine learning model, and the refined mesh with post-process, respectively.

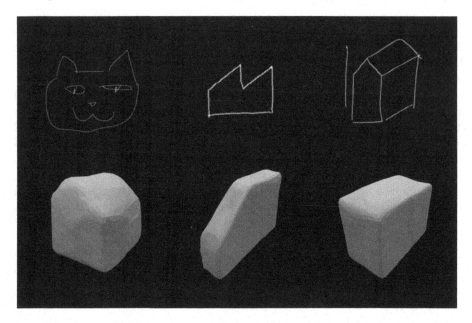

Fig. 7. Example of failed cases. The first row shows the sketch drawn by the user, and the second row shows the corresponding result generated by the machine-learning model.

middle). This also applies to a sketch that has nothing to do with a house (Fig. 7, left). That means no matter what the user draws, it attempts to generate a house that fits the silhouette as closely as possible. Though it can be a feature that inspires the designer, it may also harm this system's potential. One feasible development is to include automated and user-generated feedback that balances the machine learning model's precision, generalization ability, and uncertainty in generating 3D models.

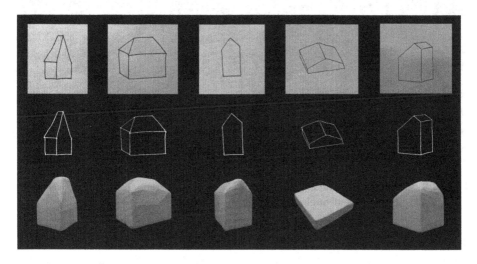

Fig. 8. Example of reconstructing 3D model from hand-drawn sketch with pen and paper. The original sketch on paper, the prepossessed result, and the generated 3D model are from top to bottom.

Another drawback of this model is the limitation on the consistent topology. Since it generates an object by deforming a template mesh, the topology will be the same as the template object. For example, if the template is a sphere (genus 0), but a torus (genus 1) is presented in the training or testing data, the generated result will be distorted.

We also tested the model generator by sketching with pen and paper. The result is shown in Fig. 8. After drawing on paper, a photo is taken and cropped so that the target shape is almost in the center of the whole image. The image will then be sent to a prepossessing pipeline, where the background is removed, and the image is converted to a similar style to the training dataset. The results show that with sufficient image pre-processing steps to clean up the image, we can extend the interface by using a mobile phone's camera to capture the sketches on paper, further developing the flexibility of the interface.

4.2   Consistency Test

In Fig. 9, we show that the reconstructed 3D meshes from different viewpoints are almost identical, indicating the robustness of the various perspectives. While the view-point consistency is preserved, the user can sketch from the preferred perspective instead of an assigned one but still get the expected 3D model.

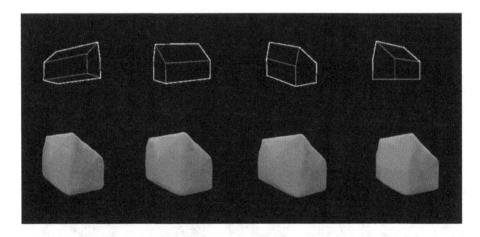

Fig. 9. Testing result of the consistency of the generated model toward the rendered 2D wire-frame of different angles.

4.3   User Interface

In Fig. 10, we show screenshots of the user interface, which includes (a) a web application demonstrating a sketch drawn by a user and translation into a 3D model and (b) a Rhino and Grasshopper script for importing the resulting model. In the web application, the user can sketch on the left side and generate the 3D model by pressing the convert button in the middle, the 3D model on the right will be updated in nearly real-time. The example shows a scenario where the user accesses the interface through a tablet and sketches with a stylus pencil. The user can redo, undo, and clear the canvas anytime. On the desktop side, the Grasshopper is connected to the same "room" as the tablet by launching a script. When the mesh is generated, the user can trigger a button in Grasshopper to import the model into Rhino automatically.

a) Web Application   b) Grasshopper Script

Fig. 10. A user is drawing on a tablet, and the result is synchronized to Rhino viewport through the Grasshopper script.

5   Conclusions

The proposed system offers a user-friendly and efficient way for architects and designers to explore their ideas in the early design process. By using deep neural networks to convert hand-drawn sketches into 3D models, the system allows users to quickly visualize their concepts in 3D. With the combination of Grasshopper, users can also perform further editing on top of the generated model.

We separate the front-end and back-end, making the system easily accessible for the end user while preserving scalability and extendability. This makes embedding the new technology into the conventional design phase more realistic.

We generated a synthetic dataset for training the neural network used in our system. This involved the randomized 3D model generation and 2D wireframe rendering, and a free-hand sketching simulation. This approach allows for creating a dataset tailored explicitly for the architectural design task, thereby opening up opportunities for future researchers to develop learning-based methods for this field. Additionally, we proposed a post-processing optimization procedure to enhance the generated 3D models and make them more suitable for architectural purposes.

Although the current model may have some limitations regarding generalization ability, it still shows the potential for creative use and can serve as a starting point for further exploration. The early translation of sketches into a 3D modeling environment could make the evaluation and simulation of design be performed earlier. It could also make the design processes more inclusive by making the 3D modeling less cumbersome to broader users. We believe that including artificial intelligence in computer-aided design will inevitably expand the landscape of architectural design processes.

References

1. Bhardwaj, N., Bharadwaj, D., Dubey, A.: SingleSketch2Mesh: generating 3D mesh model from sketch. arXiv preprint arXiv:2203.03157 (2022)
2. Cabello, R.: three.js (2010). https://github.com/mrdoob/three.js. Accessed 01 Mar 2023
3. Chang, A.X., et al.: ShapeNet: An information-rich 3D model repository. arXiv preprint arXiv:1512.03012 (2015)
4. Choy, C.B., Xu, D., Gwak, J.Y., Chen, K., Savarese, S.: 3D-R2N2: a unified approach for single and multi-view 3D object reconstruction. In: Leibe, B., Matas, J., Sebe, N., Welling, M. (eds.) ECCV 2016. LNCS, vol. 9912, pp. 628–644. Springer, Cham (2016). https://doi.org/10.1007/978-3-319-46484-8_38
5. Duda, R.O., Hart, P.E.: Use of the Hough transformation to detect lines and curves in pictures. Commun. ACM **15**(1), 11–15 (1972)
6. Dvorožňák, M., Sýkora, D., Curtis, C., Curless, B., Sorkine-Hornung, O., Salesin, D.: Monster mash: a single-view approach to casual 3D modeling and animation. ACM Trans. Graph. (TOG) **39**(6), 1–12 (2020)
7. Fan, H., Su, H., Guibas, L.J.: A point set generation network for 3D object reconstruction from a single image. In: Proceedings of the IEEE Conference on Computer Vision and Pattern Recognition, pp. 605–613 (2017)

8. Fedorova, S., et al.: Synthetic 3D data generation pipeline for geometric deep learning in architecture (2021)
9. Gadelha, M., Wang, R., Maji, S.: Multiresolution tree networks for 3D point cloud processing. In: Proceedings of the European Conference on Computer Vision (ECCV), pp. 103–118 (2018)
10. Han, X.F., Laga, H., Bennamoun, M.: Image-based 3D object reconstruction: state-of-the-art and trends in the deep learning era. IEEE Trans. Pattern Anal. Mach. Intell. **43**(5), 1578–1604 (2019)
11. He, K., Zhang, X., Ren, S., Sun, J.: Deep residual learning for image recognition. In: Proceedings of the IEEE Conference on Computer Vision and Pattern Recognition, pp. 770–778 (2016)
12. Igarashi, T., Matsuoka, S., Tanaka, H.: Teddy: a sketching interface for 3D freeform design. In: ACM SIGGRAPH 2006 Courses, pp. 11-es (2006)
13. Karpenko, O.A., Hughes, J.F.: SmoothSketch: 3D free-form shapes from complex sketches. In: ACM SIGGRAPH 2006 Papers, pp. 589–598 (2006)
14. Kato, H., Ushiku, Y., Harada, T.: Neural 3D mesh renderer. In: Proceedings of the IEEE Conference on Computer Vision and Pattern Recognition (CVPR) (2018)
15. Knyaz, V.A., Kniaz, V.V., Remondino, F.: Image-to-voxel model translation with conditional adversarial networks. In: Proceedings of the European Conference on Computer Vision (ECCV) Workshops (2018)
16. Li, C., Pan, H., Liu, Y., Tong, X., Sheffer, A., Wang, W.: BendSketch: modeling freeform surfaces through 2D sketching. ACM Trans. Graph. (TOG) **36**(4), 1–14 (2017)
17. Liu, S., Li, T., Chen, W., Li, H.: Soft rasterizer: a differentiable renderer for image-based 3D reasoning. In: The IEEE International Conference on Computer Vision (ICCV) (2019)
18. MacDonnell, J.A. (ed.): About Designing: Analysing Design Meetings. CRC Pr./Balkema, Boca Raton (2009)
19. Masry, M., Lipson, H.: A sketch-based interface for iterative design and analysis of 3D objects. In: ACM SIGGRAPH 2007 Courses, pp. 31-es (2007)
20. McCarthy, L.: p5.js (2013). https://github.com/processing/p5.js. Accessed 01 Mar 2023
21. Nan, L., Wonka, P.: PolyFit: polygonal surface reconstruction from point clouds. In: Proceedings of the IEEE International Conference on Computer Vision, pp. 2353–2361 (2017)
22. Park, T., Efros, A.A., Zhang, R., Zhu, J.-Y.: Contrastive learning for unpaired image-to-image translation. In: Vedaldi, A., Bischof, H., Brox, T., Frahm, J.-M. (eds.) ECCV 2020. LNCS, vol. 12354, pp. 319–345. Springer, Cham (2020). https://doi.org/10.1007/978-3-030-58545-7_19
23. Ping, G., Esfahani, M.A., Chen, J., Wang, H.: Visual enhancement of single-view 3D point cloud reconstruction. Comput. Graph. **102**, 112–119 (2022)
24. Rivers, A., Durand, F., Igarashi, T.: 3D modeling with silhouettes. In: ACM SIGGRAPH 2010 papers, pp. 1–8 (2010)
25. Serriano, P.: Form Follows Software, pp. 185–205. Indianapolis (Indiana), USA (2003). https://doi.org/10.52842/conf.acadia.2003.185
26. Thai, A., Stojanov, S., Upadhya, V., Rehg, J.M.: 3D reconstruction of novel object shapes from single images. In: 2021 International Conference on 3D Vision (3DV), pp. 85–95. IEEE (2021)
27. Wang, J., Lin, J., Yu, Q., Liu, R., Chen, Y., Yu, S.X.: 3D shape reconstruction from free-hand sketches. arXiv preprint arXiv:2006.09694 (2020)

28. Wang, N., Zhang, Y., Li, Z., Fu, Y., Liu, W., Jiang, Y.G.: Pixel2Mesh: generating 3D mesh models from single RGB images. In: Proceedings of the European Conference on Computer Vision (ECCV), pp. 52–67 (2018)
29. Yang, H.B.: Architectural sketch to 3D model: an experiment on simple-form houses (2020). htttps://github.com/Petingo/Architectural-Sketch-To-3D-Printing. Accessed 06 Oct 2022
30. Zhang, S.H., Guo, Y.C., Gu, Q.W.: Sketch2Model: view-aware 3D modeling from single free-hand sketches. In: Proceedings of the IEEE/CVF Conference on Computer Vision and Pattern Recognition (CVPR), pp. 6012–6021 (2021)
31. Zhang, X., Zhang, Z., Zhang, C., Tenenbaum, J., Freeman, B., Wu, J.: Learning to reconstruct shapes from unseen classes. In: Advances in Neural Information Processing Systems, vol. 31 (2018)

Towards Human-AI Collaborative Architectural Concept Design via Semantic AI

Shuyao Dai[1(✉)] ⓘ, Yang Li[2] ⓘ, Kazjon Grace[1] ⓘ, and Anastasia Globa[1] ⓘ

[1] School of Architecture, Design and Planning, The University of Sydney, 148 City Road, Darlington, NSW 2008, Australia
sdai8012@uni.sydney.edu.au, {kazjon.grace, anastasia.globa}@sydney.edu.au
[2] Department of Architecture, National University of Singapore, 4 Architecture Drive, Singapore 117566, Singapore
l.y@u.nus.edu

Abstract. As artificial intelligence (AI) methods advance quickly, more and more researchers are becoming interested in how to incorporate them into architectural design. Co-creation between humans and machines is also gaining popularity, which lends credence to the idea that AI can aid in the creative phases of design. The research presented in this article develops an AI-assisted method for generative design. It envisions a pipeline that iteratively kneads from semantics to two-dimensional (2D) images to three-dimensional (3D) models and back again by combining a semantic AI model (CLIP) with differentiable rendering. It also enables conceptual form exploration in Rhino3D with the help of a neural network built on a Text2Mesh tool. The real-time, conceptual, iterative interplay between human designers and AI collaborators could be facilitated by this pipeline. We also conducted a case study on early concept exploration for a museum to validate our approach, showcasing its potential in practical design scenarios.

Keywords: Artificial intelligence (AI) · design methods · concept design · semantic AI · architectural design · computational architecture · co-creative systems · differentiable rendering · 3D machine learning

1 Introduction

Artificial intelligence (AI) is rapidly transforming many disciplines but has yet to significantly affect architectural design. Recently developed AI models based on contrastive training over language-image pairs such as DALLE-2 [1] and IMAGEN [2] emerge with the certain revolutionary potential to change how AI and designs. This revolutionary innovation provides the ability to convert conceptual ideas between natural language and (2D) images [1, 3]. These AI models display the ability to deeply understand human language and associate it with images in both literal and abstract ways [1]. This suggests the possibility that designers may be able to semantically interact with computational "drawings". In this research, we develop an early, exploratory prototype of how that might work, and use it to speculate on the future potential of these AI models in the context of augmenting architectural design.

© The Author(s), under exclusive license to Springer Nature Switzerland AG 2023
M. Turrin et al. (Eds.): CAAD Futures 2023, CCIS 1819, pp. 68–82, 2023.
https://doi.org/10.1007/978-3-031-37189-9_5

Architectural design relies on three-dimensional (3D) representations, especially during the concept design stage. At this stage, design ideas are ambiguous and require multiple expressions and iterative exploration. Specifically, it is impossible to separate an emerging idea from its multi-modal expression across diverse (and often subtly conflicting) texts, sketches, and models [4]. Architects use sketches, floor plans, sections, and elevations to design and represent 3D spaces in two-dimensional (2D) media. Fragmentary and seemingly ambiguous language is used to describe the design elements and concepts, such as a "transparent" facade, or a garden "with a Zen feel". In this space of multi-modal, abstract representation, repeated iterations are required to find solutions to design problems [5], and those designers who are open to making significant changes to their concepts during these iterations often produce more creative results [6]. There is a mismatch between the write-prompt-get-image interaction model of generative AI systems and the process of architectural design in practice. We are exploring how these new AI techniques can be applied to support the design process.

As an example of such a system, Mid-Journey [7] is a very successful AI model that can turn text prompts into images. Architects, both students and practitioners, have been exploring the use of "semantic" AI technologies for design concept ideation. These technologies can transition between linguistic and visual representations, providing a new avenue for Human-AI collaboration in the creative process. However, most current AI systems for design concept ideation are limited to 2D images and do not support 3D modelling. This proof-of-concept study focuses on the possibility of taking semantic AI into the 3D architecture design context by integrating CLIP [8], VQGAN [9], and a 3D mesh library for machine learning called Kaolin[1] into the conceptual design stage. Design is an exploratory process that involves iterative iterations, and this study aims to establish a framework for AI's involvement in augmenting designers during conceptual decision-making.

This research integrates semantic AI into the early stages of architectural design to propose a new method of human-computer co-creation. Prototypes and case studies validate this approach, establishing a foundation for its integration into interactive design systems. The approach showcases AI's potential as a design assistant and its role in shaping the future of architectural design.

2   Background

This study draws from two relevant fields of research: digital architecture, which explores the use of digital tools in architectural design, and semantic AI, which focuses on algorithms that can understand the meaning of language and translate it into 3D form ideation.

---

[1] https://github.com/NVIDIAGameWorks/kaolin.

2.1 Digital Tools for Architectural Design

Architectural form is a nuanced and intricate concept that holds significant importance within the realm of architectural design. It is an essential and integral component of an architect's design practice. According to Edmund Bacon, it is the point of contact between mass and space [10]. Gestalt psychology suggests that the mind simplifies the visual environment to enhance understanding, and when presented with a composition of forms, individuals tend to reduce complexity by focusing on the simplest and most regular shapes [10]. Architectural design appears to conform to this phenomenon, as demonstrated by existing paradigms that trace progress from primal geometries to complex forms [4]. During the early stage of concept ideation, the search for architectural solutions involves a process of iterative exploration between simple forms and slightly more developed ones [5]. This study divides the form ideation process into primal form, exploring the simplest shapes, and detailed form, adding indicative yet preliminary details. The building's form constrains its spatial organization and functionality, and the primal form stage conceptualizes rough form ideas regarding cultural context, functionality, and other key factors [4]. The iterative exploration of the simplest forms during the early stages of design is a critical process for determining the top-level relationships within a building's design, knowing as massing design. The design of details often builds upon this preceding step, as with adding shape complexity like doors, windows, and facade details to a basic box.

Architects in the detailed form stage usually create multiple versions of a chosen conceptual form, often using 3D models to compare and assess them, adopting certain features and discarding others based on rapidly evolving aspects [11, 4]. Schøn defines a "design move" as a prerequisite for an idea to become "seen" and be evaluated, which involves developing a proposed concept to a level where it can be adequately perceived [12]. Thus, the aim of detailed form is to quickly test concepts as architecture is complex and unpredictable, with countless potential details and features that remain unknown at the start of the design process [13].

Alternative approaches exist for form finding beyond the simple-to-complex progression, such as emergent forms derived from natural laws. However, regardless of the chosen approach, designers rely on a similar process that involves reification, reflection, interpretation, and refinement, to ensure the controllability of human in that process. In our prototype, we operationalize conceptual form as an iteratively modifiable 3D model. This model is represented by a mesh that can be manipulated by an AI based on prompts. We do not intend this to be a complete definition that captures all aspects of conceptual form, but as an approximation that serves our explorations for now.

2.2 Semantic AI

Semantic AI is derived from the classical machine learning field of natural language processing, known as NLP. Semantic models extract the "meaning" of an artefact from one representation (such as a written phrase in French), allowing it to be transferred to another representation (such as a spoken phrase in Chinese). Most notable for our study is the development of semantic models of images, which started with automated captioning

[14] but have evolved into the capacity to generate images from text. Compared to previous computer vision systems which were trained for the prediction of fixed categories of images, semantic AI enlarges the space of possibilities to (in theory) the whole of written language. CLIP [8] is one of the most well-known published AI models for a task like this and was trained on 400 million text-image pairs scraped from the Internet. This pretrained model can perform zero-shot prediction, meaning it can classify images across a very large number of hypothetical experiments without any (further) training examples. This reduces the high cost of AI model training in data, computational power, and time, making it more feasible for individuals and smaller organizations. New "semantic interaction" applications of this technology are becoming rapidly available, like CLIP-Draw [15], Mid-Journey [7], and Stable Diffusion [16]. The ambition of some of these is to go beyond being digital tools and become co-creators, expanding the imaginative powers of their human collaborators. In Fig. 1, we demonstrate how easily Mid-Journey can be used to generate potential design inspiration by only using a simple prompt (in this case, "courtyard house with triangle form").

Fig. 1. Images generated by Mid-Journey from prompt "courtyard house with triangle form."

2.3 Differentiable Rendering

Differentiable rendering is a critical technology that makes machine learning over non-pixel-based representations possible. It comprises a series of techniques for "reversibly" rendering of 3D models into 2D images, forming stylized images from virtual 3D environments [17]. Traditional rendering pipelines, rasterization, and ray tracing make it difficult to correlate the image results to the rendering parameters like camera location, lighting, texture, etc.(ibid). With differentiable rendering, this process is mathematically reversible (i.e., it is possible to take the 2D images back to 2D vectors or 3D objects) using only processes to which differentiable calculus can be applied. This has no impact on the final product but allows a machine-learning technique called "gradient descent" to be applied. Gradient descent is a search algorithm that uses calculus-based methods to "backpropagate" the current error from the output (the generated image) back to the parameters used to define it (in this case, the 2D and 3D forms), which can then be updated. Undifferentiable processes require other (much slower) search and optimization processes, rendering the advantages of the deep learning revolution largely useless. Measuring the "quality" of a rendered output in a differentiable way effectively means that modern machine learning can be used to improve that quality. This opens the possibility of 3D modelling and editing based on semantic AI.

2.4 ML for Architectural Design

Currently, machine learning in architecture is still in a research stage with a wide range of explorations, without a clear consensus on its applicability and limitations [18]. Parametric design, typically relies on designer-driven approaches to modelling problems, while the statistics-based (learning) ML algorithms permit the exploration of more data-driven approaches [19]. During a teaching practice, we delved into the possibility of incorporating ML into a parametric system. This involved utilizing ML either as a means of interpreting natural language and converting it into parameters for the system, or as a component to facilitate the generation of certain forms within the system. Other design researchers have looked to adopt existing deep learning models from the computation vision domain into a design context [20]. For instance, the pix2pix model, a popular architecture based on a Generative Adversarial Network (GAN), has been used to automatically generate floor plans and transform plan styles [19]. ML also offers new potential representations for 3D scenes, such as neural radiance fields (NeRFs), which directly generate the results of rendering a 3D object or scene from any angle, without any 3D model data at all [21]. Another area of research in generative AI involves using variational autoencoder (VAE) for form blending [22] and structure optimization [23]. Co-creation is a comparatively underexplored theme as it presupposes the capacity of AI to engage in creativity and meaningful interaction, which demands the utilization of more advanced technologies.

3   New Methods for Semantic AI Augmented Design

Our approach involves the use of prompts to aid architects in controlling meshes during the conceptual ideation stage of design, which is characterized by ambiguity, uncertainty, and evolving goals. The conventional approach to this process involved physical sketching or model-making, whereas digital models are increasingly used for exploring design concepts. Our approach, illustrated in Fig. 2, exemplifies the integration of semantic AI into the ideation stage of design. This involves four distinct steps: primal form exploration, re-representation, semantically guided optimization, and detailed form exploration. The initial and final stages of the process are driven by humans, while the two intervening stages are computational. We have developed proof-of-concept implementations that taken together demonstrate each stage of this process, seen in Fig. 3.

In architectural design, the primal form stage, commonly known as massing, aims to explore potential building forms while considering factors such as size, volume, aesthetics, and functional groupings. In the subsequent stage, the 3D model resulting from massing comprises multiple primal forms amenable to manipulation.

Our optimization algorithm relies on the use of per-vertex colour and normal vectors. To achieve significant changes in massing objects using CLIP-based guidance, it is necessary to re-mesh the represented 3D mesh to increase its topological density. Due to the substantial size of the solution space, the computational complexity of the task

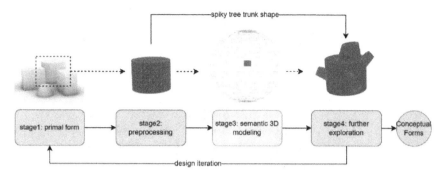

Fig. 2. A worked example of using semantic AI to generate 3D conceptual forms. In the first step, a designer produces a combination of primal forms. In the second step, it is converted for use with our AI. In the third step, it is augmented through an optimization process with a semantic objective. And in the final step, it is converted back into a form that can be further manipulated by the designer. These images are indicative, see Figs. 4, 5, 6, 7, 8, 9, 10, 11, 12, 13, 14 and 15 for actual output.

is challenging. As a result, the careful selection of the division and parameterization of massing models is of utmost importance. In our approach, the neural network's represented meshes are predetermined, with a fixed number of vertices. This allows for various resolutions but necessitates the determination of the resolution prior to semantic exploration (e.g., to control the proportion and distribution of details). Based on our practical experience, we have observed that the selection of mesh subdivisions plays a significant role in influencing the transformation process between semantic language and mesh objects.

We tested two re-meshing methods - dynamic relaxation algorithm for even triangle distribution and a regular horizontal/orthogonal grid - and compared their effectiveness. When dealing with organic shapes, the dynamic relaxation algorithm for re-subdivision proves to be advantageous, as it facilitates easier soft form adjustments. Conversely, for more geometric forms, the regular horizontal/orthogonal grid subdivision method is more beneficial, as it conforms more accurately to the target object's geometric topology. While it may be theoretically possible to achieve any form with a high enough mesh density, the associated computational and time costs would be significant. Therefore, designers must exercise caution in selecting appropriate subdivision approaches and densities to guide the addition of details more efficiently.

In a design scene comprising multiple model blocks, each requiring exploration and refinement, we suggest segmenting the scene into easily describable chunks before undertaking semantic exploration. Despite the absence of an absolute standard for segmentation, we will delve deeper into this topic in our case study. As each object will be optimized separately, it is imperative to normalize them to facilitate learning. Continuous zooming and movement after receiving results can pose a challenge for existing semantic AI models. Therefore, we preserve this transform information with prompts of each object to restore them back to the 3D scene later.

In the third stage (Fig. 3), integrating text2mesh [24] allows us to encode the mesh deforming as an optimization problem. Two attributes per vertex are optimized: the vertex

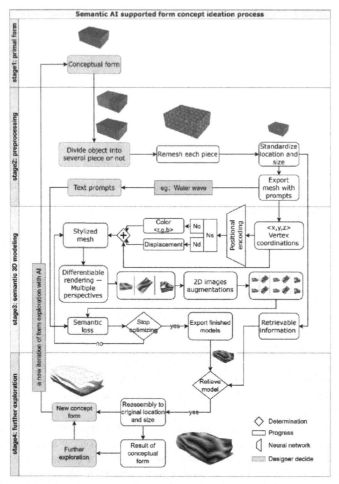

Fig. 3. Details of the generation process we use to implement the four-step process described in Fig. 2. Given the early stage of much research in this area, it is very likely that this somewhat involved process can be refined and simplified in future and present it as what is currently needed to reify our vision, rather than what might be optimal.

colour (RGB) and the displacement (the moving distance along the normal direction). We obtain multi-angle 2D representations of the 3D object through differentiable rendering from various angles. The resulting images are used to calculate the loss between the prompt and the 3D model via the CLIP neural network [8]. Subsequently, the mesh can be deformed iteratively to reduce the error in approaching the user-provided prompts.

In the fourth stage (Fig. 3), the AI modified 3D object is put back into the scene where it can be manually edited again. The previously stored relevant change information can be used to retrieve the original position, size, and orientation of the object in the scene. Architectural designers can choose to reintegrate, utilize, or draw inspiration from these AI augmented forms, which can also serve as an alternate starting point for further iteration of AI supported exploration.

Our current experiments do not incorporate this human-AI collaborative design step, and we present the AI-authored objects as a proof-of-concept of the approach. Future research would be needed to develop user interfaces for designer-AI collaboration to further explore and iterate on the form.

4   Co-creative Semantic AI Experiments

The figures presented below illustrate the outcomes of 700 iterations that were optimized using semantic AI. Figure 4 shows that the input mesh is orthogonally and equally subdivided, and the results show that although the overall feeling is very close to the described prompt, the details are not architecturally distributed with expected proportions. To address this issue, the basic object needs to be cut into pieces and individually subdivided in a way that matches the expected distribution of detail in the output.

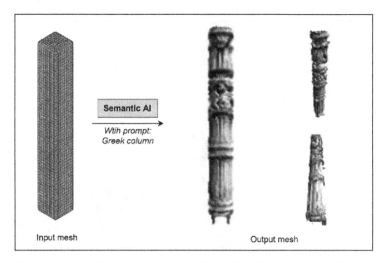

Fig. 4. Test on orthogonally and evenly subdivide mesh with prompt input "Greek column".

In Fig. 5, we manually divided the object according to the classical column ratio and only increased the mesh density at the stigma and base. By using the same prompt, the outputs show much better control of the distribution of details, and it also works for multiple objects at the same time. On the one hand, this approach provides designers with greater control over the level of detail in their designs. On the other hand, it also highlights the current limitations of AI in discovering such relationships on its own.

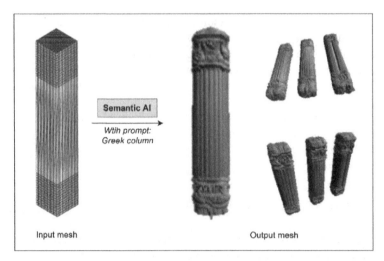

Fig. 5. Test on portion-controlled mesh with prompt input "Greek column".

We compared the impact of mesh topology on the outcome of our study by contrasting dynamic subdivision that creates regular triangles with orthogonal subdivision. Figure 6 shows the former and Fig. 7 the latter, both using the prompt "high-rise buildings with glass curtain walls". It can be clearly seen for a typically orthogonal shape like a high-rise building, that the subdivision approach produces more architectural effective detail. Conversely, the use of dynamic subdivision with an ununiformed mesh topology result in distorted generated details, often resembling a melted façade (Fig. 6).

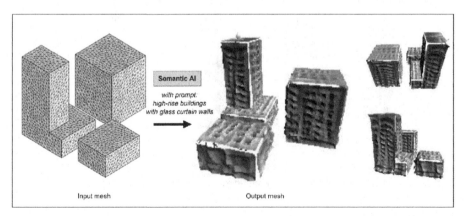

Fig. 6. Multiple dynamic subdivided meshes (based on dynamic relaxation trying to make all the triangular faces on the surface as identical as possible) with prompt "high-rise buildings with glass curtain walls". Note the non-planar orientation of windows to match the mesh.

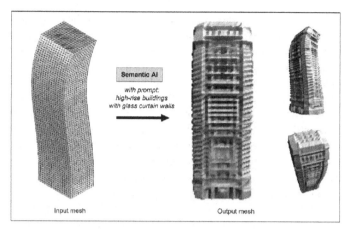

Fig. 7. Single orthogonal subdivided mesh with prompt "high-rise buildings with glass curtain walls". Note the planar mesh detail, even though the form is curved (contrast with Fig. 6).

5 Case Study: Museum Form Exploration with Semantic AI

In order to adequately showcase the potential of AI collaborative concept design, it is crucial to demonstrate its intended application and interactive collaborative capabilities in a real scenario. The aim of this case study is to generate a variety of formal design concepts for the new Lijiang Museum project in Yunnan Province, China. The project site is a rectangular plot of land situated in a suburban area of Lijiang. The design requirements call for a three-story cuboid form with dimensions similar to those of the site. To explore the form of the museum, we will utilize semantic concepts extracted from the cultural features of Lijiang City.

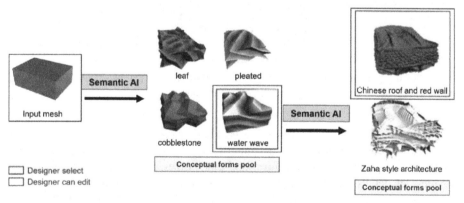

Fig. 8. An example process of how a designer collaborates with semantic AI, in which the designer can either select and edit the output of AI or treat it as another input of iteration.

The initial form (illustrated in Fig. 8 as the input mesh) represents a generic cuboid-mesh object that meets the design requirements and is commonly used in the initial design stage. During the conceptual or massing stage, the density of the mesh subdivision can be kept low to allow the machine to modify the form abstractly without being overly focused on details. Upon generating the initial base form, sub-concepts were applied to regenerate the design inspiration with specific styles as depicted in Fig. 8's middle step. The resultant output images of the cuboid, altered by different prompts, are illustrated in Fig. 9. By utilizing the triangular division of the cuboid mesh, the AI was able to modify the planar surface to an array of rippled forms. The outcome highlights the considerable potential to materialize semantic content into a tangible design form.

We selected four different concept forms (Fig. 9) using different prompts that best fit the local environment and culture such as "water wave", "pleated", "cobblestone", and "leaf". These four prompts provide significantly different formal textures on surfaces followed by semantic logic. While the tests were conducted in a preliminary manner, they demonstrated the potential to actualize verbal concepts into tangible forms. After refinement, we determined that the form generated by the "water wave" prompt was most suitable for further sub-concept exploration. As an initial architectural massing, the ripples on the surface helped to partially define the internal space separation and building stories. This provided designers with recognizable cues.

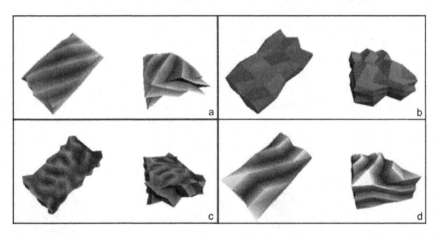

Fig. 9. The top left (a) uses the prompt "pleated". The top right (b) uses "cobblestone". The bottom left (c) uses "leaf". The bottom right (d) uses "water wave".

The machine exports a colour in each vertex to express the connection between form and prompt. Figure 10 illustrates the importance of colour mapping. It significantly influences our perceptions of the form and emphasizes the change of the form and provides rich details, which cannot be substituted by a non-textured form. Additionally, the colour mapping enhances the continuity of the ripples. After carefully analysing the previous test output, we have chosen a specific mesh outcome to further enhance the primary conceptual form with stylistic sub-concepts.

Following the generation of four distinct preliminary forms, we proceeded to refine the mesh subdivision depicted in Fig. 10 to incorporate additional details into the current form, using three new stylistic prompts that were tested as shown in Fig. 11. This represents a proof-of-concept for the effective utilization of semantic AI in iterative human-AI collaboration. Based on the initial iteration, we plan to further investigate the potential of AI in seamlessly blending an additional concept into the existing one.

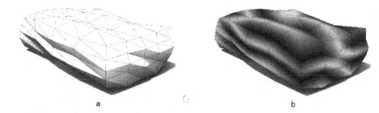

Fig. 10. Comparison between the plain mesh (a) and the colour-mapped mesh (b), both generated by the same CLIP-guided generative process.

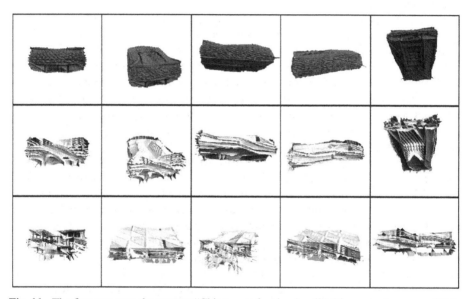

Fig. 11. The first row uses the prompt "Chinese roof and red wall". The second row uses "Zaha style architecture". The third row uses "Bauhaus style architecture".

The first row of results from Fig. 11 portrays the main entrance of the museum, highlighting the rhythm of the roof. The examples in the second and third rows demonstrate the fluidity of curvature and industrial expression, respectively. We can observe distinct stylistic changes and 3D morphological alterations, although they are occasionally interspersed with glitch-art style algorithmic additions. Although some angles are semantically closer to the prompt, which relates to the problem of pre-positioning camera angles from which to render. It is important to acknowledge that from certain perspectives, geometry may not be entirely optimal. This is because during the optimization process, the loss is averaged across different perspectives. As a result, certain angles may produce better 2D rendering results, which are then used to improve the 3D model.

However, overall formal exploration is interesting in terms of shape and colour. The base massing of the former iteration (Fig. 10) was not destroyed by the secondary form. The second concept serves to enhance the original massing by introducing unique and characteristic details.

6 Discussion

Our approach is presented as a proof-of-concept rather than a new standard for AI-augmented architectural design. It serves to demonstrate the potential of 3D semantic AI in this field, and the specific details of our algorithm reflect the current requirements for its successful application. We hope and expect that further innovation will reduce the complexity required, and accordingly present our specific approach as one data point in an as-yet underexplored field. We hope it is of use in facilitating further development in AI-mediated design.

From our exploration, we can say confidently that semantic AI offers radical new opportunities for human-machine co-creative exploration of form. This new approach to computer-aided design allows designers to translate vague and abstract language content directly into design objects. Language is composed of abstract symbols and is characterized by a high degree of both succinctness and room for interpretation. One would expect most designers to have a relatively common basic conception of most terms, but also that those conceptions would vary near-infinitely in their specifics. This adds a new way for architects to manipulate sketches and language, both of which are key components of the conceptual design process.

This semantic AI approach is not intended to produce finished designs, in the way that other generative AI approaches have been applied to create images, texts, or other media at the click of a button, but rather as a way to help a designer rapidly explore possibilities. Put simply, we want to generate compelling abstractions, providing just enough detail to aid further exploration, but leaving room for interpretation and the serendipitous pursuit of productive tangents. We hypothesise that tools based on this approach have great potential in shaping a positive future for human-machine co-creation. In addition, we can also see this as a pivot point in the role of the computer: a semantic AI can be thought of as more assistant than a tool. Being able to understand semantics and re-create and integrate elements at the design level capabilities we typically reserve for our collaborators, and if they become embedded in future tools then it will be necessary to re-think the way those tools themselves are designed, particularly the way designers might interact with them.

However, as an approach to exploring design possibilities, our early prototype has many limitations and shortcomings. First, the machine learning algorithm in this work-study is based on the vertex colours of the mesh, so the number of points and faces largely determines the resolution of exploitable details. Accordingly, if you want to get relatively detailed objects, you need to increase the number of vertexes. This is in contradiction with the typical goal of minimizing mesh density, particularly during the early stages of design. The number of objects in an architectural design scene is huge, and if each object uses a high-poly model, it will be computationally prohibitive. At the same time, high-density models are difficult to edit, which makes the results only useful as a source of inspiration or expression, as it would be difficult to use them directly in the design process.

A further limitation is the amount of time (10–15 min per 500 epochs) required to generate a 3D model. Although lower-quality results can be provided incrementally during the training process, its efficiency as a computer-aided tool is still limited by current hardware capabilities. Synthesis-by-optimization tools, which search for a solution using a pre-trained model as a guide, are inherently slower than pre-trained generative models, which can infer a solution in one shot. Another limitation is that our mesh-editing process is limited to movements perpendicular to the mesh surface – along the vertex normal relative to the object centroid. This limits the expressive capability of our prototypes: it is possible to allow all vertices in the mesh to move freely, but the search space for such optimization is currently prohibitively large.

Our approach is an early prototype, but we feel its potential outweighs its limitations. We have not yet completed the development of our massing exploration system, which will leverage similar semantic AI techniques for generating basic forms such as those used as inputs in this study. We believe that further integration of similar approaches in these two stages can improve our proposed method. We also believe that an effective assistant for the exploration of architectural form needs to be capable of two-way interaction: not just to be able to generate form, but to respond to and manipulate form created by the designer. This will require not just semantic AI models like CLIP, which have an impressive broad but also non-designer's lay understanding of terminology, but the integration of architectural reasoning to some extent. If a stronger integration of architectural reasoning into these models can be achieved, along with the development of usable interfaces for human-AI collaboration in this medium, then the potential for semantic AI in design is great. Our future research will investigate this combination.

References

1. Ramesh, A., Dhariwal, P., Nichol, A., Chu, C., Chen, M.: Hierarchical Text-Conditional Image Generation with CLIP Latents (2022)
2. Saharia, C., et al.: Photorealistic text-to-image diffusion models with deep language understanding (2022)
3. Wang, J.J.: Dall-e: creating images from text, 1–12 (2021)
4. Metz, T.: Building Meaning: An Architecture Studio Primer on Design, Theory, and History (2022)
5. Rezaei, M.: Reviewing design process theories: Discourses in architecture, urban design and planning theories (2021)
6. Suwa, M., Gero, J., Purcell, T.: Unexpected discoveries and S-invention of design requirements: important vehicles for a design process. Des. Stud. **21**, 539–567 (2000). https://doi.org/10.1016/S0142-694X(99)00034-4
7. Daniel, Max, Jack, Thomas, Red, Sam, Nadir, Sebastian: Midjourney. https://www.midjourney.com/home/#about
8. Radford, A., et al.: Learning transferable visual models from natural language supervision (2021)
9. Esser, P., Rombach, R., Ommer, B.: Taming transformers for high-resolution image synthesis. In: Proceedings of the IEEE Computer Society Conference on Computer Vision and Pattern Recognition, pp. 12868–12878 (2021). https://doi.org/10.1109/CVPR46437.2021.01268

10. Ching, F.D.K.: Architecture - Form, Space & Order (2015)
11. Fawcett, P.: Architecture Design Notebook (2003)
12. Schön, D.A., Wiggins, G.: Kinds of seeing in designing. Creativity Innov. Manage. **1**, 68–74 (1992). https://doi.org/10.1111/j.1467-8691.1992.tb00031.x
13. Bachman, L.R.: Two Spheres: Physical and Strategic Design in Architecture (2012)
14. Pan, J.Y., Yang, H.J., Duygulu, P., Faloutsos, C.: Automatic image captioning. In: 2004 IEEE International Conference on Multimedia and Expo (ICME), vol. 3, pp. 1987–1990 (2004). https://doi.org/10.1109/icme.2004.1394652
15. Frans, K., Soros, L.B., Witkowski, O.: CLIPDraw: exploring text-to-drawing synthesis through language-image encoders (2021)
16. Rombach, R., Blattmann, A., Lorenz, D., Esser, P., Ommer, B.: High-resolution image synthesis with latent diffusion models (2021)
17. Nguyen-Phuoc, T., Li, C., Yang, Y.L., Balaban, S.: Rendernet: a deep convolutional network for differentiable rendering from 3D shapes. In: Advances in Neural Information Processing Systems, 2018-Decem, pp. 7891–7901 (2018)
18. Chaillou, S.: Artificial Intelligence in Architecture: From Research to Practice (2022)
19. Haviland, W.A.: II. Architecture. Excavations in the West Plaza of Tikal, pp. 3–30 (2019). https://doi.org/10.9783/9781949057027-006
20. Eltarabishy, S.: Towards Data-Driven Design: Leveraging open data and deep learning in symbol spotting for furniture layout planning (2017)
21. Mildenhall, B., Srinivasan, P.P., Tancik, M., Barron, J.T., Ramamoorthi, R., Ng, R.: NeRF: representing scenes as neural radiance fields for view synthesis. In: Vedaldi, A., Bischof, H., Brox, T., Frahm, J.-M. (eds.) ECCV 2020. LNCS, vol. 12346, pp. 405–421. Springer, Cham (2020). https://doi.org/10.1007/978-3-030-58452-8_24
22. De Miguel Rodríguez, J., Villafañe, M.E., Piškorec, L., Sancho Caparrini, F.: Generation of geometric interpolations of building types with deep variational autoencoders. Des. Sci. (2020). https://doi.org/10.1017/dsj.2020.31
23. Petrov, M., Wortmann, T.: Latent fitness landscapes - exploring performance within the latent space of post-optimization results. SimAUD 2021 (2021)
24. Michel, O., Bar-On, R., Liu, R., Benaim, S., Hanocka, R.: Text2Mesh: text-driven neural stylization for meshes (2021)

Use of Language to Generate Architectural Scenery with AI-Powered Tools

Hanım Gülsüm Karahan[1]($\boxtimes$) (iD), Begüm Aktaş[1,2] (iD), and Cemal Koray Bingöl[3] (iD)

[1] Architectural Design Computing Graduate Program, Istanbul Technical University, 34367 Istanbul, Turkey
{karahan21,aktas15}@itu.edu.tr

[2] Department of Architecture, Altınbaş University, Istanbul, Turkey

[3] Department of Management in the Built Environment, Delft University of Technology, Delft, The Netherlands
c.k.bingol@tudelft.nl

Abstract. The quality of communication with a computer impacts how the designer performs during the design process. Today, Artificial Intelligence (AI) empowers the designer by expanding the solution space using the expertise from previous knowledge. However, the developments in AI-powered design tools mainly focus on visual and spatial enhancements. In the last decade, AI-powered design tools mostly experimented with image transformation models (GANs) to provide fast insights to designers using learned experiences, simulations, or datasets. The studies on the design process using verbal language with the help of AI are limited. Therefore, designers' capacity to communicate with intelligent machines would lead us to envision the future of AI-powered design tools.

In design practice, designers develop individual and contextual studies through digital tools. This study investigates the process of architectural visual generation and verbal communication to describe architectural images by architecture graduates with prior experience or no experience in prior with Midjourney. The research focuses on the designers' semantic language during the design process with the AI-powered tool and analysis of the verbal part of the communication. The results of this study show that participants' first impressions of the image and how they express their impressions through description do not correspond with how Midjourney interprets those descriptions. Furthermore, architects' image generation process using the tool is nonlinear. As architects develop a deeper understanding of changing modes of interactions, they are more likely to benefit from AI-powered tools as collaborative entities.

Keywords: Artificial Intelligence · Design Cognition · Human-Machine Interaction · Digital Design

1 Introduction

Designing is a complex and temporal activity that requires generating, transforming, and refining images of different aspects of that still non-existent artifact and making representations of it, enabling communication and examination of the ideas involved

M. Turrin et al. (Eds.): CAAD Futures 2023, CCIS 1819, pp. 83–96, 2023.
https://doi.org/10.1007/978-3-031-37189-9_6

[1, 2]. In the design process, a sketch is a reflection of the guiding mental image, but it cannot be identical to it, and this difference is precisely what makes it a precious instrument for the designer. By making a sketch, the designer supplies the mental image with the assistance of an optical image, which has all the properties of such visual perception [3]. The interaction of arguments in the design process, sketch represents the visual perception and exploration process of finding solutions to the design problem and reasoning about it. Working in some visual medium-drawing, in Schön's examples-the designer sees what is "there" in some representation of a site, draws in relation to it, and sees what he or she has drawn, thereby informing further designing [4]. Therefore, seeing triggers new ideas constantly and changes things accordingly to what has been drawn on paper. This conversational structure of seeing-moving-seeing represents an iterative process where every move feeds the next moves or vice versa to construct new meanings.

The discovery of unintentional ideas in this process comprises the dual nature of visual thinking. Visual thinking has the power to reveal both intended and unintended ideas for the design process. The interaction of arguments of "see-move-see" is a representation process of the goal image and realization of the idea in the context. In all this "seeing," the designer visually registers information and constructs its meaning, identifies patterns, and gives them meaning beyond themselves. Words like "recognize," "detect," "discover," and "appreciate" denote variants of seeing, as do such terms as "seeing that," "seeing as," and "seeing in." This process of seeing-drawing-seeing" is one example of what Schön means by designing as a reflective conversation with the materials of a situation [4]. Therefore, visual thinking can coalesce abstract and perceptible ideas as one or analyze them separately. In addition, in order to see, we had to think, and we had nothing to think about if we were not looking [5]. It is now well established from various studies that sketching is not just a representation of an idea; but a process of seeing, visual reasoning, and imagination. Sketching is not merely an act of representing a preformulated image; in this context, we often deal with a search for such an image [2]. Also, what the viewer "sees" in the picture is already the outcome of that organizational process [5]. In this process, in the interaction of arguments, sketching reveals the design knowledge of a designer.

2   Reasoning Through Visual and Language

When reflecting on the nature of thinking, most people associate it, primarily with words, with language when visual thinking is considered; however, we tend to concentrate on visual and almost forget thinking, which fades into the background [6]. Physical actions refer to drawing and looking, but yet. Perceptual actions refer to the interpretation of visual information [7], where interpretation is associated with thinking and language. Descriptionalists believe that mental images represent the mode of language rather than pictures [2]. Thinking is mostly associated with language to identify how it is produced and develops over time. Language has syntactic, semantic, and pragmatic rules to serve different purposes of humans or machines. We use language to conceptualize reality in the physical world, and our thinking is directly related to how we perceive the physical world through language. Figure 1 shows us a sentence's mental tree consisting of rules

[8]. This structure can grow with symbols such as "if, then," which explains a reasonable story.

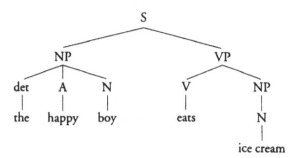

Fig. 1. A mental tree of a sentence (S: sentence, NP: noun phrase, VP: verb phrase, det: determiner, A: adjective, N: noun, V: verb), [8, p. 96].

According to Chomsky [9], the deep structure, the underlying relations of words with an abstract order, is not expressed but is only represented in mind, whereas surface structure is the aspect of syntactic description. Of interest here is that deep structure definition gives room for implicit meanings for words that extend the context of the subject. Therefore, although the language has some strict rules, it also has the power to enrich the thinking process with its profound structure aspects. However, words containing meanings may evoke different meanings in someone's mind when s/he extracts them. This situation affects our communication in both ways. It makes us think through various perspectives of the relevant concept, which may result in creative thinking through intended and unintended consequences. Sometimes, various meanings undermine the conversation when participants do not construct the desired meaning from the word or sentence.

Magnifying the number of decisions in the thinking process is one of the unique features of visual thinking. It brings out a different aspect of thinking: the ability to see infinitely and select the one feature designers to desire among numerous possibilities. In this sense, visual thinking is unbounded compared to language, which has syntactic and semantic rules. At every iteration, we can construct new rules and see new elements within a shape, and this process can be continuous. Figure 2 shows us that these branches can be multiplied by the rules we set every time we see them, and they do not have to follow specific rules, whereas the tree diagram of language has to. Our minds can cut a shape into different parts and create new forms or combine them to create whole new shapes while looking at it.

The flexibility and continuity of visual thinking expands the design space with many unintentional consequences of our moves. Although words can have different meanings, and meanings can change depending on who will reconstruct the idea, we are still restricted by the descriptions and their leading memories encoded in our minds. Language mediates as a conduit between how we think in our mind and how we externalize it. However, this externalization process may also provide a space for creativity since interventions made through the process cannot be defined enough. The notion of

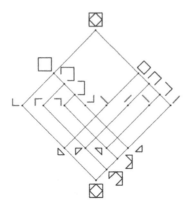

Fig. 2. A tree diagram of two squares [10].

"deep and surface structure" is common in language and visual thinking. The idea of abstract meaning only represented in mind can also be traced in visual thinking. In visual thinking, we try to externalize our ideas by employing visual elements within a constant reflective process. Here, we can emulate "visual elements" to "surface structures" and "ideas" to "deep structures." Visual thinking is also related to reflective thinking since seeing gives constant feedback to reveal the deep structure of thought.

Visual Reasoning over Digital Tools. A primary concern of cognition, particularly design cognition, is to first better understand the design to improve the design process and, secondly, to produce tools to assist designers in improving design outcomes. It is better to develop the tools for design and designers considering cognitive actions like the "seeing-moving-seeing" and contain an appreciative system comparable to a designer's appreciative system to support the evolution of the design problem [4]. The drawings are a representation of the evolving design. In that sense, the computational tools help the designers starting from the early stages to assist the evolution of the design process. Nowadays, the development of autonomous architectural tools is becoming an established research area in architecture; plan and space recognition [11]; reducing design repetitions [12]. Autonomous tools are also investigated in the field of urban design; generating urban morphologies [13]; generating building footprints and volume [14]. Bank et al. [15] also integrated GAN into their architectural design courses to explore the opportunities for designers and their role in defining the relevant domains for design possibilities, and to introduce machine learning algorithms to concept modeling processes for architectural students. Bolojan et al. [16]'s study is to apply Neural Language models (NLM) that are machine learning techniques for text-to-image synthesis by e DALL-E, VQGAN/CLIP, and Diffusion models (DMs). Text prompts with a visual reference are utilized to generate adopting John Gero's schema of 'design prototypes' [17], a new approach to architectural design was pursued, employing interconnected deep learning models. They stated that this line of work supports UN SDG #9, pushing forward the current technological capabilities of the AEC industry by offering innovative workflows.

3 A Protocol Study with AI-Powered Design Tool

The experiment consisted of two stages with the same rules and steps. Nine architects participated in the experiment individually through an online video call where each experiment was recorded as a video and audio. One out of nine architects declared that he had experienced the AI-Powered tool Midjourney (detailed explanation at Sect. 3.1) before. Then, participants (P) were provided with two architectural images (the original images): the exterior image of Ronchamp by Le Corbusier and the interior image of Therme Vals by Peter Zumthor (Fig. 3). First, the image of Ronchamp was shown to the participants, and they were asked to make verbal descriptions (T as text) of the visual in a way they felt comfortable without including the name of the building and the architect. We allow participants to use an English dictionary since English is not their native language. Each participant sent their keywords or sentences through chat to allow us to copy them to the Midjourney bot in Discord. Each participant determined their thinking duration to describe images. They were also informed that there was no limitation on how to describe the images or the number of words when asked if they should describe tangible features or the feelings the image evokes. After the first results, all participants were informed by the short presentation about how Midjourney works and how they can enrich their descriptions. Every image was shown to participants to allow them to evaluate the image and make changes accordingly. Then, each participant was given three more chances to use the Midjourney bot, including writing texts or using "Upgrade, Variations" options to reach an image similar to the first image generated. The same steps were applied for the image of Therme Vals with the chance of using the Midjourney bot three times. Each participant used the Midjourney bot seven times in total. At the end of each stage, participants were asked to decide on one of the generated images they found closer to the original image (Table 1).

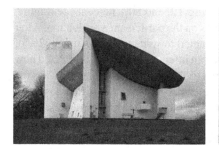 

Fig. 3. The original images: The exterior image of Ronchamp by Le Corbusier (left) and the interior image of Therme Vals by Peter Zumthor (right) [20].

3.1 Analyzing Tool Options

Midjourney is a text-based diffusion model; however, it offers "Variations/Upgrade" options. These are fully automated options; the user cannot control how variations/upgrade options will occur. It is important to note that choosing Variation/Upgrade

Table 1. Prompt examples from which participants made final choices

Ronchamp
P5,T1: *organic, extraordinary, looks like a shelter structure*
P6,T3: *curved black roof, curved gray roof, white house, futuristic, famous architect building, flue, small windows, grass, tree, stone building*

Therme Vals
P7, T2: *concrete, gray scale, pool, spa interior, rural landscape, tree, white illumination*
P8, T1: *perspective of a pool from inside looking at pool stairs and tiled blue-gray walls with a square openness in middle with green landscape seen from the middle, realistic style, sharp edges*
(P: Participant, T: Text)

options indicates that the user is satisfied with what s/he ends up with since these options do not generate something unexpected from the current outputs. However, our case study shows that these options do not always align with the user's mental imagery. First, we analyzed the path of choices and durations of decision-making processes for image generation from the protocol study. (Fig. 4). 'Thinking duration' for texts is determined from their typing period till their message timestamp. Thinking durations for 'Variations' and 'Upgrade' are counted as the duration from the output generation timestamp till the participants' response.

After the first results, only one participant decided to go "Variation" after writing the first text for Ronchamp, due to prior exposure to the tool. However, the participants, who are novel to the tool, needed more relevance with the first output via their first texts to the Ronchamp image. When we compare the thinking duration of T1 and T2, 50% of percent of participants spent less time during T2, and 50% percent spent more. Therefore, their reaction to their first image depends on the participant's evaluation. After the second result, except for P4, all participants wanted to write text. This time, 62% of participants increased their thinking duration. We can infer that participants realized how they describe the image can lead to the AI-powered tool many results, and not always the ones they imagine. Furthermore, after the first results, participants started to assess their words more carefully, whether the word has another meaning or how it matches with other words, or if two different words come together and stand for new meanings, or refer to different concepts. We had these inferences from the texts of Ronchamp, such as P9 wrote "*white fluke*" for text one and edited this as "*white fluke roof*" for text 2. P1 wrote "*with a blunt tower*" for text three and edited this as "*with a blunt tower in addition to the main building*" for text four. Another example is writing "*building as a sculpture*" in addition to "*huge curved roof, curvilinear shape, thick and twisted walls*" for text three and trying to emphasize its physical attributes with the "sculpture" term. It is clear that Variations/Upgrade options were used more for the Therme Vals image. In this session, 33% of participants went for variation for the second image generation, and this ratio climbed to 77% for the third image generation, including the upgrade option. When we analyzed all tool options, no one except one participant wanted to write text after using the Variation/Upgrade option. Of interest here is that only P4, who experienced Midjourney before, went for writing after the first image variation. This supports our idea that these options do not provide a sequential image generation process.

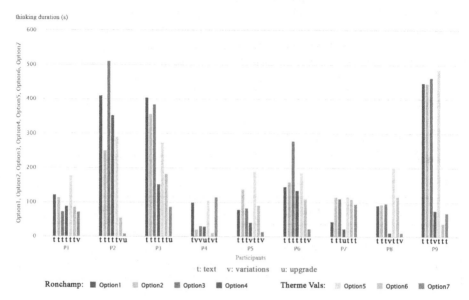

Fig. 4. Participants' option choices for description and their thinking duration(s).

3.2 Segmentation

We break down the whole process into segments. For each segment, we examine participants' cognitive actions, their number of words, their transmitted words from the previous segments, and if the transmitted words were changed or not, in terms of developing the word with new words or simplifying it by reducing the words. We code the participant's cognitive action as continuous or discrete (Table 2). If a participant keeps at least 50% of the exact words from the previous text, then we code this as 'continuous' action. In this sense, participants kept their thinking process continuously, accumulating their words gradually or changing them accordingly. If a participant's half and over half of the new text consists of new words never used before, then we code this as a 'discrete' action. In this action, participants shifted their focus and tried to find new types of descriptions. Furthermore, we code Variation/Upgrade options as continuous action if these options were used based on the last image. If not, we code them as discrete actions since the user did not correspond with what s/he imagined and decided to use Variation/Upgrade options on previous actions. Finally, we highlighted the segments to which each participant's final choice belongs so we could make inferences about their experience. When we analyzed their actions based on our coding scheme, 77% of participants made discrete actions when they wrote T2 for Ronchamp. We can infer that they developed new ideas after they evaluated their results and got information about the tool Midjourney. For instance, when we analyzed T1 and T2 for Therme Vals, none of the participants made discrete actions. In fact, only 3 participants made discrete actions during the Therme Vals session, and these actions were made for the last option. In this sense, it is clear that the more participants are exposed to the tool, the longer they would sequentially keep their thinking. 66% increased their number of words for T2 of Ronchamp, and 83% also made discrete actions. Therefore, we can conjecture that participants thought they would be more likely to get similar results if they elaborate their descriptions with more words. When we asked participants to choose one result that

Table 2. Segmentation of participants' design actions. (T: text, V: variation, U: upgrade).

○ : final choice

▬▬ : continuous action

===== : discrete action

w : word, phrase, sentence

w : transmitted word from previous session

c.w : transmitted changed word from previous session

(*continued*)

Table 2. (*continued*)

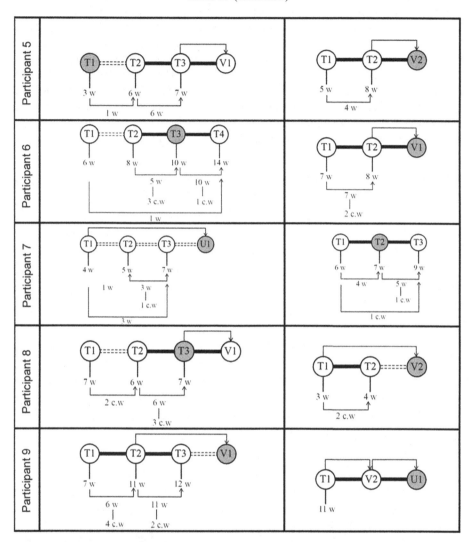

they think is the most similar to the original image, the choices were interesting. They did not always choose what they ended up with or the image they produced with more words. P5 chose one of the images he produced with T1 for Ronchamp, where he had no idea what Midjourney was. 44% of participants did not choose the last image they generated for Ronchamp. For Therme Vals, 77% chose the last image they generated. We can infer that as designers engage more with a new tool, such as Midjourney, it changes the decision processes of the designers.

We measured perceptual distance, which measures how similar two images are in a way that coincides with human judgment [18]. We used the code available on Github [19] that the author provided and we executed it on Google Colab. The results of metrics range between 0 to 1 with being 0 means they are the same (Fig. 5). The lower the number, the greater the similarity. P4 has the most similar result with 0.598 for Ronchamp and P2 has it with 0.637 for Therme Vals (Fig. 6). The results are not in correlation with educational background since P6's final choice, who is a doctorate student and a practicing architect, is rated as 6th the most similar image for Ronchamp whereas P4's, who is a master student with no practice experience, is rated as 1st one. P2's final choice, who is a practicing architect with 2 years of experience and without a graduate degree, is rated as the most similar one for Therme Vals while P9's choice, who is both an architect and software engineer with two major degrees, is rated as 7th among others.

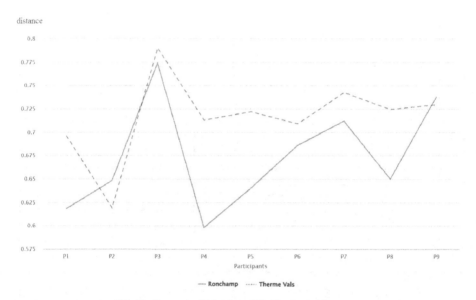

Fig. 5. Perceptual Metrics of final selected images.

P4,T1: architecture, unusual, amorphous structure, different details, unusual form representation

P2,T1: concrete, modernist building, pool, villa, purist style, modernism, open floor plan

Fig. 6. The most similar images are based on the perceptual similarity metric.

3.3   Encoding Descriptions into Categories

Participants delivered different descriptions that emphasized images' various aspects during the experiment. In order to analyze how descriptions differ, we devised seven categories to encode all the information that participants provide through text writing. These categories include type, physical attributes, style, material, feeling, reference, and render style. The six categories were shown to participants except for physical attributes to enrich their descriptions during the presentation. The "Physical attributes" category was devised after the experiment when we examined descriptions and realized many physical descriptions exist. The encoding we developed here explores architects' architectural description characteristics by analyzing what categories the majority of words belong to. Since the number of discrete actions was the highest during writing text two for Ronchamp, we wanted to demonstrate how participants' word choice changed after they evaluated their results generated with T1. To clarify how we encode descriptions, we showed some examples and the output image of texts (Table 3). Figure 7 shows the categories and how frequently they are preferred by participants for T1-T2 of Ronchamp and Therme Vals. If no words exist for each category, the user selects the Variation/Upgrade option. It is quite noticeable that participants mostly use words related to physical attributes both for Ronchamp and Therme Vals. For T1-T2 of Therme Vals, physical attributes descriptions are dominant, whereas T2 only consists of physical attributes. Furthermore, 88% expressed their feelings for text one of Ronchamp, while this ratio declined by 33% for T2 even though we informed participants they could include their feelings.

4   Discussion and Conclusion

The case study shows that participants' first impressions of the image and how they express their impressions through descriptions do not correspond with how an AI-powered tool (in our case, Midjourney) interprets those descriptions. Furthermore, Variations/Upgrade options do not always result in compatible images even though they are programmed to be as. This incompatibility may arise for two reasons: AI-powered tools are in their nascent phases to assist architects or architects need to adapt to communicate with an AI-powered tool properly. The results show that generated images do not always get better based on the number of words or how developed the words are, and participants tend to describe the images' physical attributes more than the other categories. However, in the study we conducted using Midjourney shows us that describing only physical attributes is not enough for desired results. Abstract concepts such as feelings, style and references help both the AI-powered tool and the participant to elaborate their textual definitions.

Digital external representation tools usually have been developed based on the insights of sketches. AI-powered tools such as Midjourney bring a new design space where users externalize their mental imagery with verbal descriptions. When architects were restricted to draw, they struggled to control the design process which they would amplify with "see-move-see". Instead of adopting a heuristic approach, they experienced a stochastic approach as recalling the most common and appropriate words to prevent AI from being puzzled. We see that architects still adopted a reflective thinking process

Table 3. Encoding descriptions into categories

P2: text 2-Ronchamp: chapel, France, white walls, purism, curvilinear shape, dramatic effect, modernist architecture, curved roof type: chapel physical attributes: white walls, curvilinear shape, curved roof style: modernist architecture feeling: dramatic effect,purism reference: France	
P8: text 1-Ronchamp: organic, spacious, sharp, bright, brave, geometric, plain physical attributes: organic, sharp, bright, geometric, plain feeling: spacious, brave	
P9: text 1- Therme Vals: modern half-open space, pool in the half-open space, geometrical sharpness, geometrical balance, modernist, modernist simplicity, brutal concrete, relaxing, forest view, heidegger architecture philosophy, interior half-open wet space type: modern half-open space, interior half-open wet space physical attributes: pool in the half-open space, geometrical sharpness, geometrical balance, forest view style: modernist material: brutal concrete feeling: relaxing reference: modernist simplicity, heidegger architecture philosophy	

with the cycle where they adjust their descriptions based on outputs. How Midjourney interprets the word can also give rise to creative thinking since the deep structure in language depends on the one who extracts the meaning. However, using only words and relinquishing the drawing seems not quite robust enough to externalize the mental imagery of architects. AI-powered tools should be developed to corroborate the shared mode of interaction with architects. Furthermore, the drawing process is more internal in that architects evaluate most of the ideas implicitly in their minds. However, AI-powered tools want us to externalize our thinking process explicitly through language. We can imply that exploiting this kind of AI-powered tool at its best depends on how much tacit knowledge we can transform into explicit knowledge, which can be expressed through language.

The change in the descriptions during the experiment gave us a hint of how the designers can describe architectural sceneries and how they try to understand the perspective of the AI-powered tool as it has its knowledge. This change in the mental mode is one of the most important points of the reflection process due to its collaborative nature with an AI. Even though the tool used in this experiment is not unbiased, its collaboration with the designers is unique at every sequence due to the interaction both parties create during the design process.

In the protocol study, the interaction was not direct, and the designers were getting familiar with the process. However, due to the tool's novelty, they needed some time to grasp the dynamics of the process. Despite the lack of direct interaction with the tool, the

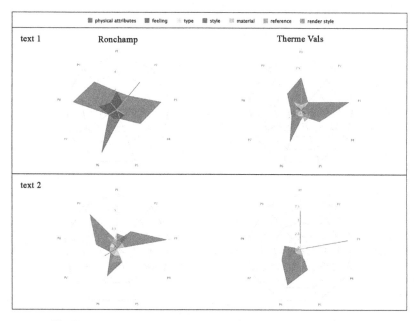

Fig. 7. The categories and how frequently participants prefer them.

design process allowed us to evaluate the wording designers use during the generative processes. One major finding was that the design process's nonlinearity was still present. Even though the AI-powered tool generated endless variations, the decision-making process was the human designer. The final decisions differed from the latest images generated and were the most similar to the original images that participants think as they were. This also shows the divergence between computational and design thinking; ambiguity and aesthetics play a different role for the human designers than the AI model, like GAN and stable diffusion. Another finding is that the wording is another skill for designers, and how they define scenery is unique to their prior architectural education, literature, and mother language. Midjourney is developed for the English language, and the study was conducted with non-native English speakers; this may have limited the participants' expression of feelings and other descriptive words even though they were allowed to use a dictionary.

The developing tools with AI models are changing how we interact with digital tools, hence the way we design using these tools. The increase of voice-commanded devices we use in our daily lives, such as smartphones and home appliances, changes how we interact with our environment. Therefore the interaction with smart design tools will affect how we interact with the new design tools. The changing modes of interactions also have the potential to transform the roles of tools in design processes. The knowledge and the support that AI-powered tools will allow designers to recognize them as collaborative entities that help designers to decide with an extended intelligence.

Acknowledgements. We would like to thank Prof. Mine Özkar for her insightful course "Computing Theories and Models in Architectural Design" given at ITU. "Reasoning through Visual and Language" section has been developed through this course.

References

1. Hay, L., Cash, P., McKilligan, S.: The future of design cognition analysis. Des. Sci. **6**, e20 (2020). https://doi.org/10.1017/dsj.2020.20
2. Goldschmidt, G.: The dialectics of sketching. Creat. Res. J. **4**(2), 123–143 (1991). https://doi.org/10.1080/10400419109534381
3. Arnheim, R.: Sketching and the psychology of design. Des. Issues **9**(2), 15 (1993). https://doi.org/10.2307/1511669
4. Schon, D.A.: Designing as reflective conversation with the materials of a design situation. Res. Eng. Design **3**(3), 131–147 (1992). https://doi.org/10.1007/bf01580516
5. Arnheim, R.: A plea for visual thinking. Critical Inquiry **6**(3), 489–497 (1980). University of California Press
6. Goldschmidt, G.: On visual design thinking: the vis kids of architecture. Des. Stud. **15**(2), 158–174 (1994). https://doi.org/10.1016/0142-694x(94)90022-1
7. Bilda, Z., Gero, J.S.: Does sketching off-load visuo-spatial working memory. Studying Designers **5**, 145–160 (2005)
8. Pinker, S.: The Language Instinct, p. 96. W. Morrow and Co., Harper Collins Publishers, New York (1994)https://doi.org/10.1057/9780230379411_23
9. Chomsky, N.: Cartesian Linguistics. Harper, and Row, New York, London (1966)
10. Stiny, G.: The critic as artist: Oscar Wilde's prolegomena to shape grammars. Nexus Netw. J. **17**(3), 723–758 (2016)
11. Huang, W., Zheng, H.: Architectural drawings recognition and generation through machine learning. In: Proceedings of the 38th ACADIA, pp. 18–20. Mexico City, Mexico (2018)
12. Uzun, C., Çolakoğlu, M.B., İnceoğlu, A.: GAN as a generative architectural plan layout tool: a case study for training DCGAN with palladian plans and evaluation of DCGAN outputs. A|Z ITU J. Faculty Architect. **17**, 185–198 (2020)
13. Boim, A., Dortheimer, J., Sprecher, A.: A machine-learning approach to urban design interventions in non-planned settlements. In: van Ameijde, J., Gardner, N., Hoon Hyun, K., Luo, D., Sheth, U. (eds.) POST-CARBON - Proceedings of the 27th CAADRIA Conference, pp. 223–232 (2022)
14. Di Carlo, R., Mittal, D., Vesely, O.: Generating 3D building volumes for a given urban context using Pix2Pix GAN. In: Proceedings of the 40th Conference of eCAADe, pp. 287–296 (2023)
15. Bank, M., Sandor, V., Schinegger, K., Rutzinger, S.: Learning spatiality-A GAN method for designing architectural models through labeled sections. In: Pak, B., Wurzer, G., Stouffs, R. (eds.) Co-creating the Future: Inclusion in and through Design - Proceedings of the 40th Conference of eCAADe, vol. 2, Ghent, 13–16 September 2022, pp. 611–619 (2022)
16. Bolojan, D., Vermisso, E., Yousif, S.: Is language all we need? A query into architectural semantics using a multimodal generative workflow. In: Proceedings of the 27th CAADRIA Proceedings (2022). https://doi.org/10.52842/conf.caadria.2022.1.353
17. Gero, J.S.: Design prototypes: a knowledge representation schema for design. AI Mag. **11**(4), 26 (1990)
18. Zhang, R., Isola, P., Efros, A.A., Shectman, E., Wang, O.: The Unreasonable Effectiveness of Deep Features as a Perceptual Metric. arXiv (2018). https://doi.org/10.48550/arXiv.1801.03924
19. Github. https://github.com/richzhang/PerceptualSimilarity. last accessed 2023/02/21
20. Archdaily. https://www.archdaily.com/84988/ad-classics-ronchamp-le-corbusier. Vals. https://vals.ch/erleben/erholung/therme-vals/. last accessed 2023/02/28

The House that Looked Like It Should Collapse. Natural Language Processing for Architectural Design

Nadja Gaudillière-Jami[(✉)]

Digital Design Unit, Technical University Darmstadt, Darmstadt, Germany
nadja.gaudilliere@gmail.com

Abstract. Machine learning (ML) has not only become increasingly popular in the fields of computer science and AI since the 1990s, it also has become increasingly used in architectural design in the past few years. While some experimentations deal with fabrication or engineering issues and performance prediction, others deal with layout generation. As in any process involving ML, layout generation necessitates a large database to train the system. These databases are currently in their majority 2D images or 3D-models. However, these databases are often either scarce or monolithic – they gather too few examples and/or examples that are very similar in regard to the vast diversity of architectural design.

The Generative Pre-trained Transformer 3 (GPT-3) is a language model developed by Open AI to produce text answers to text prompts submitted to it. With 175 billion parameters, it is one of the largest models available. It has been trained on billions of online texts, coming from databases such as Common Crawl, Books2 or Wikipedia. The texts it generates are therefore based on a very large range of information, including about architecture. While only being word-based, the amount of information it has been trained with makes the suggestions relative to architecture GPT-3 outputs worthwhile to examine. What patterns relating to architectural design emerge when training a ML algorithm on such a database, much larger than what our field is currently used to manipulating?

We have been testing and calibrating GPT-3 in order to obtain descriptions of buildings that could then be translated into drawings, i.e. design briefs. This calibration has also enabled us to assess the patterns identified by the system. We thus propose an evaluation of how design issues usually perceived by architects as key are handled by GPT-3 – location, climate, program, references, inhabitants and their peculiarities. This evaluation is based on an assessment method that can be applied to other general ML tools in architecture, including image generators. Upon performing these tests, some prompts generated particularly various and interesting answers, oscillating between rationality and fiction. The present paper uses these prompts as examples in order to deepen the assessment of GPT-3, to evaluate its potential as a design tool and to open a larger discussion about databases within ML use for architectural design.

Keywords: GPT-3 · ML databases · AI for architecture · layout generation

© The Author(s), under exclusive license to Springer Nature Switzerland AG 2023
M. Turrin et al. (Eds.): CAAD Futures 2023, CCIS 1819, pp. 97–111, 2023.
https://doi.org/10.1007/978-3-031-37189-9_7

1 Introduction

The past few decades have seen uses of machine learning (ML) soaring, resorted to in an increasing number of applications. From recognizing zip codes [1] to suggesting medical diagnosis [2], ML algorithms are now widespread. Their success has not weakened since the beginning of the latest golden age of Artificial Intelligence in the early 1990s [3]. This popularity has not receded despite ML being particularly efficient only in some fields rather than in all [4]. However, one of these fields is image recognition and generation, which upon very significant research successes these past years is now leveraged in architectural applications as well. Architecture too has indeed rushed into ML applications dedicated to the built environment and its design. A large number of research projects deal for example with using ML in order to provide robotic systems with recognition and placement capacities to in turn be able to assemble building blocks together [5, 6]. Both in the academic world and in the industry, many have also confronted the issue of layout generation, training algorithms to propose plans, often for housing, steered by various inputs, from aesthetic preferences to evaluations of the most efficient configuration [7–9]. While these works are too numerous to be listed in full here, a recent overview of works performed in the field can be found in the The Routledge Companion to Artificial Intelligence in Architecture [10]. From bronze casting [11] to structural form-finding [12], no area of digital design research seems to have escaped experimentation with ML.

However, a core issue remains to be tackled: the challenges currently posed by databases. As in any process involving ML, architectural applications necessitates a large database to train the system. These databases are currently in their majority 2D images or 3D-models. However, these databases are often either scarce or monolithic – they gather too few examples and/or examples that are very similar in regard to the vast diversity of architectural design. While teams of researchers do work towards the constitution of larger and more homogeneous databases, many of the research examples that are currently being published in the field are limited by the narrowness of the datasets available. The issue is particularly crucial, as ML can be understood as a process of pattern-seeking within these databases [13]. The type of information they contain and how they are structured is especially important to the identification of patterns. As Dominique Cardon and his co-authors put it, in order for a computational system to spot existing patterns in a database, the ideal is for the elements in this database to have been stripped as much as possible of any type of structure reminiscent of human-created patterns [13]. This way, the pattern research is not clouded by existing human patterns, and can highlight novel ones. This vision entails much work to be done on architectural databases for ML.

Exploring larger databases that are outside the realm of architecture and the patterns they might reveal seems therefore a valuable research direction, first given the size of databases available elsewhere, and secondly given the fact that they might be stripped of usual patterns crafted by architects themselves. The present research focuses on one database in particular. The Generative Pre-trained Transformer 3 (GPT-3) is a language model developed by Open AI to produce text answers to text prompts submitted to it [14]. With 175 billion machine-learning parameters, it is one of the largest language models available. More interestingly, it has been trained on billions of online texts, including tokens coming from databases such as Common Crawl, Books2 or Wikipedia. The texts it generates are therefore based on a very large range of information, which include knowledge about architecture. Considered one of the most "intelligent" systems currently available, it has in particular made progress on commonsense knowledge, something with which many AI systems have struggled so far [15]. Used for many applications, in particular fiction generation, it has been analyzed by several researchers [15, 16]. However, no research on it applied to the field of architecture exists yet to the knowledge of the author.

While only being word-based, the amount of information it has been trained with makes the suggestions relative to architecture GPT-3 outputs worthwhile to examine. What patterns relating to architectural design emerge when training a ML algorithm on such a database, much larger than what our field is currently used to manipulating? Testing GPT-3 can provide a glimpse into what a GAN with a larger database would extract as design principles. It could also enable understanding the scope of Open AI's databases regarding architecture as their systems currently become used more and more widely. Our goal is thus to get to know the system, its flaws and its proclivities, and to tune it to produce design briefs, thus exploring its potential for architectural design. An evaluation method for such large and general databases can furthermore be resorted to both for text-based and visual-based ML tools. It can also more broadly lead to a reflection on the role of databases – existing and to come – for the use of ML in architecture, and to a reflection on the place of ML systems in architectural design – word processing might help in this by putting the output at a critical distance.

2 Methods

In order to evaluate what natural-language processing tools such as GPT-3 could bring to architectural design practices, the research method relies on five steps. First, testing single-sentence prompts to get a large number of answers and identify patterns in them. Second, refining these prompts in order to obtain as precise as possible design suggestions by going through a prompt engineering process. Third, perform an analysis of the contents of the answers, a step that will provide the first results for the database assessment. Fourth, testing example-based prompts, refine them and analyze the contents of the answer: GPT-3 indeed offers much more detailed possibilities than single-sentence prompts and can thus be calibrated by providing it with a series of examples guiding it into more precise answers. In order to obtain more precise design suggestions, the approach could be applied here. This could be useful in order to steer the prompts towards precise descriptions of facades or layouts for example, but as well to assess whether the associative thinking method often considered to be at work in architectural design can function here too.

A fifth step could be the production of drawings to illustrate the suggestions. This entails an exploration of the type of drawing selected to convey best the suggestions (collage, sketch, plan, axonometry, section or other) and an exploration of how to best move from the output of the system towards a fully developed architectural proposal. Options in the method of realization of these drawings can be manual drawings, generative scripts in Grasshopper, Processing or other, or of course resorting to a text-to-image ML generator such as Disco Diffusion [17] or MidJourney [18].

The first crucial issue in testing single-sentence prompts is prompt engineering, making sure that we obtain as relevant as possible design briefs as answers. In order to do this, we therefore must first define what we consider a relevant design brief. What should an ideal design brief provide as information? We are aiming at obtaining as detailed instructions as possible on the design, be it regarding its layout, its materials, its overall form, the architectural references it is designed in relation to, and so on. While usual design briefs, for example in competitions, leave the architect to make the majority of design decisions, on the contrary in the present case we are looking for briefs that limit as much as possible those decisions.

The prompt engineering phase thus entails adjusting phrasings in order to obtain what is expected. First, the type of sentence submitted has to be fixed. Different base structures for the sentences are tested, such as:

Describe the ideal house.
What is the ideal design for a house?
How to design a house.
If designing a house, what would you do?
What should the house look like?
What features should the house have?

Additionally a series of other questions starting with what are submitted to better frame the system, for example questions on how architecture is defined within the database or how buildings considered as classics are represented within it.

What is your favorite housing building?
What does a house designed by Hector Guimard look like?
What are the most important buildings in the world?

The vocabulary used is then tuned through a series of tests, in order to figure out the influence of variations in the words chosen. For example, we assess the impact of enquiring about buildings in general or a precise building such as a house. Other examples, the impact of enquiring about a biome rather than a location, of enquiring about personal preference rather than a general answer, or of asking a question rather than giving an instruction.

While these details highlight the functioning of systems and the contents of the database, they might also play a role in the quality of answers obtained and are therefore evaluated in that regard too. Finally, once the semantic calibration is achieved, the temperature setting must also be adjusted, in order to set the diversity or consistency in answers. Part of the calibration work is therefore also to determine whether this had any influence on the relevance of the answer.

Once the prompt engineering is over, the base structure of the prompt is combined with a series of criteria and constraints in order to evaluate the importance given within the system to what is usually perceived as core design issues by architects. One by one, elements relating to the program, to the climate, to the location, to the users, to the structure, to the materials, to aesthetics and to architectural references or cultural habits are integrated in the prompts to assess the answer provided. The list of elements submitted is presented in Table 1. Finally, several criteria are combined into a single prompt to assess the answer to such indications as well.

Table 1. List of elements submitted through prompts

PROGRAM			
house	single-family house	two-family house	residential building
museum	library	factory	office building
shop	store	supermarket	department store
school	art school	high school	primary school
INHABITANTS			
adult	child	teenager	elderly person
woman	man	person	single person
family	group of friends	group of strangers	
CLIMATE			
desert	fields	temperate / mild	jungle
mountain	beach	very cold	tropical forest
ice / snow	cold	very hot	cliff
forest	hot	wet	
LOCATIONS			
city / urban area	the moon	island	asia
rural area	mars	north america	africa
the north pole	space	south america	australia
the south pole	venus	europe	Tunisia
Brasil	France	China	South Africa
Venezuela	Germany	Japan	Nigeria
USA	Denmark	South Korea	Egypt
Canada	England	North Korea	Kenya
Mexico	Russia	India	Berlin
New York	Paris	Toronto	Frankfurt
San Francisco	Lyon	Montreal	Munich
Washington	Marseille	Quebec	Copenhagen
Manchester	London	Odensee	Aalborg
Moscow	Edinburgh	Irkoutsk	Saint Petersburg

The evaluation of the answers is first conducted through a quantitative assessment. The total number of prompts, the different semantic structures of prompts and their topics are counted. The mentions of the assessed design criteria are as well, both when integrating the criteria in the prompt and when spontaneously appearing in the answer. The number of answers providing specific design guidelines is also evaluated. Alongside specific features of the answers of which the frequency might be relevant to determine. The type of answer to each prompt is documented as well: generalist, absurd, inaccurate or functional. Generalist answers contain elements that could be part of any answer, such as:

Prompt: *What is the ideal design for a house in Venezuela?*
Answer: *There is no definitive answer to this question, as everyone's ideal design for a house in Venezuela would likely be different. However, some common features that could be included in an ideal design for a house in Venezuela might include plenty of natural*

light, comfortable indoor and outdoor spaces, and a layout that makes efficient use of the available space.

Inaccurate answers contain irrelevant information, such as:

Prompt: *Describe architecture.*
Answer: *The term "architecture" can refer to a number of different things, but generally it is used to describe the overall design of a system. This can include the physical layout of the hardware, the organization of the software, or the overall structure of a network.*

Absurd answers contain elements that make only little sense rationally, such as:

Prompt: *Two parents and their two children want to build a house. They hate each other. What should their house look like?*
Answer: *The house should look like a big, ugly tree.*

Functional answers contain only rational elements, such as:

Prompt: *How to design a museum.*
Answer: *There is no one-size-fits-all answer to this question, as the design of a museum will vary depending on its purpose, location, and the type of collections it houses. However some tips on designing a museum include considering the needs of both visitors and staff, creating a layout that is easy to navigate, and using lighting and signage to highlight key areas.*

This first evaluation of results is intended to provide a broad overview of the system, indications on the database contents and on the considerations on architecture found in it that GPT-3 bases its answers on.

Specific prompts yielding promising results are then studied further as examples, as they provided particularly interesting insights into the structuration of GPT-3. The examples balance prompts derived from existing project briefs, in order to evaluate the potential of the answers within a conventional practice of architecture, and prompts derived from absurd situations, in order to evaluate the potential of the answers in provoking the imagination of the user towards original architectural proposals.

3   General Assessment

Overall, 274 prompts have been submitted to GPT-3, the answers provided by the system gathered and evaluated. Figure 1 presents the different semantic types of prompts that have been submitted to and their repartition across these 274 submissions.

Results of the prompt engineering research phase show that in order to obtain the most precise design briefs, the prompt structure *"what features should this building have?"* is particularly efficient. *"What should the building look like?"* is the second most efficient prompt structure, and provides quite relevant briefs for a design process too. Finally, *"what is the ideal design?"* also yields particularly interesting results, both in terms of design specifications and in providing insights into the information about architecture in general stored in the database. Semantic variations on vocabulary are however more erratic, as in some cases the answer radically changes with a variation while in others not at all. However, details of some phrasings that might appear completely innocuous

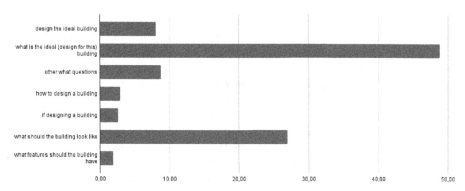

Fig. 1. Types of prompts submitted – semantics.

for humans sometimes lead GPT-3 to some drastic changes in the answers. As we have not yet determined the exact reason for these occasional changes, this is a potential lead for further research.

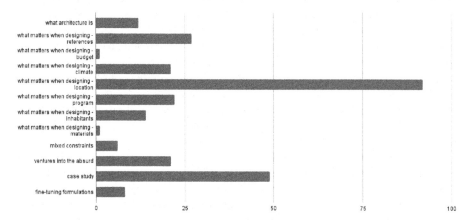

Fig. 2. Types of prompts submitted – topics.

Figure 2 shows the repartition of topics in the prompts submitted. The strong predominance of location is explained by the fact that we have sought to test the sensibility of the system to that parameter through many examples differing in scale and in geographical location (see Table 1 for detailed examples). Many architects link the location of a given building to a large range of other issues such as cultural or climatic issues. However GPT-3 does rarely create these links, scarcely demonstrating sensibility to cultural issues in the prompts dedicated to location importance testing. The location prompts very often simply recommend "plenty of natural light, a spacious layout, and a modern design" – an answer as conventional as it is general.

Out of the 274 prompts, 118 provided answers with specific design elements, such as:

Prompt: *Describe the ideal french building.*

Answer: *The ideal french building would be a tall, elegant structure with a symmetrical design. It would be made of stone or brick, with large windows and a steeply pitched roof. The front door would be grand and imposing, and the interior would be filled with light and airy rooms.*
Prompt: *What is the ideal design for a house for an elderly person?*
Answer: *There is no one-size-fits all answer to this question, as the ideal design for a house for an elderly person will vary depending on the specific needs and preferences of the individual. However, some features that may be beneficial for an elderly person include single-story living, wide doorways and hallways, and grab bars in the bathroom.*
Prompt: *What is the ideal design for a house in Japan?*
Answer: *There is no one ideal design for a house in Japan, as people's preferences vary greatly. However, some common features in Japanese homes include tatami mat flooring, shoji screens, and sliding doors. These elements create a sense of space and harmony that is often lacking in Western-style homes.*

Nevertheless, it has proven quite hard to obtain very detailed design briefs with single-sentence prompts. These design-specific answers however do hold enough elements to start sketching out designs. While many decisions are left to the user, on the contrary to what we stated was the goal in the Method section, many of the design-specific answers provide some inspiration and a starting point. Tests with text-to-image tools and example-based prompts are furthermore a good lead to expand on the level of detail expected in the answers, be it by finding a way to produce images with short answers or by finding a way to lengthen the answers and the design specifications they suggest.

Figure 3 presents the frequency of mention of various design issues across the 274 prompts. More precisely, it shows the number of unexpected mentions of these issues – each time GPT-3 has taken these issues into account in the answer provided while the issue had not been mentioned in the prompt beforehand.

The predominance of layout mentions (120 out of 274 answers, the most significant number in the quantitative assessment of mentions) is in part explained by the fact that the prompt engineering has been accomplished towards obtaining precisely that kind of design specification. One can also observe in the results a strong focus on aesthetics and form as well as on light, which are the second and third most often mentioned issues in the answers. It is also quite interesting to observe that several spontaneous mentions of available resources and of sustainability issues in general are generated by GPT-3. The issue of resources in particular is strongly linked to prompts about locations. This often replaces warnings about extreme environments such as the Arctic, the Sahara or the Moon. While a human might have suggested to not go live in such places, GPT-3 simply states resources might be difficult to reach in such locations. This is a trace of what has already been highlighted in existing assessments of GPT-3: its incapacity to not answer rather than falling into absurdity [15].

Prompt: *Describe the ideal building.*
Answer: *The ideal building would be one that is both aesthetically pleasing and functional. It would have plenty of natural light, ample storage, and be designed with the user in mind. The building would be energy efficient and made from sustainable materials.*
Prompt: *What is the ideal design for a house in Kilpisjärvi?*

Answer: *There is no definitive answer to this question as everyone's preferences are different. However, some factors to consider when designing a house in Kilpisjärvi would include the climate (which can be very cold and snowy), the surrounding landscape (which is mostly mountainous), and the availability of resources (such as wood for building and heating).*

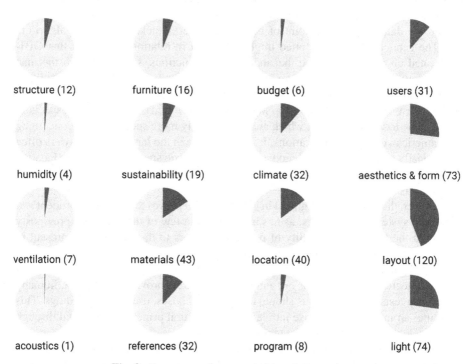

structure (12)   furniture (16)   budget (6)   users (31)

humidity (4)   sustainability (19)   climate (32)   aesthetics & form (73)

ventilation (7)   materials (43)   location (40)   layout (120)

acoustics (1)   references (32)   program (8)   light (74)

Fig. 3. Frequency of unexpected mentions per issue.

A key question in the study of such ML word-generating systems is to assess how biased they are. Prompts submitted to assess whether GPT-3 establishes a gender difference do show some biases:

Prompt: *What is the ideal design for a house for a woman?*
Answer: *There is no definitive answer to this question as everyone's needs and preferences are different. However, some features that might be included in an ideal design for a woman's house could include plenty of storage space, a cozy and inviting living room, and a private outdoor space.*
Prompt: *What is the ideal design for a house for a man?*
Answer: *There is no definitive answer to this question as everyone's ideal design for a house will be different. However, some features that could be included in an ideal design for a man's house might include a large garage for storing cars or tools, a comfortable den or man cave where he can relax and watch TV or play video games, and a backyard with a grill or fire pit for entertaining.*

Prompt: *What is the ideal design for a house for a person?*
Answer: *There is no one answer to this question as everyone's needs and preferences are different. However, some features that might be included in an ideal design for a house could be plenty of natural light, ample storage space, a comfortable and inviting layout, and a peaceful and private outdoor area.*

This is however not particularly surprising, as many experiments and studies have shown before that such systems are often biased, reflecting our own biases as they spread through the databases to the point of becoming patterns picked up by the algorithm [19]. The gender difference spotted might also exist in relation to the fact that GPT-3 is trained on a large database that includes many fictions, some in which the man den – a particularly specific feature in the answers shown above – is certainly made into a feature of houses, be it sarcastical or not. Answers provided to the prompts also display a few cultural clichés and a strong occidental focus. The latter is especially visible in the location prompts, which are often slightly more specific when concerning north-american or european locations. It is also visible in the fact that the answers often automatically suggest occidental modern features when no specification on the location is given, rather than anything else. This however is not astonishing either though, for the same reasons.

However, these trends are light, as in general GPT-3 proves to be a rather prudent and consensual system. Disclaimers, as has been seen in a few of the examples previously quoted, on the lack of universality of potential answers to the prompts are present in 145 out of 274 answers. Many of the answers are very conservative, with only a few unexpected pointers spiking out of the overall smooth answers. One particular statement about architectural design stands out amongst the answers provided by GPT-3. In many cases, it suggests that inhabitants should design their houses, users their buildings. This constitutes an interesting glimpse into larger architectural principles conveyed through a large-scale database of generalist texts.

It will come as no surprise that beyond the peculiarity of entrusting building designs to their users, GPT-3 is rather liberal and rationalist, on top of being pretty conventional. This however raises the issue of the kind of use an architect can make of systems trained on similar databases, in particular if the architect doesn't share that ideology or culture. Hijacking or tweaking computational design tools to bring them closer is nevertheless a key feature in digital design practices [20], and might be useful in that perspective.

4 Detailed Prompt Examples

In order to assess further the possibilities of tweaking prompts to obtain more challenging or more specific answers from GPT-3, three examples have been selected to be studied in greater detail. Peculiar design cases have been used to formulate a number of prompts, varying in their detail and formulation, to generate a number of answers in relation to these design cases. This is the case in the present study in which the temperature settings matter the most. While setting temperature to 1 increases the diversity of answers to one same prompt, 0 fixes it to always the same answer, with in-between values increasing step by step the diversity of answers. It however also seems that in-between values stress one or the other of the answers characteristics. Setting the temperature to 1 indeed appears to produce answers with a more rational and careful analysis of the prompt, and setting it to 0.5 appears to produce wilder answers, with more specific yet slightly strange design elements integrated in the answers. This provides a first hint towards how to provoke more interesting answers despite the overall conventionality of the system.

The first example is based on an existing brief the author has been confronted with in their practice as architect. Working on the renovation of an apartment for a young couple, it was highlighted that they were particularly fond of Frank Lloyd Wright's work. Upon discussing the many books the apartment would have to hold, they also expressed the following wish to describe what they hoped for: "we would like to live in a library". Prompts have thus been derived from these project characteristics, which also provided a further occasion of examining how GPT-3 would perform confronted to specific architectural references. Answers provided by the system in this example are the closest to prior and general observations, as well as the least provocative answers. Upon pushing a little bit further the confrontation of different styles and references, a few suggestions came up in the answers, but nothing especially design-specific nor interesting.

Prompt: *Two parents and their three children want to build a house. They are Japanese. They live in the United States. They want to build their home in the forest. They like to cook together, to swim and to fish. Two children get along but the third is always angry. They have a lot of money. They enjoy Frank Lloyd Wright's architecture. What should their house look like?*
Answer: *The house should look like a Frank Lloyd Wright house.*
Prompt: *A family wants to build a house. They like the architecture of Alvar Aalto and of Tadao Ando. What should their house look like?*
Answer: *There is no definitive answer to this question, as it depends on the specific preferences of the family. However, if the family likes the architecture of both Alvar Aalto and Tadao Ando, their house could potentially incorporate elements from both styles. For example, the house could have clean lines and a minimalistic aesthetic like that of Ando, while also incorporating organic curves and natural materials like those used by Aalto.*

This was intended as a means of exploring further the ability of GPT-3 to produce associative thinking, which is considered to be at the core of architectural design by many authors in design theory. But GPT-3 does not appear to perform very well in this case, neither in retrieving design specification from reference architectures nor in handling specifications about many criteria at once. The latter had been highlighted in assessments of the system for other applications: "It's like GPT-3 has a limited short-term memory, and has trouble reasoning about more than one or two objects in a sentence" [15].

The second example is intended as a further venture into the absurd, considered to be one of the key aspects in using GPT-3, as it is both characteristic of what a text generator might suggest that a human would not, and a source of inspiration for users resorting to such tools. It is therefore a deeper exploration of possibilities to extract more striking and provoking briefs that might be closer to inspiration than to actual design briefs, but might nevertheless be interesting. Several of the answers obtained were very rational, confirming the effects of the temperature settings highlighted above. However some prompts yielded more surprising answers, and all over the answers were particularly specific regarding design instructions, in particular relating to color and shapes, which had appeared a lot less prominently when submitting other prompts. It also provided some curious associations, for example between the color of the house and the color of the moon itself, the first one being suggested to be white because that is the color of the second.

Prompt: *Two parents and their two children want to build a house. They want to live on the moon. What should the house look like?*
Answer: *The house should be white, because the moon is covered in dust. It should have big windows, because there is no atmosphere on the moon, so the stars are always visible. The house should be round, because there are no straight lines on the moon.*

While testing prompts relating to the combination of several design criteria, including information about relationships between the users of a potential space, one prompt in particular provided surprising and rather poetic results. Exploring further this peculiar prompt and the set of answers generated by GPT-3 showed that it provided a particularly representative example of the overall behavior and potential of the system. Answers to this series of prompts are typical of how the system handles many of the issues documented, while still provoking imagination further than more conventional assessments. In short, they are one the most interesting parts of this research.

Prompt: *Two parents and their two children want to build a house. They hate each other. What should the house look like?*
Answer 1: *The house should look like it is about to fall apart.*
Answer 2: *A house with four separate wings, each with its own entrance, would be ideal for this situation. That way the parents and children can have their own space and avoid each other as much as possible.*
Answer 3: *The house should look like a big, ugly, gray box.*

The examples cited above are demonstrative of the three types of proposal the system provided in answer to the prompt: the ugly house, the rational and separating house, the house that falls apart. While the most conventional of these proposals is an echo of the overall general characteristics observed in the previous section of this paper, the others are more poetic and underline the power of fiction for design briefs. We observe here again strong associations in the answers, for example between hate and mess, a feature that this time appears to rely more clearly on semantic fields than in previous cases. This is not surprising as the tool that is being assessed is a word-generator and semantic fields are thus a key feature in it, but in this last series of answers it is the source of more remarkable elements. The most extreme prompts, relating to living on the moon or to hate relationships between users, thus seem to provoke the most striking associations, by having the system resort not only to the elements in the database that directly relate to architecture, but also to others.

Prompt: *Two parents and their two children want to build a house. They hate each other. What should the house look like?*

Answer: *The house should look like a big mess, with clothes and toys strewn everywhere, dishes piling up in the sink, and the furniture covered in dust. The walls should be scuffed and the floors should be cluttered.*

The house that falls apart, in some specific answers, can be interpreted as a striking reference to House VI by Peter Eisenman and its slits – a canonical building in the history of XXth century architecture. While not sharing the same purposes, the two designs do pose the question of what matters most when designing a space, from the most down-to-earth issues to the most prospective. We thus reach a question that also arises when venturing into the resort to optimisation processes for space design, for what does an optimized architecture means exactly [21]? Our research on GPT-3 thus finally recalls one of the key issues in experimenting with computational design tools: they question the epistemology of architecture of their users, their ambitions for the discipline.

Prompt: *Two parents and their two children want to build a house. They hate each other. What should the house look like?*
Answer: *The house should look like a house divided in half. Each [member of the] family should have their own side of the house with a dividing line down the middle.*

5   Conclusion

The research presented here has highlighted two distinct spaces existing within the GPT-3 database that might be leveraged for architectural design. The first one, a series of conventional information on architecture, might be used in machine learning systems associated with B.I.M tools, for applications in the built industry such as automated layout generation – this set of information could be leveraged by having GPT-3 generate other text formats than design briefs. The second one, a series of much more imaginative and unexpected associations and proposals, might be used as inspiration.

While the author of the present research still considers the assessment provided here on the nature of architectural assumptions in a database as large as GPT-3 as worthy of interest as initially, the apparition of MidJourney, Disco Diffusion or Dall-e might render the whole workflow slightly out of date and put the relevance of the research in question. However AI architectural design tools are currently envisioned as mainly an inspirational tool, and it has been shown in the present research that well tuned, GPT-3 works well for that purpose albeit providing inspiration through another medium than image generation. The initially envisioned workflow with a script is thus still a worthwhile expansion of the present research. The method of evaluation applied here is furthermore applicable and relevant for MidJourney and others image generators, as it will provide equally significant insights into the structuration of their databases.

Other aspects of this research could also be expanded. We have seen that the assessments, choices of issues to examine and start prompts are based in large part on the author's decisions, relying on the training as architect of the author and on discussions with fellow architects on the topic. A larger study could be interesting, highlighting through broader exchanges with architects, including outside Europe, in order to vary the prompts submitted and observe results. A series of more hierarchized prompts could result in more details in the answers provided, thus obtaining more specific design briefs. The example-based prompts and the image generation also remain to be developed.

Acknowledgments. Contribution to this research has been made thanks to funding from the Architecture of Order Research Cluster and the Hessian State Ministry of Higher Education, Research and the Arts: State Offensive for the Development of Scientific and Economic Excellence (LOEWE).

References

1. LeCun, Y., et al.: Backpropagation applied to handwritten zip code recognition. Neural Comput. **1**(4), 541–551 (1989)
2. Kononenko, I.: Machine learning for medical diagnosis: history, state of the art and perspective. Artif. Intell. Med. **23**(1), 89–109 (2001)
3. McCorduck, P.: Machines Who Think, 2nd edn. A. K. Peters Ltd, Natick, MA (2004)
4. limits of ML
5. Loing, V., Marlet, R., Aubry, M.: Virtual training for a real application: accurate object-robot relative localization without calibration. Int. J. Comput. Vision **126**(9), 1045–1060 (2018)
6. Belousov, B., et al.: Robotic architectural assembly with tactile skills: simulation and optimization. Autom. Constr. **133**, 104006 (2022)
7. Chaillou, S.: Artificial Intelligence and Architecture: From Research to Practice. Birkhäuser, Berlin, Boston (2022)
8. Song, L., Mao, J., Zhuo, Y., Qian, X., Li, H., Chen, Y.: Hypar: Towards hybrid parallelism for deep learning accelerator array. In: 2019 IEEE International Symposium on High Performance Computer Architecture (HPCA), pp. 56–68. IEEE (2019)
9. Ahmed, S., Weber, M., Liwicki, M., Langenhan, C., Dengel, A., Petzold, F.: Automatic analysis and sketch-based retrieval of architectural floor plans. Pattern Recogn. Lett. **35**, 91–100 (2014)
10. As, I., Basu, P.: The Routledge Companion to Artificial Intelligence in Architecture. Routledge, Abingdon (2021)
11. Steinfeld, K., Tebbecke, T., Grigoriadis, G., Zhou, D.: Artificiale Rilievo GAN-generated architectural sculptural relief. In: Gengnagel, C., Baverel, O., Betti, G., Popescu, M., Thomsen, M.R., Wurm, J. (eds.) Towards Radical Regeneration. DMS 2022. Springer, Cham (2023). https://doi.org/10.1007/978-3-031-13249-0_12
12. Bleker, L., Pastrana, R., Ohlbrock, P.O., D'Acunto, P.: Structural form-finding enhanced by graph neural networks. In: Gengnagel, C., Baverel, O., Betti, G., Popescu, M., Thomsen, M.R., Wurm, J. (eds.) Towards Radical Regeneration. DMS 2022. Springer, Cham (2023)
13. Cardon, D., Cointet, J.-P., Mazières, A.: La revanche des neurones. Réseaux **211**(5), 173–220 (2018)
14. Brown, T.B., et al.: Amodei: language models are few-shot learners. In: Proceedings of the 34th Conference on Neural Information Processing Systems (NeurIPS 2020). Vancouver, Canada (2020)
15. https://lacker.io/ai/2020/07/06/giving-gpt-3-a-turing-test.html
16. Patti, K.: I forced a bot to write this book: A.I. meets B.S., Andrews McMeel Publishing (2020)
17. https://colab.research.google.com/github/alembics/disco-diffusion/blob/main/Disco_Diffusion.ipynb
18. https://www.midjourney.com/home/
19. Eubanks, V.: Automating Inequality: How High-Tech Tools Profile, Police, and Punish the Poor, 1st edn. St. Martin's Press, New York, NY (2017)

20. Gaudillière-Jami, N.: Automatiser l'architecture? Savoir-faire et calculabilité dans les pratiques des courants computationnels en architecture, 1965–2020 (2022)
21. Gaudillière-Jami, N.: The optimisation game. Computational architecture and AI: from symbolist case study to machine learning data. In: Deep City Symposium Proceedings, EPFL, Lausanne (2021)

Architectural and Social 'Word Forms' Slippage with Deep Learning

Immanuel Koh[(✉)] [ID]

Singapore University of Technology and Design, Singapore, Singapore
immanuel_koh@sutd.edu.sg

Abstract. The paper aims at contributing to the exploration of co-creative design processes afforded by the 'slippery' interplay of 'word forms' as found in human participatory designs and AI deep generative modelling. Conceptualised as an AI-augmented form-ideating social platform, it leverages the unique, and often implicit, relationship between the architect's use of textual description and visual representation of three-dimensional building forms, regardless of its site, program, and budget specificities. The paper draws from similar features found in existing platforms such as those for image sharing (Pinterest), image searching (Google Images), image breeding (Artbreeder), and text-based image generation (DALL·E 2, Stable Diffusion or Midjourney). However, unlike the large-scale pre-trained deep learning models and generic image databases provided by these platforms, the implicit disciplinary graphic and written languages of architects used in communicating forms suggests a need to not only facilitate the users in finding forms, but also the option to create their own datasets of 'named' forms and to then train their own AI models accordingly. Towards this end, the siteless, scaleless, grayscale, isometric, captioned (2-word phrases) thumbnail freehand sketches of three-dimensional building forms are adopted in this study as an architecturally representative and convenient training dataset. The paper will present the methods: extracting captioned sketches (OCR), training different deep generative models (GANs), vectorising captions (GloVe) and sketches (ResNet-18), and performing semantic clustering on captions and sketches (K-Means & PCA); followed by the results which illustrate the platform's key functionalities: inferring captions of input sketch (Form2Word), searching sketches with input captions (Word2Form), and searching or creating similar sketches with input sketch (Form2Forms). The paper thus contributes to an alternative co-design framework for architectural form ideation by reconfiguring the creative affordance between texts and images with the deep learning models.

Keywords: Co-Creation · Creative AI · Architectural Language · Architectural Intention · Text2Image

M. Turrin et al. (Eds.): CAAD Futures 2023, CCIS 1819, pp. 112–125, 2023.
https://doi.org/10.1007/978-3-031-37189-9_8

1 Introduction

1.1 Motivation

Architects not only make sketches to communicate their ideas (to themselves and/or to others), but often incorporate words to accompany these sketches. Whether which precedes which during the creative ideation process depends on individuals. That is, whether it is the word that translates into its sketch such as in a Text2Image generative model, or it is the sketch that translates into its word such as in an Image2Text generative model. One thus assumes there exists an implicit causation, or at least correlation, between any given words and sketches. However, one would only need to recall the 'word paintings' by Belgian Surrealist René Magritte to see that the pairing of words and images is anything but a linear translation, demonstrated most vividly by his 1929 painting titled 'The Treachery of Images'. The semiotic interplay between a painting of a pipe and the inscription 'Ceci n'est pas une pipe' ('This is Not a Pipe' in English) testifies to a dynamic layer of generativity that destabilises straightforward translation. Yet, this is only one among several examples of 'word images' slippage found in Magritte's illustrated text 'Les Mots et les images' ('Words and Images' in English) published in the same year.

Such slippage between words and images can be a means towards serendipitous co-creation, and further extended via the social and the computational. In this paper, the social refers to the collective interactivity among human participants who search, collect and exchange 'word forms' on an online social platform, while the computational refers to the personal curatorial generation of new 'word forms' by training one's own deep generative learning models. Both leverage the co-generativity from the implicit 'word forms' slippages found in the collective and intuitive subjectivity of the human architects, and the vectorial and high-dimensional objectivity of the machines. A final 'dimensional' slippage (or leap) is to be found when the architect begins to 'architecturalised' these pairs of two-dimensional forms and one-dimensional word into three-dimensional spatial forms of interiority. One is thus forced to generate the third dimension and manifest the 'word forms' as legitimate architectural drawings (plans/sections/elevations) and physical models. Therefore, the research project can also be understood as a pedagogical tool for students of architecture in training their spatial abilities and expanding their capacity for spatial imaginations.

1.2 Background

Previous works within architecture discipline have toyed with the generativity of pairing words and forms. Blanciak (2008), in his book 'Siteless: 1001 Building Forms', freely creates grayscale sketches of forms and pairs each with a name, as if to reinforce the intended form or to legitimise its architectural relevance. Years later, these 2D sketches became physicalised 3D ceramic pieces by artist Molanphy (2015) in his artwork called 'tripelgänger'. Mari and Yoo (2012); Di Mari (2014) formalizes words as a catalogue of 'spatial verbs' to 'operate' on forms, not unlike the explanatory formal diagrams from OMA (Office for Metropolitan Architecture) or BIG (Bjarke Ingels Group). Pedagogically, Gerber (2020) published a workbook to train and test spatial ability among architecture students, thus bringing back into focus the importance of seeing, thinking, and imagining forms. Recent understanding of form-finding in computational architectural design is often characterised more by an optimisation outlook, and less as a socially curatorial (e.g., Pinterest) and collaboratively generative one (e.g., Artbreeder). For the latter, since the invention of generative adversarial networks or GANs (Goodfellow, et al., 2014), alongside its subsequent adaptations for image synthesis (Radford et al., 2016) and GAN-based Text2Image synthesis (Reed et al., 2016), more powerful deep generative models have emerged such as DALL.E (Ramesh et al., 2021), Imagen (Saharia et al., 2022) and Stable Diffusion (Rombach et al., 2022). However, as these new deep models are trained with increasingly far larger number of parameters and training datasets by commercially and well-funded organisations like OpenAI, Google, Meta or Stability AI, their affordance for personal and bespoke creative use also somewhat diminishes, especially when the domain tasks and datasets concerned differ significantly from these pretrained generative models. In addition, the ways in which one can interact with these AI models are inevitably dictated by their AI developers, whether it is limited by typing a string of text on specific platforms like Discord or clicking buttons on their dedicated web applications. Moreover, the cost of non-free tier subscriptions further de-democratizes their access and issues of copyrights remained unaccounted. In view of these issues, gaps and opportunities at the intersection of AI and architecture, the paper aims to explore a simple (yet bespoke) ideation scenario and propose a feasible framework/platform for a more personal and pedagogically-driven alternative in generative architectural form-finding.

2 Methods

2.1 Workflow

The following sections will describe the different stages involved in the design and development of the proposed platform. All the codes were written on Google Colab; all the models were trained using Google's GPU with a free subscription account; all the files were saved on Google drive and code libraries used include tensorflow, keras, torch, pandas, numpy, sklearn, matplotlib, PIL, cv2, and pytesseract. Figure 1 shows a high-level diagrammatic representation of the proposed workflow.

2.2 Dataset Preparation

The paper has chosen the content of the book 'Siteless: 1001 Building Forms' (Blanciak, 2008) as a familiar, yet simple, dataset to demonstrate the proposed framework. The original sketches exist in a printed physical book form which necessitates the scanning of every page prior to its procedural digital extraction. Each page contains 12 entries and shares the same 4×3 grid layout where each entry is a sketch with a given name below it and a chronological number index at its top-left corner. Simple image cropping with Pillow library and optical character recognition (OCR) with Pytesseract library are used and the results saved as 1001 number of 122×122 grayscale images for the former and as a table with 1001 rows and 2 columns (i.e., first word as an adjective and the second as a noun) on an Excel sheet for the latter.

Data Preparation

Implementation

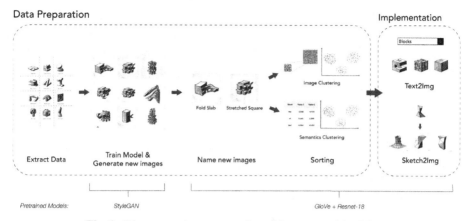

Fig. 1. Diagrammatic representation of the proposed workflow.

2.3 Training Models

With the 1001 training data points now available, the experiment proceeds with the selection of a suitable generative model architecture. Beginning with the training of a vanilla generative adversarial network (GAN) from scratch, the project eventually adopted transfer learning using a pretrained style-based GAN or StyleGAN (Karras et al., 2020) model. The advantage of the former is the ability to leverage the lower-level features that are already learned by the deep neural networks, thus making it feasible to train a new yet performative GAN model even with a small dataset of only 1001 images. The model is trained for 3000 epochs with different distribution of the original dataset, yielding 6 models with different formal characteristics derived from an unsupervised image clustering procedure (to be elaborated in the later section 'Form Clustering').

2.4 Naming New Forms

The StyleGAN model, used in a transfer-learning manner, while effective in generating new sketches as variations of the given training dataset of 1001 sketches, it is not designed to generate any accompanying names to describe the semantics of the sketches. A naive solution would be to use pretrained image captioning deep learning models (Vinyals et al., 2017; Yang et al., 2017) to literally describe the content of each sketch. However, unlike coloured photographic images containing real-world objects alongside their con-textual scenes that are used to train such models, the training set of grayscale sketches are abstract, conceptual, standalone, and imagined architectural forms which are loosely described by two-word phrases (adjective and noun) rather than as sentence-like cap-tions. Transfer learning of such pretrained image captioning models would likewise be ill-suited given the size and nature of the sketch dataset. Instead, a novel workflow com-bining feature vector extraction for measuring image and word similarities is proposed, to infer a two-word phrasal name for any newly generated sketch as shown in Fig. 2. Both tasks leverage learnt features from existing pretrained image-based and language-based deep models prior to a subsequent process of computing their respective centroids to identify either a new name for a given sketch, or a new sketch for a given name.

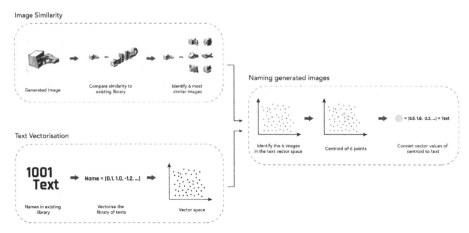

Fig. 2. Workflow for inferring a name for any given newly generated form

2.5   Form Similarity

An image can be converted to its feature vector by doing a forward pass through an image-based classification deep neural network pre-trained with a large image dataset and extracting the output of the network's earlier layer as a dense representation of the input image. A good model for such a task is ResNet or deep residual networks (He et al., 2015) pre-trained on the ImageNet dataset (Russakovsky et al., 2015) containing millions of images. More specifically, the 'avgpool' layer at the end of the pre-trained ResNet-18 is chosen for its simplicity among the 5 available versions (18, 34, 50, 101 and 152) which returns feature vectors in 512 dimensions. With all the existing sketches in the training set converted to their feature vectors, it is possible to score their visual similarity by computing their cosine similarity values. Likewise, any newly generated

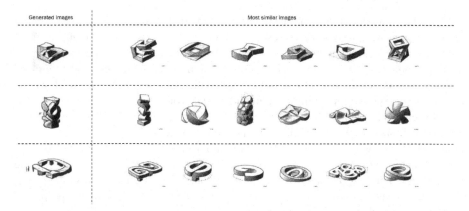

Fig. 3. Results of the six most similar looking existing forms based on a given newly generated form.

sketches from the StyleGAN model can now be measured against the existing sketches to obtain the most similar looking ones as shown in Fig. 3.

2.6   Form Clustering

With the means to now convert any given sketch into a feature vector, a preliminary EDA or exploratory data analysis (Tukey, 1977) is performed to better understand the existing visual dataset via data visualisation using K-means for clustering followed by principal component analysis (PCA) for dimensionality reduction. The K-means clustering algorithm uses all 512 dimensions of the feature vector belonging to each sketch to group the existing dataset of 1001 sketches into 6 clusters, with which they are intuitively named as 'Layered', 'Sharp', 'Tubular', 'Spherical', 'Fold' and 'Block', based on their distinct characteristics. The clustering result thus suggests underlying formal themes that could be subsequently used as constraints to separately train different thematic StyleGAN models in generating more specific type of forms. It is also a good way to identify any formal bias as a result of an imbalanced data distribution, which could in turn be rectified by further curating or manipulating the existing dataset. To effectively visualize all the 1001 sketches, PCA is used to compress the 512 dimensions into just 2 dimensions for plotting as shown in Fig. 4.

Fig. 4. A plot showing the distribution of existing forms based on a compressed version of their feature vectors

2.7   Word Similarity

A word can be converted to its vector representation called word embedding. Words with similar meanings are locally clustered in this vector space and thus provides the means to score each word by simply computing their cosine similarity values. Given the small corpus of only 1001 two-word phrases, it would be ineffective to train a new word embedding neural network model from scratch. The approach is thus to use existing word embedding models such as the Google word2vec model (Mikolov et al., 2013) trained with billions of words and containing 3 million words and phrases, or the Standford GloVe model (Pennington et al., 2014) also trained with billions of words and containing a 400,000-word vocabulary. The smaller size of the latter with 100-dimensional vectors is chosen among the 4 available ones (50, 100, 200, and 300). The two-word (adjective and noun) naming format of the sketches, however, demands a less direct use of the chosen 100-dimension GloVe model which only allows vectorization per word. Therefore, two different methods have been tested to determine an appropriate solution for the task by visually evaluating their respective word form accuracy, which will be elaborated in the later section 'Word Form Accuracy/Slippage'.

2.8   Word Clustering

During the training of the first StyleGAN model using all 1001 sketches, newly generated forms are dominated by blocky-looking outputs. In fact, even without visually inspecting the sketches themselves, a basic text frequency count procedure performed on their corresponding names is already showing a high frequency of words related to blocky-looking forms such as 'towers'(105), 'slab'(47) and 'blocks'(37). However, a more effective approach is to use word embeddings instead. Since the names come in the format of two-word phrases, each word is first converted into a word embedding of 100 dimensions before being concatenated into a 200 dimensioned vector. A similar preliminary exploratory data analysis is then performed to better understand the existing textual dataset via data visualisation using K-means for clustering followed by word clouds for retrieving the top occurring words. Two different k values have been tested (6 and10) and the results are interestingly different. On the one hand, it is beneficial to have distinctly classified names with fewer clusters, but on the other hand, overlapping meanings found in more clusters might provide a greater ambiguity for drawing inspiration.

2.9   Word-Form Accuracy/Slippage

In order to generate a name for any newly generated sketch based on the proposed workflow (Fig. 1), it is necessary to first search for the 6 most similar sketches according to their proximity to the given sketch in the 512 dimensioned image vector space. This is followed by obtaining their respective word embeddings in the 100 dimensioned text vector space, and then computing the centroid of their text vectors, before converting the centroid vector into a word based on the closest available word within the 400,000 words vocabulary of the GloVe model. As aforementioned in the section 'Word similarity', it is less trivial to do so when the names are not in the form of single words, but as two-word phrases. The first method is to split the two-word name (adjective + noun) belonging to each sketch and vectorise them into two separate lists of 100 dimensions receptively. Therefore, for the sake of clarity here, it means that each existing named sketch is represented by one 512 dimensioned image vector, one 100-dimensioned adjective text vector and one 100-dimensioned noun text vector now. The same procedure proceeds in computing and converting the centroid into an adjective and a noun, which are combined to form a final two-word phrase as the name of the given newly generated sketch. Figure 5 shows some samples of the outputs with this first method.

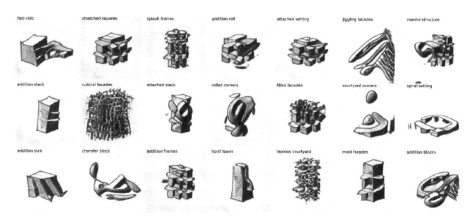

Fig. 5. Results from the first method showing convincing inference of names for the newly generated forms

The second method likewise vectorises each two-word name into two separate lists of 100 dimensions, but subsequently concatenates them into a single 200 dimensioned vector representing the two-word phrase directly. That is, each existing named sketch is now represented by one 512 dimensioned image vector and one 200 dimensioned adjective-noun texts vector. With this approach, it is no longer possible to leverage the 400,000 single words vocabulary of the pretrained GloVe model since there exists no 200 dimensioned vector space for two-word phrases. To circumvent this reduction in vocabulary size, it becomes necessary to perform name permutation, while preserving the order of adjectives and nouns. Although this increases the existing 1001 phrases to 195,639 unique phrases, the names of all newly generated sketches will be limited by those found in the original book itself. In fact, the result from this second method seems inaccurate. Eventually, the first method has been adopted and incorporated into the proposed creative AI social platform for 3D architectural form ideation.

3   Results

3.1   Application

The proposed application is a creative social platform for architectural form-finding that combines the non-human generative power of deep neural networks in computer vision and natural language processing with the participatory interaction of collective human creativity. There are three main featured functions, namely, searching with Word2Form (or Text2Image) and Form2Word (or Image2Text) and creating with Form2Forms. User will only have to sign up for a personal account with a username and a password. Once logged in, the users can explore by scrolling through their social feeds that include posts from other users whom they follow, recommended posts and posts featured by the platform itself. Each post consists of the user profile, the form, a two-word phrasal name for the form, and optionally an extended text describing the user's own formal and architectural interpretation. Each user account consists of a short account description, user posts and personal library of forms. The library serves as a source of datasets for the user to curate and train new thematic GAN models. Those newly generated forms could then be further curated, saved, and posted onto the feed for sharing with others.

3.2 Word2Form

The platform provides user the ability to search for any form by typing a two-word phrase (adjective + noun) describing the desired formal characteristics. The 6 most representative forms will be returned as the search results. Behind the interface, the two-word phrase is split, and each word is checked to ensure it is one of the 400,000 words vocabulary available in the pre-trained GloVe model. If not, the user will be prompted to type in another available adjective or noun or both. Thereafter, each of the two words is converted into its 100 dimensioned text vector before being concatenated as a single 200 dimensioned phrase vector to search for the 6 closest forms as computed in the 200-dimensional space. The available forms are taken either from the existing dataset of 1001 forms or other expanded versions made by other users. Figure 6 shows some examples of the search results.

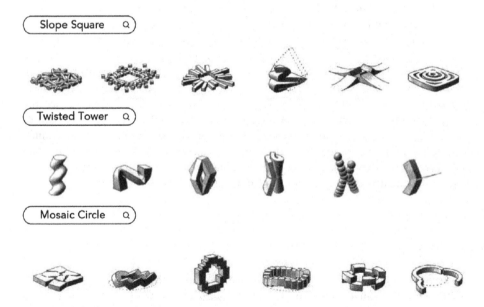

Fig. 6. Three sets of samples showing the search results from the Word2Form function.

3.3 Form2Word

The platform also provides user the ability to name any given form by uploading a form. Behind the interface, the name is obtained by the first method described in the earlier section 'Word-Form Accuracy/Slippage'. In fact, by changing the proposed pre-trained models used (ResNet-18 and GloVe), different results will be obtained due to factors such as, how other image-based models would extract features differently and how other word embeddings would contain different vocabulary set.

3.4 Form2Forms

Apart from searching words with forms, and vice versa, users can directly search for similar-looking forms by either uploading a form or selecting any existing form from any users' libraries. Although it is as trivial as computing the cosine similarity scores within the shared image feature vector space to obtain the search result, it is however crucial that the form created is of a similar style. For example, if a cross-hatching style is used, instead of the flat shading style found in the existing 1001 sketches, the model will misinterpret the hatching as string-like secondary forms despite maintaining relative accurate inference for the overall massing as shown in Fig. 7.

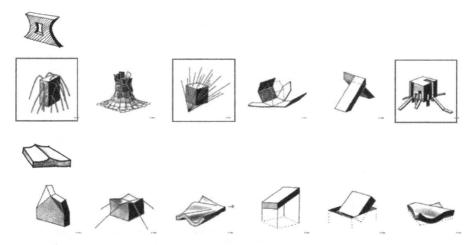

Fig. 7. Two input forms with different shading styles yielding significantly different accuracy (e.g., misinterpretation for those highlighted in red boxes) from the search results of the Form2Forms function.

A more interesting option available on the platform is for the user to generate completely new forms by uploading or curating a new dataset to train his or her own thematic StyleGAN model via transfer learning, similar to what has been described in the earlier section 'Training Models'. The user could then curate a collection of forms based on the output of the newly trained model.

3.5 Spatial Imagination in 3D

Every form on the platform can potentially be interpreted architecturally, that is, translating as two-dimensional drawing of plans or elevations or sections, modelling as three-dimensional CAD geometry, and fabricating as physical artefact. In fact, the proposed creative AI social platform aims to simultaneously couple and decouple architecture disciplinary language, both linguistically and visually, thus enriching the interpretative capacity and design affordance of the forms. Figure 8 shows examples of such potential translations.

Fig. 8. A selection of nine newly generated forms is translated into their respective physical clay massing models. These are in turn analysed as buildings with interior spaces using the architectural representation technique of the sectional drawings.

4 Conclusion

The paper has presented the end-to-end design and development of an AI-augmented form-ideating social platform using an all-familiar data source of a book and the mode of architectural representation of a grayscale thumbnail sketch. Critically, it identifies the inherent associations and slippages between words and forms used in architecture and proposes a serendipitous co-creative space of encounters among non-human word-forms and human interaction for generative designs with artificial intelligence. While the preceding section on 'spatial imagination in 3D' attempts to illustrate the interpretative capacity afforded by common architectural techniques of 2D CAD drawings and physical model making, one of the future works would be to explore a more seamless workflow in digitally generating 3D massing models from those generated 2D sketches. Given the same low-resolution, exact aspect ratio of 1:1, and similar isometric view of these grayscale sketches, one potential approach would be to perform a reprojection and voxelization procedure to infer their 3D models. Since this research was conducted in 2021 and thus preceded the more recent text2image-based deep learning models, future work will also thoroughly investigate and evaluate their performance for the same architectural ideation task. As of the time of this writing, experiments have already been underway to evaluate whether the same functionalities (i.e., word2form, form2word and form2forms) can also be achieved by popular text2image models like Midjourney, Stable Diffusion and DALL·E 2, all of which only emerged around mid-2022. As shown in Fig. 9, 10, 11 and 12, these large generic models have in fact failed in their conceptual alignments when translating between architectural forms and texts (specifically for Form2Word and Word2Form). In other words, given the same Blanciak's dataset, our models outperform theirs -- a reassuring sign that this line of research to effectively appropriate deep learning models for the specificity of architectural ideation is still promising.

Fig. 9. Form2Word: Given the sketch 'mosaic slab' (leftmost image) taken from Blanciak (2008), when running an image2text deep learning model, it returns the caption "a black and white photo of a clock tower". Form2Forms: Four outputs from DALL·E 2's 'generate similar images' feature after uploading the same sketch 'mosaic slab' (leftmost image). The first model fails and the second model seems to succeed in understanding the visual language and formal intention for this architectural ideation task. However, the latter still lacks the conceptual diversity as compared to Fig. 6.

Fig. 10. Word2Form: Four outputs from DALL·E 2's text2image feature after typing the text 'mosaic slab' taken from Blanciak (2008). The model failed to understand the disciplinary language and formal intention for this architectural ideation task.

Fig. 11. Word2Form: Four outputs from Midjourney's text2image feature after typing the text 'mosaic slab'. The model failed to understand the disciplinary language and formal intentions for this architectural ideation task, despite being less literal than DALL·E 2 as shown in Fig. 10.

Fig. 12. Word2Form: Four outputs from Midjourney's text2image feature despite after typing a more detailed string of text "'mosaic slab', grayscale, isometric, captioned (2-word phrases) thumbnail freehand sketches of three-dimensional building forms". The model failed to understand the disciplinary language and formal intention for this architectural ideation task.

References

Blanciak, F.: SITELESS: 1001 Building Forms. MIT Press, Cambridge, MA, USA (2008)

Shore – Brian Molanphy. https://brianmolanphy.com/shore/ (2015). Accessed 28 Oct 2022

Di Mari, A., Yoo, N.: Operative Design: A Catalogue of Spatial Verbs. BIS Publishers, Amsterdam (2012)

Di Mari, A.: Conditional Design: An Introduction to elemental Architecture. Uitgeverij Bis, Amsterdam (2014)

Gerber, A. (ed.): Training Spatial Abilities: A Workbook for Students of Architecture. Basel (2020)

Goodfellow, I.J., et al.: Generative Adversarial Networks. arXiv:1406.2661 [cs, stat] [Preprint]. http://arxiv.org/abs/1406.2661 (2014). Accessed 28 Oct 2022

Radford, A., Metz, L., Chintala, S.: 'Unsupervised Representation Learning with Deep Convolutional Generative Adversarial Networks', arXiv:1511.06434 [cs] [Preprint]. http://arxiv.org/abs/1511.06434 (2016). Accessed 28 Oct 2022

Reed, S., et al.: Generative Adversarial Text to Image Synthesis. arXiv:1605.05396 [cs] [Preprint]. Available at: http://arxiv.org/abs/1605.05396 (2016). Accessed 28 Oct 2022

Ramesh, A., et al.: Zero-Shot Text-to-Image Generation. arXiv:2102.12092 [cs] [Preprint]. http://arxiv.org/abs/2102.12092 (2021). Accessed 28 Oct 2022

Saharia, C., et al.: Photorealistic Text-to-Image Diffusion Models with Deep Language Understanding. arXiv. https://doi.org/10.48550/arXiv.2205.11487 (2022). Accessed 28 Oct 2022

Rombach, R., et al.: High-Resolution Image Synthesis with Latent Diffusion Models. arXiv. https://doi.org/10.48550/arXiv.2112.10752 (2022). Accessed 28 Oct 2022

Karras, T,. et al.: Analyzing and Improving the Image Quality of StyleGAN. arXiv:1912.04958 [cs, eess, stat] [Preprint] http://arxiv.org/abs/1912.04958 (2020). Accessed: 28 Oct 2022

Vinyals, O., et al.: Show and tell: lessons learned from the 2015 MSCOCO image captioning challenge. IEEE Trans. Pattern Anal. Mach. Intell. **39**(4), 652–663 (2017). https://doi.org/10.1109/TPAMI.2016.2587640

Yang, L., et al.: Dense Captioning with Joint Inference and Visual Context. arXiv:1611.06949 [cs] [Preprint]. http://arxiv.org/abs/1611.06949 (2017). Accessed 28 Oct 2022

He, K., et al.: Deep Residual Learning for Image Recognition. arXiv:1512.03385 [cs] [Preprint]. http://arxiv.org/abs/1512.03385 (2015). 28 Oct 2022

Russakovsky, O., et al.: ImageNet Large Scale Visual Recognition Challenge. arXiv:1409.0575 [cs] [Preprint]. http://arxiv.org/abs/1409.0575 (2015). 28 Oct 2022

Tukey, J.W.: Exploratory data analysis. Addison-Wesley Pub. Co., Reading, Mass. http://archive.org/details/exploratorydataa00tuke_0 (1977). Accessed 28 Oct 2022

Mikolov, T., et al.: Efficient Estimation of Word Representations in Vector Space. arXiv:1301.3781 [cs] [Preprint]. http://arxiv.org/abs/1301.3781 (2013). Accessed 28 Oct 2022

Pennington, J., Socher, R., Manning, C.: Glove: Global Vectors for Word Representation. In: Proceedings of the 2014 Conference on Empirical Methods in Natural Language Processing (EMNLP). Proceedings of the 2014 Conference on Empirical Methods in Natural Language Processing (EMNLP), Doha, Qatar: Association for Computational Linguistics, pp. 1532–1543 (2014). https://doi.org/10.3115/v1/D14-1162

Performance-Based Design

Enabling Flexible Architectural Design Re-representations Using a Phenotype-Based Strategy

Ban Liang Ling[(✉)] [ID] and Bige Tunçer [ID]

Singapore University of Technology and Design, Singapore, Singapore
`banliang_ling@mymail.sutd.edu.sg`

Abstract. Architectural design is driven by qualitative variables like spatial aesthetics and geometry appreciation. At the early explorative stage, flexible re-representation of information is used to discover design solutions. However, in computational design, current data management methods are constrained by genotype and performance representations. These strategies do not enable a similar level of flexibility. As such, this paper presents a phenotype-based strategy that allows design logic to be transferred between computational design processes. Across different design spaces, similar phenotypes can be bridged based on their appearance. This will be demonstrated using room footprints and daylight distribution results which are represented using Elliptical Fourier descriptors and Gabor filters. This enables flexible re-representations as the algorithm can now appreciate geometrical properties. Examples to demonstrate design logic transfer, which are guided by the phenotype space, will be covered in two different scenarios: 1) in a search scenario—where the search direction of an unexplored design space is influenced by learnings from an explored design space, and 2) in a scenario where all design spaces are explored—the data structure and possibilities afforded to new search scenarios.

Keywords: Phenotype-Based Strategy · Gabor Filters · Elliptical Fourier Descriptors · Flexible Re-Representations

1 Flexible Re-representations

Architectural design involves the proposal of geometrical solutions to address various problems. Problems that have been clearly defined nevertheless remain fluid throughout the design exploration process, because the problem and solution landscape influence each other in the early phase of design. As the design progresses, new problems are illuminated, and initial problems might become irrelevant. The level of priority for each problem might also change over time [15]. Designers have been observed to re-interpret design briefs by adopting new goals or by re-organizing the provided information [8]. The key to such re-representations is the human ability to recognize patterns from any presented information. In architectural design, an essential piece of information is the solution geometry. Designers can identify geometrical patterns within a current design

M. Turrin et al. (Eds.): CAAD Futures 2023, CCIS 1819, pp. 129–144, 2023.
https://doi.org/10.1007/978-3-031-37189-9_9

scenario or across past and current design scenarios [8]. These patterns help with performance prediction and can lead to alternative proposals [7]. To replicate pattern recognition using computational methods, geometry representation is based on quantifiable parameters like size, position, and how the geometry was constructed. Algorithms will then identify patterns according to these parameters or according to their response to evaluation criteria. Generating a feasible computational solution to a problem, that would benefit from recognizing a geometrical pattern and re-representing it in a similar fashion as a human would, would therefore rely on the representation of parameters and evaluation criteria.

When presented with a cube geometry, humans can describe its construction in many ways. Figure 1 shows two parametric descriptions, logic A generates a cube using 3 steps, whereas logic B generates a cube using 2 steps. The computer sees both possibility spaces (A & B) as different solution spaces due to their different construction methods. However, humans can recognize patterns based on the final geometry, a parameter that is unique and more primal than the construction method. In this case, humans can propose alterations that veer off logic A and B (e.g., chamfers and hole punching), or humans can identify different pathways that lead to similar solutions. On the other hand, algorithms require the construction method as an input and as an identity. Pattern recognition in a computational setting would therefore be constrained by logic A or B (Fig. 1). If logic A is in effect, only 4-sided extrusions can be constructed, and performance patterns are identified based on steps 1–3 of logic A.

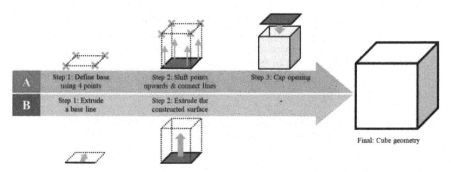

Fig. 1. Construction logic A and B both build towards a cube using different methods

The above describes a parametric approach using the construction logic as a representation. The construction parameters are compared against a performance score to identify high performing patterns. This approach has been shown to improve solution diversity [2] and overall performance [6]. While such methods are successful, common drawbacks include the long computational runtime and the specialized expertise required for a meaningful parametric search [3]. One reason is because computers are unable to appreciate the generated geometry. When the computer is assigned logic A, it only explores the possibility space of logic A. However, human designers can easily identify patterns, using them to expand possibilities.

Developing a geometry-based representation is therefore essential for computers to identify geometrical patterns. Geometry representation contributes towards parametric design in two ways: 1) Performance patterns are related to geometry instead of construction parameters. This means that similar looking solutions, which will otherwise be flagged as diverse according to their differing parameters, can be identified. Instead of evaluating each solution, a diverse spread of solutions can first be identified before being evaluated. 2) Parametric design can recognize patterns across different construction logic. Instead of reading the cube as a string of values based on logic A or B (Fig. 1), a geometry-based representation can provide the cube with a unique identity. Computational learnings from each possibility space will then be related to a unique geometry-based identity and these can be brought across logic spaces.

1.1   A Phenotype-Based Strategy

There exist three methods for solution representation: 1) genotype—describes the actions that lead to solution formation, 2) phenotype—describes the overall geometry, and 3) performance—describes the value of the solution [11]. Figure 2 uses logic A of Fig. 1 as an illustration. Firstly, the genotype representation refers to the position of each point and the upward shift height. This representation is common in computational design, where various construction methods have been used as a solution identity [2]. Secondly, the phenotype representation captures the "look" of a solution. This refers to the uniqueness of a geometry and the expression of key geometrical parameters. Since the final geometry is represented, the phenotype representation will be independent of its construction method. Examples include computer vision being employed to quantify the composition of shapes [11], and the expression of surface patterns [16]. Lastly, the performance representation refers to a figure that captures the solution value according to an evaluation metric. Commonly used in performative design, this representation can sieve out high performing solutions [2].

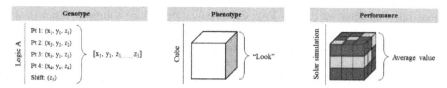

Fig. 2. Types of solution description: Genotype describes the construction actions, phenotype describes the geometry, and performance describes the value

In typical parametric search scenarios, search directions are driven by genotype- or performance-based methods [10]. Transferring learnings into a new scenario requires a human to analyze and translate lessons into a new parametric model. Compared to existing search methods, a phenotype-based strategy can span across scenarios due to its focus on the look of solutions [11]. Using Fig. 1 as an example, logic A generates cubes by shifting 4 points and logic B generates cubes by extruding a base line. There would exist common looking solutions between both, and a phenotype-based strategy can recognize them to bridge across A and B possibility spaces. This is similar to search conducted by a human designer, who can identify the common looking solutions. Lessons learnt from logic A, where a longer length of the East facing side corresponds to better daylight performance, can be extended into logic B. This way, the search commences with logic B solutions that have longer East facing lengths.

1.2 State of the Art in Parametric Design

Compared to a genotype representation that has many-to-one descriptions (Fig. 1 demonstrates this using a cube which can be described based on logic A and B), a phenotype representation describes unique geometrical properties. This means that a phenotype representation is independent of its genotype and performance values. It also means that if a possibility space of all solution properties is constructed, the phenotype representation will always take up a unique position within this space. An example is an image-based representation, whereby solutions are exported as images, and these are then used to generate unique identities. Ling and Tunçer propose the phenotype as an intermediate layer of geometrical considerations, such that various genotype logic can be consolidated based on "looks" [11]. They further demonstrate that Zernike polynomials can describe qualitative features like geometry composition. Based on these descriptions, similar looking solutions can then be identified. Stuart-smith and Danahy present results of Gabor filters that were used to measure the textures of roof patterns [18]. The resultant values formed a metric to analyze the visual character of generated roof designs. Both computer vision methods use image-based representations to provide a targeted description of solution phenotypes. While Zernike polynomials describe compositions, Gabor filters describe surface textures.

Other image-based methods involve the training of Variational Auto-encoders (VAE) to generate solutions [1, 13]. These methods represent the phenotype as images, attempting to identify high dimensional patterns in the latent space. Ampanavos and Malkawi [1] trained a VAE with bird's eye view depth maps, using optimized forms as an input for the algorithm to identify high performing patterns in relation to site conditions. They demonstrate that after training, a VAE can produce optimized solutions faster than a typical optimization run. Mirra and Pugnale [13] introduce another strategy that has similar aspirations. After training a VAE on common shell structures, a floorplan boundary can be used to generate plausible shell structures. These generated alternatives are presented to a designer, who will then interact and evaluate solutions. Both strategies encode geometrical properties (heights or boundary) within an image, and this enables the algorithm to focus on key properties. Furthermore, the phenotype representations also allow computers to "see" presented data, expanding their previously limited scope within genotype representations.

The above methods present the usage of phenotype representations as a measurement method, or as a training data set. These demonstrate the potential of phenotype representations in encoding geometrical information. While the encoding of heights allowed VAEs to generate solutions, the encoding of surface textures enables a measurement of aesthetic diversity. However, these applications of phenotype-based methods are limited to one design scenario. For example, if Ampanavos and Malkawi alter the site boundary, the VAE might misrepresent data and produce infeasible solutions. Even if a new VAE is trained, the training would be conducted from scratch. New sets of training data would be required and this leads to an overfitting or an underfitting of input data. This contrasts human designers who can easily identify patterns from a previous site and apply them under new conditions.

To address this limitation, this paper presents a phenotype-base strategy that enables computers to perform flexible re-representations. The case study involves room level spaces whereby Elliptical Fourier descriptors (EFD) are used to represent the room footprint [4]. Figure 3 demonstrates an approximation of a freeform shape within ten "shrink-wrap" steps. At each step, four values are generated, and each value represents the incremental morph that an ellipse must undertake. The values of a subsequent step relate to earlier steps, and they all combine into a unique identity for the footprint. This incremental morph was favored as it captures macro-level geometrical features at earlier steps. If the freeform shape is compared to a horizontal rectangle, the first few "shrink-wraps" would produce similar values. To differentiate between both shapes, later steps would have different morph values. This means that footprints with similar macro-level features can be identified yet be differentiated when the micro-level features are considered. This relates to daylight performance as the macro-level features contribute most to daylight patterns.

Fig. 3. Elliptical Fourier Descriptors used to approximate a shape boundary.

2   Measuring Daylight Phenotypes

In building performance, daylight measurements look at the frequency of achievement of an illuminance threshold. Designers map these frequencies onto a floorplan to study the spatial performance with respect to daylight performance. The illuminance and spatial considerations form two decision spaces that designers alternate between.

2.1   Daylight Performance Representations

A common strategy is Climate-Base Daylight Modelling (CBDM), where illuminance levels are measured and reported as an annual frequency. Examples include spatial daylight autonomy (sDA) and useful daylight illuminance (UDI). In this study, UDI (300–3000 lx, Singapore climate) is used as the performance indicator. UDI is preferred because it considers overlit and underlit spaces as undesirable. This provides meaningful daylight patterns near the East facing windows and dark corners. UDI patterns also capture the size and proportions of comfortably lit spaces.

In typical optimization strategies, the area-weighted average is used as a representation of daylight performance [9, 10]. However, this average score might overlook climate and spatial considerations. Figure 4 details these considerations in the footprint pheno-type, the UDI phenotype, and the UDI area-weighted average. In search algorithms, the genotype is manipulated based on the average value (Fig. 4 bottom row). Looking at the right most column of Fig. 4, the rectangular option achieves the highest UDI score (52.4%). However, this comparison is solely based on daylight performance. If the design intent is for a bedroom layout, then the shifted and L-shape options might be better performers. For example, the areas of near 0% UDI (blue areas) can be used as a wardrobe space, and this avoids shadow casting into the main room space. As such, a new design direction related to having kinks might become useful. Human designers can easily change their spatial evaluations, leading to new performance metrics in comparing solutions. On the contrary, search algorithms are fixated on the UDI average calculated across a floorplan. This suggests that the UDI phenotype can be used to guide design decisions related to spatial planning.

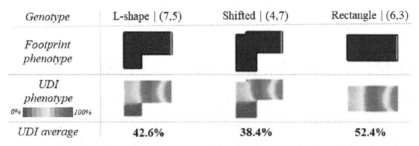

Fig. 4. Comparison across footprint phenotypes, UDI phenotypes, and UDI average

To address the issue of long simulation runtime, surrogate modelling (SM) has been proposed as an alternative performance heuristic [13]. Muthumanickam et al. train a surrogate model to take in window and room dimensions before the algorithm predicts the daylight performance of a solution [13]. In essence, a performance landscape is first built based on discovered solutions. This landscape relates input genotypes to performance, and it predicts the performance of yet-to-be discovered solutions. Designers would then interact with this landscape (using window and room dimensions), exploring geometrical properties and their relationship with performance values. SM provides a performance representation where predictions are carefully presented as absolute values and as a part of the larger landscape. This allows designers to understand the overall performance of the possibility space.

Although SM reduces computational runtime, its prediction is still related to a single representational value. A floorplan of daylight pattern is reduced into a single number (from row 3 into row 4 in Fig. 4) and the focus is on the final value. In computational design, this property limits the exploration of the design space into a quantity-driven approach, whereby the agreement of more data towards a goal suggests the existence of a design direction. This means that the most populous data trends become design pathways. However, the success of this approach depends on the definition of parametric models and performance metrics. When the performance representation is constrained, current search methods might lead to an early-stage design fixation [7].

When looking at daylight distributions, different patterns can lead to the same average score. Figure 5 demonstrates this with four room shapes that are all within an UDI average range of 59–61%. Although they all have similar UDI average, the difference in daylight patterns suggest various spatial potential for each room. However, existing computational search logic works within fixed performance representations and the diverse patterns in Fig. 5 are identified as similar. While designers can rectify the underlit spaces by shifting the room boundary, a search algorithm only identifies rectifications based on the UDI average. To the computer, an improvement step would require further trial-and error of genotypes to find a better UDI performer.

Fig. 5. Various daylight phenotypes can have similar UDI averages.

2.2 Gabor Filters to Describe UDI Patterns

UDI measures the frequency that a measurement point is within a comfortable range. Sharp edge boundaries usually exist along the room boundary while value gradients might exist near the windows. These are key geometrical parameters for computer vision to pick out. Gabor filters are chosen due to their ability to recognize texture by highlighting changes in color frequency in different pixel directions [5].

Fig. 6. Gabor filters and their impact on UDI phenotypes

By changing hyperparameters, various filters are constructed for textures recognition (top row in Fig. 6). The response to these filters (bottom row in Fig. 6) is represented as a list of numbers that combine into a unique identity. In a daylight distribution, this texture recognition is useful for identifying daylit patterns. Since the hyperparameters check for color frequency in specific angles, this means that Gabor filters can also relate macro-level patterns (second and last columns in Fig. 6) while differentiating solutions with micro-level differences (third and sixth columns in Fig. 6).

3 Driving Design Search

Five different design spaces were set up to study daylight performance within a room space. Figure 7 details the names of each design space, alongside the generated room footprint—all rooms have equal area and are rotated true North. Each parametric model has two genes which exert different geometrical influence in each design space. The east most boundary is always set as a full height window, while the rest of the boundaries are set as generic walls. Honeybee [14] was used to measure UDI (between 300–3000 lx) in the Singapore climate. Each image of the footprint and UDI pattern is exported as a JPEG image (100 × 100 px). UDI patterns are then adjusted to have a white (0%) to black (100%) scale. This ensures that areas with 0% UDI would be a part of the image background and areas with 100% UDI will be more prominent. EFDs are calculated up to 25 iterations and normalized such that results are rotational invariant. Gabor filters are set at angles of 0°, 90°, 180° and 270° with two variations of sigma (1 & 3) and frequency (0.05 & 0.25). These values were chosen for both techniques such that macro- and micro-level geometry properties can be captured.

This study demonstrates re-representation across different design spaces based on two scenarios: 1) in a search scenario—where one design space has been explored, and another new design space is still unknown, a transfer of information might facilitate a more efficient search process. 2) when all five design spaces have been explored—the data structure used for storage and future expansion strategies. To address these two scenarios, the design spaces represent five different solution spaces. Similar footprints exist across some design spaces, while others have different footprints. Lessons from one design space can be transferred to another by a human designer, and this study aims to replicate this ability using a phenotype-based strategy.

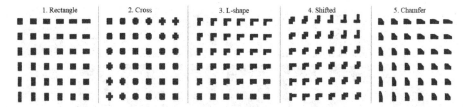

Fig. 7. The set-up and overall look of solutions generated using five different parametric logic.

3.1 Logic of Phenotype Space

Phenotype-based representations connect across all five design spaces in terms of the footprint and the UDI patterns. This allows the computer to make geometry-based associations which are independent of the genotype. The aim is to enable algorithms to recognize a geometrical pattern and re-represent it in a similar fashion as a human would. Therefore, the data space of both footprint and UDI phenotypes are presented to demonstrate the ability of computer vision in explaining geometry logic.

Based on all five design spaces (Fig. 7), Figs. 8 and 9 are principal component analysis (PCA) results for the footprint phenotypes (using EFD) and the daylight distributions (using Gabor filters) respectively. For the EFDs, the first three PCA descriptions account for a total of 78% of total data variance. Plotting the footprints on these three axes, the square footprints were found to sit at the origin. Extending along PCA 1, which has ~50% of the total variance, the geometrical logic of a compact square shifting towards long rectangular shapes is captured (row 1 of Fig. 8). PCA 2 describes the change from a square towards L-shapes that have equal length legs (row 2 of Fig. 8). The negative space bounded by the L-shape increases in area along PCA 2. PCA 3 starts off with a square shape and morphs towards cross shapes with equal length legs (row 3 of Fig. 8). These results demonstrate that EFDs can describe various geometrical morphs and its data space provides a good organization of shapes. Shapes with similar properties are placed within the same vicinity, whereas shapes which have mixed properties are spaced accordingly based on existing data in the PCA space.

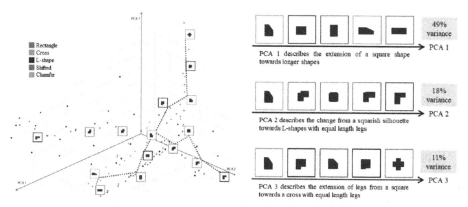

Fig. 8. PCA of room footprint based on Elliptical Fourier descriptors—footprint images placed at PCA location (left), description of PCA axis (right)

For the Gabor filters, the first two PCA axis has a total of 94% variance (Fig. 9). One reason for the 90% variance contributed by PCA 1 is due to the simple room layouts, where windows are only on the East most wall. This affects the daylight patterns, and PCA 1 captures the logic of a squarish daylight distribution morphing towards distributions which are stretched horizontally (row 1 of Fig. 9). Due to the existence of smaller windows in the cross and chamfer design spaces, some solutions have an arched daylight distribution. This is captured in PCA 2, where it starts off with a straight-line distribution and shifts towards arched daylight distributions. These results demonstrate that the applied computer vision methods can produce unique identities that describe the desired logic of phenotypes. Furthermore, nearby solutions also have similar footprints or daylight distributions. In other words, the EFD and Gabor filters provide good descriptions of the footprint and daylight distribution phenotypes.

Figures 8 and 9 demonstrate that a collection of differently generated solutions can be organized in a common data space. Furthermore, the organization is based on geometrical properties as demonstrated by the different PCA axis. This represents a scenario where many design spaces have been discovered and a phenotype representation is used to re-represent all the different genotype values into a common data space. Since the discovered solutions will always take a unique position within the phenotype data space, new additions will enhance the existing collection. For example, if a new design space constructs the cross shape using a different parametric logic, the footprint phenotype can recognize the similarity and gain a new design pathway. Due to the similar cross shape, the current data structure can also appreciate the geometrical information and connect both cross shape parametric spaces (Sect. 4).

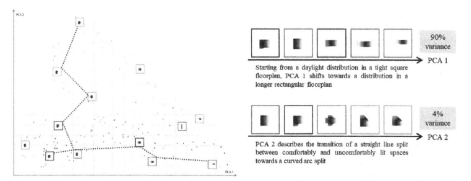

Fig. 9. PCA of daylight distributions based on Gabor filters. Distributions are adjusted to a white (0%) to black (100%) scale for clarity in numerical sense.

Expanding on the possibilities of phenotypes, Fig. 10 displays the results of a K-means clustering of Gabor filters (UDI phenotypes). Within the three clusters, solutions from different design spaces are selectively shown. Cluster 1 are UDI patterns which are horizontally long with a band of good UDI in the middle of the floorplan. In cluster 2, there is a larger portion of good daylight, and this stretches vertically into the floorplan. Lastly, cluster 3 has a distinctively square UDI pattern. The clustering is meant to demonstrate how a desirable UDI pattern can be used to identify similar looking ones within the UDI phenotype space. For example, a designer might be working with horizontal rectangular office layouts and is now tasked to add meeting rooms. An optimal orientation might have been discovered earlier and this UDI pattern should remain even with the new meeting rooms. A search within the UDI phenotype space can suggest alternatives like the cross shape or L-shape layouts (cluster 1 of Fig. 10). Otherwise, there could exist some UDI goal and the designer is interested in possibilities to achieve this goal. By searching for an average UDI of 55%, diverse solutions in cluster 2 can be presented.

3.2 Actions in Genotype, Phenotype, and Performance Space

The above strategies describe a scenario where information on discovered solutions are already available. Genotype, phenotype, and performance values can then be used together or separately to explore the combined design space. In the proposed phenotype-based strategy, EFD and Gabor filters help connect and organize these solutions. The top half of Fig. 11 describes this relationship. While genotype representations describe

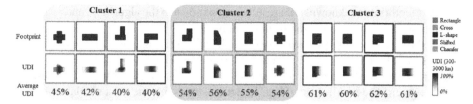

Fig. 10. Clusters of UDI patterns identified using K-means clustering.

parametric logic, performance representations describe possibilities to achieve them. The phenotype space focuses on the "look" of solutions, and it represents a space where each unique solution will have a unique identity. This imitates a human designer's perception on geometry. Although the designer might propose various methods to construct a geometry, the properties of the final outcome remain constant across all methods of construction. This human quality allows designers to flexibly re-represent information and learn quickly from past design explorations.

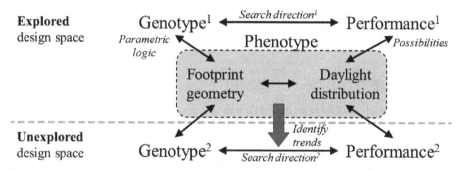

Fig. 11. Relationship between different data spaces: 1) Genotype, phenotype, and performance space. 2) Identifying trends from explored design space to influence unexplored design space.

In the bottom half of Fig. 11, an unexplored design space is introduced. Despite the new genotype values, the phenotype becomes a common data space between both design spaces. This means that trends identified from the footprint or daylight distributions can influence the search direction of a new unexplored design space. Instead of a random start for the new search, an informed starting point could be identified using the phenotype.

4   Design Space Re-representation

In a search scenario, there exists at least one design space that is explored with a full set of genotype, phenotype, and performance data. Figure 12 demonstrates this using the rectangle design space as an explored design space. The outlines of solutions are colored based on their UDI average, and the best and worst solutions have a red and blue background respectively. In this case, the L-shape design space is assumed to be unexplored. This means that for the unexplored case, only the parametric logic and the generated footprints are known. UDI results are unknown to the algorithm, and available data consists of genotypes and footprint phenotypes.

Based on the best solution (red background) in the rectangle design space, an EFD comparison on footprint similarity is made across the L-shape footprint space. Similarity results are colored on a green to red scale. Similar footprints are red, while different footprints are green. The middle grid of Fig. 12 demonstrates that the most similar solutions exist on the bottom row, and this is highly similar to the L-shape UDI performance landscape. Furthermore, the most different footprint (green background) is also one genotype value away from the actual worst performer.

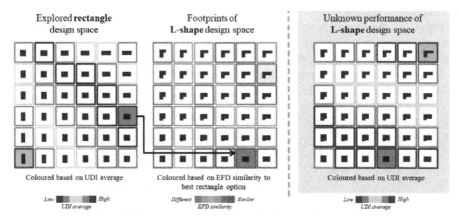

Fig. 12. Transfer of trend from an explored rectangle design space to an unexplored L-shape design space using EFD similarity to obtain an informed start point.

Firstly, this demonstrates a successful influence on the start point of a search. Instead of a random start point, the algorithm can now refer to the rectangle design space and begin at an informed position within the L-shape design space. The footprint geometry acted like a bridge between both parametric logics to enable this re-representation. Secondly, the success is due to the similarity in room proportions and its impact on daylighting levels. An L-shape is like a long rectangular shape, whereas there are squarish footprints that exist across both design spaces. Nonetheless, the ability to estimate performance, based on just the footprint, speeds up design search. Thirdly, this transfer in logic considers geometrical information instead of genotype information. Comparing the genotype spread of both design spaces, location of squares is along the diagonal for the rectangle design space but across the bottom row of the L-shape design space. This demonstrates an appreciation of footprint geometry by using EFD. Figure 13 presents logic transfer over the other three design spaces.

4.1 Imitating Human Design Re-representation

Beyond the estimation of a new start point, footprint phenotypes can also be used to strategize design exploration. The chamfer design space, which has a poor estimation on the best solution (Fig. 13), is used as an example. If a transfer of logic landed on a poor start point, the phenotype space is used to reinforce or discredit the move. In a search scenario, the genotype and footprint geometry are available at the start. Therefore, a K-means clustering of the footprint phenotype can be performed. This algorithmic step imitates a designer choosing the next search step. Instead of a blind guess, the designer selects a footprint that is significantly diverse from the initial guess. K-means clustering imitates this property by identifying diverse clusters from the footprint EFD (first column in Fig. 14).

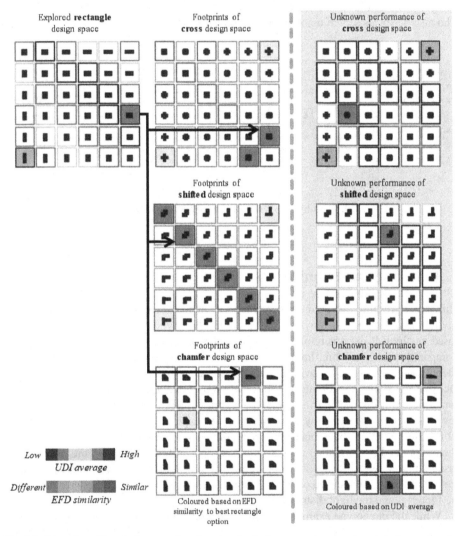

Fig. 13. Transfer of trends from explored rectangle design space to unexplored design spaces to identify an informed start point.

Next, a representative solution from each cluster is selected. Accordingly, this represents a diverse solution which is most similar to a previous best performer, or one which is most diverse based on the current design space (second column in Fig. 14). After the selection is made, an initial spread of daylight results will be achieved. This means that the UDI score and the Gabor filters from the chamfer design space will progressively become available. These daylight distributions can then be compared to the existing databank of phenotype results and identify a similar footprint within the chamfer design space clusters. The constant reference to the phenotype space is summarized in Fig. 14. Compared to traditional computational search methods, a phenotype-based strategy guides search based on the geometrical properties. This imitates a human designer who also performs search using modelling as a key strategy.

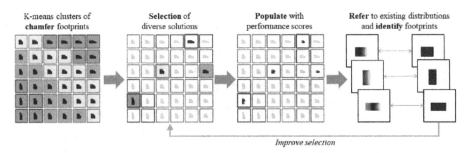

Fig. 14. Step-by-step refinement of phenotype-driven search

5   Conclusion and Limitations

This paper presented EFD and Gabor filters as methods to measure the footprint and daylight distribution respectively. The resultant phenotype space enables comparisons based on "looks" and it is reminiscent of a human designer's appreciation of geometry. Compared to an algorithm, humans can perform flexible re-representations at ease. A phenotype-based strategy provides a uniform data space based on geometrical properties, such that algorithms can imitate this ability. Re-representation is then demonstrated using two different scenarios: 1) in a search scenario—search direction of an unexplored design space is influenced by learnings from an explored design space, and 2) when all five design spaces have been explored—the possibilities afforded by the genotype, phenotype, and performance data structure. Based on both scenarios, the phenotype-based strategy demonstrates an improvement in terms of geometry appreciation. The presented results are focused on daylighting scenarios, where phenotypes are specific to the footprint and daylight distributions.

While the phenotype examples demonstrate good descriptions of geometrical properties, the current hyperparameters might not work as well in other geometry sets. One drawback might be the comparison with geometry with holes (e.g., Donut shapes) or a composition of geometry that are non-touching (e.g., two squares placed nearby each other to suggest a rectangular shape). These solutions require more than one boundary to describe and therefore fall outside of the above EFD space.

In future cases, the phenotype representation can consider the location of windows and walls. Learnings and transfers into the chamfer design space might therefore improve as the computer can now identify window sizes and proportions. The advantage of a phenotype space is the uniformity in description, and this leads to a longer timeframe of relevancy. Even in a new design scenario, solutions can still use the same phenotype description. However, this leads to an increase in data volume, and it requires a better data management system to recall relevant phenotype heuristics.

The results of this paper demonstrate the strength of a phenotype-based strategy in describing geometrical properties. This adds a layer of representation that extends beyond existing genotype and performance representations. When the appropriate computer vision method is used, it allows the algorithm to "see" geometrical properties, thereby leading to improved search or training results.

References

1. Ampanavos, S., Malkawi, A.: Early-phase performance-driven design using generative models. In: Gerber, D., Pantazis, E., Bogosian, B., Nahmad, A., Miltiadis, C. (eds.) Computer-Aided Architectural Design. Design Imperatives: The Future is Now: 19th International Conference, CAAD Futures 2021, Los Angeles, CA, USA, July 16–18, 2021, Selected Papers, pp. 87–106. Springer Singapore, Singapore (2022). https://doi.org/10.1007/978-981-19-1280-1_6
2. Beulow, P.: Paragen: performative exploration of generative systems. J. Int. Assoc. Shell Spatial Struct. **53**(4), 271 (2012)
3. Bhooshan, S.: Parametric design thinking: a case study of practice embedded architectural research. Des. Stud. **52**, 115–143 (2017)
4. Blidh, H., Witthoeft, N.: PyEFD. Github. https://github.com/hbldh/pyefd/blob/master/docs/index.rst (2021). Last retrieved 11 Nov 2021
5. Bradski, G.: The OpenCV Libary. Dr. Dobb's J. Softw. Tools 122–125 (2000)
6. Brown, N.: Design performance and designer preference in an interactive, data-driven conceptual building design scenario. Des. Stud. **68**, 1–33 (2020)
7. Crilly, N., Firth, R.M.: Creativity and fixation in the real world: three case studies of invention, design and innovation. Des. Stud. **64**, 169–212 (2019)
8. Dorst, C.K., Cross, N.: Creativity in the design process: co-evolution of problem–solution. Des. Stud. **22**, 425–437 (2001)
9. Jakubiec, J.A., Doelling, M.C., Heckmann, O., Thambiraj, R., Jathar, V.: Dynamic building environment dashboard: spatial simulation data visualization in sustainable design. Technol.|Archit. + Des. **1**, 27–40 (2017)
10. Ling, BL., Jakubiec, A.: A three-part visualisation framework to navigate complex multi-objective (>3) building performance optimisation design space. In: Proceedings of BSO 2018: 4th Building Simulation and Optimization Conference, Cambridge UK (2018)
11. Ling, B.L., Tunçer, B.: A phenotype-based representation that quantifies aesthetic variables. In: Gerber, D., Pantazis, E., Bogosian, B., Nahmad, A., Miltiadis, C. (eds.) CAAD Futures 2021. CCIS, vol. 1465, pp. 250–267. Springer, Singapore (2022). https://doi.org/10.1007/978-981-19-1280-1_16
12. Mirra, G., Pugnale, A.: Exploring a design space of shell and tensile structures generated by AI from historical precedents. J. Int. Assoc. Shell Spatial Struct. **63**, 172–188 (2022)

13. Muthumanickam, N., Duarte, J., Simpson, T.: Machine learning based surrogate model for faster daylighting estimation in building design. In: 6th RBDCC (2022)
14. Roudsari, M., Pak, M.: Ladybug: a parametric environmental plugin for grasshopper to help designers create an environmentally conscious design. In: 13th IBPSA Conference (2013)
15. Simon, H.A.: The structure of ill structured problems. Artif. Intell. **4**, 181–201 (1973)
16. Stuart-Smith, R., Danahy, P.: Visual character analysis within algorithmic design. In: 27th International Conference of the Association for CAADRIA, pp. 131–140 (2022)

Expanding Performance-Driven Parametric Design Spaces Through Data Streams

Laura E. Hinkle🆔 and Nathan C. Brown$^{(\boxtimes)}$ 🆔

The Pennsylvania State University, State College, PA, USA
{luh183,ncb5048}@psu.edu

Abstract. There has been recent interest in using sampled datasets of parametric models to create fast, accurate predictions of building performance. This rapid feedback can enhance creative processes in which a designer is exploring potential design options live, compared to waiting for the results of individual simulations for each design possibility. However, basing these predictions on an existing parametric dataset can be limiting, as the original variable structure must be maintained. Drawing from advances in data streams, this paper proposes and evaluates strategies for adding a variable to the parametric design space without needing to re-simulate the entire dataset to update the prediction. Several approaches are tested to expand the design variables for three situations: a daylit room, a massing model for energy prediction, and a stadium geometry. The main strategy, an online updating method, significantly reduces the number of new simulations required with the new variable to achieve reasonably accurate performance prediction, compared to re-simulating the entire dataset with the new variable included. However, in certain cases this approach only slightly outperforms creating a simple linear model with just the new datapoints. Future work can continue to investigate how to best enable the addition of variables to a parametric design space and corresponding performance prediction model.

Keywords: parametric design · online learning · surrogate modeling · design space exploration · model updating

1 Introduction

Parametric modelling allows designers to explore many design possibilities rapidly within a design space. For these parametric models, design space exploration can help in early design to find solutions that meet desired performance objectives. Because many simulation engines have been integrated into visual programming environments, parametric models can also be used to generate simulation data and train prediction models that provide real-time performance feedback during exploration. Parametric models paired with prediction models are particularly useful as a collaborative decision-making tool. However, in early design, flexibility is crucial to achieve performance goals [1], as well as qualitative goals desired by architects. A surrogate modeling approach with a previously simulated parametric dataset inherently limits the range of possible designs

© The Author(s), under exclusive license to Springer Nature Switzerland AG 2023
M. Turrin et al. (Eds.): CAAD Futures 2023, CCIS 1819, pp. 145–158, 2023.
https://doi.org/10.1007/978-3-031-37189-9_10

based on the initial parametric definition and corresponding parameters in the prediction model. While surrogate modeling is attractive due to its accessibility, it has the potential to limit creativity: if design variables are added to the parametric model, the old prediction model becomes obsolete, and updating the prediction model requires re-running simulations to generate new data. This can become a time-consuming process depending on the objective function, and it no longer tracks with the pace of design practice.

However, if simulation results are viewed as a data stream rather than a fixed dataset, related methods can incorporate new variables from the parametric model into the prediction model. In this paper, we explore statistical methods from the domain of big data streams to reduce the number of simulations required to update a prediction model. Rather than re-running all simulations with the new variable, we implement an online updating algorithm that leverages the existing prediction model to make new predictions for the expanded parametric design space. This research quantifies the tradeoff between the number of new simulations required and prediction model fit, and it contextualizes the results in terms of computational time saved. We apply the online updating algorithm to three parametric design examples with different scales: predicting spatial daylight autonomy of a sidelit room, energy consumption of a residential building, and embodied carbon of a stadium's structural system. The goal is to allow for the addition of creative new design directions to existing performance-driven parametric design spaces without starting from scratch. By making existing performance-driven parametric design spaces more flexible, this study addresses computation barriers for designers, with the goal of increasing adoption of surrogate modeling in early design.

2   Literature Review

2.1   Performance Feedback in Early Design

Due to time constraints in practice, rapid performance feedback is often necessary to incorporate technical objectives while conceptual design options are considered. If used properly, it can enable designers to make informed decisions in the earliest stages when the design is most flexible. Many researchers have enabled real-time or near-real-time feedback in the form of interactive optimization [2], adaptive sampling [3], and prediction models or surrogate models [4]. Interactive optimization allows a user to provide input to an automated process that is seeking better designs. In this vein, Danhaive and Mueller developed a performance-driven sampling algorithm that uses filtering to introduce performance bias but still balance diversity in the generated samples [5]. However, these examples do not necessarily accommodate new variables mid-design. Often, adding a variable to the design space requires starting from scratch to build up new simulations for a predication model. Therefore, designers are encouraged to finalize certain variables early on, which limits design options and can lead to design fixation [6]. Alternatively, designers are forced to decouple their design decisions from the simulation feedback, since it is not accurately depicting the variables they are considering. In this work, we propose treating simulations in the early design process as a data stream rather than a fixed dataset to improve their flexibility.

2.2 Online Updating Methods

Data streams are common in many data science applications. For example, an airline may be making predictions about customers, when a new data field is added on a customer form. The new variable may be helpful in making more accurate predictions about customer behavior, but it is desirable to adjust an established prediction model built on a large volume of customers to incorporate the new variable, rather than build a new prediction model based only on the datapoints containing the newer field. One potential strategy to address this problem involves online updating. Online updating methods for regression are limited, but they have been primarily developed for linear or generalized linear models, decision trees, and ensemble methods. Schifano et al. [7] developed an iterative estimating algorithm for linear models and estimating equations that update as new data arrives. The very fast decision trees (VFDT) algorithm was developed in 2000 [8] to build decision trees online as accurate as conventional systems, and it was later adapted for regression. A variety of ensemble methods have been developed [9], including bagging and boosting algorithms [10, 11] and ensembles of model trees [12], among others.

While most methods do not address the case of adding a new variable to the data stream, [7] was extended to accommodate new variables mid-stream [13]. The Wang et. al [13] online updating algorithm was selected for this study to leverage the initial model information and reduce the number of simulations required to update the predictions when a new variable is added to the design space. It is worth noting several differences between the applications. For example, in typical data streams the new variable may have been latent but never recorded, whereas in a parametric model a new variable may substantially change the response. However, by testing out the approach on surrogate modeling problems in architectural design, this paper seeks to 1) quantify the number of samples required to update the model to achieve a desired accuracy, and 2) determine the requirements for the added variable for this method to be successful in an early design setting. The number of new samples required to achieve useful accuracy based on the online updating method is then compared to the simulations required to create a simple linear regression and neural network based on the new data points, or rerun simulations on the whole dataset from scratch.

This implementation is potentially useful in several early design scenarios. For example, researchers have proposed reusable performance-driven parametric design spaces for common design problems, such as a shoebox energy calculation or sidelit room [14]. By applying the identified online updating algorithm, designers can customize reusable performance-driven parametric design spaces. This not only reduces the number of simulations required, but also improves parametric modeling time. The second relevant design scenario is the need to quickly add a variable to a complex, custom parametric model. For example, a stakeholder may be interested in manipulating an aspect of the building that was not considered in the initial prediction model, and it might be infeasible to quickly adjust it and provide timely feedback. In an ideal scenario, a fast-updating scheme could occur during the course of a workday or even a meeting when collaborating designers are exploring performance feedback for a design space together but seek more flexibility. While online updating methods are only one possible solution, this paper tests their applicability.

3 Methods

To evaluate the suitability of these methods for adding variables to a prediction model, several architectural case studies were developed. For each case study, we first determined the appropriate number of samples required to train a linear model to predict an architectural performance metric based on several parametric variables. We did this by incrementally sampling and setting a threshold for the change in the mean squared error. Then, the dataset was manipulated to test each variable as the possible "added" variable. This gave an understanding of the impact of the correlation between the added variable and existing variables, along with the importance of the added variable with respect to initial variables on the performance of the online updating algorithm. Data without the added variable was used to train the initial linear model. The initial linear model coefficients were stored to be updated. Next, new data was generated to include the added variable, and the correlation between the added variable and initial variables was calculated. The new data was then used in the online updating algorithm to update the initial linear model coefficients and incorporate the added variable. An illustration of this process is provided in Fig. 1.

The updated model was then evaluated against multiple baselines, including a "scratch" model, linear model without the online updating algorithm, and neural network, all seeking to make an accurate prediction of the performance metric now that the new variable affects the design space. The results are presented in terms of error of the prediction versus number of samples, which informs a discussion of computation time saved and its implications for the early design process.

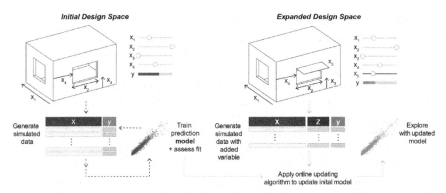

Fig. 1. An example of adding a variable in an early design setting and updating the corresponding prediction model

3.1 Selected Online Updating Algorithm

This section describes the details of the updating process of the selected algorithm by Wang et al. [13]. The notation is consistent with [13]. In general, online updating algorithms fit statistical models to blocks of data from the data stream. To avoid storing data and the prediction model, the model coefficients and variance estimates are updated at each new block. At the crux of the selected method is the assumption that the true model contains the new (referred to as "added" in this application) variable(s), and the initial model coefficients are biased. To leverage information from the initial model, this method corrects the bias in the cumulative model coefficients and corresponding variance estimates for the initial variables. This method also assumes that the added variable is completely missing before it becomes available.

In the following equations, $\hat{b}_{k+1}$ and $\hat{\beta}_{k+1}$ are the estimator for the new block $k+1$ restricted to the initial variables where the new variable was included in the input and the estimator for the new block $k+1$ where the new variable was excluded from the input, respectively. Note that $\hat{b}_{k+1}$ and $\hat{\beta}_{k+1}$ are both vectors of the same dimensionality. Then, $\hat{\delta}_{k+1}$ is the difference between the two estimators, aka bias.

$$\hat{\delta}_{k+1} = \hat{b}_{k+1} - \hat{\beta}_{k+1}. \tag{1}$$

Once the bias is calculated, we define $\tilde{b}_k$ as the cumulative estimator for all blocks 1 to k that includes the initial variables. Next, we denote $\tilde{\beta}_k$ as the difference between $\tilde{b}_k$ and the bias.

$$\tilde{\beta}_k = \tilde{b}_k - \hat{\delta}_{k+1}. \tag{2}$$

The updated cumulative estimator $\tilde{\beta}_{k+1}$, is obtained recursively through Eq. 3, which is a function of variance of estimators that we already know. This provides an updated model that efficiently leverages information from the initial model.

$$\tilde{\beta}_{k+1} = \left[\mathrm{Var}^{-1}\left(\tilde{\beta}_k\right) + \mathrm{Var}^{-1}\left(\hat{\beta}_{k+1}\right) \right]^{-1} \left[\mathrm{Var}^{-1}\left(\tilde{\beta}_k\right)\tilde{\beta}_k + \mathrm{Var}^{-1}\left(\hat{\beta}_{k+1}\right)\hat{\beta}_{k+1} \right]. \tag{3}$$

3.2 Data Preparation for Case Studies

This method is tested on three different design scenarios and corresponding parametric models (see Fig. 2). The goal was to test on problems that span in scale, complexity, and reusability. The first example is a sidelit room, which was modeled in Grasshopper within Rhinoceros [15]. This problem was selected to explain the method and because it could be adapted for many projects in the earliest stage of design when high-level decisions are made. The problem has seven design variables including room depth, sill height, head height, orientation, number of panels, panel width, and wall thickness (Fig. 2). The performance response was spatial daylight autonomy (sDA), which was simulated using Climate Studio [16]. The path-tracing settings were set to 100 rays emitted for each sensor at each pass, and 6 ambient bounces before discarding a ray. To determine the number of samples required to train a linear model, data points were randomly sampled at an interval of ten from a pre-computed dataset with 1000 points. This dataset was generated with the Latin Hypercube sampling method. Next, a linear model was trained, and the process continued until the change in the mean squared error was less than 0.001. Ultimately, 130 points were required to train a linear model with a mean squared error of 0.111.

Fig. 2. Design spaces and included variables for the daylit room (left), residence (middle), and stadium (right)

The second dataset was retrieved from the UCI machine learning repository. It was generated by Angeliki Xifara and processed by Athanasios Tsanas in [17]. There were eight variables in the original dataset: relative compactness, surface area, wall area, roof area, overall height, glazing area, area distribution, and orientation. There were two response variables, cooling load and heating load, which were combined for this study as a single response. The design space was grid sampled and the data was simulated using Ecotect. For this dataset, the added variable was removed from the initial dataset by setting it to one of the values in the parameter grid. It was then "added" to the design space by using the remaining data for which it varied.

A third dataset is based on a parametric stadium model similar to Old Trafford stadium in Manchester, England. There were eight variables in the design space that define the overall stadium geometry, including north-south column line, east-west column line, roof change in y-direction, roof change in x-direction, roof level overhead bleacher back, roof level overhead bleacher front, truss top height, and main frame depth (Fig. 2). The response was the embodied carbon of the structural system. As the geometry changes, structural loads are applied and the structural elements are sized. Material quantities are then multiplied by corresponding embodied carbon coefficients to calculate the overall embodied carbon of the structure. The structural analysis and sizing for this process were completed with the Karamba plug-in [18]. As with the first dataset, the points were randomly sampled from a large dataset at an interval of ten until the change in the mean squared error was less than 0.01. The initial linear model required 130 points to achieve a mean squared error of 12.80.

4   Results

This section reports the model accuracy and additional computation required for each individual case study. Overall results are then summarized and analyzed in the discussion. The first step in all three cases is to check the correlation of the variables, which helps determine when the method is likely to be effective. Figure 3 shows the correlation coefficients between all variables for all three datasets. Collinearity is not an issue for the daylit room dataset or the stadium embodied carbon dataset. However, there are a number of highly correlated variables in the energy dataset, mainly due to the relationship between compactness and surface area. The selected online updating algorithm requires the *squared* correlation coefficient between the added and initial variables to be less than 0.5. Therefore, glazing area, glazing distribution, and orientation were selected as potential added variables, and the remaining were used as initial variables. All variables were demonstrated as added for the daylit room dataset and embodied carbon stadium dataset. In the case of a new design application, potential collinearity is unknown and would have to be checked—this is addressed in the discussion.

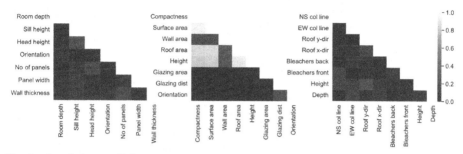

Fig. 3. Correlation coefficients between all variables for the daylit room dataset (left), residential energy dataset (middle), and stadium embodied carbon dataset (right)

4.1 Spatial Daylight Autonomy of a Sidelit Room

Figure 4 compares performance of the online updated model, a new linear model, a new neural network model, and a naïve "scratch" linear model for the room case study. The "updated model" is the updated linear model using the online updating algorithm. The linear model and neural network model were trained using only the new data (including the added variable), and the "scratch" linear model was trained using new data (including the added variable) with the same number of samples as the original model. As more samples are added to the updated linear model and linear model, the MSE converges to the scratch model MSE. However, the neural network MSE is several hundred times greater than the updated linear model and linear model within this sample range. For this reason, the neural network was excluded from the plot in order to compare the other two models.

Figure 4 shows that the two variables on the left, room depth and panel width, do not directly benefit from the online updating algorithm, likely due to their high partial R^2. The importance of the added variable relative to the initial variables affects the effectiveness of the online updating algorithm [13]. However, the partial R^2 of the number of panels variable is relatively high compared to room depth, and the online updating algorithm helped reduce the number of simulations required to update the prediction model. For example, if the designer desired a MSE of 0.01, 21 fewer simulations are required to update the prediction model compared to the standalone linear model (Table 1).

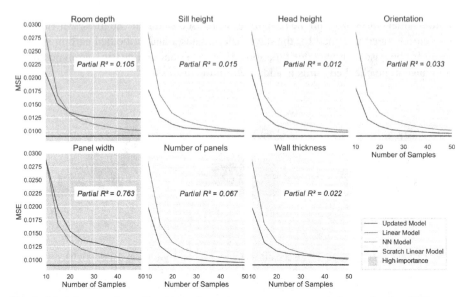

Fig. 4. Model comparison of number of samples versus mean squared error for the sidelit room example

Table 1. Estimated number of samples required per MSE for the 'number of panels' variable

MSE	Updated model samples	Linear model samples	Delta
0.0175	12	15	3
0.0150	13	18	5
0.0125	15	23	8
0.0100	31	52	21

4.2 Energy Consumption of a Residential Building

Figure 5 compares model performance for the energy dataset. Similar to the daylit room dataset, the two variables with lower importance relative to the other variables included in the initial model benefited from the online updating algorithm. Table 2 shows that, depending on the desired MSE, up to 27 simulations can be saved when adding the glazing distribution variable to the initial prediction model. This is also true of the orientation variable. However, Fig. 5 also shows that if glazing area were the added variable, the online updating algorithm would only be effective if the desired MSE was above ~42. The partial R^2 indicated that glazing area is several times more important than glazing distribution and orientation, which likely lead to lower effectiveness.

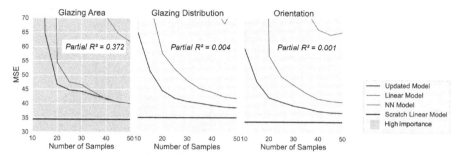

Fig. 5. Model comparison of number of samples versus mean squared error for the residential energy consumption example

Table 2. Estimated number of samples required per MSE for 'glazing distribution' variable

MSE	Updated model samples	Linear model samples	Delta
55	14	22	8
50	16	28	12
45	20	36	16
40	36	63	27

4.3 Embodied Carbon of a Stadium

Figure 6 displays the results of adding a variable to the stadium dataset. This is the most complicated of the models, as it contains the most potential variables for addition. As the embodied carbon calculation requires structural analysis and sizing to obtain material quantities before converting to embodied carbon, it is also potentially discontinuous and highly nonlinear. Thus, it may be difficult to predict accuracy with few datapoints. The goal of including this case study is to test the method on a custom model of a signature structure, as opposed to the previous two datasets, which were meant to be general across many different buildings.

The results show that the NS column line variable and EW column line variable did not directly benefit from the online algorithm, as their importance was very high relative to the other variables included in the initial model. The addition of these variables changed the linear model coefficients from the initial model in a way that could not be corrected through the online updating algorithm. It is possible that adopting a weighting scheme, such that the new model with the added variable is weighted more than the initial model, would improve the effectiveness [13]. Nevertheless, the other six variables show that the adoption of the online updating algorithm reduces the number of simulations required to update the prediction model. The partial R^2 for the depth variable indicates that it is relatively important, and with the online updating algorithm, it is possible to save up to 26 simulations depending on the desired MSE (Table 3) when it is added to the model. Among the variables that benefit from the online updating algorithm, as the desired MSE decreases, the number of simulations saved increases.

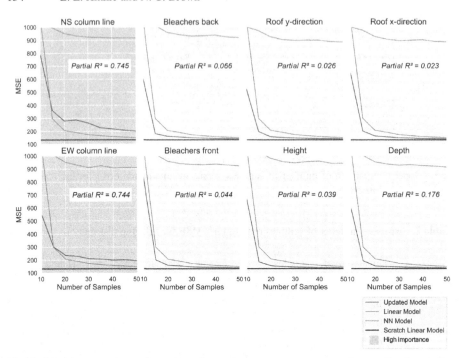

Fig. 6. Model comparison of number of samples versus mean squared error for the stadium example

Table 3. Estimated number of samples required per MSE for 'depth' variable

MSE	Updated model samples	Linear model samples	Delta
300	14	15	1
250	14	18	4
200	15	22	7
150	24	50	26

5   Discussion

To consider how useful these updated models would be for informing conceptual design, the MSE was converted to root mean squared error (RMSE) and normalized by the response range. This is meant to give an idea of how accurate a prediction is likely to be if a designer is manipulating the model and receiving live feedback. Figure 7 provides the number of simulations required to update the prediction model for different levels of desired accuracy. One variable is selected to demonstrate the accuracy for each problem. The bars compare the number of simulations required for the online updating algorithm to the linear model, which in all selected cases takes more simulations to achieve the same accuracy.

While the computational time saved depends on the simulation type, resolution, and model size, implementing an online updating algorithm has the potential to reduce the updating time in several useful increments. For example, in some cases, a variable could be added and the prediction model updated within the span of a meeting, if a design team is together exploring options and decides they need to consider a new variable. In other situations, treating design data as a data stream and implementing online updating algorithms could allow a designer to expand existing performance-driven design spaces without slowing down the design process by updating it over the span of a meeting, break, or workday, rather than rerunning all simulations overnight. All surrogate models have some prediction error for such early design scenarios, as do the simulation models themselves. The desired degree of accuracy is left up to the designer.

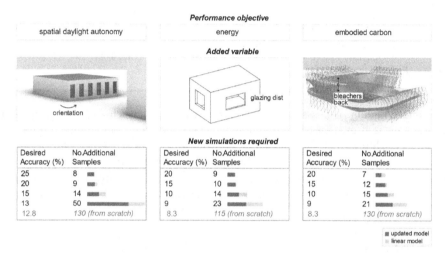

Fig. 7. Example of estimated accuracy versus number of samples required. Accuracy is in terms of normalized RMSE.

During the experiments, several factors emerged that impact the effectiveness of the updating algorithm in an early design setting. The squared correlation coefficients between the added variable and initial variables must fall below the prescribed threshold [13], and if the added variable is very important relative to the initial variables, which was measured by partial R^2 in this study, the online updating algorithm may not be useful. However, implementing a weighting scheme between the old and new predication infrastructure may still prove the algorithm effective. The algorithm described below demonstrates how these considerations could be applied in practice. Once a new variable is modeled and s simulations are added, a series of checks are conducted to update with the fewest number of simulations.

Algorithm 1: Update prediction model	
0	Generate s simulations
1	Train linear model
2	Calculate MSE
3	While MSE > threshold:
4	Calculate squared correlation coefficients $\mathbf{A}^2$
5	If any entry of $\mathbf{A}^2 > 0.5$:
6	Add new variable to model in terms of highly correlated variable and remove highly correlated variable OR exclude new variable from model // out of scope of paper
7	Return updated prediction model
8	Train linear model and calculate partial R^2 for new variable, b
9	If b is "high":
10	Go to 12
11	Run online updating algorithm
12	If MSE < threshold:
13	Return updated prediction model
14	Generate s more simulations // lower s if MSE is close to threshold
15	End
16	Return prediction model

Another finding of this study is that parametric models such as linear regression can be valuable if extremely high accuracy is not needed and computational time is a concern. For all three datasets, the neural network MSE did not approach the accuracy of the scratch model within a reasonable number of samples. Although nonparametric models such as neural networks have proven to be very accurate for many building design problems [19], they require substantially more data [20] and are thus not as useful for a quick updating scenario.

6 Conclusion

This paper has shown how online updating algorithms can be used to update a performance prediction for a parametric design space when a new variable is added. The amount of computation required with this method is compared to several other possibilities, including new linear models or neural network models created with just the new data points. The initial results show that for some of the cases, an online updating algorithm can save considerable simulation (up to 90–95%) compared to re-running all simulations with the new variable. However, creating a new linear model with the new points was almost as accurate as the updating algorithm in some cases, requiring very few more simulations to achieve similar results.

Future work might consider other methods for updating a performance-based parametric model that achieve even greater accuracy. The authors have attempted several other strategies for this application, including incremental learning algorithms and statistical imputation strategies. Data streams and online updating algorithms are the most promising so far, but more significant computational gains are desired. Additionally, extensions of this method may include piece-wise linear regression models to account for non-linearities. It may also be possible to expand a design space for multiple variables at once, which could be tested in another study. For example, if the relationship between two objectives are known, one could be added as a variable to predict the other. Otherwise, separate models could be updated simultaneously using respective simulation engines. Nevertheless, this paper provides an initial step towards faster updating of expanding parametric design spaces, based on thinking of the model as a data stream with potentially more variables rather than a fixed dataset.

Acknowledgments. We thank Jonathan Broyles for his efforts in creating the stadium model in Grasshopper for Rhinoceros.

References

1. Wortmann, T., Costa, A., Nannicini, G., Schroepfer, T.: Advantages of surrogate models for architectural design optimization. Artif. Intell. Eng. Des. Anal. Manuf. AIEDAM **29**(4), 471–481 (2015). https://doi.org/10.1017/S0890060415000451
2. Brown, N.C.: Design performance and designer preference in an interactive, data-driven conceptual building design scenario. Des. Stud. **68**, 1–33 (2020). https://doi.org/10.1016/j.destud.2020.01.001
3. Westermann, P., Evins, R.: Adaptive Sampling For Building Simulation Surrogate Model Derivation Using The LOLA-Voronoi Algorithm. https://doi.org/10.26868/25222708.2019.211232
4. Westermann, P., Evins, R.: Surrogate modelling for sustainable building design – A review. Energy Buildings **198**, 170–186 (2019). https://doi.org/10.1016/j.enbuild.2019.05.057
5. Danhaive, R., Mueller, C.T.: Design subspace learning: structural design space exploration using performance-conditioned generative modeling. Autom. Constr. **127**, 103664 (2021). https://doi.org/10.1016/j.autcon.2021.103664
6. Tseng, I., Moss, J., Cagan, J., Kotovsky, K.: The role of timing and analogical similarity in the stimulation of idea generation in design. Des. Stud. **29**(3), 203–221 (2008). https://doi.org/10.1016/J.DESTUD.2008.01.003
7. Schifano, E.D., Wu, J., Wang, C., Yan, J., Chen, M.-H., Yan, J.: Online updating of statistical inference in the big data setting HHS public access. Technometrics **58**(3), 393–403 (2016). https://doi.org/10.1080/00401706.2016.1142900
8. Domingos, P., Hulten, G.: Mining High-Speed Data Streams (2000)
9. Sakr, S., Zomaya, A.Y. (eds.): Encyclopedia of Big Data Technologies. Springer, Cham (2019). https://doi.org/10.1007/978-3-319-77525-8
10. Bifet, A., Holmes, G., Pfahringer, B., Gavaldà, R.: Improving adaptive bagging methods for evolving data streams. In: Zhou, Z.-H., Washio, T. (eds.) ACML 2009. LNCS (LNAI), vol. 5828, pp. 23–37. Springer, Heidelberg (2009). https://doi.org/10.1007/978-3-642-05224-8_4
11. Bifet, A., Holmes, G., Pfahringer, B.: Leveraging bagging for evolving data streams. In: Balcázar, J.L., Bonchi, F., Gionis, A., Sebag, M. (eds.) ECML PKDD 2010. LNCS (LNAI), vol. 6321, pp. 135–150. Springer, Heidelberg (2010). https://doi.org/10.1007/978-3-642-15880-3_15
12. Krawczyk, B., Minku, L.L., Gama, J., Stefanowski, J., Woźniak, M.: Ensemble learning for data stream analysis: a survey. Inf. Fusion **37**, 132–156 (2017). https://doi.org/10.1016/J.INFFUS.2017.02.004
13. Wang, C., Chen, M.H., Wu, J., Yan, J., Zhang, Y., Schifano, E.: Online updating method with new variables for big data streams. Can. J. Stat. **46**(1), 123–146 (2018). https://doi.org/10.1002/CJS.11330/FULL

14. Hinkle, L.. Pavlak, G., Brown, N.. Curtis, L.: Dynamic subset sensitivity analysis for design exploration. In: 2022 Annual Modeling Simulation Conference, pp. 581–592 (2022). https://doi.org/10.23919/ANNSIM55834.2022.9859293

15. "Rhinoceros." Robert McNeel & Associates (1993). https://www.rhino3d.com/. Accessed: 25 Feb 2021

16. "Climate Studio." Solemma LLC

17. Tsanas, A., Xifara, A.: Accurate quantitative estimation of energy performance of residential buildings using statistical machine learning tools. Energy Build. **49**, 560–567 (2012). https://doi.org/10.1016/J.ENBUILD.2012.03.003

18. Preisinger, C., Heimrath, M.: Karamba – A toolkit for parametric structural design. Struct. Eng. Int. J. Int. Assoc. Bridg. Struct. Eng. **24**(2), 217–221 (2014). https://doi.org/10.2749/101686614X13830790993483

19. Østergård, T., Jensen, R.L., Maagaard, S.E.: A comparison of six metamodeling techniques applied to building performance simulations. Appl. Energy **211**, 89–103 (2018). https://doi.org/10.1016/J.APENERGY.2017.10.102

20. Shaikhina, T., Khovanova, N.A.: Handling limited datasets with neural networks in medical applications: a small-data approach. Artif. Intell. Med. **75**, 51–63 (2017). https://doi.org/10.1016/J.ARTMED.2016.12.003

A Design Ranking Method for Many-Objective Evolutionary Optimization

Likai Wang[1][(✉)] ⓘ, Do Phuong Bui Tung[2] ⓘ, and Patrick Janssen[3] ⓘ

[1] Xi'an Jiaotong – Liverpool University, Suzhou, China
likai.wang@xjtlu.edu.cn
[2] National University of Singapore, 21 Lower Kent Ridge Road, Singapore
[3] Packhunt, Amsterdam, Netherlands

Abstract. This study presents a design ranking method for evolutionary optimization that is aimed to address design optimization problems with many performance-related evaluation metrics. The application of the method consists of three strategies. First, all evaluation scores are expressed as percentages that indicate the proportion of the design achieving acceptable performance. Second, related evaluation scores are grouped, and for each group, a combined score is calculated using a weighted product approach. Third, design populations are evolved using the Pareto optimization of the combined evaluation scores. The combination of the three steps helps designers to define and organize the design evaluation metrics and can also produce optimization results revealing meaningful information. A case study is presented to demonstrate the efficacy of the proposed design ranking method. The relevance of the proposed method to performance-based evolutionary optimization research is also discussed.

Keywords: Performance-based Design · Design Evaluation · Many-Objective Evolutionary Optimization · Architectural Design Optimization

1 Introduction

Evolutionary optimization has been widely considered an effective tool that can help designers improve the performance of the design and achieve sustainable development goals [1, 2]. While the utility of evolutionary design optimization has been demonstrated by many studies, its applications to real-world design scenarios are still rather limited. One reason is various challenges relating to design ranking methods when there are many performance objectives.

In the research literature, evolutionary design optimization is often simplified to include only one or two performance objectives [3], which is unlikely to reflect actual design challenges in real-world projects. Even for research focusing on multi-objective optimization, the number of objectives is typically below four. In contrast, the number of objectives in real-world projects is often higher. However, formulating an effective design ranking method to handle many objectives could be difficult for most designers with limited expertise in computational optimization. At the same time, a poorly-defined method could introduce bias and produce misleading optimization results. Therefore,

© The Author(s), under exclusive license to Springer Nature Switzerland AG 2023
M. Turrin et al. (Eds.): CAAD Futures 2023, CCIS 1819, pp. 159–173, 2023.
https://doi.org/10.1007/978-3-031-37189-9_11

an appropriate ranking method needs to be developed to guide the optimization process and produce results with meaningful and rich design information [4].

Compared with the existing multi-objective optimization research in architectural and urban design, many-objective evolutionary optimization [5] and the corresponding design ranking methods are relatively under-researched in architecture. In response, this study focuses on the development of a ranking method for the many-objective evolutionary optimization for performance-based urban massing designs. To demonstrate our proposed method, a case study is presented. The case study is derived from a real-world residential precinct design project with many objectives and constraints.

To put this study into context, a brief overview of the ranking methods applied in relevant studies is first presented. Afterward, the proposed method and the implementation are elaborated, followed by the case study. We conclude the paper by discussing its relevance to the research field.

1.1 Ranking Methods in Evolutionary Optimization

Evolutionary optimization, or computational design optimization, aims to identify high-fitness designs from a design space, and a critical factor in the optimization process is the ranking of the design population, which guides the evolutionary algorithm when searching the design space. Moreover, the ranking of the designs also reflects the design objective and/or design intent of the design task. As a result, an appropriate design ranking method can guide the optimization toward the desirable design direction.

In the research literature, three design ranking methods are commonly used [6]. The single-objective ranking method ranks designs based on one design evaluation metric, which is typically applied with a single-objective evolutionary algorithm, such as standard genetic algorithms. The Pareto-ranking method ranks designs based on multiple design evaluation metrics. This method is often combined with multi-objective evolutionary algorithms, such as non-dominated sorting genetic algorithms (NSGA). The weighted ranking method combines multiple evaluation metrics into a single combined score, typically using weighted sum or weighted product functions. As a result, designers can use single-objective evolutionary algorithms to solve such problems. The three methods have certain strengths and weaknesses when applied to architectural and urban design tasks.

Regarding the single-objective ranking method, as it only considers one design evaluation metric, the optimization is typically more efficient, and the high-ranking design can reflect implications directly related to the objective. This method is often applied to problems with a well-defined design objective [7]. However, in architectural design, the design problem is often ill-defined, and designs often need to be evaluated against more than one factor. As a result, ranking designs based on one evaluation does not always meet practical needs in architectural design.

Regarding the Pareto ranking method, designs are ranked by multiple objectives separately, and it is considered more applicable to architectural design. In many research studies, this method, combined with Pareto optimization, has been widely used, and the optimization result can display the trade-offs and compromises between design options [8]. Nevertheless, as Pareto optimization works most efficiently with two or three performance objectives, relevant studies typically avoid using design problems with more performance objectives [9]. With a higher number of performance objectives, the search mechanism of NSGAs will force the optimization to search for more non-dominated designs to expand the Pareto front. As a result, this mechanism undermines the search efficiency of the optimization and makes the optimization process becomes less effective in improving fitness [10]. Moreover, as multi-objective optimization often produces a large number of non-dominated designs, this could also cause cognitive overload for designers to analyze and interpret the optimization result.

Regarding the weighted ranking method, it solves the problem of many performance objectives by combining multiple evaluation scores into a single score, using weighted sum or weighted product functions [4]. This method drives the optimization process to search for designs that can maximize the combined evaluation score. In addition, designers can adjust the weights to prioritize certain performance objectives over others, which gives designers more control in guiding the optimization search process. Although this method can handle many objectives, it has the disadvantage of making it difficult for designers to understand the design implications of the optimization result. As evolutionary optimization is often used to explore trade-offs between performance objectives, multiple design options need to be identified that can illustrate such trade-offs. However, when using this method, optimized designs tend to blend features responding to different design objectives, making it difficult to distinguish any trade-offs. Moreover, prior knowledge is also required for designers to define reasonable weights and scales for each item in the function [10], and inappropriate settings of the weight and scale can make optimization produce invalid or misleading results.

In addition to the design ranking method, design evaluation methods have also received insufficient attention. In many studies related to performance-based design optimization, designs are often evaluated by factors such as solar irradiation, daylight accessibility, and unobstructed views. When evaluating these factors, researchers and designers often directly average or aggregate the simulated results to represent the fitness and performance of the design, while the optimization is typically aimed to maximize or minimize this value. However, using this method often produces unbalanced designs with a high overall score while still leaving a significant proportion of the design unusable. In contrast, in real-world architectural design projects, minimizing the parts that cannot perform to acceptable levels plays a more critical role in determining the overall quality of the design.

The above-mentioned issues undermine the efficacy of evolutionary optimization in real-world performance-based design tasks. As a consequence, it is critical to developing a design ranking method that can handle many performance objectives as well as can guide evolutionary optimization to produce useful results. This study aims to address these challenges by proposing a hybrid design ranking method.

2 Method

The proposed design ranking method is developed for an optimization system focusing on residential precinct design projects in Singapore. In this system, several performance factors, such as sky exposure, wind permeability, solar heat, and noise, are considered. In addition, indoor and outdoor spaces are also considered in this system. In total, 11 different performance-related metrics are included in the design evaluation. At the same time, this system has been developed for designers with limited computational design expertise. Hence, it is critical that the optimization can provide meaningful results and cognitively manageable information.

2.1 Overview of the Design Ranking Method

In order to make the proposed ranking method usable by designers, two factors are considered. First, the method can help the designer with minimum expertise in evolutionary optimization effectively select and manage different evaluation metrics before running the optimization. Second, the ranking method should guide the evolutionary optimization to produce results that can reveal clear and rich information related to the performance objectives. The proposed ranking method uses three strategies (Fig. 1):

- First, in order to make evaluation scores commensurate, all evaluations are calculated as percentages rather than as absolute values. These percentages generally indicate the proportion of the design that achieves desirable performance.
- Second, the evaluation metrics are divided into a small number of evaluation groups (for example two or three groups). The evaluation metrics in each group should be related to one another so that the group is meaningful to the designer.
- Third, for each evaluation group, an integrated performance objective is defined by calculating a combined score using a weighted product or weight sum approach. In addition to performance objectives, penalty functions are adopted to embed other functional constraints and imperative requirements into the design evaluation.

The Pareto optimization is then applied, using the combined evaluation score for each group. In the case study, two groups were defined: evaluation metrics related to the building facades and those related to the ground plane. The rest of this section elaborates on the details of the method and techniques mentioned above.

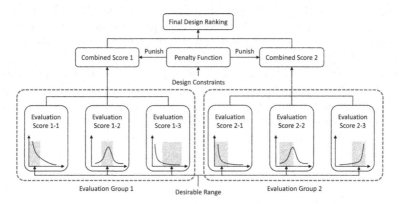

Fig. 1. Decomposition of the design ranking method

2.2 Design Evaluations

In architectural and urban design, the proportion of a design that achieves an acceptable performance is often a critical metric when evaluating the overall design performance. Other metrics, such as the overall average performance or the maximum performance can be misleading and lead to inappropriate results. Similar to "Liebig's law of the minimum", or "Liebig's Barrel", the performance of a design can be significantly impacted by its weakest part. In this regard, when optimizing a design, we prioritize reducing those parts with low performance over improving those parts that already have good or acceptable performance. Accordingly, in the proposed design ranking method, all design evaluation metrics are expressed as a percentage indicating the proportion of the design that is able to achieve desirable performance (Fig. 2). At the same time, using percentage value also normalizes the evaluation metrics, making them commensurate.

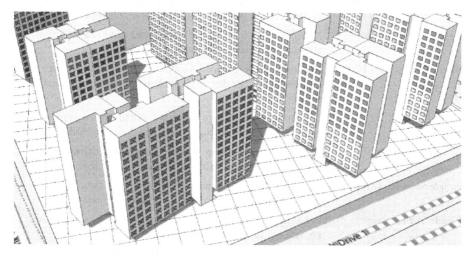

Fig. 2. Counting of the windows that cannot achieve acceptable performance (windows with a cross)

When using this design evaluation metric, a critical step is defining the desirable range for each evaluation metric. For the case study, the optimization system is applied to Singapore residential precinct design, and the desirable range for each evaluation is predefined based on currently used design rules. These ranges are relatively stable across different projects. For example, in façade sky exposure, 50 to 100 % is set as the desirable range, and windows that have a sky exposure value outside this range are considered unacceptable. Nevertheless, designers can modify these ranges according to the design conditions and brief.

2.3 Evaluation Groups

As mentioned above, there are typically many performance objectives in real-world design projects. Thus, in the proposed design ranking method, we consider two dimensions that can affect the overall quality of a residential precinct: the performance of the ground plane and the performance of the building façade. For the ground plane, it is relevant to residents' outdoor activities, which are largely affected by solar exposure, wind, and the urban heat island effect. Regarding the building façade, it is pertinent to residents' indoor comfort, which can be affected by solar heat gain, views, daylighting, and noise. In this regard, we group the design evaluation metrics according to these two dimensions (Table 1). These evaluation groups allow the optimization to produce results that show the trade-offs between the ground plane and the building façade. As such, it provides a broader view of the design problem and prevents designers from being distracted by overly detailed design information.

Table 1. Evaluation metrics

Ground Solar Exposure	Maximize the percentage of ground area that receives solar exposure below a certain maximum desirable level
Ground Sky Exposure	Maximize the percentage of ground area that has a sky exposure above a certain minimum desirable level
Ground Urban Heat Island	Maximize the percentage of ground area that has a temperature difference between day and night below a certain maximum desirable level
Ground Wind	Maximize the percentage of ground area that has a wind permeability above a certain minimum desirable level
Ground Irradiance	Maximize the percentage of ground area that receives solar irradiance below a certain maximum desirable level
Facade Solar Exposure	Maximize the percentage of windows that receives solar exposure below a certain maximum desirable level
Facade Sky Exposure	Maximize the percentage of windows that have a sky exposure above a certain minimum desirable level
Façade Irradiance	Maximize the percentage of windows that receive a solar irradiance below a certain maximum desirable level
Façade Noise	Maximize the percentage of windows that receive direct noise from the surrounding roads below a certain maximum desirable level
Façade Unobstructed View	Maximize the percentage of windows that have an unobstructed view above a certain minimum desirable level
Façade Visibility	Maximize the percentage of windows that have a direct scenic view

Note that different design tasks may require different grouping strategies. Nevertheless, in order to make the optimization solvable, it is recommended to keep the number of groups below four. However, it is also not recommended to combine all the design evaluation metrics in one group, as this can hinder designers to distinguish the trade-offs in the optimization result.

After determining the evaluation groups, the next step is to define performance objectives by combining the evaluation metrics in each group. In the proposed design ranking method, a weighted product approach is applied to combine the scores calculated by different evaluation metrics. The weighted product approach has the advantage that the combined score is not affected by the magnitude of each evaluation metric, and the change in each evaluation score can equally influence the aggregated value. Moreover, while all the evaluation metrics are normalized in the proposed method, the variable range in each evaluation metric can still differ significantly. For example, some metrics can change from 70% to 85%, and others change from 10% to 90%. Thus, if using a weighted sum approach, the impact of the metric with a smaller variable range would be reduced.

2.4 Penalty Functions

In architectural design, certain functional requirements are imperative when determining the feasibility of a design, such as gross floor area and site efficiency. Unlike performance objectives, designs that fail to meet these requirements cannot be compensated by the improvement of the building performance. As a result, penalty functions are used to punish designs that fail to meet such functional feasibility requirements. In the proposed design ranking method, users can formulate various penalty functions based on different functional requirements. These penalty functions are then applied to the combined scores of every evaluation group as a soft constraint [11]. This results in the overall fitness of infeasible designs being reduced.

In terms of the implementation, we considered offering more flexible ways of applying penalty functions. For example, we considered that it might be possible to apply penalties to specific evaluation scores. However, in the end, such approaches were rejected since they would result in additional complexity for the end users and would require too much expertise.

2.5 Optimization

Pareto optimization is applied using the evaluation groups defined by the design ranking method. The combination of evaluation groups and Pareto optimization enables the designer to see a broad picture of the design problem without being distracted by subtle mutual restraints within the facade or the ground plane. At the same time, the existence of multiple design evaluation metrics can still produce a variety of design options, with varying performance strengths and weaknesses for each performance objective.

2.6 Implementation

In order to facilitate ease of use and to streamline the design process, the proposed design ranking method is implemented within an integrated design optimization system [12]. This system consists of an evolutionary component, embedded with a standard NSGA, in Rhino-Grasshopper and an evaluation server that conducts all the design evaluations (Fig. 3). In this system, designs are generated in Rhino-Grasshopper and sent to the evaluation server, in which designs are evaluated against every evaluation metric. With all evaluations completed, the evaluation result (combined evaluation score) is sent back to the evolutionary component in Rhino-Grasshopper for evolving the design population. Using the evaluation server reduces the simulation time by taking advantage of cloud parallel computing and ray-shooting simulations [13]. In addition, the evaluation server also provides designers with various interactive interfaces for configuring the evaluation settings and viewing the optimization results (Fig. 4).

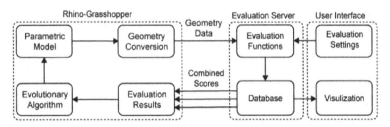

Fig. 3. System framework

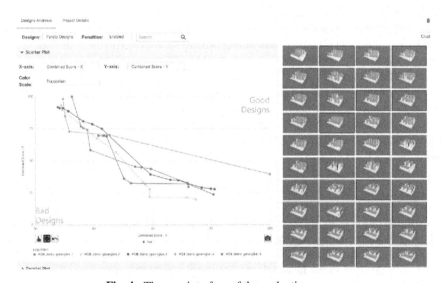

Fig. 4. The user interface of the evaluation server

In the evaluation server, designers first have to define the evaluation setting by selecting evaluation metrics for each evaluation group. Regarding advanced users, they can also modify the desirable range for any evaluation metric. After the evaluation setting is defined, other optimization parameters in the evolutionary component can be set on Rhino-Grasshopper, such as population sizes and numbers of generations, and launch the optimization. The optimization process uses the defined evaluation settings to rank the designs generated in Rhino-Grasshopper. After the optimization has been executed, the evaluation server (Fig. 4) provides various visualization diagrams to assist designers in analyzing the optimization result and extracting information from the optimization result.

3   Case Study

To demonstrate the efficacy of the proposed design ranking method, a case study is presented that focuses on a residential precinct design optimization task. The residential precinct design consists of a stand-alone carpark building and multiple residential towers. In this case study, a parametric model using a skeletal strategy is used to generate design variants [14].

3.1   Evaluation Settings

In this case study, not all the evaluation metrics available in the web application are selected. It is because some of the evaluation metrics have similar effects on the optimization, and as a result, selecting these evaluations can introduce bias against other evaluations. The selected evaluation metrics are displayed in Fig. 5. Nevertheless, designers can also change the selection of the evaluation metrics to investigate the impact of the precinct configurations on other performance factors.

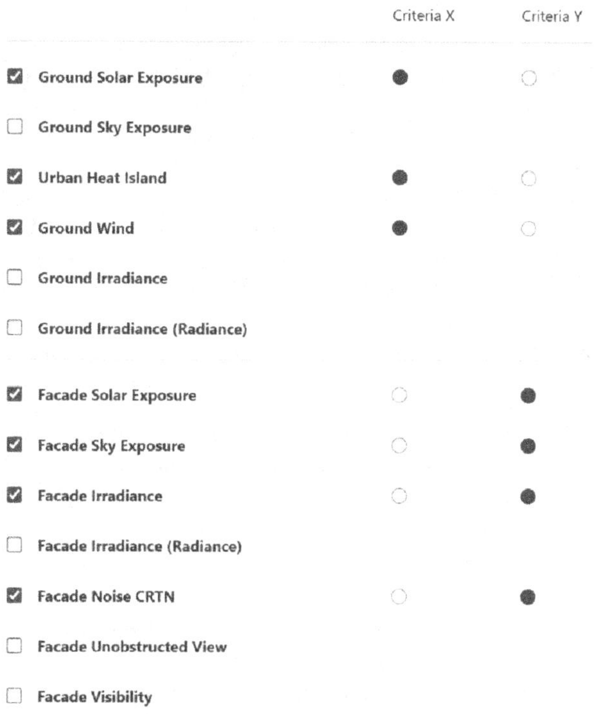

	Criteria X	Criteria Y
☑ Ground Solar Exposure	●	○
☐ Ground Sky Exposure		
☑ Urban Heat Island	●	○
☑ Ground Wind	●	○
☐ Ground Irradiance		
☐ Ground Irradiance (Radiance)		
☑ Facade Solar Exposure	○	●
☑ Facade Sky Exposure	○	●
☑ Facade Irradiance	○	●
☐ Facade Irradiance (Radiance)		
☑ Facade Noise CRTN	○	●
☐ Facade Unobstructed View		
☐ Facade Visibility		

Fig. 5. Selected evaluation metrics

Moreover, a penalty function is applied according to the difference between the target flat number required in the project brief and the actual flat number in the generated design. In this design task, there are three different types of flats, and therefore, the penalty is calculated based on the aggregated differences of these three types of flats. In the optimization process, a heavy penalty is applied to the design with a large difference between the actual flat number and the target flat number, making this design less likely to be chosen in evolutionary selection.

3.2 Design Generation and Optimization

A parametric model developed using a skeletal strategy for design generation is used for the case study and generates precinct configurations based on a hierarchical schema. [14, 15]. The parametric model first generates a plot for the carpark building alongside the boundary of the building plot. Afterward, the remaining building plot is further divided into two subregions for generating residential towers. In each of the subregions, a set of parallel skeletal lines are generated, which serves as an organizational device to control the orientation and density of the residential towers. Based on the skeletal lines, residential towers are then generated. The use of the parametric model can produce a variety of residential precinct configurations with different strengths and weaknesses in regard to different evaluation metrics (Fig. 6).

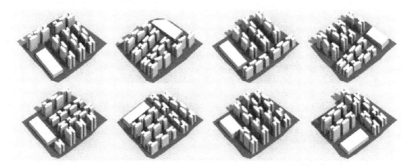

Fig. 6. Random sampling of the generated precinct configuration [14]

In this case study, a design population with 40 individuals is used, and the population evolves over 12 generations. The fitness progress trendlines show that the optimization is sufficiently converged.

3.3 Result

Figure 7 demonstrates the live designs with a high fitness value in the population in the objective space, which also displays the Pareto front for the two combined evaluation scores. The ground plane scores are plotted along the X-axis, and the building façade evaluation scores are plotted along the Y-axis. The Pareto front indicates the performance trade-off between the ground plane and the building façades. The use of two evaluation groups facilitates designers to understand the compromise among different design options with regard to these two design dimensions (ground versus façade). In contrast, if all the evaluation metrics are treated separately and equally, the trade-offs between each pair of evaluation metrics can be difficult to comprehend for the designer and can lead to cognitive overload.

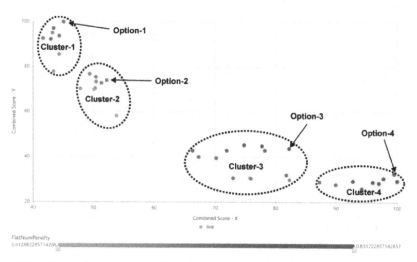

Fig. 7. The visualization of the optimization results in the objective space (x-axis: ground plane evaluations, y-axis: building façade evaluations)

With regards to the optimization results, four clusters of designs can be identified in Fig. 7. In order to compare their morphological features, we choose one representative design from each cluster (Fig. 8). The designs shown in Fig. 8 display the transition from the design with good façade performance to that with good ground performance. The design with good façade performance typically has north-south facing residential buildings, whereas the design with good ground performance has more residential buildings with an east-west orientation. At the same time, there is also a shared feature among these designs: the standalone carpark building at the southwest corner. With the above analysis, designers can gain a broader understanding of the ground-façade relationship.

Apart from the morphological features, the difference in the evaluation metrics also reveals the relationship between the building configuration and its performance. On the one hand, some evaluation metrics are significantly affected by the building configurations, including ground solar exposure, ground wind, façade solar exposure, and solar irradiation. On the other hand, evaluation metrics, such as ground urban heat island, façade sky exposure, and façade noise (CRTN), are less sensitive to the difference in the building configuration. This analysis further helps designers understand the impact of the building configuration on various performance factors.

In addition to the above analysis at the macro design level, designers can further explore detailed and subtle design variations and trade-offs at the micro design level. With the information provided by the design clustering, designers can easily filter out design clusters and directions that are not viable for further development while selecting more promising ones for more detailed examination. In this regard, we assume that the designer will analyze the second cluster further.

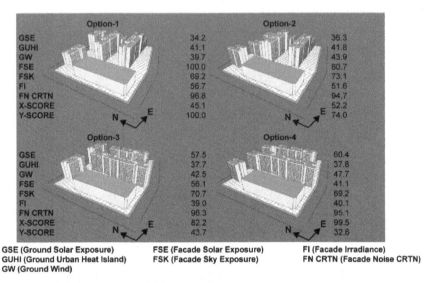

Fig. 8. The building configuration and evaluation results of the selected designs

Figure 9 illustrates six selected design options from the second cluster. On the one hand, option-6 displays a remarkable difference from the other five designs. The position of the standalone carpark building and the orientation of the residential tower make option-6 have better performance than the other five design options except for façade

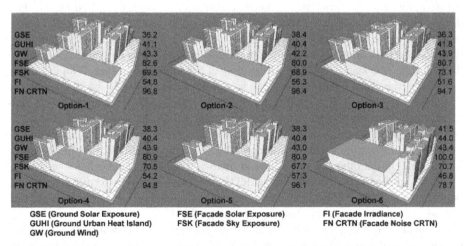

Fig. 9. The building configuration and evaluation results of the selected designs in the second cluster

irradiation and façade noise. On the other hand, for the remaining five designs, the major difference among them is the height of the residential in the north, which subtly differs in the performance of each evaluation metric of these designs.

4 Discussion and Conclusion

The above case study demonstrates how the proposed design ranking method can facilitate the application of evolutionary optimization and performance-driven design to architectural and urban design tasks. First, dividing all evaluation metrics into groups makes the optimization more computationally feasible and makes the optimization result easier to analyze. Second, combing multiple evaluation metrics using the weighted product function allows the optimization process to improve the overall performance of the design, while also allowing designers to compare these metrics across different design options. This helps designers better understand the impact of subtle design variations on the performance as well as the performance strengths and weaknesses of different building configurations. Third, by focusing on reducing the underperforming parts of the design, the optimization can produce precinct designs with acceptable performance across most flats.

The above case study also demonstrates a procedure for applying the proposed ranking method to the design process. In the beginning, the combination and grouping of the evaluation metrics can be seen as a "packing" process that encapsulates performance objectives into an organizational structure. The designer then runs the optimization to discover high-performing design options. The designer's analysis of the optimization results can then be viewed as an "unpacking" process, in which the designer compares design options at the macro and micro levels to extract design information. This two-way process has two key benefits. On the one hand, it makes complex architectural design optimization problems manageable. On the other hand, it ensures that optimization results contain sufficient design information for designers to understand the design problem.

With regard to the generalizability of the proposed method, the major issue is that the evaluation metrics used in the case study may not be suitable for all other building or urban design tasks. Therefore, it is essential for the system configurator to define feasible evaluation grouping criteria based on the characteristics of each design task. For example, other possible grouping criteria for common building or urban design tasks include: 1) factors related to a building versus factors related to the surrounding environment, 2) factors related to a building's indoor space versus factors related to a building's façade, 3) factors related to a building's performance versus factors related to subjective design intent. Hence, using the proposed design ranking method requires the system configurator to effectively formulate the design problem and apply an appropriate grouping criterion to the design optimization task. Thus, design domain knowledge and having a comprehensive understanding of the design problem in the architecture or urban design becomes critical when using this method.

In terms of application, the proposed method also offers a relatively simple and intuitive way to help designers, especially computational novices, handle complex performance-related design problems with many objectives. In contrast to building customized optimization workflows using native Grasshopper components and plugins, the proposed method is significantly easier. This is achieved by limiting the designers' inputs to specifying design requirements and customizing evaluation metrics and penalty functions. For most designers with basic or minimum expertise in evolutionary optimization, greater flexibility often results in mistakes that may make any optimization result invalid. In contrast, the proposed method minimizes the possibility of such mistakes and allows designers to apply performance-based design optimization with minimum expertise. Moreover, if a specific need arises, it is still possible for system configurators to modify aspects of the optimization system.

Lastly, the application of the proposed design ranking method also highlights the importance of an appropriate infrastructure to support designers in conducting design optimization and post-optimization analysis and exploration. As mentioned at the beginning of this paper, the proposed design ranking method aims to address complex real-world design application scenarios, and the case study demonstrates that when analyzing the optimization results, there can be multiple perspectives. In this regard, using conventional data analysis tools, such as Excel and relevant python libraries, can be technically demanding for many designers. Therefore, in order to further support the application of evolutionary optimization to practice, integrated and easy-to-use tools are also crucial components to support the application. As illustrated in the case study, the evaluation server plays an essential role in supporting the proposed design ranking method. However, the development and research of such tools and systems are still rather scarce.

4.1 Conclusion

The paper proposes a design ranking method to address building and urban design optimization problems with many objectives. Despite limited flexibility, the method is aimed to help designers handle and organize various performance objectives when confronted with complex real-world design tasks. As demonstrated in the case study, the application of the method not only makes the optimization of such complex design problems computationally feasible. It also ensures that the analysis of the optimization result is cognitively manageable. Moreover, the paper also highlights the importance of developing supportive infrastructures to implement the proposed design ranking method and to apply evolutionary optimization to address real-world design problems.

Acknowledgement. This work is part of the research project: "Optimization Algorithm for Rapid Sustainable Planning and Design", supported by Housing Development Board (HDB), Singapore.

References

1. Wang, L., Janssen, P., Ji, G.: SSIEA: a hybrid evolutionary algorithm for supporting conceptual architectural design. Artif Intell Eng Des Anal Manuf **34**, 458–476 (2020). https://doi.org/10.1017/S0890060420000281
2. Cubukcuoglu, C., Ekici, B., Tasgetiren, M.F., Sariyildiz, S.: OPTIMUS: self-adaptive differential evolution with ensemble of mutation strategies for grasshopper algorithmic modeling. Algorithms **12**, 141 (2019). https://doi.org/10.3390/a12070141

3. Li, S., Liu, L., Peng, C.: A review of performance-oriented architectural design and optimization in the context of sustainability: dividends and challenges. Sustainability **12**, 1427 (2020). https://doi.org/10.3390/su12041427

4. Chen, Y., Lu, Y., Gu, T., Bian, Z., Wang, L., Tong, Z.: From Separation to Incorporation - A Full-Circle Application of Computational Approaches to Performance-Based Architectural Design. In: Yuan, P.F., Chai, H., Yan, C., Leach, N. (eds.) CDRF 2021, pp. 189–198. Springer, Singapore (2022). https://doi.org/10.1007/978-981-16-5983-6_18

5. Hisao, I., Noritaka, T., Yusuke, N.: Evolutionary many-objective optimization: a short review. In: 2008 IEEE Congress on Evolutionary Computation (IEEE World Congress on Computational Intelligence). IEEE, pp 2419–2426 (2008)

6. Liu, X., Wang, L., Ji, G.: Optimization approaches in performance-based architectural design - a comparison study. In: Proceedings of the International Conference on Education and Research in Computer Aided Architectural Design in Europe, pp. 599–608 (2022)

7. Caldas, L.G., Norford, L.K.: A design optimization tool based on a genetic algorithm. Autom Constr **11**, 173–184 (2002). https://doi.org/10.1016/S0926-5805(00)00096-0

8. Negendahl, K.: Building performance simulation in the early design stage: an introduction to integrated dynamic models. Autom Constr **54**, 39–53 (2015). https://doi.org/10.1016/j.autcon.2015.03.002

9. Emmerich, M.T.M., Deutz, A.H.: A tutorial on multiobjective optimization: fundamentals and evolutionary methods. Nat. Comput. **17**(3), 585–609 (2018). https://doi.org/10.1007/s11047-018-9685-y

10. Cao, K., Huang, B., Wang, S., Lin, H.: Sustainable land use optimization using boundary-based fast genetic algorithm. Comput Environ Urban Syst **36**, 257–269 (2012). https://doi.org/10.1016/j.compenvurbsys.2011.08.001

11. Yang, D., Wang, L., Guohua, J.: Embedding design intent into performance-based architectural design—case study of applying soft constraints to design optimization. In: Hybrid Intelligence, pp. 165–174 (2023)

12. Wang, L., Janssen, P., Do, T., et al.: COMPARING DESIGN STRATEGIES a system for optimization-based design exploration. In: CAADRIA 2023 (2023)

13. Wang, L., Janssen, P., Do, T., et al.: A rapid design optimization framework. In: Co-creating the Future: Inclusion in and through Design - Proceedings of the 40th Conference on Education and Research in Computer Aided Architectural Design in Europe (eCAADe 2022), pp. 619–628 (2022)

14. Wang, L., Janssen, P., Chen, K.: EVOLUTIONARY DESIGN OF RESIDENTIAL PRECINCTS: a skeletal modelling approach for generating building layout configurations. In: POST-CARBON, Proceedings of the 27th International Conference of the Association for Computer- Aided Architectural Design Research in Asia (CAADRIA) 2022. Pp. 415–424 (2022)

15. Wang, L., Janssen, P., Chen, K.W.: Evolutionary Optimization of Benchmarks: Parametric Typologies for Generating Typical Designs. In: Gero, J.S. (ed.) Design Computing and Cognition'22, pp. 699–717. Springer International Publishing, Cham (2023)

Optimization Strategies of Architecture and Engineering Graduate Students: Responding to Data During Design

Stephanie Bunt[1]([✉]) [iD], Catherine Berdanier[2] [iD], and Nathan Brown[1] [iD]

[1] College of Engineering, Department of Architectural Engineering, The Pennsylvania State University, 104 Engineering Unit A, University Park, Pennsylvania, PA 16802, USA
s.bunt@psu.edu

[2] College of Engineering, Department of Mechanical Engineering, The Pennsylvania State University, 137 Reber Building, University Park, Pennsylvania, PA 16802, USA

Abstract. Both architects and engineers increasingly use design optimization in the early stages, but it is unclear how designers' disciplinary background may influence their optimization strategies. In considering designs with multiple conflicting objectives, large datasets of options are often produced, which can be difficult to navigate. Architects and engineers may engage with optimization tools and their feedback differently based on their background, which can affect collaborative efforts and influence design outcomes. In this study, graduate architecture and engineering students with experience in optimization responded to a design task with both quantitative and qualitative goals. The task required participants to establish and explore their own parametric design variables, producing large datasets with numerical and visual feedback. Screen recordings of the design sessions were analyzed to characterize optimization events initiated by the designers, revealing when and how often they ran optimization routines and how they reviewed the optimization feedback. The study showed that the architecture students tended to use optimization later and iterate less than the engineering students, who relied on quantitative data more often to edit their design space and justify their decisions. Future efforts to incorporate design optimization into graduate education should be cognizant of these differences, especially in multidisciplinary settings that encourage architects and engineers to mutually engage with data during collaborative design.

Keywords: Multi-objective optimization · 3D parametric design · Disciplinary design strategies · Design study

1 Introduction

Although architects and engineers both contribute professional expertise in designing our built environment, they often use different tools, which can hinder cross-disciplinary considerations. However, optimization tools embedded in 3D modeling environments allow designers to consider many numeric and geometric objectives simultaneously,

M. Turrin et al. (Eds.): CAAD Futures 2023, CCIS 1819, pp. 174–189, 2023.
https://doi.org/10.1007/978-3-031-37189-9_12

which can support integrated design decisions [1, 2]. The ability to interactively create, manipulate, and analyze datasets with multi-objective feedback is advantageous in navigating visual and performance implications. While considering these possible advantages, it should be acknowledged that architects and engineers may use these tools in different ways—interdisciplinary environments do not simply merge the professions. In their design training, architects and engineers engage with design data in different forms, ranging from open-ended analyses to strictly defined problems with clear parameters and constraints. Understanding how developing designers make decisions in data-rich digital environments, and particularly how architects and engineers might show different optimization strategies, is a first step in facilitating better collaboration between the disciplines when optimizing. While it has been shown that parametric design has distinct design thinking characteristics [3, 4], optimization approaches are still being explored in these environments [1, 5–7]. This research thus asks: **How does the disciplinary background of architecture and engineering design students relate to their optimization strategies during conceptual design?**

This paper presents an initial study which prompted architecture and engineering graduate students with experience in optimization to develop an atrium roof for a fictional university in the Southwestern United States. They were asked to account for daylight, solar radiation, and/or structural performance, along with the contextual appearance of their design. Participants developed a 3D parametric model for geometry and used optimization tools to account for the specified objectives. All variables were created by the designers, making them responsible for defining the structure of the optimization problem within an architectural design prompt. A survey of participants focused on educational experience. Screen recordings from the design sessions were collected and analyzed to capture significant events, assessing when and how frequently the designers navigate between performance feedback and design development. It was expected that while there would be recuring and similar behaviors exhibited by the designers, the focus of their optimization efforts would align with typical disciplinary characteristics, such as greater comfort with numerical feedback for engineers and more internalized decisions for the architects.

2 Background

2.1 Optimization in Building Design

Optimization, as a design strategy, enables the consideration of quantitative objectives such that designers make more informed decisions [5]. This approach is useful in building design as the requirements of our built environment become more numerous and complex [8, 9]. Many design goals for buildings are inversely related, such as daylight and energy conservation, where the use of more glass will increase daylight, but not provide an as efficient U-value compared to an insulated wall for thermal performance. Conflicting relationships between objectives can make finding optimal design solutions in the objective space challenging, particularly when the design goals become more numerous, as in full building design [6, 10].

Though formal mathematical optimization seeks a single answer, in practice, multi-objective design optimization strategies often produce dense datasets in pursuit of finding "better" solutions [11], which can be difficult for designers to sort. Optimization tools employ a range of algorithms within specialized interfaces for user interaction, such as displaying a 3D model with plots of the design and objective spaces. Figure 1 shows an example of the relationship between 3D model space, variable design space, and optimization objective space. A designer develops the parametric model with variables, and the resulting geometry displays in the 3D space. At the same time, objective performance values are generated in an objective space. When an optimization process is used, the tool will rapidly iterate through the model, changing the values of variables to minimize the resulting objectives. In situations with no clear winner, a designer will need to edit their original design and rerun the optimization or select a design based on priorities or characteristics not captured in the model. Often, selecting a final design will rely on qualitative requirements, intuition, or preferences. Optimization tools can thus help designers make informed decisions while still allowing for design freedom. However, architects and engineers are trained differently, and thus may diverge on objectives, as well as how they engage with such tools.

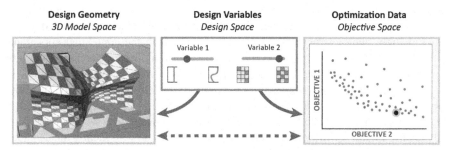

Fig. 1. Navigating between the geometric and numeric feedback in the 3D modelling, optimization design process.

2.2 Differences in Architecture and Engineering Education

It has been shown that architects and engineers tend to design differently, with architects assuming partially defined problems and engineers pursuing well-defined problems [12]. This distinction in design strategy aligns with aspects of their design education. In contemporary architecture education, architects are trained to think spatially, often using 3D modeling tools in their first or second years of training, while there is still a call to improve visualization skills in engineering education [13]. The additional years of computer-space design experience can set architects apart from engineers as spatial design thinkers. Furthermore, parametric modeling has recently been incorporated in architecture education [14, 15] and some researchers have called for more parametric modeling in architecture practice [16]. However, research has shown that AEC students generally express a larger learning curve in becoming acquainted with 3D digital modeling tools than in learning the associated design concepts [17].

Engineering education characteristically emphasizes design decision making as well, though typically centered in the first-year engineering courses and senior design. It has been documented that engineering undergraduates typically struggle to negotiate authentic problem- and project-based design tasks without "right" answers, since traditional coursework often emphasizes arriving at a correct, tangible solution, following established problem-solving methods [18]. Focusing on a final solution is imperative in our built world where design decisions have monetary and life-safety consequences. However, there has been a shift towards project-based learning in engineering education as engineers also need to be adept communicators, team members, and lifelong learners [19, 20]. An additional complication is that architects and engineers can define the outcome of a "design" differently: in practice, engineering design can be represented by algorithms and codes, numerical simulations, spreadsheet outputs, and physical prototypes, in addition to spatial computer-aided designs.

The appropriate scoping of design tasks for education is critical, as design tasks with definitive solutions rarely allow for ambiguity of interpretation and can narrow thinking. Engineers with a low tolerance for ambiguity also create fewer novel ideas [21]. Similarly, solution-focused thinking can be a barrier for navigating optimization strategies because of conflicting qualitative and quantitative goals in building design. Selecting a "best" solution from optimized datasets relies on both informed performance feedback and the intuition of the designer.

With this background, optimization has been incorporated in the education of architects and engineers with initial positive results [22–24]. Researchers have looked at how designers make decisions in a parametric modelling space, recognizing the difference between choices made by the designer's knowledge versus decisions made by algorithms [25, 26]. Because variables are incorporated into a parametric model, there is the potential for unexpected designs to emerge. Developing a well-built parametric model, with appropriate constraints but freedom of exploration, requires effective parametric strategies [27]. Moreover, what might motivate when a designer makes decisions based their own intuition or on the suggestion of an algorithm is difficult to distinguish, particularly in the application of optimization. Disciplinary training likely influences the way that optimization occurs in designerly domains, though this relationship is not yet explored. Observing how disciplinary background influences optimization strategies can inform how the professions approach complicated datasets created during the process. To consider these questions, this research used a design study which asked graduate architecture and engineering students with experience in optimization to respond to a building design task. Their behaviors with the optimization tools were then compared.

3   Methods

3.1   Study Setup

This study asked participants to respond to a conceptual design task with clear design goals using optimization techniques. The design interface provided numeric and visual feedback, while the optimization tool produced datasets with a range of possible solutions. In observing how the participants interacted with the tool and responded to feedback data, this study established patterns of behavior in relation to the disciplinary background and experience of participants.

Participant Recruitment, Background, and Selection. Participants were recruited from the graduate programs of the architecture and architectural engineering departments at a large university in the Northeastern United States. After expressing interest, potential participants completed an intake survey which collected data about educational background, previous professional work experience, parametric modeling skills, and their understanding of optimization. To qualify for inclusion, participants were required to have completed coursework in multidisciplinary building design and have at least 6 months of experience with parametric modeling and optimization. These prerequisites established their ability to adequately respond to the study task. However, the participants were graduate students and not yet experts in their field, which was taken into consideration when developing complexity in the design task and managing for design fatigue. The study was approved by the Institutional Review Board and participants were financially compensated for their time after completing their design session.

Ten total participants (five from architecture and five from architectural engineering) were included in the study. Although this number does not establish statistical significance, the goal is to identify behavioral differences as an initial investigation into optimization techniques. Each participant represents a dense collection of data with 3+ hours of recorded files per person. This work follows methods evident in qualitative research to establish a framework for studies with more participants in the future.

Design Session. To begin, participants watched a video brief which explained the design task and introduced the provided base file with site context. The video ensured that participants received a standard level of detail along with illustrations to demonstrate context. participants were allowed to take notes or sketch while watching the video. They could return to their paper and pencil tools at any point in the session.

Next, participants were situated in front of a workstation that was pre-loaded with unobtrusive eye-tracking software and hardware (i.e., no headgear), which was then calibrated to each participant. The researcher sat to the side and observed a mirrored computer screen to not intrude on the participants' space. The researcher collected memos during this time, making notes about design choices, behaviors, and sketching. Although the researcher answered participant questions about the study's instructions, the researcher did not provide feedback on the designs or optimization strategies. Once the participants were satisfied with a design, they wrote a design statement justifying their final solution and submitted 3–5 screenshots of their design to the researcher. Figure 2 shows the study events and data that was collected to compare behaviors based on disciplinary background experiences.

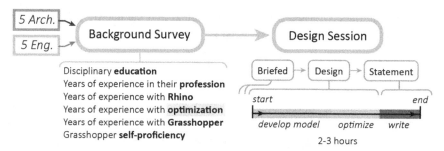

Fig. 2. The study sequence, including a background survey, design session, and data collection.

Design Task and Tools. Participants were asked to develop an atrium roof for a fictional university in the hot climate of Arizona. The site was chosen for its generally known sunny conditions, which can also be readily found online. The participants were shown site context and instructed to address two of three optional performance criteria: maximizing daylight, minimizing solar radiation, and maximizing structural performance. Their designs also had to consider contextual appearance.

Participants used Grasshopper, a 3D parametric tool, to develop the geometry of their design in Rhinoceros, a 3D digital environment. Optimization was conducted using plugins built for Grasshopper. These tools are established environments for design optimization, being used in past research [28]. Although participants could use any available optimization tool, the participants chose Galapagos [29], Design Space Exploration [30], or Octopus [31]. They were provided a file with existing site context and a script with pre-built quick calculations for the objectives, as opposed to more detailed simulations. Daylight and solar radiation calculations accounted for area, materials, and geometry of the surface panels. Structural performance was measured by reducing elastic energy using the structural analysis plugin Karamba 3D [32]. Figure 3 shows a sample of the

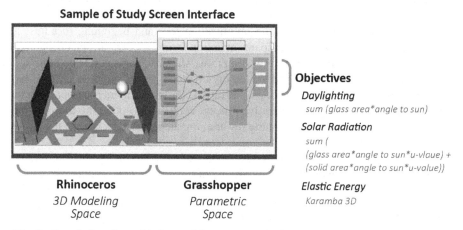

Fig. 3. Sample interface with the model space, parametric space, and objective calculations.

environment interface with the provided site context and parametric canvas with the objective generators.

3.2 Data Collection and Evaluation

The following three data streams were collected to answer the research question:

Background Survey and Analysis. The survey collected data about participants' educational background and experience with the study tools, which is used to compare participant experience and training to designerly behavior when engaging with optimization. A summary of background characteristics is given in the results.

Screen Recordings and Analysis of Observational Screen Capture Data Analysis Methods. Screens were recorded using screen capture software from Eyetracking, Inc. The screen capture hardware and software were non-intrusive, meaning they are not wearable and do not interfere with a Participant's natural behavioral tendencies and processes. To analyze the recorded screen capture data from the $N = 10$ participants, methods consistent with observational qualitative analysis using content analysis methods [33] were employed, relying on a modification of an *a priori* framework of design behaviors informed from the FBS literature [34]. This was then honed to describe the significant observable design events that are captured via screen recording. The coding schema was discussed with the broader research team and validated in early rounds of analysis to ensure that it was comprehensive to define and verify behaviors in the parametric space, such as the placement of the first component in the parametric model space and interaction with the optimization tools. The coding scheme and definitions are presented below:

(1) Activate Objective Feedback: Plugging geometry into the objective generators indicated a shift from focusing on visual model development to model performance. The importance of defining numerical objectives and appropriately incorporating them into design decisions has been established in previous assessments of optimization [35].
(2) Preparing Optimizer: Opening the optimization tool and beginning to adjust its settings signified a shift in participants attention from their own design decisions to engaging feedback from the optimization tool. All participants performed this act.
(3) Run Optimizer: Running an automated optimization process showed that the participant began generating data for observation. Early and numerous optimization runs indicated an integrated, iterative process compared to plugging a model into the objective generators and running fewer automated processes later in the design session.
(4) Review Results: Viewing either a data visualization in the tool or cycling through designs it produced indicated the beginning of this action. While some participants

only glanced at the results, others considered the options for an extended period of time.

To analyze the data, the researcher watched the recordings while qualitatively "coding" the design behaviors occurring over time through descriptive content analysis methods using a post-positivist paradigm [33]. The occurrences of significant events were plotted on session timelines like the one shown in Fig. 4.

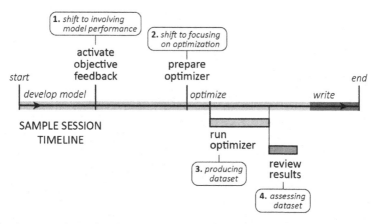

Fig. 4. A sample session timeline labeling the key events identified in each session.

Design Statements. The participants also submitted a 150–250 word design statement, written to the fictional client, that presented the suggested solution. There were no explicit requirements of the design statement, and thus they varied in content. Thematic analysis methods were employed to characterize the dominant themes in the written design prompts, using an emergent coding scheme to understand the patterns, again relying on conventional qualitative content analysis methods, this time employing an emergent coding approach [33]. Four primary topics were noted in the statements: (1) the potential *users*, such as students; (2) the Participant's *design vision*, which could be as explicit as "I Wanted the roof to look like a tree" or abstract as in wanting the space to be well shaded; (3) stating which of the three *objectives* were considered; and (4) referring to *optimization* or improving design performance. Comprehensively, the characteristics of the design statement were compared to the design behaviors and educational background of the participants to draw initial observations about the difference between architecture and engineering students when optimizing.

4   Results and Discussion

This study's primary research question asked, **how do the disciplinary background of architecture and engineering design students relate to their optimization strategies during conceptual design?** To create an authentic design challenge, the study's design task allowed the participants to develop their own geometries with variables that they defined. This unrestricted parametric space prompted solutions of different quality and parametric range. Although evaluation of the final designs falls outside the scope of this research, the range of design spaces developed by four of the participants are presented in Fig. 5. While some participants built models with greater geometric variation, others focused on controlling smaller changes in the design space. The emphasis of this work, however, responds to the research question and considers differences of the designers' design strategies, not their final designs.

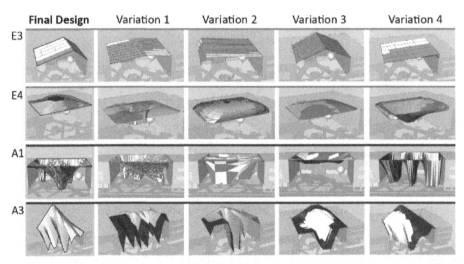

Fig. 5. Sample of two engineers' and two architects' designs, showing their final design and two of the variations of their parametric model.

From the results of all participants, three influential dimensions of data emerge: 1) years of experience and confidence in the study's tools, 2) patterns in the participants' design sessions, and 3) characteristics of the designers' final design statements. Collectively, these data suggest disciplinary strategies that increase or limit the inclusion of optimization feedback in design, which, when utilized effectively, can positively influence overall design performance. As a result, introducing design students to both qualitative and quantitative design goals throughout their education may equip them to more wholistically incorporate dataset feedback and optimization suggestions in their design decisions.

4.1 Participant Background

The participants' background information is summarized in Fig. 6. The Grasshopper self-proficiency is illustrated by a scale of 0 to 5. The architects had considerably more experience with Rhino, and some more experience with Grasshopper, but all reported experience levels were considered adequate to authentically respond to the design task. The architects also had greater professional experience, with participant A1 reporting the greatest number of years. The distinction between participants' years with optimization were closer. This aligns with expectations from disciplinary backgrounds since architects are trained early in their education to use 3D modeling tools. Their expressed confidence with Grasshopper also reflects a comfort with the design tool, perhaps as a result of their design training. To consider the influence of these experiences, the background differences were included in the context of the design session behaviors.

	Years of Experience				Grasshopper
	Professional	Rhino	Optimization	Grasshopper	Self-proficiency
Engineering Graduate Students					
E1	1	1	1	3	●●●●○
E2	1	1	3	3	●●●○○
E3	0	1	1	1	●○○○○
E4	0	3	2	3	●●●●●
E5	1	1	1	1	●●○○○
Architecture Graduate Students					
A1	10	3	2	3	●●●●●
A2	4	5	2	5	●●●●●
A3	1	7	2	7	●●●●○
A4	2	7	4	5	●●●●●
A5	6	5	3	5	●●●●○

Fig. 6. Background information on the graduate student participants

4.2 Session Sequences

The design sessions were plotted to detail important events in the sequence. The results from the engineering graduate participants are presented in Fig. 7 and the architecture graduate participants in Fig. 8. The timeline labels when the participants plugged their model into the objective calculators (measured as a percentage of the session), when they started to prepare the optimization tool, and how long the session lasted. It also shows when the designer ran an optimization tool (indicated in yellow-green) and for how long they reviewed the results, if at all (indicated in brown). E2 and A2 did not review the datasets. Along the right side of the figure, the participants number of optimization runs, years of experience with the 3D modeling tool, and self-provided proficiency with Grasshopper are provided.

Disciplinary Differences in Design Sessions. The architecture students seemed to engage less significantly with the feedback data from optimization compared to the engineers and from the session plots, they ran the optimization tool a fewer number of times than the engineers. The architects also tended to wait until later in the design session to engage with the numeric feedback of the optimization tool, with the exception of Participant A1, who plugged their model into the objective algorithms earlier than all of the other participants. This designer also spent the longest amount of time considering the results of the optimization tool compared to the other participants. A primary difference between this participant compared to the others is years of professional experience, as was presented in Fig. 8, where Participant A1 worked at least 4 more years in their disciplinary professional setting than the other participants. Although the architecture students tended to include optimization later in their process compared to the engineering students, years of work experience and maturity of a designer can influence optimization strategy, incorporating algorithmic thinking early in an optimization, conceptual design task. Additional research should be conducted to better determine the relationship between professional experience and optimization strategies.

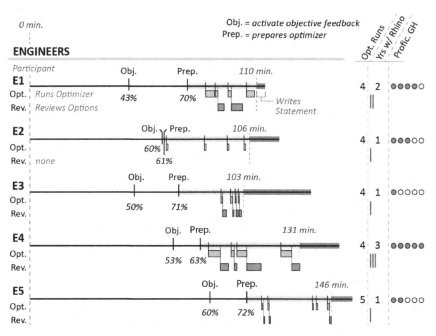

Fig. 7. Session plots for the engineering participants, showing key events in their design process.

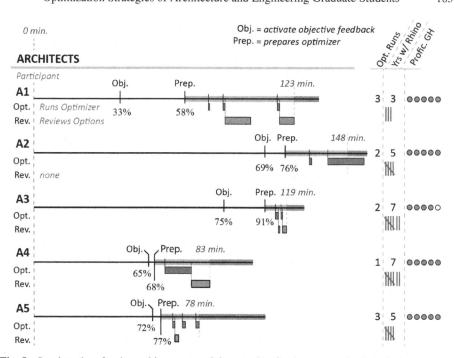

Fig. 8. Session plots for the architecture participants, showing key events in their design process.

4.3 Design Statements

Figure 9 summarizes the mentions of design topics within the participants' statements, emphasizing when the topics were mentioned most with a yellow-green box.

	Statement Mentions			
	Users	Design Vision	Objectives	Optimization
Engineers	2	3	5	2
Architects	2	5	3	4

Fig. 9. The mentions of design topics within the participants final design statements.

Disciplinary Differences in Design Statements. As was expected from the hypothesis, the architecture participants consistently referred to a design vision, stating what they imagined for the design, while the engineers all stated their quantitative objective goals. However, not all participants mentioned the users of the atrium space or referred to optimization in their statements. Despite the engineers incorporating optimization feedback earlier in their design sessions compared to the architects, they were not as explicit in mentioning optimization. Meanwhile, the architects, who engaged with optimization feedback later in their design sessions, mentioned the study's focus of optimization more consistently. From their disciplinary education in design studios, they may be more practiced at reflecting a design task's requirements, despite the spread of project-based learning in engineering [18]. Alternatively, architects may be less formal in their use of the term "optimize." While engineers may view optimization and dataset processing as an inherent aspect of design and not important to mention to the client, the architects may have viewed optimization and dataset parsing as an influence on their design process, despite incorporating it less.

4.4   Implications for Disciplinary Education and Approaches to Optimization Feedback

Disciplinary background clearly related to different patterns in optimization behavior, but professional years of experience or greater comfort with the tools may have also played a role. During the design sessions, the more professionally experienced architects spent more time designing without optimization feedback for a greater percentage of time, rather than on optimization feedback. Alternatively, the engineers incorporated optimization feedback data into their design process more often. However, when writing about their final design, the architects mentioned optimization more consistently than the engineers. Other details of the design statements supported expected characteristics of the profession, such as the engineers stating their chosen objectives while the architects described their design visions.

With these results, it may be that participants' disciplinary training influenced their preparedness to navigate large datasets produced by optimization, but not recognize the role of optimization on their final design. While the engineers relied on optimization feedback iteratively to develop their design, they did not all include it in their statement. In contrast, the architects consider optimization later, often after large geometric decisions were already set, yet they mention optimization more readily. Although architecture pedagogies tend to approach design decisions conceptually, rather than requiring numerical feedback, there may be advantages to incorporating quantitative approaches to design in the context of optimization. As quantitative metrics can increasingly be simulated and optimized, preparing architecture designers to navigate dataset feedback as a part of their education may be valuable. Figure 10 shows a summary of these relationships.

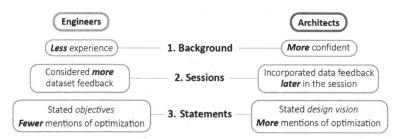

Fig. 10. Summary of the results, emphasizing the different characteristics of the disciplinary backgrounds, design sessions, and design statements.

4.5 Study Limitations

To elicit authentic design behavior from the participants, the design task tried to provide an approachable scenario with challenging goals and accessible expectations. However, limiting the design space to the framework of the computer programs may have affected the designers' natural design process. Working within the constraints of cognitive fatigue was also central to the quality of data collection for this study. As has been noted previously, studying design behavior inherently impacts dimensions of the process, but this alone does not discredit the research, as the data must be considered within its context [36]. Additionally, this study did not assess the overall quality of the final designs and leaves the dimension of design efficacy for future investigations. Based on the participant background qualifications to participate in the study, it was established that the designers could develop an adequately performing solution in response to the design task. Future work intends to investigate design efficacy, but it is valuable to first establish differences in optimization behavior.

5 Conclusion

This study considered the optimization strategies of architecture and engineering graduate student designers compared to their disciplinary background when responding to a conceptual building design task that produced large datasets of solutions. Although the architecture students incorporated optimization and dataset feedback later in their design process and with less frequency compared to the engineering students, this was not true for one architecture student who had more at least 4 more years of professional work experience compared to the other participants. It may be that both disciplinary background and experience influence optimization behaviors. These initial findings may help instructors approach optimization curriculum with attention to the students' disciplinary processes. It lays a groundwork for understanding design behavior when students construct and use different performance feedback tools in combination with optimization tools. This work also establishes methods for investigation large datasets produced by optimization that can be applied to future investigations of optimization efficacy of both design students and design professionals.

Funding Acknowledgement. This research is supported by the National Science Foundation under Grant #2033332. Any opinions, findings, and conclusions or recommendations expressed in this material are those of the authors and do not necessarily reflect the views of the NSF.

References

1. Wortmann, T.: Efficient, visual, and interactive architectural design optimization with model-based methods opossum-surrogate-based optimization for architectural design view project Informed design group view project (2018)
2. Haymaker, J., et al.: Design space construction: A framework to support collaborative, parametric decision making. J. Info. Technol. Constr. **23**, 157–178 (2018)

3. Oxman, R.: Thinking difference: Theories and models of parametric design thinking. Des Stud **52**, 4–39 (2017). Sep.
4. Wortmann, T., Tunçer, B.: Differentiating parametric design: digital workflows in contemporary architecture and construction. Des Stud **52**, 173–197 (2017). Sep.
5. Yang, D., Wang, L., Ji, G.: Optimization-assisted building design - cases study of design optimization based on real-world projects. In: the 40th Conference on Education and Research in Computer Aided Architectural Design in Europe, pp. 609–618 (2022)
6. Brown, N.C., Mueller, C.T.: The effect of performance feedback and optimization on the conceptual design process. In: International Association of Spatial Structures Annual Symposium (Sep. 2016)
7. Geyer, P., Beucke, K.: An integrative approach for using multidisciplinary design optimization in AEC. In: the International Conference on Computing in Civil and Building Engineering (2010)
8. Gerber, D.J., Lin, S.-H.E.: Designing in complexity: simulation, integration, and multidisciplinary design optimization for architecture. SIMULATION **90**(8), 936–959 (2014). Aug.
9. Mueller, C.T., Ochsendorf, J.A.: Combining structural performance and designer preferences in evolutionary design space exploration. Autom Constr **52**, 70–82 (2015)
10. Wortmann, T., Nannicini, G.: Introduction to architectural design optimization. In: Karakitsiou, A., Migdalas, A., Rassia, S., Pardalos, P. (eds.) City Networks 128, pp. 259–278. Springer (2017)
11. Machairas, V., Tsangrassoulis, A., Axarli, K.: Algorithms for optimization of building design: A review. Renew. Sustain. Energy Rev. **31**, 101–112 (2014). Mar.
12. Cross, N., Roozenburg, N.: Modelling the design process in engineering and in architecture. J. Eng. Des. **3**(4), 325–337 (1992). Jan.
13. Contero, M., Naya, F., Company, P., Saorin, J.L., Conesa, J.: Improving visualization skills in engineering education. IEEE Comput Graph Appl **25**(5), 24–31 (2005). Sep.
14. Hsu, C.-Y., Ou, S.-J.: Innovative practice of sustainable landscape architecture education—parametric-aided design and application. Sustainability **14**(8), 4627 (Apr. 2022)
15. Gallas, M., Jacquot, K., Jancart, S., Delvaux, F.: Parametric modeling: an advanced design process for architectural education. In eCAADe **33**, 119–127 (2015)
16. Stals, A., Jancart, S., Elsen, C.: Parametric modeling tools in small architectural offices: Towards an adapted design process model. Des Stud **72**, 100978 (2021). Jan.
17. Diao, P.-H., Shih, N.-J.: Trends and research issues of augmented reality studies in architectural and civil engineering education—a review of academic journal publications. Applied Sciences **9**(9), 1840 (May 2019)
18. Lucas, B., Hanson, J.: Thinking like an engineer: using engineering habits of mind and signature pedagogies to redesign engineering education. Int. J. Eng. Pedagogy **6**(2), 4–13 (2016)
19. Dym, C.L., Agogino, A.M., Eris, O., Frey, D.D., Leifer, L.J.: Engineering design thinking, teaching, and learning. J. Eng. Educ. **94**(1), 103–120 (2005). Jan.
20. Smith, K.: Teamwork and Project Management, 2nd edn. McGraw-Hill, New York (2004)
21. Toh, C.A., Miller, S.R.: Choosing creativity: the role of individual risk and ambiguity aversion on creative concept selection in engineering design. Res. Eng. Design **27**(3), 195–219 (2016). https://doi.org/10.1007/s00163-015-0212-1
22. N. C. Brown, S. Bunt. Optimization tools as a platform for latent qualitative design education of technical designers. In: National Conference on Beginning Design Students (2022)
23. de Oliveira, C.J., Steffen, L.O., Vasconcellos, C.A.de M., Sanchez, P.F.: Structural topology optimization as a teaching tool in architecture. Revista Ensino de Engenharia **37**(3) (2018)
24. Pasternak, A., Kwiecinski, K.: High-rise Building Optimization - A Design Studio Curriculum, pp. 305–314 (2015)

25. Yu, R., Gu, N., Ostwald, M.: Comparing designers' problem-solving behavior in a parametric design environment and a geometric modeling environment. Buildings **3**(3), 621–638 (2013). Sep.
26. Yu, R., Gero, J., Gu, N.: Architects' cognitive behaviour in parametric design. Int. J. Archit. Comput. **13**(1), 83–101 (2015). Mar.
27. Lee, J., Gu, N., Williams, A.P.: Parametric design strategies for the generation of creative designs. Int. J. Archit. Comput. **12**(3), 263–282 (2014). Sep.
28. Shi, X., Yang, W.: Performance-driven architectural design and optimization technique from a perspective of architects. Autom Constr **32**, 125–135 (2013). Jul.
29. Rutten, D.: Galapagos: On the Logic and Limitations of Generic Solvers. Archit. Des. **83**(2), 132–135 (2013). Mar.
30. Brown, N.C., Jusiega, V., Mueller, C.T.: Implementing data-driven parametric building design with a flexible toolbox. Autom Constr **118**, 103252 (2020). Oct.
31. Vierlinger, R.: Multi objective design interface. Masters Thesis. University of Applied Arts Vienna, Vienna (2013)
32. Preisinger, C., Heimrath, M.: Karamba—a toolkit for parametric structural design. Struct. Eng. Int. **24**(2), 217–221 (2014). May
33. Hsieh, H.-F., Shannon, S.E.: Three approaches to qualitative content analysis. Qual Health Res **15**(9), 1277–1288 (2005). Nov.
34. Gero, J.S., Kannengiesser, U.: The situated function–behaviour–structure framework. Des Stud **25**(4), 373–391 (2004). Jul.
35. Geyer, P.: Component-oriented decomposition for multidisciplinary design optimization in building design. Adv. Eng. Inform. **23**(1), 12–31 (2009). Jan.
36. McGrath, J.E.: Method for study of groups. In: Groups: interactions and performance, pp. 31–37. Prentice-Hall Inc., New Jersey (1984)

Wieringa Surface: The Implementation of Aperiodicity into Architectural Acoustics

Ross Cocks, John Nguyen[(✉)], and Brady Peters

John H. Daniels Faculty of Architecture, Landscape, and Design, University of Toronto, 1 Spadina Crescent, Toronto, ON, Canada

{rossjory.cocks,johnnie.nguyen}@mail.utoronto.ca,
brady.peters@daniels.utoronto.ca

Abstract. At the intersection of order and disorder exists a balance that is referred to as aperiodicity. This mathematical concept is exemplified by the famous Penrose tiling methods, which display local five-fold symmetries that quickly dissipate globally. Architectural designers have previously exploited this property through floors tiling's, sculptures, and building facades. However, like mathematicians in the latter half of the 20th century such as Nicolaas Govert de Bruijn and R.M.A. Wieringa, designers in the early 21st century are now looking to the third dimension. In acoustics, surfaces with complex sectional profiles have been demonstrated to have scattering properties where the relationship between depth and width of undulations relate to the amount and frequency of the sound scattered. Mathematical sequences have been demonstrated to have optimal sound scattering properties, however, these mathematical sequences are periodic, and can therefore create unwanted acoustic effects. This research proposes combining the concept of aperiodicity with the predictability of a mathematical formula and defined through a parametric model, where the acoustic performance of these aperiodic surfaces will be analyzed through simulations. The contributions of this research are two-fold: first, we have developed a novel approach to creating a 3D object using parametric modeling to generate a Wieringa Surface, which when orthographically projected to 2D creates a Penrose Tiling; and second, we have carried out preliminary simulations that suggests there is a potential for this complex mathematical surface to exhibit sound scattering properties without periodic effects. This paper documents the parametrization of a specific case of the "Cut and Project Method", a mathematical method by which a slice of a multi-dimensional periodic space is selected through a cutting window and then projected to a lower dimension to produce aperiodic forms.

Keywords: Aperiodic Tiling · Mathematics and Architecture · Wieringa Surface · Penrose Tiling · Architectural Acoustics · Sound Scattering

1 Introduction

Architecture and acoustics share many common roots in mathematics. These disciplines rely on many of the same mathematical models and methods for the prediction, morphology, and optimization of behaviors. This is why the specialization of architectural

© The Author(s), under exclusive license to Springer Nature Switzerland AG 2023
M. Turrin et al. (Eds.): CAAD Futures 2023, CCIS 1819, pp. 190–203, 2023.
https://doi.org/10.1007/978-3-031-37189-9_13

acoustics utilizes mathematics to predict the auditory behavior of a space, explore new geometrical morphologies, and optimize acoustic-related scenarios. This paper documents the development of a parametric model for a mathematical surface that was observed by RMA Wieringa and described by N.G. de Bruijn. Although de Bruijn speculated about its applications in architecture, a supplementary hypothesis is presented for architectural acoustics. This approach was informed by previous works and observations made by Alan Mackay, while a formulated parametric model provides a novel computational approach to the established mathematical method. Iterations of the mathematical surface are tested through simulations and analyzed for performance. A case study consisting of three different geometrical scattering surfaces was conducted, where physical models for each of the surfaces was casted to demonstrate de Bruijn's initial architectural application ideas.

2 Background

2.1 Mathematical Definitions

In this section, we clarify all mathematical terminology used throughout this paper. "Euclidean space" is the vector space of all n-tuples of real numbers, $(x_1, \ldots, x_n)$. All geometries discussed within this paper reside in Euclidean space. When the word "space" is used singularly, then it refers to when $n = 3$, while the word "plane" is used for when $n = 2$. A "topological disc" refers to any set whose boundary is a single simple closed curve. A "plane tiling" is a countably infinite arrangement of closed topological disks on a plane without any gaps or overlaps. The individual components of a plane tiling are called "tiles". When the words tile(s) or tiling(s) are used singularly, then this refers to the aforementioned mathematical objects that reside on the plane unless otherwise specified. A tiling is "periodic" when there is an ordered copying and translation of a specific set of finite tiles. Thus, periodic tilings have translational symmetry, however they may, or may not, have rotational and/or reflection symmetry. Periodic tilings are said to exhibit "periodicity." If there is no periodicity, then there is no translational symmetry, and a tiling is said to be "non-periodic". In this paper, we will use the word "aperiodic" as a synonym for "non-periodic" and will use the two terms interchangeably. Aperiodic tilings are said to exhibit "aperiodicity"; however, aperiodicity can have alternate meanings within other mathematical contexts. A well-known example of aperiodicity is the Penrose tilings (Fig. 1) developed by the mathematical physicist and Nobel laureate, Sir Roger Penrose. (Penrose, 1974; Penrose, 1979).

2.2 The Fourier Transform and Aperiodicity

Mathematically, both periodic and aperiodic tessellations can be constructed in any dimension. In the field of crystallography however, tessellations of unit crystal cells were thought to only be periodic due to observed physical properties and experimental results. This imposed periodicity on structure meant that a crystal needed to be spatially arranged to follow the crystallographic restriction theorem which states that, if a set is periodic and the discrete group of translations of that set has more than one center

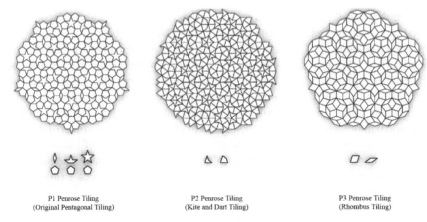

P1 Penrose Tiling
(Original Pentagonal Tiling)

P2 Penrose Tiling
(Kite and Dart Tiling)

P3 Penrose Tiling
(Rhombus Tiling)

Fig. 1. Penrose Tiling variations

of rotation, then the only rotations that can occur are by 2, 3, 4, and 6. (Senechal, 2006) Theoretical crystallographer, Alan Mackay, (1981) was one of the first to consider spatial arrangements outside the periodical discourse. Using the vertices of Penrose's Rhombus tiling (P3), Mackay generated a sphere packing of both the plane and space and put them through a mathematical transformation known as the Fourier Transform (Fig. 2). The Fourier Transform decomposes a function and reformulates it into one that depends on spatial or temporal frequency. It is extensively used in the analysis of wavelike phenomena such as acoustical decomposition and diffraction patterning. (Bracewell, 1986) Mackay realized that his patterns displayed clear strong and weak modulations of the transform. There was a clearly defined ten-fold symmetry diffraction pattern after the transform was performed indicating that there is an ordering within the aperiodic arrangement despite the lack of translational symmetry. This construction allowed Mackay (1982) to theorize that arrangements of unit crystal cells could be both periodic and aperiodic. A year later at the National Bureau of Standards (now National Institute for Standard and Technology) in the city of Gaithersburg, Maryland in the USA, Mackay's theories were experimentally verified by Daniel Schectman's discovery of quasicrystals, a form of solid-state matter that exhibits aperiodicity. (Shechtman et al., 1984).

2.3 Architectural Acoustics

In architecture, the prediction of acoustic performance is complicated by the interaction of numerous competing factors, including space dimensions, proportions, geometry, material properties, and surface details. An important determinant of acoustical behavior is the design of architectural surfaces that affect the propagation of sound reflections through space. Such reflections may cause flutter echoes or comb filtering if they are not carefully controlled. (Long, 2014) In proper acoustical design, sound reflections can be effectively tamed by evenly dispersing the reflected sound energy. (Jaramillo and Steel, 2015) Several studies have examined the behavior of sound reflections off different surface geometries and how these reflections contribute to shaping the spatial auditory

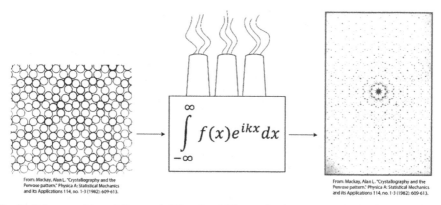

Fig. 2. Sphere packing of Penrose's Rhombus Tiling and subsequent diffraction pattern (Note: diffraction image was color inverted)

atmosphere. (Patel, 2017) A key challenge associated with the design of effective surface diffusers is making sure that sound energy is dispersed uniformly throughout the sound field while keeping the sound energy within the room and not absorbing it, thus maintaining an appropriate reverberation time. (Kuttruff, 2014) The diffusion of sound is enhanced when every position in the diffuse sound field receives the same density of reflected energy. A common practice among diffuser designers is to repeat a limited number of basic shapes on a periodic basis. (Ermann, 2015) This results in repetitive energy loops that reduce the diffuser's ability to uniformly disperse sound. (Tyler, 2016).

2.4 Schroeder Diffusers

In the 1970s, Manfred Schroeder (1975), utilized number theory to determine the optimal diffusion for a surface profile. A Schroeder diffuser consists of a series of wells of the same width and depth and can be arranged in a single or multi-dimensional configuration (see Fig. 3), and since their introduction, Schroeder's diffusers have been widely adopted in technical and architectural acoustics. (Cox and D'Antonio, 2003) The performance of Schroeder diffusers is limited by their ability to respond to the targeted sound wavelength, and depends on their surface design, arrangement, and depth profile. Aesthetically, Schroeder's irregular well arrangements can present a challenge for designers, and their protruding profiles make these products limiting for many wall applications. Contemporary architecture faces an important challenge in the absence of innovative acoustic designs that is complementary to established workflows. (Peter and Nguyen, 2021) A number of Architects are researching and developing new strategies overcome or complement the shortcomings of Schroeder diffusers through broadband absorption using Helmholtz resonators and Acoustic Metamaterials. (Setaki et al., 2016; Kladeftira et al., 2019; Nguyen et al., 2022; Cop, Nguyen, and Peters, 2023).

2.5 The Wieringa Roof

In 1981, the mathematician N. G. de Bruijn wrote a seminal paper on the algebraic theory and construction of Penrose in which he showed that they can be obtained from a periodic

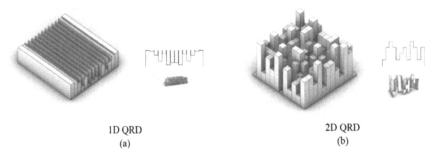

1D QRD
(a)

2D QRD
(b)

Fig. 3. Two types of Schroeder Quadratic Residue Diffusers (QRD)

5D space. (De Bruijn, 1981) Additionally, an observation made by his colleague R. M. A. Wieringa states that Penrose's Rhombi tiling can also be viewed as an orthogonal projection of a 3D surface containing a single rhombus (Fig. 4). De Bruijn named this surface a "Wieringa Roof" (WR) as he thought it could be used in architectural settings, particularly ceilings. (De Bruijn, 1981) Using an indexing system based around his construction of Penrose's P3 tiling, de Bruijn details a method to align and orient the rhombi of this 3D surface. This paper uses a similar method of indexing to orient and align the rhombi that are raised from 2D to 3D, as described by mathematician, and historian, Marjorie Senechal (1996). It is important to note that a WR is not a 3D tessellation of space, but the lifting of a planar tiling into a third dimension. The research described in this paper develops a parametric model for the WR that results in a generalization of the morphology of the surface. Due to this generalization, we designate any surface that can be orthographically projected from space to a plane to form Penrose's Rhombus tiling to be henceforth referred to as a Wieringa Surface (WS).

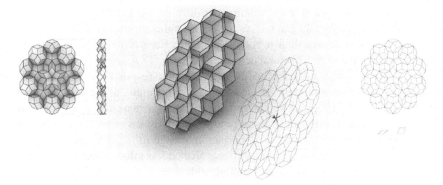

Fig. 4. Wieringa Roof morphology and projection to Penrose's P3 Tiling

2.6   Wieringa Surface for Architectural Acoustics

The application of aperiodicity into architectural acoustics is a promising endeavor, as aperiodic formations appear to have an advantage over periodic structures. (Karaiskou et al., 2020) Using a single asymmetric base shape, optimized for acoustical performance, D'Antonio and Cox (2017) have demonstrated curved diffusers that minimize the effects of periodicity. They reported that it is possible to increase the repeat distance, and hence improve performance, all the while creating a novel aesthetic for architects. In the design of Rafael Moneo's L'Auditori Barcelona, the Fibonacci sequence, an instance of 1D aperiodicity, was implemented to provide diffraction that progresses from low to high frequencies. The implementation of this sequence as done to avoid echoes and coloration. (Arau-Puchades, 2012) In 2018, Lee et al. verified that Penrose-tiling-type diffusers have only a very small dependence of incidence angle on scattering coefficients, which suggests that while there is promise for 2D aperiodic scattering surfaces, 3D aperiodic surfaces should be investigated. Rima Ajlouni (2018) has proposed that "quasi-periodic geometry" could be used for designing better surface diffusers as this means periodicity is removed from all directions and a limited number of manufactured shapes are used. An inadequacy of Rima's work is the lack of any acknowledgement or explanation for foundational governing mathematics that yields aperiodic forms and approaches the topic through deconstructing existing patterns found in historical documents.

Mackay's recognition that a clear diffraction pattern can result from the Fourier Transform of a sphere packing of the vertices of Penrose's Rhombus tiling, despite its lack of translational symmetry is a mathematical indication that there is an underlying order to its aperiodicity. The fact that aperiodic tessellations can be formulated from higher dimensional periodic spaces demonstrates that disorder and randomness are not the governing principles of aperiodic patterns. As mathematical models can explain this order (Fig. 5), predictions can be made on the implementation of aperiodicity into acoustics. We hypothesize that Wieringa Surfaces can provide a non-repeating scattering surface, perhaps comparable to the celebrated Schroeder QRDs, on the basis that its morphology lacks periodicity but is not random, a clear diffraction pattern results from similar geometries, its aperiodicity can be mathematically modelled from higher dimensional periodic spaces, and that its aesthetics are desirable for architectural implementation.

$$\sqrt{\frac{2}{5}} \begin{bmatrix} 0 & \sin(\frac{2\pi}{5}) & \sin(\frac{4\pi}{5}) & -\sin(\frac{4\pi}{5}) & -\sin(\frac{2\pi}{5}) \\ 1 & \cos(\frac{2\pi}{5}) & \cos(\frac{4\pi}{5}) & \cos(\frac{4\pi}{5}) & \cos(\frac{2\pi}{5}) \\ 1 & \cos(\frac{4\pi}{5}) & \cos(\frac{2\pi}{5}) & \cos(\frac{2\pi}{5}) & \cos(\frac{4\pi}{5}) \\ 0 & \sin(\frac{4\pi}{5}) & -\sin(\frac{2\pi}{5}) & \sin(\frac{2\pi}{5}) & -\sin(\frac{4\pi}{5}) \\ \frac{1}{\sqrt{2}} & \frac{1}{\sqrt{2}} & \frac{1}{\sqrt{2}} & \frac{1}{\sqrt{2}} & \frac{1}{\sqrt{2}} \end{bmatrix} \begin{bmatrix} x_1 \\ x_2 \\ x_3 \\ x_4 \\ x_5 \end{bmatrix} = \begin{bmatrix} b_1 \\ b_2 \\ b_3 \\ b_4 \\ b_5 \end{bmatrix}$$

Fig. 5. The rotational matrix used in the construction of Wieringa Surfaces

3 Methods

In view of the complexity of the geometry, it was decided that the best approach would be to create a parametric model for quick surface iterations while using digital simulation to study the acoustic performance of the structure. A total of twelve case iterations, based off six parameters (Fig. 6), were used as the basis for testing and fabricating Wieringa surfaces. The parameters were: lattice size, α (alpha), β (beta), γ (gamma), δ (delta), and index order. The selected iteration for fabrication had a lattice size of 2, $\alpha = -0.25$, $\beta = -0.25$, $\gamma = -0.25$, $\delta = 0.5$, and had an indexing order of 1,2,3,4. For fabrication, the objective was to preserve aesthetic values while indicating a significant departure from the canonical Wieringa surface, as well as displaying all possible variations of rhombi orientations. The fabrication and testing required the model to be scaled to obtain a desired height, width, and radius that could perform acoustically. The fabricated model spans $20.76'' \times 20.73'' \times 4.45''$, the deepest well depth being $4''$. It was determined that computer numerically controlled (CNC) milling would provide the best surface fidelity since it contains a minimum amount of surface difference from the 3D model.

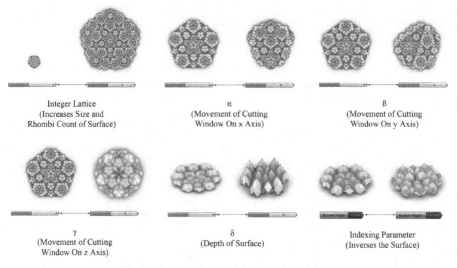

Fig. 6. Six parameters of the Wieringa Surface

3.1 Parametric Model

The parametric model developed in this research paper is based on the Cut and Project method. (Whittaker and Whittaker, 1988; Levine and Steinhardt, 1986; Socolar and Steinhardt, 1986) This method is well-developed in mathematics and essentially involves capturing a slice of a higher dimensional periodic space through a constructed cutting window and projecting the slice into a lower dimensional space, resulting in an aperiodic tessellation. The higher dimensional space used in this parametric model was a

5D periodic integer lattice, while the cutting window was a rhombic triacontahedron. Throughout extensive background research no evidence of a publicly available parametric CAD script was found. As of October 14, 2022, a public Wolfram Demonstration Project by Adam P. Goucher is available which can create a single canonical Wieringa surface arrangement with 5 degrees of sizing, however the resulting surface is small and cannot be converted to any CAD format. Likewise, there are several parametric tiling programs that allow for a P3 Penrose tiling to be constructed, however these programs are primarily for visual presentations and are not readily available to be placed within a CAD software. There exists a grasshopper plugin named Parakeet (Mottaghi and Khalil-Beigi, 2022) which can produce a P3 Penrose tiling, however the plugin only offers a single patch of Penrose's Rhombus tiling and does not allow for exploration of other portions of the tiling. Additionally, the plugin has no way of indexing the vertices of the tiling in order to systematically lift and orient the vertices into a third dimension, allowing for the construction of a Wieringa Surface.

3.2 Simulation

The software package AFMG Reflex (AFMG, 2022, October 13) was used to test the acoustic scattering properties for several Wieringa surface case studies throughout this investigation (Fig. 7). AFMG Reflex was selected on the basis of its simulative accuracy and informative results as demonstrated by Shtrepi et al. (2020) and Ajlouni (2018). A rigid two-dimensional slice of a three-dimensional object is subjected to the Boundary Element Method (BEM) for the calculation of acoustic diffusion and scattering coefficients, and modelled in an environment outlined and stipulated by the ISO 17497–1 Standard. Using BEM, it is possible to model acoustic performance on the basis that the linear partial differential equations used within the successive process, mathematically describe the wave-like phenomenon. (Cox and D'Antonio, 2009) With a simple width and height value, geometry is created natively within the program. The Wieringa Surfaces were measured in accordance with the modelling capabilities of the program using slices taken from each iteration. In this experiment, a Wieringa surface was subjected to incoming sound at a 45-degree angle with respect to the surface. Although the AFMG Reflex software is limited to two dimensions, it provides a valuable understanding of how a particular surface scatters sound. (Wolfgang and Feistel 2011) We recognize that the AFMG Reflex software is limited to two dimensions and therefore has limitations on how its results can be translated to three-dimensional scattering performance; however, our simulation results do indicate that the Wieringa surface may produce sound scattering and that this performance might be controlled through tuning its geometric properties.

3.3 Digital Fabrication

CNC milling was used as it was the most sensible way of creating a mold that could then be cast using a concrete-like material "Rockite." The material Rockite produces a smooth finish that will reflect most sound waves unlike wood or plastics and isolates the scattering performance away from sound absorptive properties. In the selected iteration, a block with dimensions of $22.5'' \times 22.5'' \times 5''$ represented the material stock to be

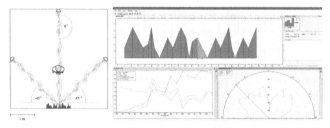

Fig. 7. AFMG Reflex simulation setup and result output

milled. A negative mold was produced into which the Rockite surface would be cast. The plugin RhinoCAM v. 2020 was used to simulate and aid the milling process (Fig. 8a). It was determined that white Styrofoam would be the most economical, convenient, and easy to dispose of material for the mold. Taking advantage of the foam's lighter weight, it was taped to the CNC bed for stability and tests were performed on feeds and speeds. The first pass of the CNC mill was a horizontal roughing at 14000 RPM with a 0.5″ square carbide end mill in the clockwise direction and was controlled at a step over rate of 90%. Feed rates varied between 350–500 in/min. A second pass was then performed to clear the flat tops with the same speed and feed as the horizontal roughing. A 0.5″ ball carbide end mill was then used for parallel finishing at 14000 RPM. A speed of 300 in/min and a step-over control of 5% were used to achieve flat faces for the rhombi with minimal striations (Fig. 8b). For each block of Styrofoam, the total milling time resulted in approximately 4 h due to the low step-over rate in the parallel finishing, however this reduced post-process sanding (Fig. 8c).

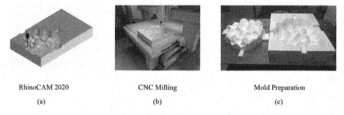

RhinoCAM 2020   CNC Milling   Mold Preparation

(a)   (b)   (c)

Fig. 8. Milling and processing of Styrofoam molds

4   Results

4.1   Acoustic

In total, twelve iterations of the Wieringa surface were developed using our parametric model. A 2D slice of the 3D geometry was taken at the centroid of each surface and modeled in AFMG Reflex to evaluate scattering properties. In the absence of a change in any other factors, an increase in the delta parameter (Fig. 6) improved acoustic performance for both normalized diffusion coefficients and scattering coefficients (Fig. 9a). This is expected as the increase in delta parameter increases the peaks and valleys of the Wieringa surface and thereby allowing for deeper and larger acoustic well depths to occur. This pattern is best observed when the simulated sound source was placed at 45 degrees to the normal. When alpha and beta parameters (Fig. 6) were altered while other variables are left constant there is an observable pattern in the graphs (Fig. 9b). Overall deviation between iterations is insignificant, but local variability between iterations exists.

Consequently, because orientation is not a factor in determining the acoustic performance of a WS, less planning would be devoted to the fabrication of these surfaces since both a positive and negative cast will produce similar acoustical results. This is a great advantage for mass production, as the cost of fabrication is normally proportional to the complexity of the surface's profile. Considering several iterations of the WS' and comparing them to the 1D and 2D Schroeder QRDs, it is evident that the WS' are significantly less effective at lower frequencies (Fig. 9c). For the normalized diffusion coefficient, both 1D and 2D QRDs outperform the WS' for $x < 400\,Hz$, while for the scattering coefficient, QRDs outperform the Wieringa surfaces as much as $x < 500\,Hz$. In spite of this, the WS' are comparable to calculable periodic surfaces at $x > 500\,Hz$ (See Fig. 9c). This would suggest that WS are not sufficient for acoustic scattering at frequencies $x < 500\,Hz$, however at $x > 500\,Hz$ are comparable (and at some frequencies superior to) both forms of QRDs. As previously written, these simulations were carried out using two dimensional BEM techniques and therefore the results cannot be directly translated to the actual three-dimensional geometry of the Wieringa surface. However, we speculate that these results indicate that there might also be similar acoustic properties when these surfaces are simulated using three-dimensional software.

4.2 Fabrication Process

The material Rockite was selected for to its suitable properties to be casted and cured as a solid mass that can absorb negligible acoustic energy. However, working with concrete requires the understanding of physical limitations that must be overcome if there is to be functionality within the fabricated model. The first model was a complete negative out of a stock material, meaning that all the material removed from the stock material was to be replaced by Rockite as a positive molded cast (Fig. 10). In this first experiment, the milled white Styrofoam came out with crisp edges and flat rhombi due to the 5% step-over of the drill bit in the parallel finishing. This was unexpected as it was assumed that white foam would not have the rigidity to create a clean cast due to the forces exerted on it by the CNC mill. White Styrofoam produced a defined, rigid, and affordable mold was produced. A coating of Vaseline was applied accordingly as a releasing agent. The finishes on the model were smooth, the lines between the rhombi were crisp, there was no observable cracking, and the integrity of the model was strong. The only setback of the model was its extreme weight. The mold itself weighed 15.9 kg. Because our purposes of the Wieringa surface was acoustics, we knew that a solid mold cast of Rockite was not the final iteration in the fabrication process. With the insights that we had gained with the first cast it was decided that a compression mold would be the next experiment. A negative mold would be milled out and then a positive of that negative would also be milled, however the positive would be offset by 0.5″. It was hypothesized that this would give a sufficient the wall thickness to uphold the complex surface but also be resilient to

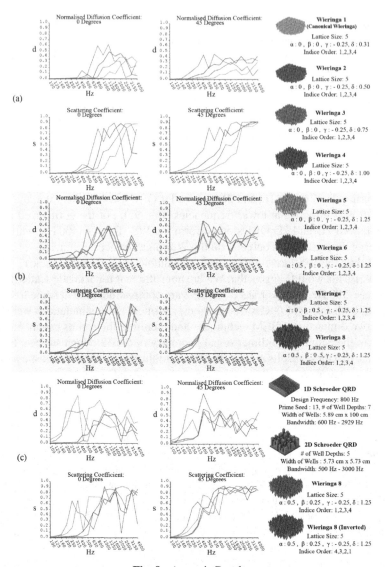

Fig. 9. Acoustic Results

handling, mounting, and small tensile forces placed upon on it. Guides were etched into both the positive and the negative molds in an effort to prevent uneven accumulation of material. While the Rockite mixture, Vaseline application, temperature, and curing time were all equal to the first cast, the second cast was much more difficult to release. Upon release there was noticeable air bubble pockets that had formed at the valleys of the surface which created unwanted aesthetics, flaking, and cracking. It was hypothesized that there was not enough release area within the initial compression of the mold to allow air to escape. Additionally, it was noted that agitating the bass of the mold would be an added part to the fabrication process. Despite these drawbacks, the mold itself was significantly lighter then the first mold, all the while keeping its strength integrity. This allowed us to produce a third mold which we kept the idea of a compression mold, while addressing issues of air bubbles and the release of those bubbles. Fortunately, with the added agitation to the mixture placed in the mold, a redesigning of releasing areas, and a well-developed casting routine, our third iteration was a success. Air bubbles had not created the unwanted aesthetics, flaking, and cracking. Additionally, both the structural integrity and the molds fidelity were retained, thereby establishing a clear lineage of experimentation that ultimately led to a successful prototype (Fig. 10).

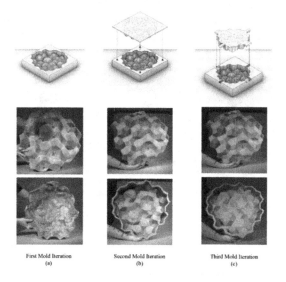

First Mold Iteration   Second Mold Iteration   Third Mold Iteration
(a)   (b)   (c)

Fig. 10. Diagram Explaining Molding Process Discussion

5 Discussion

The formation of a parametric model of the Wieringa surface is an exciting step in design and is a reaffirmation of the intimate relationship that mathematics has with architecture. The age of computation, along with the democratization of mathematical knowledge and tools can now allow us to probe into spaces once deemed unobservable. There are many opportunities for designers to investigate properties and aspects of 5D space with the Wieringa Surfaces and based on the acoustic findings, as well as a freely distributed grasshopper script, we conjecture that they will be beneficial to structures, aesthetics, and acoustics as de Bruijn had once hoped. Beyond the mathematical contributions, we have investigated the acoustic properties of these surfaces through simulation. Due to the scope of this experiment, initial acoustic analysis was limited to 2D analysis and provided ideas as to what can be achieved with the now parameterized Wieringa Surface. While there is confidence in the initial results through AFMG Reflex, it is noted that the

Wieringa Surface is a complicated 3D surface topology. Further research implementing physical testing will need to be conducted to verify results obtained. Additionally, assessments of sound quality within an architectural setting will be required to fully test the postulated acoustic application. While a concrete molding process was investigated in this paper it is not limited to this fabrication option for architectural acoustics. Although not implemented in the fabrication process of this paper, the use of modularity and aggregate assembly has been built into the code for future experimentation and fabrication. The use of Styrofoam as a material is destructive, alternatively, a nondestructive option for molding would involve silicone casting and release. This would be beneficial in a fabrication setting that calls for precise repeatability and mass production. It is important to note that while Wieringa surfaces have an uncountably infinite number of morphologies, there are other aperiodic geometries that extend beyond the plane that can also be explored and exploited (Fig. 11).

Fig. 11. Casted Tile B and A interlocking, respectively

6 Conclusion

The formation of a parametric model of the Wieringa Surface is an exciting step in design and provides opportunities for the exploration of new architectural knowledge. We have discussed the mathematical background to provide other architectural designers with the requisite information to holistically explore these amazing surfaces as well as provide a rational foundation for the hypothesis that such surfaces have useful scattering and acoustic properties. The acoustic simulations conducted revealed that Wieringa Surfaces have the potential for tunable acoustic performance when properly optimized. Our Fabrication experiments were performed to demonstrate avenues of physical construction and knowledge gained from these experiments have been outlined for future refinement. Wieringa Surfaces are a testament to the intimate relationship that mathematics has with architecture, acoustics, and design, and we conjecture that these surfaces will be beneficial to structures, aesthetics, and acoustics as de Bruijn had once hoped.

References

Penrose, R.: The role of aesthetics in pure and applied mathematical research. Bull. Inst. Math. Appl. **10**, 266–271 (1974)

Penrose, R.: Pentaplexity a class of non-periodic tilings of the plane. The mathematical intelligencer **2**(1), 32–37 (1979)

Senechal, M.: What is a quasicrystal. Notices of the AMS **53**(8), 886–887 (2006)

Mackay, A.L.: De nive quinquangula: On the pentagon snowflake. Sov. Phys. Crystallogr **26**, 517–522 (1981)

Mackay, A.L.: Crystallography and the Penrose pattern. Physica A **114**(1–3), 609–613 (1982)

Bracewell, R.N., Bracewell, R.N.: The Fourier transform and its applications, vol. 31999. McGraw-Hill, New York (1986)

Shechtman, D., Blech, I., Gratias, D., Cahn, J.W.: Metallic phase with long-range orientational order and no translational symmetry. Phys. Rev. Lett. **53**(20), 1951 (1984)

Patel, R.: Architectural Acoustics: A Guide to Integrated Thinking, RIBA, London, UK (2017)

Jaramillo, A.M., Steel, C.: Architectural Acoustics. Routledge, Abingdon, Oxon (2015)

Long, M.: Architectural Acoustics, 2nd edn. Elsevier/Academic Press, Boston (2014)

Kuttruff, H.: Room Acoustics, 5th edn. CRC Press, Boca Raton, FL (2014)

Ermann, M.A.: Architectural Acoustics Illustrated. Hoboken, N.J (2015)

Schroeder, M.R.: Diffuse sound reflection by maximum-length sequences. J. Acoust. Soc. Am. **57**(1), 149–150 (1975)

Cox, T.J., D'Antonio, P.: Acoustic Absorbers and Diffusers:Theory, Design and Application, 2nd ed. Taylor & Francis, London (2009)

Peters, P., Nguyen, J.: Parametric acoustics: design techniques that integrate modelling and simulation. In: Proceedings of Euronoise Congress (2021)

De Bruijn, N.G.: Algebraic theory of Penrose's non-periodic tilings of the plane. Kon. Nederl. Akad. Wetensch. Proc. Ser. A **43**(84), 1–7 (1981)

Senechal, M.: Quasicrystals and geometry. CUP Archive (1996)

Karaiskou, A., Tenpierik, M., Turrin, M.: Fine tuning of aperiodic ordered structures for speech intelligibility. In: Proceedings of the Symposium on Simulatiuon In Architecture + Urban Design (2020)

Arau-Puchades, H.: The Refurbishment of Tonhalle. Building Acoustics **19**(3), 185–204 (2012)

Ajlouni, R.: Quasi-periodic geometry for architectural acoustics. Enquiry The ARCC Journal for Architectural Research **15**(1), 42–61 (2018)

Setaki, F., Tenpierik, M., Timmeren, A., Turrin, M.: New Sound Absorption Materials: Using Additive Manufacturing for Compact Size, Broadband Sound Absorption at Low Frequencies (2016)

Kladeftira, M., Pachi, M., Bernhard, M., Shammas, D., Dillenburger, B.: Design Strategies for a 3D Printed Acoustic Mirror. In: Heusler, M., Schnabel, M.A., Fukuda, T. (eds.) Intelligent & Informed – Proceedings of the 24th CAADRIA Conference – Volume 1 (2019)

Nguyen, J., Cop, P., Hoban, N., Peters, B., Kesik, T.: Resonant hexagon diffuser: designing tunable acoustic surfaces by combining sound scattering and Helmholtz resonators. In: Hybrids & Hacceities - Proceedings of ACADIA 2022 (2022)

Cop, P., Nguyen, J., Peters, B.: Modelling and simulation of acoustic metamaterials for architectural application. In: Gengnagel, C., Baverel, O., Betti, G., Popescu, M., Thomsen, M.R., Wurm, J. (eds.) DMS 2022. Springer, Cham (2023). https://doi.org/10.1007/978-3-031-13249-0_19

Whittaker, E.J.W., Whittaker, R.M.: Some generalized Penrose patterns from projections of n-dimensional lattices. Crystallographica: Foundations of Crystallography **44**(2), 105–112 (1988)

Levine, D., Steinhardt, P.J.: Quasicrystals. Phys. Rev. B **34**(2), 596 (1986)

Socolar, J.E.S., Steinhardt, P.J.: Quasicrystals. II. Unit-cell configurations. Physical Review B **34**(2), 617 (1986)

Mottaghi, E., KhalilBeigi, A.: Parakeet. Computer software. Parakeet I Food4Rhino, March 6 (2022). https://www.food4rhino.com/en/app/parakeet

Urban Models and Analysis

Cost-Effective 3D Urban Massing Reconstruction of Public Aerial LiDAR Scans

Jinmo Rhee$^{(\boxtimes)}$ (iD) and Brad Williams

Carnegie Mellon University, Pittsburgh, PA 15206, USA
{jinmor,bwillia2}@andrew.cmu.edu

Abstract. While highly developed metropolitans such as Los Angeles and New York City facilitate the availability of high-definition and costly digital urban models for designers and planners, second-tier American cities lack similar resources for digital design environments. This unbalanced availability of digital urban contexts discourages the utilization of 3D models for urban research planning in less developed cities. In an attempt to make digital urban models more widely available, this paper introduces an approach on publicly accessible aerial LiDAR scans to automate inexpensive 3D reconstructions of existing building mass at the urban scale. The automation includes collecting raw data from an aerial LiDAR scan database and processes their point clouds for architectural representation. With the understanding that a building is represented as a composite of roof planes, building forms are identified and created via a point cloud plane detection and geometrical approximation. Implementing the automation in the widely used design software, we suggest an automation framework of a cost-effective method for building massing in urban scale. By applying the automation framework to five different urban scenarios, we can evaluate the quality of the reconstruction of building masses and limitations. The reconstructions are quite swiftly executed with reasonable accuracy in urban scale.

Keywords: Urban Modeling · Building Forms · LiDAR · Reconstruction · Affordability

1 Introduction

Urban data has long been a core material in urban planning, research, and design. Recently, due to the advancements of survey and software technologies, working with such data is no longer limited to institutes or government authorities, and individual researchers can construct, process, manage, and distribute their own urban data. In light of this development, the demands for urban data are becoming varied; different and specific types of urban data are in demand based on particular urban situations. In response to this demand, large metropolitan areas such as New York City and Los Angeles have constructed their own digital three-dimensional massing models of their cities. Since these models are typically generated from high-resolution aerial LiDAR (Light Detection and Ranging) or satellite images from SAR (Synthetic Aperture Radar), the individual building massings within these models are highly detailed.

M. Turrin et al. (Eds.): CAAD Futures 2023, CCIS 1819, pp. 207–218, 2023.
https://doi.org/10.1007/978-3-031-37189-9_14

Although there have been recent advances in automation methods for generating building mass models for entire countries [1], such approaches still involve substantial computational costs, which can be a significant barrier for individuals or private companies seeking to undertake such tasks. The high cost is a prohibitive constraint for smaller, second-tier American cities—smaller than larger metropolitan cities in economy, population, and territory—to develop their own 3D building data. Therefore, these cities rely mostly on LiDAR data from federally produced surveys, which cover the entirety of the United States, but have lower resolution than the aforementioned regional custom surveys. Although the federal data is publicly accessible, it requires additional expensive processing to make it useful towards urban planning, research, and policy making.

Open 3D data—e.g., OSM3D (Open Street Map 3D)—and CityGML[2] can be alternative ways to obtain building mass data in smaller, second-tier cities. Urban planners, researchers, designers, and architects have also relied on open 3D data [3–6]. However, in both CityGML and OSM, several metropolitan areas—e.g., New York City, Montreal, and Berlin—have detailed information of building components. However, most second-tier city areas provide mere simple extrusions of building footprints without detailed descriptions of roof shapes. Alternative methods to obtain building mass are either manually modeling buildings in a targeted site, or employing GIS services. These methods can provide the lacking details but are limited in coverage from a cost perspective.

There have been technical efforts to reconstruct building mass using aerial LiDAR data of roof shapes. Certain techniques demonstrate great details of reconstructing building mass such as roof parapets, chimneys, staircases, etc. However, these techniques are not compatible with popular software used in design and planning; they are conceptual studies and have yet to be distributed in the format of stand-alone or plug-in software.

In this research, we have developed a framework to automate the reconstruction of building mass from public lidar data, in order to provide reasonably detailed building massings in urban-scale models of second-tier American cities. We have focused on the framework to have light computations within a widely used 3D modeling software, enabling users to automatically generate building massings at the urban scale without large costs and heavy computation powers.

2 Background

LiDAR is a technique for determining ranges of targeted objects from a laser emitting source. The source, a scanner, repeatedly emits a laser and measures the time of flight for each instance in order to compute varying distances from its location. A single range measured from the scanner is represented as a 3D point, and the composite of all ranges measured is collected as an aggregation of points called a point cloud. Point clouds are one of the most popular representations of 3D objects or spaces. The resolution of the object is dependent on the number of points used to represent it. A high resolution is time intensive to create and requires greater computational resources to store and represent the data, increasing scanning costs. Conversely, a low resolution is more efficient but is less accurate in representing the scanned object.

LiDAR is widely used to survey 3D space and cities due to its inherent laser precision. USGS (United States Geological Survey) executes a federal scale land survey using

LiDAR attached to an aircraft. The government regularly records and traces the entire country using LiDAR. The results of the survey are published on their on-line repositories in different forms of data such as LAS (LASer point cloud data), DEM (Digital Elevation Model, or DSM (Digital Surface Model).

Modeling 3D buildings from LiDAR data can be understood in three different ways: geometrical, vision-based, and learning-based. The geometrical approach is reconstructing building mass by tracking the roof shapes using geometrical algorithms from parsed point clouds. For example, segmented points of the roof shapes are reconstructed using α-shape, extracting outlines of the shape, constructing topological graphs, and geometrically simplifying the polygons of the roof shape [7, 8]. Intersections and boundaries extracted from the points can be used for reconstructing building mass through binary space partitioning—set operation in 3D space [9]. Clustering algorithms and Voronoi neighborhood techniques are used to segment the points by the plane of roofs and construct adjacency of the planes [10].

The geometrical approach demonstrates a reasonable quality of reconstructed mass using comparatively low computation power than vision-based or learning-based. The approach does not require heavy usage of the graphic processing unit (GPU). If the input LiDAR file has high resolution, the approach can efficiently yield well-detailed building masses of multiple buildings in an area. However, the approach can have limited generalization in the reconstruction; since the approaches rely on sensitive parameters for computing geometry, they have shown limitations when it comes to reconstructing the diverse building forms in different cities, requiring manual changes of the parameters.

Vision-based approach is using computer vision technology to detect geometrical features and construct building mass using the features. Computer vision is used to trace the contours of buildings in rasterized point clouds with the color schemes of elevation—the higher altitude of the point, the lighter color [11] and support the detailed segmentation of point clouds and regularization of the roof shapes [12]. These approaches mostly perform well in generating clean mass of diverse forms of buildings in a large-scale area, unless the input image data for vision technology has high-resolution.

The learning-based approach uses a complex probabilistic model to reconstruct building mass by training the point cloud data. The model can work as a supportive or generative tool for the reconstructing building mass. For example, a machine learning model can provide the information of a building's program which will be used to decide the types of roof planes' heights [13]. Deep learning models supported the segmentation of roof shapes [14] or directly converted point clouds to polygon meshes [15]. These approaches can result in the accurate reconstruction but require high computation power and huge cost for training the models, heavily relying on GPU.

Although there are many different ways to reconstruct building mass from LiDAR data, we used the geometrical approach, focusing on the cost-effectiveness of the reconstruction. For this, we developed a framework of reconstructing building mass in larger areas with slow computation.

3　Reconstructing Building Mass

We developed a cost-efficient framework to reconstruct building mass from the federal data of aerial LiDAR. Figure 1 illustrates diagrams of the framework. The framework consists of two parts: front- and back-end. In this paper, we define the front-end as an interface through which users communicate with the software or system, and the back-end as the software or system where data processing occurs without user visibility. For the front-end, we developed custom components in Grasshopper—an algorithmic design plug-in of Rhinoceros 3D—enabling designers to control the reconstruction parameters such as an offset tolerance from a potential plane and minimum points for constituting a plane. The parameters are exported to the back-end where a compiled script receives the parameters, imports point cloud data, and runs a random sample consensus (RANSAC) algorithm [16]. The script returns points segmented by shared planes which are further processed by custom components in Grasshopper to reconstruct the polygon meshes of building masses.

　　Within the context of this framework, the user is required to download a publicly available LiDAR file and subsequently specify the path to this file in Grasshopper. By adjusting the relevant parameters, the user can then generate a reconstructed building mass. This process typically takes a few tens of seconds. The subsequent sections of this paper will provide a detailed explanation of the underlying process that drives the functionality of this framework (Figs. 2 and 3).

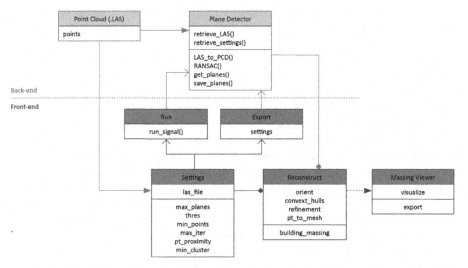

Fig. 1. The pipeline of a cost-efficient reconstruction of building form point cloud data

　　The first step of the reconstruction process is importing the raw LiDAR file consisting of millions of points into Grasshopper. The points are visualized as a point cloud in Grasshopper in gray. By running the backend script, the points are segmented by planes using RANSAC. Points have different colors according to the planes where they belong; if points have the same color, they belong to the same plane. When the plane information

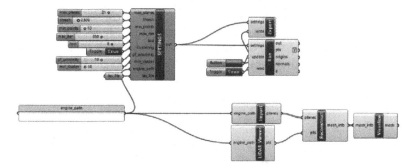

Fig. 2. Custom components as an interface in Grasshopper

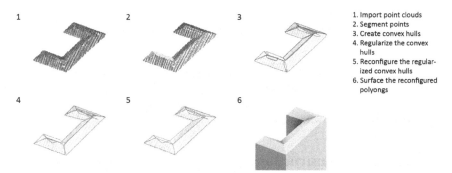

1. Import point clouds
2. Segment points
3. Create convex hulls
4. Regularize the convex hulls
5. Reconfigure the regularized convex hulls
6. Surface the reconfigured polyongs

Fig. 3. The process of reconstructing building mass from LiDAR in Grasshopper

is retrieved in Grasshopper, convex hulls of the segmented points are generated. The convex hulls are planar to each sampled plane.

In order to trace the fitted lines of building roofs, polyline regularization and reconfiguration are required. The regularization is reducing the polyline points aggregated to each other within a range R1 considering slopes on the points. Figure 4 illustrates how the regularization works. Firstly, the red point is selected and other points within the range R1 are selected; in this case, one white point is in the range. By comparing the slope changes at the red point, the approximation algorithm decides whether the white point is removed or not. If the slope change at the white point is smaller than at the red one, the white will be removed.

Fig. 4. Regularization process of convex hulls

The reconfiguration is matching vertices of different polylines within a range R2. For instance, Fig. 4 illustrates three different vertices of three different polylines within 0.5 m. The location of the vertices will be updated into the average point of the vertices. By the reconfiguration, the polylines can be welded without any gap between themselves (Fig. 5).

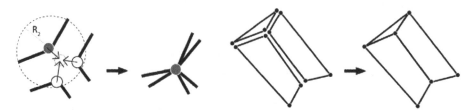

Fig. 5. Reconfiguration process of the regularized convex hulls

The final step of the reconstruction is surfacing. The polyline will be projected to a plane and cut at the level of the terrain. Lofting the roof and projected polylines, the outline of the building mass is generated. Making Delaunay meshes, the polylines will be triangulated and filled with mesh faces.

4   Evaluating Reconstructions

To evaluate our reconstruction technique across urban data, we applied the framework to five cases of different building types with two alternative methods—ContextCapture from Bentley Systems [17] and manual modeling—and observed the results of the automation. The case A, a university arts building, is a single structure with a continuous pitched roof. The case B, a university college complex, is a few structures with multiple, discontinuous, pitched roofs. The case C, a residential neighborhood, contains a few streets of single-family homes with pitched roofs. The case D, a mixed-use urban block, is a collection of flat roof buildings with varying footprint shapes. The case E, a downtown district, is a high-density collection of large commercial buildings and a large variety of heights. See detailed results in Fig. 6.

Our reconstruction results visually indicate that the quality of reconstruction achieved using the proposed framework surpassed that of ContextCapture. However, it is worth noting that the reconstruction quality was still inferior to that achieved through manual modeling. The proposed reconstruction method also outperformed ContextCapture in terms of reconstruction speed. Specifically, the proposed method was able to generate a reconstructed building mass in approximately 30 s, which is faster than ContextCapture's processing time of approximately 1 min. It is important to note that the results presented here do not necessarily suggest that our reconstruction method is superior to Bentley Systems' software. A more comprehensive analysis of these results will be provided in the subsequent Discussion section.

Moreover, we observed that pitched roof buildings exhibit better reconstruction success over flat roofs (Fig. 7). The plane-detection in the reconstruction technique allowed

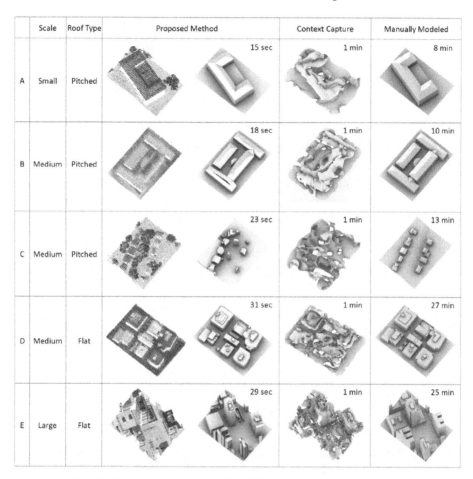

	Scale	Roof Type	Proposed Method	Context Capture	Manually Modeled
A	Small	Pitched	15 sec	1 min	8 min
B	Medium	Pitched	18 sec	1 min	10 min
C	Medium	Pitched	23 sec	1 min	13 min
D	Medium	Flat	31 sec	1 min	27 min
E	Large	Flat	29 sec	1 min	25 min

Fig. 6. Five cases of reconstructing different types of building masses

the individual segments of the pitched roofs to be reconstructed first, which then form a composite representation of the overall roof shape. Flat roofs on the contrary, are detected wholly as a single plane, and their subsequent reconstruction relies on one remeshing computation. Since our process utilizes convex-hulls for surface construction from a plane of points, flat roof geometrical features such as reentrant corners and courtyards are built over. Pitched roof buildings were successfully reconstructed when continuous, as shown on the university arts building. As with the university college complex however, pitched roofs displayed reconstruction errors when a roof ridge was interrupted by an intersecting roof hip. This is due to the fact that the converging valley lines of the hip create an averaged point location, which deviates from the adjacent ridge.

We also observed that larger scale buildings appear more defined than smaller scale buildings. This is directly related to the resolution of the urban point cloud and uniform spacing of points across the city, independent of building types. The downtown buildings benefit from large roof planes which are dense with points and easy to detect, while

Fig. 7. Reconstruction quality comparison between pitched and flat roofs

the small residential buildings struggle to contain enough points to properly represent their roofs. Though the reconstruction technique was able to track most of the planes successfully, the extent of the planes from the given points was often not enough to represent a large enough roof. The results of the residential neighborhood therefore appear more fragmented and incomplete compared to other building types (Fig. 8).

Fig. 8. Reconstruction quality comparison between large and small buildings

Another factor we observed between different urban settings was the amount of noise in the point clouds and its influence on detecting building roofs. Any points that were not used to represent building mass were considered to be noise. The most obstructive form of noise was from surrounding trees and vegetation. In the arts building, a few trees were easy to parse out from the building, because they were adjacent to one another with little overlap. Additionally, the size of the trees was much smaller than the building in question. In the residential neighborhood however, the noise from trees dominated the data to an obstructive extent. The trees were dense, overlapping the houses, and much larger than structures; significantly, the trees were often taller than the houses, which prohibited the LiDAR scans to obtain points of the structures below. For all other urban scenarios, tree noise nor additional noise was present and did not cause error (Fig. 9).

In all urban contexts, we focused the study and analysis solely on the building forms and simplified the terrain. In each sample of data, the points representing the ground plane were normalized as an average elevation and the resulting mesh reconstruction was a simple ground plane. While the point clouds contained scans of the surrounding ground and landscapes, we limited the scope of this study to building forms which resulted in this simplification of ground geometry. Furthermore, the ground data was noisy and interfered with the building data. Simplifying the ground eliminated errors associated with this noise.

Fig. 9. Reconstruction quality comparison between different tree noise cases

5 Discussions

By applying the framework to five different urban conditions, we can evaluate the performance of the proposed method and limitations. Generally, our results demonstrate that the proposed reconstruction method is capable of generating reasonably accurate building masses at an urban scale in less than 30 s, effectively eliminating lengthy wait times associated with traditional manual modeling approaches.

However, the evaluation of reconstruction performance does not necessarily imply that the proposed method is inherently superior from a technical perspective. It is reasonable to infer that the observed differences between the proposed method and ContextCapture may be attributed to underlying differences in the systems and methodologies used for reconstruction. ContextCapture requires numerous inputs such as point cloud data from multiple angles, as well as measurement information such as photographs and survey points. However, the publicly available lidar data used in our study often had lower resolution, with missing or limited scan angles.

The proposed method aims to address the unbalanced accessibility and usability of urban modeling information. Our evaluation demonstrates that the proposed method not only leverages lightweight computation and widely used software but also is well-suited to working with public data that may have limited attributes. This highlights the potential of our approach to provide an efficient and effective solution for generating building mass models in urban modeling applications.

Nevertheless, compared to more computationally expensive methods such as deep learning, our automation approach faces challenges in terms of reconstruction accuracy. The most significant factor determining a successful result is consistency of resolution in point clouds. In some cases, the publicly available LiDAR data did not provide sufficient points to define a building's structure adequately. For example, the single-family houses in the residential neighborhood had less than 30 points to define a roof pitch in some cases. Conversely, some buildings had an excess of points, which could positively impact computational processing and speed. The commercial buildings in the mixed-use district for example, were relatively simple in form and nature with large, flat roofs defined with an excess of 1,000 points contributing to one flat plane. This imbalance between insufficient and excessive points can make the reconstruction process more challenging.

Another key challenge of the proposed method is parameter tuning. The first significant parameter affecting accuracy was the threshold for detecting a viable plane. The threshold parameter is an offset tolerance from a potential plane for which points can be considered part of the plane. This parameter is necessary because geometric planes are infinitely thin, but scanned LiDAR points are never perfectly coplanar. Therefore, we must introduce some offset to account for the inconsistency between geometrical definition and reality.

Smaller buildings achieved better reconstruction results from smaller tolerances, while larger buildings achieved better results from larger tolerances. This is largely due in part to the complexity of what each roof is composed of. Smaller roofs, such as a single-family home, are simple single planes. Larger commercial buildings, however, have elements such as parapets, flues, and pavers, among many roof-hosted items. Since we conceptually want to reconstruct one, simple, flat roof, increasing the offset tolerance to include all of these elevational variances was beneficial for simplified reconstruction.

Similar to tolerance, the second significant parameter affecting performance was the amount of minimum points needed to constitute a plane. While a plane requires only 3 points to be defined, we must tell the RANSAC algorithm what the minimum number of points within a detected plane is in order to be substantial enough to be returned for reconstruction. A low minimum points parameter will return many small and random planes, while a large minimum points parameter will struggle to find enough planes to form a reconstruction. Due to the uniform density of points, it became necessary to lower the amount of minimum points required for a plane for single-family roofs, because a typical roof plane might be no more than 100 points.

However, in a downtown urban scenario, there are many insignificant planes comprising 100 points, such as roof-top mechanical equipment. Therefore, increasing the minimum point parameter filtered out unwanted artifacts and returns only significant architectural planes.

6 Conclusion

In this research, we demonstrated a framework for reconstructing building mass from publicly accessible lidar data of second-tier American cities. This framework is appropriate for cities which do not provide detailed massing models of their buildings, or only have models of simple footprint extrusions. The framework was implemented in a widely used design software with light and swift computation of geometry including six steps: 1) importing point clouds 2) segment points by identifying planes 3) creating convex hulls 4) regularizing the hulls 5) reconfiguring the hulls and 6) surfacing the hulls.

By applying this framework to five different urban scenarios, we were able to discover that the geometrical approach in this paper is sensitive to the certain features and parameters in the reconstructions such as building scale, roof shapes, and RANSAC parameters. We might simplify the parametrization in future versions of this framework. Currently, the algorithm requires the user to adjust several parameters to achieve varying levels of results. This was helpful in our process to explore the relationship between the algorithm's mechanics and the point cloud data we were concerned with.

However, the tuning of all parameters for different urban conditions was time consuming and counterproductive towards a swift and accessible framework. Therefore, in future versions, we may compose a set of predefined parameter configurations for common urban scenarios, such as the scenarios explored in this paper. The user might select one option from a list of possible use cases. For example, a user might select "residential," or "large commercial," before executing the reconstruction. Since American cities are commonly zoned according to allowable building scale, this approach might be applicable for most American use cases.

Another approach for the future direction is developing a hybrid framework by supplementing the parameter-sensitive steps. By providing additional guidance for the reconstruction such as footprint information from computer vision techniques and roof type detection from learning-based models, the quality of the reconstructed masses can increase.

References

1. Peters, R., Dukai, B., Vitalis, S., van Liempt, J., Stoter, J.: Automated 3D reconstruction of LoD2 and LoD1 models for all 10 million buildings of the Netherlands. Photogram. Eng. Remote Sens. **88**(3), 165–170 (2022). https://doi.org/10.14358/PERS.21-00032R2
2. Gröger, G., Plümer, L.: CityGML – Interoperable semantic 3D city models. ISPRS J. Photogrammetry Remote Sens. **71**, 12–33 (2012). https://doi.org/10.1016/j.isprsjprs.2012.04.004
3. Kang, J., Körner, M., Wang, Y., Taubenböck, H., Zhu, X.X.: Building instance classification using street view images. ISPRS J. Photogramm. Remote. Sens. **145**, 44–59 (2018). https://doi.org/10.1016/j.isprsjprs.2018.02.006
4. Boeing, G.: OSMnx: new methods for acquiring, constructing, analyzing, and visualizing complex street networks. Comput. Environ. Urban Syst. **65**, 126–139 (2017). https://doi.org/10.1016/j.compenvurbsys.2017.05.004
5. Moosavi, V.: Urban morphology meets deep learning: Exploring urban forms in one million cities, town and villages across the planet. ArXiv (2017)
6. Fleischmann, M., Feliciotti, A., Kerr, W.: Evolution of urban patterns: urban morphology as an open reproducible data science. Geogr. Anal. **54**(3), 536–558 (2021). https://doi.org/10.1111/gean.12302
7. Boltcheva, D., Basselin, J., Poull, C., Barthélemy, H., Sokolov, D.: Topological-based roof modeling from 3D point clouds. J. WSCG **28**, 137–146 (2020). https://doi.org/10.24132/JWSCG.2020.28.17
8. Xie, L., et al.: Combined rule-based and hypothesis-based method for building model reconstruction from photogrammetric point clouds. Remote Sens. **13**(6), 1107 (2021). https://doi.org/10.3390/rs13061107
9. Jung, J., Jwa, Y., Sohn, G.: Implicit regularization for reconstructing 3D building rooftop models using airborne LiDAR data. Sensors **17**(3), 621 (2017). https://doi.org/10.3390/s17030621
10. Sampath, A., Shan, J.: Segmentation and reconstruction of polyhedral building roofs from aerial Lidar point clouds. IEEE Trans. Geosci. Remote Sens. **48**(3), 1554–1567 (2010). https://doi.org/10.1109/TGRS.2009.2030180
11. Bauchet, J.-P., Lafarge, F.: City Reconstruction from Airborne Lidar: A Computational Geometry Approach. Presented at the 3D GeoInfo 2019 – 14thConference 3D GeoInfo (2019). https://hal.inria.fr/hal-02295768. Accessed 30 Aug 2022
12. Cheng, L., et al.: Integration of LiDAR data and optical multi-view images for 3D reconstruction of building roofs. Opt. Lasers Eng. **51**(4), 493–502 (2013). https://doi.org/10.1016/j.optlaseng.2012.10.010

13. Park, Y., Guldmann, J.-M.: Creating 3D city models with building footprints and LiDAR point cloud classification: a machine learning approach. Comput. Environ. Urban Syst. **75**, 76–89 (2019). https://doi.org/10.1016/j.compenvurbsys.2019.01.004

14. Alidoost, F., Arefi, H., Hahn, M.: Y-shaped convolutional neural network for 3d roof elements extraction to reconstruct building models from a single aerial image. ISPRS Ann. Photogrammetry, Remote Sens. Spatial Inform. Sci. **V-2–2020**, 321–328 (2020). https://doi.org/10.5194/isprs-annals-V-2-2020-321-2020

15. Zhang, L., Zhang, L.: Deep learning-based classification and reconstruction of residential scenes from large-scale point clouds. IEEE Trans. Geosci. Remote Sens. **56**(4), 1887–1897 (2018). https://doi.org/10.1109/TGRS.2017.2769120

16. Fischler, M.A., Bolles, R.C.: Random sample consensus: a paradigm for model fitting with applications to image analysis and automated cartography. Commun. ACM **24**(6), 381–395 (1981). https://doi.org/10.1145/358669.358692

17. https://www.bentley.com/software/contextcapture/

Fine-Grained Long-Term Analysis of Resurgent Urban Morphotypes

João Paulouro$^{(\boxtimes)}$ ⓘ and José Nuno Beirão ⓘ

CIAUD, Research Centre for Architecture, Urbanism and Design,
Lisbon School of Architecture, Universidade de Lisboa, Lisbon, Portugal
joao.neves@edu.ulisboa.pt, jnb@fa.ulisboa.pt

Abstract. We present the methodology employed for the study of an architectural type within the Lisbon Metropolitan Area spanning over eight hundred years with two chief epochs of exponential growth and commonly referred today as the gated community. Covering the use of remote sensing, georeferencing, data scraping, contemporary census, early modern period population counts, spatial disaggregation, spatial analytics, network analysis and new network-based socio-urban measures, the methodology offers a single semi-automated design for the reproducible urban analysis of a metropolitan area, culminating with the production of eight urban models covering periods from the late XVI to the early XXI centuries. Evaluation carried out on the models through statistical analysis and probabilistic unsupervised learning methods shows the resurgence of morphological and socio-urban patterns separated in time by centuries of urban development.

Keywords: Urban Morphology · Gated Enclaves · Space Syntax

1 Introduction

One of the more interesting socio-urban dichotomies present in contemporary urbanities is the concurrent manifestation of globalisation practices that have seen cities turned into veritable test-beds of multicultural integration on the one hand, and a rise in socio-urban fragmentation and community fortification on the other. In previous research we have argued that, following Bateson's theories on unstructured liminality, such a dichotomy confronting globalisation with confinement is a sign of a schismogenic urban development, leading to a feedback loop caused by the urban environment whereby, as Madsen & van Naerssen [9] have noted, "multiculturalism is no longer the postmodern thrill of encountering differences and creating new forms of hybrid cultures, but rather as something that separates". A separation that finds its exemplar paradigm in the gated enclave—a type that can be simply defined in space as a single walled collection of long-term habitation possessing an exclusive open sky private commons.

© The Author(s) 2023
M. Turrin et al. (Eds.): CAAD Futures 2023, CCIS 1819, pp. 219–235, 2023.
https://doi.org/10.1007/978-3-031-37189-9_15

Our research shows that this dichotomy however, appears to be not so much related to the individual characterisation of enclaves, but to how they function in synergy. Sadly, whilst contemporary community fortification has enjoyed growing prominence in urban research (as the work of David Harvey, Stephen Graham, Saskia Sassen, or Setha Low attest), the same has rarely translated into scientific theory founded on empirical evidence and reproducible models (the work of Laura Vaughan being a notable exception). Whilst the reasons for this handicap are numerous we can distil them into three main limitations that can be roughly categorised as technological, aetiological, and epistemological, or, in cruder terms, referring to the research instrument, the phenomenon's origin, and its analysis. These have manifested themselves through often incomplete data, hindering the observation of the phenomena; a focus on modest cross-sectional studies leading to limited temporal inferences of the morphotype; and finally, a predominance within the fields of architecture and urban planning of, at worst, what C. P. Snow termed the 'literary intellectuals' in his Rede Lecture, averse to technological innovation or, at best, what Ihde termed the 'Idealists' [6], who consider technology as subsidiary to science, leading in our view to a focus on localised case studies with limited insight on their combined behaviour on the city as intrinsic components of a complex system. In what follows, we will briefly summarise these three limitations in Sects. 1.1–1.3, demonstrate the general methodology applied to address them within the Lisbon Metropolitan Area (AML) in Sect. 2, introduce one of the algorithms applied in Sect. 2.4, followed by some initial results for discussion.

1.1 Observing Enclaves in Place

It is hardly surprising that following a literature review on gated enclave research, it is the technological limitation through data-gathering and observation instruments that is most prevalent, as only recently has the necessary data and technology become publicly available. From the onset of research into the gated phenomenon, which for the sake of brevity we can place as Blakely and Snyder's 1997 *Fortress America* [2], data acquisition has focused almost exclusively on secondary sources, raising questions of bias, imprecision or incompleteness. Blakely and Snyder's research for instance, along with studies that capitalise on their findings, is based on data from household polls carried out by Community Associations Institute members representing the gated enclave home-owners themselves and therefore can hardly be considered impartial. To our knowledge, the scale of their study is comparable only to Atkinson's [1] report on England's gated enclaves and Landman's [8] national survey of gated enclaves in South Africa, both written in the first decade of this millennium and relying on postal surveys of local planning authorities. Whilst targeting public authorities may offer greater impartiality in the survey responses, it is not void of issues, as Landman herself acknowledged. For one, certain local authorities have a vested interest in concealing the true scale of gated enclaves within their territories, but mostly they are simply not aware, due to either inadequate public surveys or a lack of consensus on what constitutes a gated enclave. These limitations were promptly encountered by Raposo [13] during her 2002 study of gated enclaves within the AML, leading her to dismiss public authorities as a viable source of data and turning instead to a survey of real estate publicity. Replacing the interlocutor however, does not solve the underlying issue, namely the reliance on secondary sources for data on an object which is neither easily defined nor discernible. The reason is simple: a survey of real estate publicity reveals only the realtor's desired portrayal of their product to his target audience, leading to certain developments being characterised as gated enclaves when in fact they're not and other developments clearly concealing the fact. Disconcertingly, this particular methodology failed to identify a high number of enclaves, putting the total number initially at 97 developments, significantly below our own survey results that put the number at 328 for the same period.

1.2 Observing Enclaves in Time

The aetiological limitation, involving the search for the origins and ontology of the phenomenon, is patent in the bulk of studies on gated enclaves. In the majority of cases this has led to a misconception that gated developments are a new phenomenon originating in the US during the 1970s. This incredibly short-sighted assumption which was particularly prevalent in early social criticism of gated enclaves has, thankfully, been put to rest by an increasing number of researchers in the past two decades suggesting other comparable examples in history, from the gated commons of the Western European bourgeoisie starting in the seventeenth century, through the ideal communities of the Romantic era, to the various 'total' or 'disciplinary' institutions defined by Goffman and Foucault. These early types however, were all exceptions to the residential norm of the time. They were not the rule, nor did they ever threaten to become the rule. Their influence over the city was relatively minor, in terms of both the population they housed and the urban space they occupied. All these types had their versions in Lisbon, but none can compare with the scale and extent to which contemporary gated enclaves have begun to dominate the landscape since the late 1990s. As our survey will show, to find another single architectural type that had a comparable impact on the enclosure of open space within the city and subsequent formation of private communities, we have to travel back centuries, to before the modern age of company compounds and social housing, before the enlightenment and industrial age that led to the romantic gardens, to what is commonly referred to as a much darker age in Europe and yet was responsible for originating many of the spatial and social institutions that last to this day: to the late medieval age of monastic enclosure.

1.3 Observing Enclaves in Synergy

Where we're lucky to obtain a study relying primarily on spatial primary sources, incorporating the entire geographical extent of the affected territory and applying longitudinal analysis to assess its development, we are nevertheless often faced with a final constraint which we tentatively term epistemological as it goes to the heart of the entire research process, from data gathering, through treatment, analysis, and ultimately, interpretation. Essentially, this is chiefly a result of one's definition of the city and consequently, of how it should be studied. The majority of studies on gated enclaves have so far presented their results as aggregate statistics varying in scale from the metropolitan level to the municipality. Such methodologies have the unfortunate consequence of interpreting gated enclaves as singular independent entities, or, at most, a conglomerate of entities with unknown relations. Consequently, such results only inform a part of reality, ignoring the intricate connections that play a vital part in urban and complex systems where the whole is greater than the sum of its parts. We would argue this has been the case with a majority of studies on architectural typology and is often repeated when researching the city as a singular object, as in Busquet's et al. [3] momentous morphological study of urban grids, employing a methodology summarised as a "series of abstract reflections that are open to varied interpretations". One of the fallacies of such morphological studies is that virtually their entire preoccupation and object of analysis is not the phenomenon they often seek to study (a hypothetical fragmentation of the city caused by an architectural type), but what they consider to be the causation of the phenomenon. They seek answers not in the affected territory, but through observations into what they assume shapes it. This constitutes an inversion of the hypothetico-deductive scientific method, eschewing measurement-based deductive logic in favour of pure inductive reasoning coupled with posterior measurements on a sample of hypothesised generators of the problematique. Such use of latent variable models via direct observation of disparate observable quantities is common in several fields where the object of study cannot be directly observed, either due to ineffective instrumentation or incalculable measurements. These limitations in instrumentation and computational power however, no longer apply to urban phenomena. Furthermore, the use of classical urban morphology consisting primarily of typo-morphological inventories of a limited number of urban elements, devised primarily through figure-ground representations based on *Nolli* maps with limited applicability, producing findings with non-objective interpretation and restricted reproducibility, is simply not compatible with our understanding both of the city as the most complex artificial entity known to man and with the various mature scientific sub-fields that have arisen in the past century specifically to study such complex phenomena.

2   Methodology

Our research on the development of gated enclaves within the AML, home to approximately 2.8 million people, has sought to address these three limitations by augmenting the quality and breadth of previous urban observations, substantially increase the temporal delimitation of the survey, whilst preserving the intricate connections and relations that play a vital part in complex systems. This was accomplished through the development of a methodology targeting bias and incompleteness of previous surveys whilst adhering to the definition of the urban as a complex artificial organism, equipping the researcher, one hopes, with the tools that allow for fine-grained long-term urban analysis, from data acquisition, through treatment, analytics, and interpretation. It does this by addressing all three limitations that have plagued a majority of studies on the problematique through an exhaustive aerial survey of the region, a secondary survey targeting the initial development of a comparable type, and the generation of digital twin models allowing for the analysis of the totality of surveyed elements and their effects on the socio-urban landscape.

2.1 Searching in Place

Our response to the first limitation—quality, integrity, and completion of observations—was implemented through an aerial survey of gated enclaves in the AML undertaken between 2016 and 2017. Rather than relying on existent or incomplete public records or publicity material, our intention was to survey *de facto* gated enclaves through a spatial definition stipulated as: clearly cordoned long-term habitations or collection of habitations, possessing a common open sky area, accessible by the majority of the inhabitants whilst barred to the outside public. The survey, based on ortho-rectified aerial imagery provided by Google, was carried out on a GIS platform through manual examination of the built landscape occupying a virtual square grid consisting of 11 688 0.25 km^2 cells, providing a viewing scale of approximately 1:2 000 which we deemed sufficient to detect any potential gated enclaves for further inspection and final validation through street-level imagery or direct field observation. Whilst the use of remotely sensed data is more popular in other fields of study, the increase in freely available high resolution imagery has made it a viable means of observation in architecture and urban studies. This resulted in the identification, georeferencing, and characterisation of 770 inhabited gated enclaves occupying a total area of 2 260 ha.

2.2 Searching in Time

The multiple conjectures on the origin of contemporary gated enclaves have led to numerous branches within the tree of self-segregated habitation being pinpointed as proto-enclaves or the origin of this type, particularly during the late nineteenth century. We find most of the evidence severely flawed, principally because we can find alternative earlier examples that are at once more complete, mature, and achieved a far greater distribution, in number and in population, within the walls of cities, their fringes and across different continents, nations, and cultures (such as the 10th century Chinese *fang* or the 17th century Latin American reductions). As such we proposed to study the hypothesis that, whilst the notion of self-segregation and potentially its basic morphological representation are at least as old as cities themselves, it would be the development of coenobitic monasticism and its dissemination within Europe's nations and overseas colonies that constitutes the typological aetiology of contemporary globalised gated enclaves as repeatable, persistent and adaptable architectural objects as defined by Moneo and Habraken [11]. This was tested within the geographic delimitations of the AML across an extended temporal delimitation covering the late medieval, through the early and late modern periods, to the current era. The identification of past and present monastic structures within the AML required the consultation of sources ranging from sixteenth century manuscripts to the latest data from the national Architectural Heritage Information System and a welcome recent survey of religious houses within the Lisbon municipality [12]. The precise location, footprint, territorial walls, and entries of the religious houses were subsequently determined through a combination of aerial imagery, street-view photography, and period sources. This resulted in the identification, georeferencing, and characterisation of 153 monastic houses at different points in history, with 140 active at its height during the decade of 1750, occupying a total area of 432 ha.

2.3   A Model for Informed Urban Analysis

The survey results are sufficient to carry out certain morphological classifications of the enclaves or simple aggregate statistical analysis based on administrative boundaries. In order to perform a fine-grained cross-sectional study at the scale of the smallest element, undertake comparative analysis with different past configurations of the city, and unmask morphological and behavioural patterns within the immaterial in-between spaces connecting urban elements, these results were introduced within a digital model of the city augmented with ancillary data acquired primarily from the 1801 and 2011 census. This data was then converted into a population dot map produced by its spatial disaggregation into a continuous surface of geo-referenced demographic information through statistical sampling based on Mennis' [10] Intelligent Dasymetric Mapping, where each geographic point is equal to one person in space, thereby responding to possible modifiable area unit problems and ecological fallacies that may arise (see Fig. 1).

In total, eight geographic digital models of the AML were produced: focusing on the monastic houses and covering the XVII, XVIII pre- and post-earthquake, and XIX centuries: and focusing on contemporary gated enclaves and covering the periods of 1991, 2001, 2011, and 2016. Between the dissolution of the monasteries in 1838 to the mid-1980s, no *comparable* occupation of land by alternative self-segregated typologies took place. As such, these two models can be seen to represent the final epoch of a form of self-segregated living that arose in Lisbon since the Second Crusade of the XII century, and its contemporary spatial equivalent. The analysis of the models—entailing the formulation of a series of methods and algorithms—works within a typical closed control research system, receiving its input from the data collection stage, calibrating the algorithms and finally constructing the specific methods and subsequent geospatial execution. These methods can summarily be partitioned and sorted by ascending scale of analysis into:

Descriptive which uses field and satellite imagery observations to qualify and quantify the principal observable and qualitative features of gated enclaves,

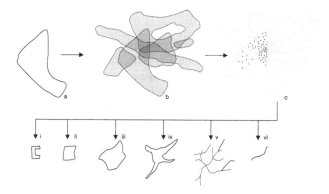

Fig. 1. Multi-spatial demographic metrics. The original census subtract (a) is intersected with various land-use and building footprints (b) which result in a disaggregated population point map containing socio-economic attributes through a process of empirical sampling (c). The resulting point map is then processed to inform multiple spaces, such as individual buildings (i), blocks or gated enclaves (ii), geospatial buffers or clusters (iii), isochrones (iv), sub-networks (v), and street segments (vi).

thereby affording an initial set of indicators and classification of these structures.

Demographic which relies on population statistics and ancillary socio-economic measures obtained primarily from census and scraped public data, disaggregated and combined with quantitative metrics, to inform the spatial and structural patterns revealed within the city.

Morphological which, in conformity with the postulate that the shape of things have an effect on, and are affected by the environment in which they are located, we look at the finer details of the geometry of each individual enclave urban footprint, predominantly through shape metrics stemming from the field of landscape ecology.

Zonal which seeks to chart the extent and geometry of the specific zones within the metropolitan area that are affected by gated enclaves or demonstrate shared configurational attributes and is primarily implemented through the computation of density clusters and isochronic polygons.

Configurational which seeks to examine the structural configuration of the city as a single organism, implemented primarily through space syntax and derived network measures of the vehicular and pedestrian urban structure.

Each method provides its own set of measures and indicators targeting their disciplinary paradigm which are in turn aggregated to explore their effects at multiple scales within areal or linear geometries. In what follows, we present an example of measures and algorithms devised for the configurational method, which is our preferred means of addressing the third limitation.

2.4   Configurational Analysis

If gated enclaves constitute an architectural type, than their morphological characterisation would have us identify such a type in dramatically different moments
in history, spread throughout considerably disparate geographies. And yet, whilst
this characterisation may aid us in identifying the elements of a typological set,
it does not necessarily define nor validate the typology. One of the fundamental
properties of a building type is the measure of its reproduction to the extent
that it has an effect on a much larger relative scale of increasing complexity.
As such, its validation and ultimate definition should derive from measurable
observations of its impact with reality, that is to say, the behavioural, social,
economic, or environmental concerns that are defined by and define the type
and which constitute our principal source of observations once we move away
from the purely formal definition of the element. Consequently, our response to
the final limitation is ultimately built upon an early development of complex systems studies through morphic language and graph theory, establishing a number
of methodologies which Karl Kropf terms the 'configurational approach', and
whose most mature conception is arguably Hillier & Hanson's [5] space syntax.

In addition to computing these syntactic measures, we proceeded to devise
a set of custom indicators, using street segment geometry from standard space
syntax angular segmentation. We were interested in augmenting the information derived from Hillier's methodology by introducing disaggregated population, commerce, and service sector counts onto the network. Our logic is that, as
syntactic measures are often used to portray estimates of pedestrian movement,
we can gain accuracy and specificity if we weight these measures with a social
metric residing within the area of influence of each street segment. As such, an
integrated street segment in a highly populated area would receive a higher value
than a similarly integrated segment in a less populated area. This also has the
advantage that the population coefficient can be qualitative, as it is based on
census surveys, and therefore associated with other demographic attributes. Furthermore, such a process can be replicated for different time periods since such
population counts have been taking place since at least medieval times.

Segment Population (SP), *Segment Service* (SS) and *Segment Commerce*
(SC), are measures derived from a method we propose to introduce demographic
and socio-economic geospatial information into angular street segment analysis.
These measures represent, respectively, the resident population, number of service sector establishments, and number of commerce establishments closest to
each street segment, calculated by Euclidean distance through a Voronoi tessellation of the urban mesh defined by street network segments and intersected
either with a disaggregated dot distribution map of the population (allowing for
further implementation of demographic measures), or the geospatial point locations of the relevant objects being measured (in this case service or commerce
establishments). They can be used either independently or combined with angular segment analysis, producing population- or trade-weighted syntax measures.
When implemented on the street network, we have found our SS and SC measures to be highly correlated with the space syntax measure of *choice* as well as

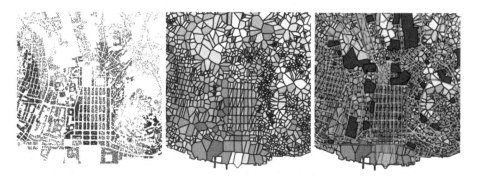

Fig. 2. Integrating population in network analysis. From left to right: disaggregated 1801 Lisbon population dot map; resulting population by street segment voronoi cells; integration of voronoi values on network.

pedestrian count from a modest dataset. Our SP measure however, was found to be inversely correlated to pedestrian movement, presumably an effect of the commuter town to central city pendular effect on street occupation and land value. Together, they afford our network model with some much needed local socio-economic differentiation.

Naturally, one can simply combine demographic data, in its original geographic tract unit with the aggregated street segments intersecting the unit. However, in our case, we aspired to attribute a coefficient to each individual segment, acting therefore on the original data from the space syntax analysis, in the case of combined measures, and on the original street segment geometry alone, in the case of pure population measures. To achieve this, we take advantage of a population dot map produced by the spatial disaggregation of census data into a continuous surface of georeferenced demographic information as previously described in Sect. 2.3. This time, the disaggregated population is assigned to the relevant street segment by partitioning the street network into a set of cells in such a way that the borders of each cell are at equal distance to the next closest segment through the use of Thiessen polygons, more commonly known as Voronoi diagrams. This is achieved through a series of spatial operations summarised below.

Given a space syntax network relation N, an operator pt which returns the vertices of a vector, an operator sp_x which splits a street segment vector geometry g into a series of vectors k at x meter intervals, and an operator vor which returns a two-dimensional Voronoi diagram from the supplied vertices, we first derive the relation R of Voronoi regions R_k for a street network diagram through the following expression:

$$R = \Pi_{\mathrm{vor}\,g}\sigma_{(\mathrm{pt}(\mathrm{sp}_6\,g))}(N) \tag{1}$$

Following that, deriving the sample of a population within the sphere of influence of a network segment part k is a simple operation of counting the number of previously disaggregated population points in D within each Voronoi region R, such that the relation A containing the a attribute count of elements closest to a street segment part k can be had calculating:

$$A = \Pi_{\rho_a \text{ count } d_g \in R_k} \sigma_a \left({}^{R \bowtie D}_{g \cap g} \right) \tag{2}$$

We nevertheless need to account for the fact that our street segments are generally split into multiple lines (to afford length consistency and diminish edge effects), and that a Voronoi cell may be intersected by more than one street segment. This means that the population of a Voronoi cell belonging to a specific street segment, may be acting upon either a large or small percentage of that segment. Depending on the size of the region, the length of the segment will also differ. Finally, if other street segments are intersecting the cell, they may be more affected by the population depending how much of the cell they intersect. We therefore cannot simply attach the standardised population from a cell to the intersecting segment. Rather, we need to allocate the population to the intersecting segments according to their relative importance within the cell. We approach this issue by taking the sum of the length of the intersection of individual parts of the segment intersecting the Voronoi cell as S, and by taking the sum of the parts of the intersection of all the segments intersecting the Voronoi cell as C, such that:

$$S = \Pi_{\gamma n, \gamma r, \rho_s} \sum \|g\| \left({}^{N \bowtie R}_{g \cap g} \right) \tag{3}$$

$$C = \Pi_{\gamma r, \rho_c} \sum \|g\| \left({}^{N \bowtie R}_{g \cap g} \right) \tag{4}$$

Thus far we have created a Voronoi diagram where each cell represents the total number of attribute elements closest to a street segment, be they people, commerce, services, or any unique georeferenced element. To be able to apply these attributes to the network we normalise the sum of the attribute values contained in the set of Voronoi cells representing each unique network segment into a uniform range. Finally, the relation N_w representing the population weighted street network is derived as the sum, for all segments parts f of a unique street segment $n \in N$, of the ratio derived from the division of s by c, calculated in Eqs. (3–4), multiplied by a scaled attribute $a \in A$, to construct the network diagram of attributes, or multiplied by a scaled attribute $a \in A$ and a respective syntactic measure $n_a \in N$, to construct the space syntax attribute-weighted network diagram, such that:

$$N_w = \Pi_{\rho_a \left(\sum_{f \in n} \frac{s}{c} \cdot a \right), \dots, \rho_s \left(\sum_{f \in n} \frac{s}{c} \cdot a \cdot n_a \right)}^{(N \bowtie A \bowtie S \bowtie C)} \tag{5}$$

This final relation essentially distributes a series of attributes (consisting of population, commerce, service, *integration*, *choice*, and weighted combinations of these attributes) through the street network, allowing us to visualise the variation of effect the network has on these properties, establish correlations between street segment measures and gated enclaves, or identify socio-urban patterns (see

Fig. 2). These are further explored through the use of what we term zonal methods which assume that distance and interaction are inversely proportional and therefore hold distance as the primary metric of classification. This premise was used to implement three techniques to determine the spatial extent and morphology of zones deeply affected by gated enclaves, through buffer, isochrone, and density based cluster analysis.

We generated isochrones at a variety of temporal distances for each entry point of each gated enclave within the metropolitan area, allowing us to visualise in one stroke how accessible these enclaves are, how far their accessibility extends, to which urban areas this access is granted, and what urban pockets are excluded. In addition, we generated similar isochrones for the walled perimeter of the enclaves. This in turn allowed us to visualise the reach and area of influence that these walled obstacles and blind façades generate on urban mobility. The isochrone maps we've produced for the AML covering different eras from the fifteenth to the twenty-first century, all focus on answering the same key question, which is to uncover how the spatial-temporal spheres of influence of gated enclaves manifest themselves in the city. Combined with disaggregated or segment data, they become isochronic aggregators of spatio-temporal properties focusing on the space of public movement. For both isochrone and density clusters this can be achieved by calculating the total public street segment attributes within the area of influence as summarised in Eq. 6, where P represents the target area's street segment population estimate, W represents the target area's street segment geometric intersection ratio, and M the original target street segment attributes.

$$A = \Pi_{\left(\frac{P_s \cdot W_w}{\max(P_p)}+1\right) \cdot M_x} (P \bowtie W \bowtie M) \tag{6}$$

3   Results

At the small printed scale, the isochrone maps provide a good macro perspective of the region's enclaves, showing how these are distributed along the system's network, and where and how a single architectural type is responsible for substantial occupation at strategic points within the grid. Figure 3 depicts the area of influence created by gated enclave footprints, represented through isochrones calculated for 5-, 10-, and 15-min intervals for the walled footprint of all 768 gated enclaves found to be active in 2016 within the AML for three different scenarios. These areas of influence can be directly compared to our 1821 model of the same geographic extent, revealing a staggering increase in scale, extent, and coverage of the influence areas. The sum total area of the 15-min pedestrian isochrones increases seven-fold, in line with the upsurge in number of enclaves. And while this area of influence is now more fragmented due to the higher number of enclaves in the contemporary model, the inverse is true once we account for contemporary modes of transportation, where we observe between five isochrone regions covering 37% of the area's landmass for a 15-min vehicle isochrone during peak times to a singular region covering 67% of the metropolitan area's landmass

Fig. 3. AML delimitation in 2016 depicting gated enclaves, in black, and isochrones of enclave walls for the, from left to right, pedestrian network, high traffic, and low traffic vehicle network.

during low traffic. Moreover, it is not just the areas of influence that occupy a significant portion. In certain municipalities for instance, the land occupied by the walled area of gated enclaves exceeds that occupied by all other residential types combined. Whether we focus on the isochrones or the enclave footprints, the growth in urban occupation during both the medieval and contemporary periods is vertiginous.

If we add the street network graduated according to the space syntax measure of global angular segment *choice*, which has been shown to be a faithful measure of pedestrian movement, we achieve a model that is designed to portray the physical and cognitive impact of the enclave walls on the public street network, rather than the accessibility of the enclave entry points, and consequently, is particularly suited as a street-perspective counterpart to spatial analysis. What our observations on the dynamics of the bulk of monastic houses in the region have shown, is that the traditional view that these structures functioned as population attractors playing a vital part in the foundation of new settlements, is not portrayed in either their materiality which is designed to deter entry, the low relative population of their particular locations within the region, the morphology of their footprints which minimise contact with the network, and through their interaction with the local settlements, which portrays a constant that we have found repeatedly in our observations within the AML, century upon century, and produce one of the great contradictions of this typology, one that harkens back to the early prototypical forms of gated enclaves springing in the Nitrian desert at the doorsteps to Alexandria nearly two millennia ago. It is the need to be at once close to population centres whilst yearning to remove themselves from them. This contradiction has been a constant of gated enclaves and its inhabitants, from the earliest forms of asceticism, eremitism, and cenobitic monasticism—they are all a reaction to the city, and therefore, inexplicably linked to it.

Such observations however, continue to be somewhat subjective, hindered as they are by the limitations of the human eye and cartographic visualisation.

Fig. 4. Angular integration in the 2016 AML network and its enclaves. Left: r = n. Middle: r = 2400. Right: Kernel density of NAIN for closest street segment to all gated enclaves (GE), vs all residential buildings (AML).

These become harder to sustain the more complex the network becomes and the more built structures we have to examine. It is necessary therefore to leave the domain of cartographic representations of space and enter the fields of statistical analysis, data visualisation, and probabilistic pattern recognition, where we can handle a much greater quantity of data, visualise multiple scales and measures simultaneously, while generating interactive reproducible computational workflows.

A simple example can be had through a kernel density plot comparing the behaviour in probability density of normalised angular *integration* (NAIN) for the closest street segment to gated enclave walls, to that of all other predominantly residential buildings in the AML. This results in a similar, yet negatively shifted (by approximately 12%), multimodal distribution, whereupon the mean syntactic measure for NAIN for the entire universe of surveyed gated enclaves is lower than that for the remaining residential built types occupying the AML region (see Fig. 4). It seems therefore that gated enclaves are associated with lower levels of street network *integration*. However one views gated enclaves, as urban attractors or detractors, generators or impediments of local socio-urban development, in the AML they are consistently associated with a decreased level of urban *integration* in the street network, creating a disruption in the wider street grid system in addition to the obstructions already created by their walls. This reduction, by operating within the medium and global foreground grid, takes on a life of its own, disseminating its impact across the network to affect a much wider area than other forms of residential constructions are typically capable of, but also to subvert the effects of what are traditionally recognised to be socio-urban attractors.

Integration however, is only part of the story surrounding gated enclaves. At the local scale, where we study the behaviour of through-movement within the local grid systems surrounding gated enclaves through segment angular *choice*, the mutually exclusive constant of enclaves being at once part of the city and apart from it becomes apparent, with reproducible results that are at once more complex to interpret, and pose some uncomfortable questions. It is here that we observe a high frequency of zero-*choice* street segments within the areas of influence of both medieval and contemporary enclaves. These tend to represent the dead-end streets artificially created by the gates. But even if one were to remove the gates, open up the enclaves so to speak, these would remain dead-end streets, reducing *choice* and flow within the network. Little is achieved by opening up enclaves in this way. We also observe differences among these measures depending on whether the foci are the gates, the walls, or the surrounding extramural local network grids. Gates are the most integrated and have the least *choice*, walls have higher levels of *choice*, whilst the perimeter has the highest SP values. This describes a reproducible environment amongst gated enclaves of a strong residential area on its periphery, exhibiting high street population SP values, surrounded by a network with stifled through-movement potential, and exhibiting high relative *choice* but low commercial SC and SS values, leading to enclave gates that are globally integrated within the city whilst locally marginalised. To put it in another way, gated enclaves not only segregate through walls. We have found that they also segregate through the topology of the surrounding road network and the morphology of their urban footprints, and they do so consistently and repeatedly, across several scales, epochs, and diverse locations within the Lisbon Metropolitan Area.

4   Conclusion

If the measure of successful methodology is the questions it unearths rather than those it answers, than we have, unfortunately, met our goal. For the initial dichotomy that informed our problematique has led us to identify a second dichotomy. Why is it that enclave gates, being the area of highest relative *integration*, are also the area with the lowest relative *choice*? It helps to understand what exactly normalised angular *choice* measures in space syntax. It is often used as a predictor of movement, and for good reason, but it is nevertheless based on the graph theory measure of betweenness centrality, which measures the degree to which a node is likely to lie in between others, and therefore more connected to others. The most connected node was initially termed 'the relayer' by Marvin Shaw, representing the point with most potential for control of the system's information flow, precisely because it was the point through which most information passed. The entry points of gated enclaves may be consistently well integrated within the network, but they nevertheless remain isolated from it— they receive little to no information from outside their exclusive private network, and as such, they complete the apparently mutually exclusive constant of being at once part of the city and apart from it. Conversely, the low *integration* and high *choice* present within the sub-grids emanating from the streets lining the exterior of the enclave walls, exacerbated by high SP and low SC and SS values, suggest that whilst gated enclaves are located adjacent to, or artificially create, connection routes between the various sub-centres in their vicinity, the potential for connection is not realised. The 'relayer' has essentially been cut. As such, where G represents the gates, P represents the perimeter of the urban footprint, i and c their respective *integration* and *choice* values, a standard gated enclave type E can be summarised as $E = G_i > P_i \wedge G_c < P_c$.

Combined with the withdrawn circular morphology of enclaves and the exposed structure of the adjacent residential background grid, what we obtain, in laymen's terms, is the prime routes for a series of local highways connecting all neighbouring sub-centres. But this comes at a cost. These local highways are in turn affected by the hermetic qualities of the adjacent enclave walls or blind arcades, which act as both a physical deterrent, on the enclave side, and a cognitive deterrent, on the city side, to local and micro socio-economic activity, depopulating the streets from pedestrians and further increasing the vehicle-centric focus of the road. What we end up with is not only a hermetic occupation of varying extents of land on the part of gated enclaves, but equally the suppression of adjacent pedestrian traffic and socio-economic micro-structures on the very routes that are most appropriate to sustain them within the local background grid subsystems.

The streets that most suffer from socio-urban desertification throughout the Lisbon Metropolitan Area are the streets that were best placed to accommodate it. It is the paradox that emanates from betweenness centrality. For an element to be located within a privileged position of a network system in such a way that communication favours running through it, to be in such a position of control, the element must also be exposed to the system and therefore exposed to attacks by forces within the system. The most effective way of disrupting a connected system is by targeting its most central and exposed node. This is precisely what is taking place with gated enclaves in the AML. The danger here is that removing highly connected nodes can have serious consequences—in the extreme it can split the global system into multiple independent components. And yet there are hundreds of gated enclaves, and thousands of circulation routes surrounding them all of which are adding to the disruption and segregation of the system. In this sense, one need not wait for some kind of post-enclave counteraction emanating from a feedback loop due to symmetric schismogenesis. The system can be brought down, and is being brought down, by the gated enclaves themselves.

This brings us to a delicate final question. Are gated enclaves a foreign, generic, and universal product of the foreground grid injected into the culturally specific background grid of the city? An initially top-down process descending from global capital and used as mechanisms of neoliberal localisation, in turn reinforced bottom up through a global scalar structuration often termed *glocalization*? Or are they instead, ultimately a universal product, but generated predominantly within the background grid in a bottom-up democratic process, revealing a basic and primordial characteristic of human dwelling and socialisation? The political and socio-economic background of a series of intensified periods of gated enclave development during the past millennia both in Lisbon and other cities we have studied, combined with a predominance of street segments with zero-*choice* values within the immediacy of enclaves, and further taking into account Coates findings of zero-*choice* as a possible indicator of autopoeitic urban genesis in his 1970s Damascene models [4], leads us to suggest that gated enclaves, and their interaction with the city, may represent a basic characteristic of urban development. That perhaps the cities-first social model of trade and interaction that Jacobs [7] favoured should be seen as cities-last, with trade and *integration* a consequence of posterior top-down correction of an unstable, defensive reality.

Acknowledgements. This work is financed by national funds through FCT - Fundação para a Ciência e a Tecnologia, I.P., under the Strategic Project with the references UIDB/04008/2020, UIDP/04008/2020, SFRH/BD/118696/2016.

References

1. Atkinson, R., Flint, J., Blandy, S., Lister, D.: Gated communities in England: final report of the gated communities in England 'New Horizons' project. DCLG, ODPM, London (2003)
2. Blakely, E., Snyder, M.: Fortress America: Gated Communities in the United States. Brookings Institution Press & Lincoln Institute of Land Policy, Washington, Cambridge (1997)
3. Busquets, J., Yang, D., Keller, M. (eds.): Urban Grids: Handbook for Regular City Design. ORO Editions, San Francisco (2019)
4. Coates, P.: Programming Architecture. Routledge, London (2010)
5. Hillier, B., Hanson, J.: The Social Logic of Space. Cambridge University Press, Cambridge (2005)
6. Ihde, D.: The historical-ontological priority of technology over science. In: Durbin, P.T., Rapp, F. (eds.) Philosophy and Technology, vol. 80, pp. 235–252. D. Reidel Publishing Company, Dordrecht (1983). https://doi.org/10.1007/978-94-009-7124-0_16
7. Jacobs, J.: The Economy of Cities. Vintage Books, New York (1969)
8. Landman, K.: A national survey of gated communities in South Africa. Technical report, CSIR Building and Construction Technology, Pretoria (2008)
9. Madsen, K.D., Van Naerssen, T.: Migration, identity and belonging. J. Borderlands Stud. **18**(1), 61–75 (2003). https://doi.org/10.1080/08865655.2003.9695602
10. Mennis, J., Hultgren, T.: Intelligent dasymetric mapping and its application to areal interpolation. Cartogr. Geogr. Inf. Sci. **33**(3), 179–194 (2006). https://doi.org/10.1559/152304006779077309
11. Moneo, R.: On typology. Oppositions: J. Ideas Criticism Archit. **13**, 22–45 (1978)
12. Mégre, R., Silva, H., Lourenço, T.B.: As casas religiosas de Lisboa. Universidade Nova de Lisboa, Lisboa, FCSH (2018)
13. Raposo, R.: Novas paisagens: A produção social de condomínios fechados na área metropolitana de Lisboa. Ph.D. thesis, Universidade de Lisboa, Lisboa (2002)

Street Voids: Analyzing Street-Level Walkability Based on 3D Morphology and Remotely Accessible Urban Data

Elif Ensari[1]([✉]) [iD], José Beirão[2] [iD], and Mine Özkar[3] [iD]

[1] New York University, New York, NY 10012, USA
ee2143@nyu.edu
[2] CIAUD, Research Centre for Architecture, Urbanism and Design, Lisbon School of Architecture, Universidade de Lisboa, Lisboa, Portugal
jnb@fa.ulisboa.pt
[3] Istanbul Technical University, Istanbul, Turkey
ozkar@itu.edu.tr

Abstract. Urban planning and design approaches that aim to leverage walking as a sustainable means of transportation require a thorough understanding of the built environment. Information regarding density, diversity, accessibility, and attractiveness of streets is critical to assess walkability, yet it is also resource-intensive to acquire through traditional methods. We present a computational analysis method that captures and aggregates information on walkability indicators encapsulated in the 3d morphology, street-view imagery, and POI data of streets, using a 3d component called the Street Void [1]. This component builds on the Convex and Solid-Void models [2] which are 3d representations of open-urban spaces informed by the interrelationships between topography, surrounding buildings and other immediate physical boundaries, and facilitates the quantitative evaluation of walkability attributes. The method is unique in that it allows for the walkability evaluation of urban open spaces in the micro level, with a semi-automated algorithm and utilizing remotely accessible urban data. We present the implementation of this analysis on four neighborhoods of Istanbul and Lisbon, demonstrating insight drawn from its quantitative output. The research interconnects knowledge in the domains of computational design, behavioral psychology, urban management, and planning; with the contribution of a novel quantitative analysis of streets to inform urban decision-making processes.

Keywords: Urban Walkability · Urban Morphology · Computational Design

1 Introduction

The morphology of the built environment shapes urban life and captures dynamic and highly representative information necessary to understand the urban context, in order to inform its design, planning and management processes. The quantitative evaluation of the urban built environment, especially when the third dimension is accounted for, can reveal

© The Author(s) 2023
M. Turrin et al. (Eds.): CAAD Futures 2023, CCIS 1819, pp. 236–252, 2023.
https://doi.org/10.1007/978-3-031-37189-9_16

how the street network, topography, buildings, and the open spaces in between work with the infrastructure, services, and the amenities in influencing the urban experience. This paper looks at the three dimensional (3d) morphological attributes of the urban built environment, combined with streetscape and activity data, and proposes a remotely applicable and semi-automated measuring method that is both scalable and less resource-intensive than on-site surveys. The purpose of the method is to evaluate urban walkability and to inform urban design, planning and management decisions for improving the walkability of the urban built environment.

The study firstly enhanced and expanded an existing urban classification and assessment model; secondly, used case studies to explore its application on walkability evaluation; and thirdly, developed urban planning and design recommendations based on the insight drawn from the statistical analyses of case study results. Through this research, we found that (1) 3d morphological attributes are effective in understanding the walkability of urban streets; (2) a combination of characteristics is more meaningful than individual ones in determining walkability (3) a 3d model-based morphological analysis method can be practically utilized in evaluating and improving walkability. Our premise is that 3d model based, automated and remotely applicable evaluations will be ubiquitous and practical in the future due to the increasing availability of 3d data and rapid advances in computation.

The following section provides a literature review on the use of urban 3d representation models, the theoretical background of and preceding work on the 3d analysis method, Convex and Street-Voids further developed in this study, as well as the history of walkability research that our study contributes to. Section 3 presents our methodology, which involves conducting Convex and Street-Void analysis, aggregation of additional urban data to the model's unit of analysis, a definition of walkability indicators based on the Convex and Street-Void model, and their refinement through four case studies. Section 4 lays out the case studies which were carried out in neighborhoods in Istanbul and Lisbon. Finally, Sect. 5 presents our conclusions.

This work was a part of the doctoral studies conducted in partial fulfillment of the requirements for a Ph.D. in Architectural Design Computing [1].

2   Background

This research engages with the broad literature of the methods and domains of application of 3d representation models of the urban built environment. More specifically, it investigates the potential role of a 3d representation model in the evaluation of walkability, intended to inform urban design, planning and management practices.

2.1   3D Representation Models of the Urban Environment

3d representation models emerged as part of computer aided design (CAD) tools and evolved from being merely geometric representations of the built environment utilized for visualization purposes to semantically rich information systems that facilitate uses from design, evaluation, and construction documentation to management [3]. While the initial scale of application for 3d representation models was limited to those of industrial products and buildings, as computation methods, processing power and data acquisition techniques advanced, they began to be applicable in the urban scale as well. On the other hand, GIS tools, linking geometry with geographic location and semantic data managed through databases have been used to represent the built environment in the urban scale for decades [4]. However, advantages for the use cases of 3d GIS and representation models over 2d GIS and other 2d representation models have been reported extensively [3].

The need for 3d representations in the urban scale stems from the necessity to account for the height component of the built environment to create accurate and photorealistic simulations of historic and new planning scenarios, useful both technically in design and engineering studies where multiple aspects of the built urban environment are evaluated, and when communicating with the public in urban renewal processes. State of the art urban 3d representation models, which include 3d GIS, City Information Models (CIM) or Digital Twins, offer these benefits, and are generated faster than ever through automation; integrate geometry with dynamic data and allow multiple users to collaborate, affording them real-time analysis and insight.

3d GIS, CIM and Digital Twin are designations corresponding to digital 3d city models with geometric representations of buildings, topography and other physical elements linked with additional data. 3d GIS adds the third, Z-dimension to the traditional GIS, commonly used for mapping elements citywide or in the larger scale in 2d, linked with their geographical locations and additional attributes which are managed via databases. BIM, used mainly in the architecture, construction, and manufacturing industry, provides the user with a 3d modeling interface that integrates objects with parametrically modifiable attributes such as dimension, material, cost, or scheduling information. CIM offers to integrate such a system within a city-scale and geo-referenced 3d model, facilitating simulation, analysis and management of infrastructure, buildings, and services by multiple stakeholders, supported by data manageable through the system [5].

Digital Twins of cities are used by city agencies in Dublin, Singapore, Helsinki, Zurich, Madrid, The Hague, Budapest, Lyon, Oslo, Porto, Athens, Antwerp, Pilsen and many others in managing transportation, energy production and consumption, air quality, noise, open government data and participatory processes with citizens [6]. Inevitably these technologies will become more widely integrated into the everyday operations of urban planning and management in the near future, despite their resource-intensiveness and cumbersome adaptation processes. The introduction of BIM into the architecture and construction industry was not effortless, yet it constitutes a good example of the adoption of new technology by an established industry, once it proved to be effective in responding to the ever-evolving conditions of the built environment. Considering the greater complexity and ceaseless transformation of cities along with the needs of their growing populations, CIM's offerings to the city agencies and their providers are not only rewarding, but essential.

The value of the city-scale parametric representation models such as CIM, and of the capacity to deal with urban morphology in 3d, is in the inclusion of information regarding the topography, building heights and relationships between layers of vertically superposed systems in the analysis of phenomena. Space syntax, a prominent spatial analysis theory that is used to evaluate street networks in the urban scale [7], has been criticized precisely due to its disregard of the third dimension [8]. Convex and Solid-Voids (CSV) originated from an interest in addressing this gap [9], as a method of not only analyzing the urban built environment in three dimensions, but also utilizing the voids delineated by the urban boundaries as units of analysis, rather than analyzing the solids constituting the urban boundaries as most evaluation methods do.

The motive behind the CSV's theoretical concept was to measure the perceptual qualities of the urban open spaces with a consistent analysis method, for which, a 3d-representation model was developed that could be algorithmically generated, informed by the topography, buildings, and other vertical boundaries defining the urban open space [10]. The smallest 3d units of this model, the Convex Voids (CV), already held the morphological information such as width, length, height, squareness, compactness and fatness. Fatness is the radius of the biggest circle inscribed in a 2D polygon; compactness is the ratio between the perimeter of a polygon and the perimeter of a circle of the same area and squareness is the ratio between area of a polygon and area of its smallest bounding square [2]. CVs' automated aggregation into Solid-Voids (SV) captured additional information on the horizontal and vertical continuity of the open space, the impact of neighboring open spaces with varying enclosure levels on each other and the level of connectivity within the street network based on the number of streets crossing each other at intersections.

This paper presents subsequent research of prior work [1] that introduced additional spatial information to the CSV model such as function, level of activeness and availability of amenities in the urban space. This information is external to the morphology of the urban open space, however, is analyzed in relation to the location of the CSV and its neighboring CSVs within the street network. The final model leverages the morphological attributes inherently analyzed with the CSV model and combines it with additional spatial information to derive street characteristics related with walkability. Additionally, a new unit of analysis was created by the aggregation of the original SV units along street segments, called the Street-Void (STV).

2.2 Urban Walkability Assessment, Urban Planning and Design

Walkability is a term used to describe the level of comfort, safety and pleasurableness of the walking experience afforded by the built environment to its inhabitants. Walkability literature covers methods developed to quantify urban street characteristics with the aim of informing health sciences [11], urban design [12], planning and policy making [13]. While efforts have been made to automate the data collection and analysis to most accurately capture the built environment characteristics that influence the actual walking experience, the problem with the existing methods is that they either disregard the 3d and the micro-scale attributes or require on-site analysis, which makes them resource intensive, non-scalable and non-dynamic.

Automation and scalability of urban analysis is important for its integration with Digital Twins or CIMs which hold the potential to make walkability evaluations of existing and simulated designs for the built environment more accessible to city administrations. However, scale and level of detail becomes critical for the automation of an analysis method, as data collection in high levels of detail is costly and cumbersome, if not impossible to automate. Walkability has been studied looking at different scales and levels of detail in the urban environment, yet only large-scale and low-detail evaluations have been automated through GIS or other algorithmic methods. Generally, the macro-scale measures that are based on indicators such as density, connectivity and land use mix can easily be digitized and calculated through GIS. As the level of detail to be measured increases, obtaining relevant data in GIS format becomes more difficult, hence such measuring methods often rely on in-person audits. While such measures can assess particular features of the built environment in detail, such as how well a sidewalk is maintained or how safe a street crossing is, in-person audits tend to be expensive, inefficient and unreliable [14].

Efforts have been made to adapt urban measures based on detailed and small-scaled features of urban design to GIS [14], and also to automate assessments in this scale using additional techniques and resources like 3d GIS and Google Street View Imagery that account for the third dimension [15]. These are valuable precedents as they introduce algorithmic methods of analysis for the streetscape attributes in the meso-to-micro scale, which were found to be highly influential in the walking behavior of urban inhabitants [15, 16]. These attributes are also easier to modify through local government-level urban design interventions, carrying potential for rapid and meaningful impact on mobility behavior. GIS and other automated methods render analysis in this scale efficient and objective as opposed to surveys and audits that are prone to human error and require extensive resources in terms of time, money and man-hours.

This study aims to inform urban design and policy, presenting an analytical method that fills the gap in walkability research with a semi-automated, remotely applicable, 3d and streetscape-level analysis. We anticipate that this approach also holds potential to identify urban vulnerabilities or opportunities through morphology and remotely accessible data in future applications. The motivation is to create an assessment that is easily adaptable to changes and that has the potential to be integrated into CIM and Digital Twin workflows, hence a parametric, GIS-based model structure was utilized.

3   Methodology

Departing from existing literature, a new categorization of built environment attributes is utilized in this study. The categories were selected to represent the most widely applied walkability measures, and are based primarily on morphological attributes, streetscape elements and amenity information. The categories operationalized are Density, Diversity, Complexity, Human Scale, Enclosure, Connectedness, Shape, Incline, Permeability/Transparency, and Infrastructure Quality.

First, a set of walkability indicators (Table 1), measurable using morphology and street view imagery were developed based on a literature review [1]. Next, through case studies of four neighborhoods, indicator values were calculated using CSV, Space Syntax and street view image analyses, as well as location based social network (LBSN) and location sharing services (LSS) data. The unit of analysis was modified from the units utilized in the existing CSV method from CV and SV, to STV. Finally, recommendations were compiled to inform urban design and planning processes, based on the statistical analyses of the results.

The methods of measurement used are based on 3d morphological analysis of streets and 2d morphological analysis of the street network carried out using the CSV and the

Space Syntax methods respectively, amenity locations present on the open map platform Google Places and streetscape features documented through automated image processing of Google Street View data. These evaluations were adopted firstly to facilitate a semi-automated and remotely applicable walkability analysis to any neighborhood, which has urban information available through government resources and open data platforms. This information includes the topography; building footprints with height information; lot outlines; public and private amenity locations; and street view imagery. The second criterion for the selection of these methods was their fitness for analysis in the street and neighborhood scale. These criteria are in line with the goal of building a detailed yet practical analysis method of the urban built environment characteristics, for which data collection, update and feed-in can be automated as much as possible.

3.1 Convex and Street-Void Analysis

CSVs are 3d representation models of the open spaces in the built environment, developed to capture morphological attributes useful for urban analysis [2, 10]. They are generated through a semi-automated workflow in which the main code operates in the Grasshopper visual programming environment [17] within the Rhinoceros3d CAD software [18] and with input files generated with QGIS [19]. Through CSVs, the urban empty space is compartmentalized as solids, and can be analyzed as object, network and field entities. Five primary entities that the method incorporates are the Convex Space (CS), Convex Void (CV), Solid Void (SV), Facade and Flow (Fig. 1) [20]. The additional Street Void (STV) unit presented in this paper is generated by the aggregation of CVs along street segments, and further adapted to analyze streetscape and use attributes external to but informed by the open space's morphology.

CVs derived from what may be termed 3d-informed convex representations of public space. These are convex spaces, but they contain associated information on terrain elevation as well as on the elevation of the physical elements that frame such spaces. Contrary to traditional Space Syntax convex spaces which are flat and resort to the 'fattest' algorithm (spaces with the greatest inscribed circle [8, 21] as algorithmic definition), 3d-informed CSs resort to a 'fattest + squareness' algorithm in aggregating triangles. The 'fattest + squareness' algorithm is able to recognize junctions as singular spaces defined by the street corners and therefore assess the specific spatial properties of such spaces. These convex spaces are informed because they derive from geographic information and therefore can have associated data, such as data required for studies in walkability. And they are 3d-informed because they contain information on elevation, slope and boundary elevations (façades and other types of spatial limits). This information is used to extrude the CS into a 3d representation called the CV. CV representations inherit the CS data or keep semantic relations to the CS layer.

CVs are aggregated into more complex spatial units called SVs. These aggregations occur whenever three conditions are satisfied: (1) continuity between spaces with a low angular horizontal deviation; (2) continuity between spaces with low vertical deviation; and (3) connectivity between spaces with similar scale of boundary share. SVs tend to cross each other at street intersections forming a network layer and can be characterized by properties such as the number of CVs defining them expressing continuity, the number of connections with other solid voids expressing connectivity and the topological distance to other solid voids expressing network depth. Such networks may be the representational basis for topological analysis. Though similar to Space Syntax and therefore amenable for showing similar mathematical properties, they are peculiar to these models and therefore require independent calculus and interpretation. However, the 3d nature of the representations are particularly fit for analyzing public space at street and neighborhood level, complementing models with the information taken from other GIS layers, as well as inherited data (e.g.: from CVs), and data taken from the bounding elements such as buildings, their façades and other private spaces and limits. The most recent functional software version generates two additional layers – façades and flows – where additional data can be included. Flows represent polylines connecting CS centroids, common edges' midpoints and neighboring spaces´ centroids forming a spatial kind of centerline representation. They allow for the calculation of both the horizontal and vertical shift between spaces, providing (or not) the required conditions to the solid void generation.

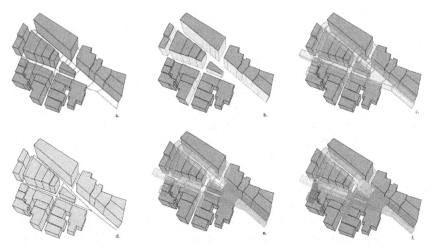

Fig. 1. Convex and Street-Void entities. a: Convex Spaces, b: Facades, c: Convex Voids, d: Flows, e: Solid Voids, f: Street Voids.

The STV was developed as an alternative to SVs, by aggregating CVs along street segments, providing a more consistent unit of analysis for research contexts. STVs only overlap at intersections, unlike SVs and inherit all data from the CVs they contain but the CV values can yield additional information based on the summary method used in this aggregation. Among the indicators developed for walkability assessment in this study, averages, weighted averages, sums and coefficients of variation of data inherited from CVs constituting each STV have been utilized. See Table 1 for the reduced set of attributes measured using the CSVs.

3.2 Street Network Analysis

Following the morphological analysis by the STVs, a 400 m radius Space Syntax analysis of each street segment was carried out and aggregated per STV to assess its connectivity within the spatial network. The indicator values of Connectivity, Node Count, Angular Connectivity, Choice, Integration and Total Depth [22] were aggregated within STVs as weighted averages based on segment lengths and utilized as part of the morphological evaluation and the classification of the street segments.

3.3 Street View Imagery Based Attribute Analysis

To collect information regarding the streetscape characteristics that has been shown in literature to influence the pedestrian experience, street view imagery from Google Street View platform was analyzed for built environment and use-indicating attributes. Street view images were captured every 15 m, facing both sides of the street segments, and the images were fed into an image-processing algorithm [23]. Next, the number of instances where the following elements were recognized by the algorithm was aggregated per STV: tree, landscape, environment, park, door, window, pavement, commercial, business, shopping, chair, bench, furniture, car, vehicle, traffic, calamity, abandoned, demolition and people.

3.4 Street Activity Attribute Analysis

The most intuitive and earliest utilized means to collect data on pedestrian behavior was counting people. Due to the resource-intensiveness and non-scalability of this method, along with the ubiquity and location tracking capability of the cellular phones today, GSM data has become one of the most practical means of tracking pedestrian behavior in open urban space. Yet this type of data was not accessible for our study, so instead, LBSN and LSS data was mapped to identify use patterns in the studied neighborhoods. Flickr, Instagram and Google Place posts were tracked within 3.5-m buffers around the analyzed STV footprints using APIs provided by these platforms and aggregated per STV. These values were used together with the people counts obtained from the street view imagery.

4   Case Studies

Neighborhood sections from Caferaga and Hasanpasa in Istanbul, and Chiado and Ajuda in Lisbon were selected for analysis based on having similar size, road density and built area densities. We selected areas that do not cut through an axis or open space perceived in continuity unless interrupted by administrative boundaries. Additionally, we assumed one neighborhood from each city to be more walkable (Caferaga and Chiado) and one less so (Hasanpasa and Ajuda), with the assumptions supported by density based on built area, diversity based on Google Place frequency, and activity on streets based on number of street sides where people were identified on street view imagery and social media activity based on Flickr posts. Incidentally, Chiado and Caferaga neighborhoods have touristic and commercial characteristics more than the other two.

The built area densities of Caferaga and Chiado considered walkable are about 44% higher and floor area densities are between 80% and 100% higher than Hasanpasa and Ajuda which are considered less walkable. Caferaga and Chiado have shorter streets, therefore smaller blocks, and more frequent intersections than Hasanpasa and Ajuda. Density of Google Place locations are significantly higher in Caferaga and Chiado (0.075/m and 0.064/m) and are very low for Ajuda (0.008/m). Density of Flickr posts are significantly higher in Caferaga (0.066), followed by Chiado (0.029) and are 94% and 80% lower in Hasanpasa and Ajuda respectively. Number of street sides where people were sighted in street view images are double as much in Caferaga (267) and Chiado (120) when compared with Hasanpasa (102) and Ajuda (38).

4.1 Statistical Analysis and Classification of Street Voids

We first conducted a descriptive statistical analysis of all attribute values and looked at the variation between the range and averages across the four neighborhoods [1]. We expected similarities between Caferaga and Chiado as both were considered walkable, and between Hasanpasa and Ajuda as both were considered less walkable. Similarly, we expected contrasting results between walkable and less walkable pairs of neighborhoods within each city. We then ran regression analyses to test the power of measured attributes in predicting pedestrian activity. Based on the results of these two exercises, we selected 22 attributes (Table 1) and ran a K-means clustering analysis [24] of their values for the four neighborhoods. We mapped the STVs in each class and explored their footprint shapes and sizes, street view imagery and the results of a second descriptive analysis of attributes comparatively between classes (Figs. 2, 3 and 4). The walkability-related characteristics of these classes of STVs were consistent and revealed changing impacts of some attributes in different combinations.

In the regression analyses, we utilized Flickr and Instagram post locations and the presence of people on street view images, some of which also appear in recent walkability literature as activity indicators [25, 26], as dependent variables. We tested them all separately, and as combined into one variable, but did not find significant correlations. However, we utilized these results in eliminating attributes with consistently low predicting power across different models of regression analyses.

The K-means analysis seeks to group attributes into a given number of clusters based on the proximity of features in the multi-dimensional space of attribute values [24]. We tested the model with 5, 6, 7 and 8 clusters to group all our STVs from the four neighborhoods within our 22-dimensional attribute space and upon mapping, observing street views and comparing descriptive statistics of the resulting clusters, we decided on the 6 clusters as showing the most consistent STV classification (See Figs. 2, 3 and 4 for examples from the 6 clusters).

Table 1. Selected attributes[1].

Characteristic	Attribute	Explanation
Density		
Physical	STV_FArea_p_STVLen	Total floor area of surrounding buildings per STV length
Use	GPlaces_pSTVLen	Number of Google Place locations that are tagged within 3.5 m of the STV footprint area divided by the length of STV
Diversity		
Morphological	STVs_#FacadesPerM	Number of surrounding Facades per STV length
Land use	GPlaces_pSTVLen	See above
Connectedness		
Space Syntax	WAv_Integration400	Wav of Integration within 400 m radius
(Human) Scale		
	STVs_Area	Footprint area of STV (not projected)
	STVs_Length	Length of STV. (Length of longest continuous street segment.)
	STVs_Width	Av width of STV. (STV area divided by length.)
	STVs_Height	Wav of heights of included CVs
	STVs_#FacadesPerM	Number of Facades per STV length

(continued)

---

[1] Attributes that were eliminated for over-representing some measures or shown to be insignificant in the case studies were: STV perimeter, enclosure, elevation change, flow length/STV area, #buildings/STV length, #buildings/STV area, total floor area/STV area, average floor area per building; weighted averages of flow incline, facade heights, facade widths and Convex Solid (CS) elevations; coefficient of variations for the values of building area, floor area, #floors, CS compactness, CS squareness, CS skyview, CS, elevation, CS diameter

Table 1. (*continued*)

Characteristic	Attribute	Explanation
	STVs_WAv_FacadeArea	Wav of building and wall façade areas surrounding STV
	STVs_Av_Floors	Av number of building floors per STV
	STVs_Av_BArea	Av footprint area of surrounding buildings per STV
Complexity		
Granularity/Articulation	STVs_#FacadesPerM	Number of Facades per STV length
	STVs_WAv_FacadeArea	Wav of building and wall façade areas surrounding STV
	STVs_PerimTArea	Perimeter of an STV divided by its Area
Other streetscape features	Green	Nss where trees, parks, natural greenery, or landscape is identifiable
	Motor_transit	Nss where cars, vehicles or traffic is identifiable
commercial amenities	GPlaces_pSTVLen	See above
Enclosure		
	STVs_Height	See above
	STVs_HeightTWidth	STV_Height divided by STV_Width
	STVs_WAv_FacadeHeightTWidth	Wav of building and wall façade height to width proportions
	WAV_CS_Skyview	Wav of included CS Sky view factors per STV
Shape		

(*continued*)

Table 1. (*continued*)

Characteristic	Attribute	Explanation
	STVs_Compactness	Ratio between the perimeter of STV footprint and perimeter of a circle of the same area
	STVs_WAv_CS_Compactness	Wav of included CS Compactnesses per STV
	STVs_WAv_CS_Squareness	Wav of included CS Squarenesses per STV
	STVs_PerimTArea	Perimeter of an STV divided by its Area
Permeability/Transparency		
	GPlaces_pSTVLen	see above
Infrastructure Quality (and Maintenance)		
	Green	see above
	Pavement	Nss where pavements are identifiable
	Motor_transit	see above
	Negative	Nss where abandonment, demolition or calamity is identifiable

Av = Average, Wav = Weighted Average, Nss = Number of street sides

4.2 Findings and Further Analysis

Each cluster, despite representing spaces from more than one neighborhood, and often from neighborhoods in both Istanbul and Lisbon, showed distinct characteristics legible on the plans, street views and in the descriptive statistical analysis.

Cluster 1 represented relatively well known, active and wide streets; most notably Moda Caddesi from Caferaga, a part of Kurbağalıdere Caddesi from Hasanpasa, Rua de Santa Catarina from Chiado and Calçada da Ajuda from Ajuda, hence, was able to identify most main streets within our selected neighborhoods. The main streets that were even more active than those identified in Cluster 1 and were almost exclusively in Chiado, were identified in Cluster 2. Cluster 3, which represented dense, diverse (high frequency of small facades hence potential for a diversity of building types and uses), well enclosed, and lower traffic STVs with the most favorable attribute values for walkability was most commonly found in Chiado as well. Cluster 4 represented large squares, with lower enclosure but high commercial and pedestrian activity, including the Praça Luís de Camões and Praça de São Paulo as well as smaller squares including Largo Barão Quintela along with their connecting streets, Rua do Alecrim, Rua das Flores, Rua da Boa Vista and Avenida Dom Carlos. The two shape attributes of STVs_Compactness and WAV_CSSquareness showed the highest median values in Cluster 4, hence revealing that these attributes were successful in identifying square-like spaces that are popular public squares. Cluster 5 represented narrow, mainly residential streets with ground-floor commercial amenities similar to the streets of Cluster 3 but were wider and commercially more active. Cluster 6 STVs captured low density, wide, little enclosed, least connected and single-use spaces, mainly in Hasanpasa and Ajuda, the less walkable neighborhoods.

Looking at the descriptive summary statistics of these 6 clusters, which had distinct characteristics relevant with walkability, we saw different attribute values to have varying influence on open spaces of different characteristics. Out of the 22 attributes, we identified 10 to be more closely linked with observable walkability characteristics: STVs_Area, STVs_Length, STVs_Width, STVs_Height, Wav_CS_Skyview, Av_Floors, BArea_pSTVLen, FArea_pSTVLen, Permeability and Negative. We had omitted the Permeability indicator which is measured based on the counts of doors and windows on street view images, as this measure proved unreliable on commercial streets where shop windows weren't identified as windows by the algorithm. However, for residential streets, the frequency of openings was accurately detected, hence we didn't disregard it in the next step. We made a second classification, this time grouping Street-Voids based on their shapes, use, and observed walkability levels. The purpose was to identify the value ranges for the finally selected 10 attributes influencing the walkability in street-like vs. square-like and commercial vs. residential open spaces. Our groups consisted of square-like, commercial and walkable, square-like, residential and walkable, street-like, commercial and walkable, street-like, residential and walkable, and square-like, residential and non-walkable STVs. We did not have many samples of commercial and non-walkable open spaces hence excluded the two groups of STVs that would have been street-like, commercial and non-walkable and square-like, commercial and non-walkable.

We found that within the urban open space network, firstly, the square-like and street-like spaces and secondly, spaces with residential and commercial uses perform differently in terms of walkability, and so can have attribute values of different levels while affording similar levels of walkability. For example, square-like and commercially active spaces in Cluster 4 have lower enclosure levels, larger buildings, wider facades (hence lower possibility for a diversity of building uses) and more frequent visual indicators of "calamity", "abandoned" or "demolition" (all negative influences on walkability) than Clusters 2, 3 and 5, yet spaces of Cluster 4 accommodate a vibrant social life both during the day and night hours and are considered highly walkable open spaces. As another example, comparing two street-like clusters of 3 and 5 which have similar densities and building morphologies, we saw higher levels of activity in Cluster 5 despite its wider streets, lower enclosure levels, and higher motor traffic. Hence, we found that commercial and mixed uses along street-like spaces as in Cluster 5, afford streets a higher level of walkability, even with lower values of walkability-supporting attributes like density and enclosure.

Based on our findings we recommend that open urban spaces first be analyzed for their shape to find whether they are square or street-like (STV_Compactness > 0.7 = square like). Since our results show the Scale and Enclosure attributes to be more influential on the walkability of street-like spaces than square-like spaces; design and planning interventions to increase building heights, human-scale elements like street furniture and green elements that constitute visual boundaries can be prioritized in improving the walkability of street-like spaces. As the second step, whether a space is commercial or not should be determined (GPlaces_pSTVLen > 0.2 = commercial). In terms of the scales of the urban spaces, we found that square-like commercial spaces can be walkable even at sizes as large as 2000 m^2; while square-like residential spaces, when they are smaller than 500 m^2. Street-like spaces are walkable if their footprints (per STV) are below 1000 m^2, regardless of their residential or commercial characteristics. For lower-level interventions than zoning and planning, design and programming of public spaces can help increase perceived levels of Enclosure and Density to make up for the larger-scaled open public spaces for improved walkability. We found that total building footprint area per STV length should be at least 45 m^2 for residential and square-like spaces and 25 m^2 for residential and street-like spaces; the total floor areas per STV area should be at least 80 m^2 for all street-like spaces and 180 m^2 for residential and square-like spaces. Refer to [1] for all analysis result values.

Fig. 2. Street View images and plan from Cluster 3. Hasanpasa (left), Chiado (center, right).

Fig. 3. Street View images and plan from Cluster 4. Chiado.

Fig. 4. Street View images and plan from Cluster 6. Hasanpasa (left), Ajuda (center, right).

5 Conclusion

CSV representations collect information on open public spaces, linked, not only with their geographical and topological location within the network of open spaces, but also with the open spaces' 3d morphological characteristics. In existing research, including those integrating GIS, the data has been either associated with points, buildings, or an abstraction of the street network, such as the street centerlines. STV data is aggregated within street-segment-based 3d representations and may combine point-based georeferenced data, hence providing a much more consistent representation for the analysis of phenomena that takes place within the public space. Furthermore, the method and representation model are able to associate vital information taken from the physical elements that shape the void, like facades along streets, and help evaluate information in relation with the morphology of the spaces, like the variation of building footprint sizes throughout the street segment.

The CSVs, being a rhizomatic network representation, assumes a particular interest for exploratory analysis but also brings an algorithmic complexity problem due to the computation being exponential depending on the size of the model. A simplification of the algorithm is under study to solve this problem. Another vulnerability of the method is that its accuracy depends on the detail and accuracy of the 3d city model fed into the generative algorithm. However, thanks to the improvement of remote sensing and street-level imaging technologies and as the 3d city models are becoming ubiquitous in urban studies and management, access to more reliable models is becoming easier.

The latent potential of the models is still to be explored in most phenomena that take place in open public spaces. Possible future applications include the study of links between the 3d morphology of open spaces with spatial, social and economic indicators relating to public health, crime, pollution, transit ridership or household income, and the provision of improvement scenarios in the forms of land use and zoning interventions.

The contribution of this research to walkability literature and the fields of urban planning, design, and management, firstly, is that it facilitates a 3d, semi-automated and remote evaluation of the open public space. Secondly, it allows for the assessment of all open public spaces, may they be streets, boulevards, intersections, or squares of different sizes. Thirdly, it proposes the need for groups of criteria in the evaluation of open spaces, depending on their morphology and use characteristics as opposed to the uniform sets of criteria applied to the evaluation of all street spaces in existing walkability studies. Overall, the research contributes to the computational design field by an analysis method that links multi-faceted big data with the 3d urban form.

The shortcoming in this walkability evaluation method is that a consistent statistical correlation could not be sought between the CSV data and walking behavior due to the lack of robust mobility data. Furthermore, STV areas do not distinguish the percentage of the surface given to pedestrians nor sidewalk width, values that are known to impact walkability. Planned future work involves testing the correlation of the developed walkability indicators with GSM-based walking data and accounting for sidewalk widths.

Acknowledgements. José Beirão is financed by national funds through FCT – Fundação para a Ciência e a Tecnologia, I.P., under the Strategic Project with the references UIDB/04008/2020 and UIDP/04008/2020.

References

1. Ensari, E.: An Uninterrupted Urban Walk: 3D Analysis Methods for Supporting the Design of Walkable Streets, (2020)
2. Sileryte, R., Čavić, L., Beirão, J.N.: Automated generation of versatile data model for analyzing urban architectural void. Comput. Environ. Urban Syst. **66**, 130–144 (2017)
3. Biljecki, F., Stoter, J., Ledoux, H., Zlatanova, S., Çöltekin, A.: Applications of 3D City models: state of the art review. ISPRS Int. J. Geo-Inf. **4**, 2842–2889 (2015)
4. Maliene, V., Grigonis, V., Palevičius, V., Griffiths, S.: Geographic information system: old principles with new capabilities. URBAN Des. Int. **16**, 1–6 (2011)
5. Beirão, J.N., Montenegro, N., Arrobas, P.: City Information Modelling: parametric urban models including design support data. In: Conferência da Rede Lusófona de Morfologia Urbana. Faculdade de Arquitectura, Universidade Técnica de Lisboa, Lisbon (2012)
6. Charitonidou, M.: Urban scale digital twins in data-driven society: challenging digital universalism in urban planning decision-making. Int. J. Archit. Comput. **20**, 238–253 (2022)
7. Hillier, B., Hanson, J.: The Social Logic of Space. Cambridge University Press (1984). https://doi.org/10.1017/CBO9780511597237
8. Ratti, C.: Space syntax: some inconsistencies. Environ. Plan. B Plan. Des. **31**, 487–499 (2004)
9. Beirão, J.N., Chaszar, A., Čavić, L.: Analysis and classification of public spaces using convex and solid-void models. In: Rassia, S.T., Pardalos, P.M. (eds.) Future City Architecture for Optimal Living. SOIA, vol. 102, pp. 241–270. Springer, Cham (2015). https://doi.org/10.1007/978-3-319-15030-7_13
10. Beirão, J.N., Chazsar, A., Čavić, L.: Convex and solid-void models for analysis and classification of public spaces. In: Gu, N., Watanabe, S., Erhan, H., Hank, M., Haeusler, W., Huang, R.S. (ed.) Rethinking Comprehensive Design: Speculative Counterculture, Proceedings of the 19th International Conference on Computer Aided Architectural Design Research in Asia CAADRIA. pp. 253–262. Kyoto (2014)
11. Chandrabose, M., et al.: Built environment and cardio-metabolic health: systematic review and meta-analysis of longitudinal studies. Obes. Rev. **20**, 41–54 (2019)
12. Ozbil, A., Gurleyen, T., Yesiltepe, D., Zunbuloglu, E.: Comparative associations of street network design, streetscape attributes and land-use characteristics on pedestrian flows in peripheral neighbourhoods. Int. J. Environ. Res. Public. Health. **16**, 1846 (2019)
13. Babb, C., Curtis, C.: Institutional practices and planning for walking: a focus on built environment audits. Plan. Theory Pract. **16**, 517–534 (2015)

14. Harvey, C.: Measuring Streetscape Design for Livability Using Spatial Data and Methods (2014)
15. Carlson, C., Aytur, S., Gardner, K., Rogers, S.: The Importance of the "Local" in Walkability. Buildings **5**, 1187–1206 (2015)
16. Learnihan, V., Van Niel, K.P., Giles-Corti, B., Knuiman, M.: Effect of scale on the links between walking and urban design. Geogr. Res. **49**, 183–191 (2011)
17. Davidson, S.: Grasshopper (2017)
18. McNeel, R.: Rhinoceros (2017)
19. QGIS Geographic Information System (2017)
20. Čavić, L., Sileryte, R., Beirão, J.N.: 3D-informed convex spaces. In: Proceedings of the 11th Space Syntax Symposium. pp. 1–15. Lisbon (2017)
21. Turner, A., Penn, A., Hillier, B.: An algorithmic definition of the axial map. Environ. Plan. B Plan. Des. **32**, 425–444 (2005)
22. Turner, A.: Depthmap 4 (2004)
23. Clarifai Inc.: Clarifai. https://clarifai.com/. Last accessed 13 May 2019
24. Hartigan, J.A., Wong, M.A.: Algorithm AS 136: a K-means clustering algorithm. J. R. Stat. Soc. Ser. C Appl. Stat. **28**, 100–108 (1979)
25. Yin, L., Cheng, Q., Wang, Z., Shao, Z.: "Big data" for pedestrian volume: exploring the use of google street view images for pedestrian counts. Appl. Geogr. **63**, 337–345 (2015)
26. Quercia, D., Aiello, L.M., Schifanella, R., Davies, A.: The digital life of walkable streets. In: WWW'15: Proceedings of the 24th International Conference on World Wide Web, pp. 875–884 (2015)

Visualizing Invisible Environmental Data in VR: Development and Implementation of Design Concepts for Communicating Urban Air Quality in a Virtual City Model

Clara Larsson[1]([✉]) [ID], Beata Stahre Wästberg[2] [ID], Daniel Sjölie[1] [ID], Thommy Eriksson[2] [ID], and Håkan Pleijel[3] [ID]

[1] Division of Informatics, University West, 461 86 Trollhattan, Sweden
clara.larsson@hv.se
[2] Department of Computer Science and Engineering, Chalmers University of Technology, 412 60 Gothenburg, Sweden
[3] Department of Biological and Environmental Sciences, University of Gothenburg, Box 461, 405 30 Gothenburg, Sweden

Abstract. As cities continue to grow, the desire to combine densification with sustainability and greenery may present a challenge to air quality, resulting from reduced ventilation caused by dense buildings and vegetation. To support the careful urban planning required, effective and interactive tools that can visualize and communicate information about air quality to stakeholders are essential. In a transdisciplinary research project aiming to explore such visualizations a prototype pedagogical virtual reality tool was developed, allowing users to explore the impact of aspects of the built environment upon urban air quality. The tool was evaluated with adolescents in upper secondary school through interviews and observations, as well as with the general public through a questionnaire study. This paper provides insights, potential solutions, and initial assessments relevant to data visualization in 3D and immersive analytics in urban planning and stakeholder communication. Identified challenges include difficulties with color association and data distinguishability, and as well as tool complexity relating to the many features requested by experts involved in a transdisciplinary project.

Keywords: Data visualization · Urban planning · Air pollution data · 3D-city model · Virtual reality · Usability study

1 Introduction

An important goal for future cities is to create sustainable, green, and dense urban environments [1]. For this end it is crucial to understand how vegetation, building structures and traffic affect air quality. Digital tools with a high degree of interactivity can promote dialogue with groups not usually involved in urban transformation processes [2] and enhance citizen involvement, thus bridging the gap between planners and citizens [3,

© The Author(s), under exclusive license to Springer Nature Switzerland AG 2023
M. Turrin et al. (Eds.): CAAD Futures 2023, CCIS 1819, pp. 253–267, 2023.
https://doi.org/10.1007/978-3-031-37189-9_17

4]. In this context 3D-city models can provide a better understanding of increasingly vertical cities [5]. Easily comprehensible visualizations that can be explored intuitively and draw public attention, promote engagement, and spread awareness about air pollution [6]. Immersive Virtual Reality (VR) enables embodied interaction, enhances the user experience, and facilitates understanding by relating new information to relevant contexts [7]. However, the range of possible interactions in immersive VR presents a challenge, in particular when combined with visualizations of realistic 3D-environments and associated variations in available views [8]. Interactive visualizations also need to be adapted to the needs of laymen [9]. Given the richness of a medium aiming to imitate reality, the discussion of additional design concepts for presenting data in specific scenarios is a valuable contribution to mapping out the full scope of the field.

The research presented in this paper was carried out within a transdisciplinary, three year-long research project[1]. The project was a collaboration between research and development competences involving environmental sciences, urban planning, interaction design and visualization. The goal of the project was to create a prototype of a virtual tool for identification, visualization, and communication of air quality where users can explore how vegetation, building formations, street widths, traffic and wind directions impact urban air quality across different scenarios.

This paper focuses on learnings from the design and development of complex data visualization in a transdisciplinary project and reports from field tests with intended end users. This includes user feedback on implemented interaction methods and visualization techniques that may inform future work. Here, we present initial results connected to the focus for this paper. Additional analysis is currently ongoing.

2 Background

A central challenge for implementing visualization as a communication tool in urban planning is visual representation [10]. To facilitate communication with the public, information must be presented in an engaging easy-to-understand manner [11, 12]. Traditionally, 2D-maps are used to analyze and communicate environmental impacts from e.g., noise or air pollution. However, for non-experts it can be difficult to connect numbers or abstract color scales in a 2D-map to a real-life scenario [13, 14]. Linked to this is the difficulty of showing different types of data together in 2D-maps to clarify correlation effects between different parameters. With increased densification cities are becoming more vertical, further limiting the use of 2D for conveying detailed information about the situation in a specific area [15].

In urban planning processes 3D-city models are increasingly used for analysis, decision making and communication. Displaying information in 3D allows for a heightened recognition factor – as maps are an abstraction of reality, a 3D-model enables an increased degree of realism [9, 16]. When visualizing abstract invisible parameters, the flexibility of 3D-media is advantageous, since a visualization grounded in reality can be augmented, for example, to display parameters that are invisible to the naked eye. Exploring these 3D-visualizations in immersive VR allows users to be able to interact

---

[1] https://www.mistraurbanfutures.org/sv/projekt/cityairsim-ska-visa-hur-trafik-gronska-och-tatt-byggande-paverkar-stadsluften.

with the model and strengthen the user experience [17]. VR interaction with a 3D-model allows the user to move around and look at a city street or proposed construction from different perspectives [9]. To easily be able to explore presented data in a naturalistic virtual environment can facilitate the understanding of what a scenario would entail in real-life [18]. A user can thereby be aided in connecting the information to a realistic situation, which can make for a memorable experience and have an impact on future actions and decisions [19]. However, it is important to point out that different techniques and tools are better suited for different aspects of communication and target groups [20, 21].

The need for a multidisciplinary approach in the development of real-time 3D-simulation tools is highlighted by Christmann [22], who emphasizes the importance of the human factor and to make data intelligible to both experts and non-experts. There has been an increase in usability studies focusing on the use of 3D-visualizations and VR within urban planning in recent years [14, 20, 22–24]. A variety of projects have been conducted using 3D-visualization to communicate air quality data, aiming to display data for analysis, and raise awareness among the public [25–27]. Teles et al. [25] developed a tool with air quality sensing data using a game-like 3D-environment. Using gamification, their aim was to engage non-experts and increase awareness of air quality. Isikdag and Sahin [26] developed a web based interactive system for visualizing air pollutant levels, focusing on high-volume geospatial data. They acknowledged the importance of visualizing information connected to city objects in different LODs. Ujang et al. [27] studied the use of a spatial 3D-city model for air quality monitoring. They concluded that visualization in 3D will improve the visual analysis for understanding the behavior of air pollutant dispersion. However, improvements could be furthered through identifications and analysis of different layers of air pollution concentrations with a topological data structure in 3D [27].

By visualizing air pollution data in a 3D-city model, an improved understanding of environmental impacts can be conveyed [28]. However, the combination of visualizations of invisible pollution data with a realistic background presents challenges that make it crucial to consider how factors of the visual expression can provide a clear contrast between the visualized data and the background in order to minimize the risk of misinterpretation to convey the information correctly and [8, 29–31]. Such factors include for example choice of shape of the visualized data, such as surfaces, lines, points, volumes, 3D-grids or heatmaps. For the information to be clarified and identified in relation to its context it is often necessary to define different visual properties for each geometric shape, such as textures, patterns, and color [32].

Traditionally, color plays an important role in cartographic visualization [33, 34]. Depending on the target audience, cultural or natural associations are important to consider [35], including divergent color vision. Today, the rainbow scale is widely used for visualizing air quality data [33–35], however this scale is also criticized for being difficult to perceptually interpret [34]. Gautier and Brédif [36] suggest that an interactive 3D-visualization of data therefore should provide the possibility to dynamically change between a rainbow color scale and alternative color scales, based on a range in luminance. Other suggestions include the use of a single-color ranging from low to high intensity, or to use other attributes, such as shapes [35].

The perception of color in a model is also linked to the perspective the visualization will be viewed in. Most 3D-city models today use birds-eye view rather than street-level view. The street-level view is however important for evaluating how complex urban changes affect life in the city [37]. Movement through the model in a street-level view perspective combined with an overall view can furthermore facilitate understanding about how e.g., a new building will affect the surroundings [37, 38].

3 Methods and Development Process

The project was conceived as an iterative, user-centered design [39], mixing methods for evaluation. The application was designed as a communication tool for understanding the impact of greenery in urban landscapes, primarily regarding air quality. Workshops with educators and urban planners were arranged to discuss needs and ideas for implementation. Early prototyping was used together with sketching to communicate complex development tasks involving data visualization, scenario design and VR interaction design. Evaluations of the material were made both in the smaller work package group as well as discussed with researchers, developers, and project partners. Throughout the development the prototype was shown to representatives of target groups. Feedback from these sessions formed a base for refinements. The visual concepts developed here also represent a continuation of previous work by the project participants [30, 31, 40] on developing methods for the visualization of invisible environmental data in 3D.

Methods were developed for visualizing simulated research results, based on an inventory of needs through discussions with representatives of the target groups. The design work focused on two areas: 1) modelling of data (nitrogen dioxide (NO_2) and particulate matter (PM_{10})), where a city model was developed with different scenarios for, e.g., buildings and street width; and 2) representation of data, where concepts for air quality were visualized. The visualizations were combined with the city model in an interactive VR-environment, facilitating interactive and contextual exploration. User tests to evaluate results were part of the refinement process to validate updates and different versions of the application. Due to the Covid-19 pandemic all user testing during the development in 2020 and 2021 was done remotely, using Zoom.

3.1 Modelling of Scenarios and Data

As a case area a district in central Gothenburg was chosen as it currently is considered for future replanning and because it is affected by heavy pollution from the nearby motorway. A baseline model, corresponding to the current status of the case area, was developed in Autodesk 3Ds Max, as well as additional building scenarios along the street (see Fig. 1A–C). These models were combined with surrounding environments, greenery, lighting, and additional environmental features, in Unreal Engine (version 4.26) to produce interactive 3D-environments corresponding to different scenarios. Interaction enabling real-time exploration in VR was also developed in Unreal Engine, as well as concepts for visualization of air pollutants.

The scenarios were designed for the user to explore the impact of different building and vegetation configurations affecting the air quality through air flow and circulation.

Fig. 1. A–C. Screenshots of scenarios from the application, **A)** today's 3-storey buildings, narrow streets, no vegetation, easterly wind, NO$_2$, a) The focused-on street, and b) the adjacent E20/E6 motorway, **B)** wide street without trees, **C)** wide street with row of deciduous trees.

Four fictitious building scenarios were designed, for 5 building variations, and combined with a narrow or wider street to create scenarios covering a diverse set of different preconditions. The different building scenarios included buildings of different heights as well as a potential blocking "barrier building" facing the motorway. Additionally, scenarios with different kinds of vegetation were developed including green walls and different placements of deciduous trees (see Fig. 2).

The visualized data was based on computer simulations and covered a volume of $512 \times 512 \times 64$ m, with one data point for each cubic meter, provided as NetCDF from project participants responsible for the data simulation. In a pre-processing step, this NetCDF data was converted to images that can be read into Unreal Engine as data textures and used in shader-based materials and particle systems (Niagara) for fast real-time visualization of the data. The application was tested and designed mainly with Meta Quest 2, connected to a desktop computer, in mind.

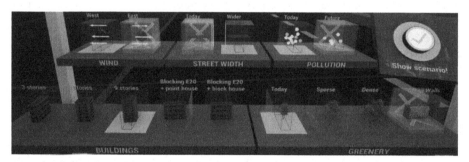

Fig. 2. Boxes that combine to select scenarios. Green is selected. Crosses indicate that boxes cannot be combined, as not all scenarios were included in the prototype (Color figure online).

3.2 Visual Exploration of Data: Visual Design and Concepts for Air Pollution

Concepts were developed for how shape, color and background conditions can be used in a virtual 3D-model for displaying NO$_2$ and PM$_{10}$. Various visual concepts for the visualization of the pollution data were developed and tested. Eventually three concepts were chosen: 1) volumetric particles, 2) large cut planes, and 3) small cut planes. Firstly, particles were spawned in the entire data-volume, colored by, and moving according to the data. Thresholds were used to hide particles outside of a certain range, e.g., to reduce clutter (see Fig. 3A). Secondly, a pair of large cut planes with a heatmap of the pollution levels, one horizontal and one vertical, were implemented to cover the entire data volume in their respective planes. These cut planes could be shown or hidden, with the possibility to move them up/down or back/forward (see Fig. 3B). Thirdly, small cut planes were designed to facilitate examination of data by pointing the hand controller in a direction and firing a cut plane. The cut planes also included a text with the current data value at the center of the cut plane. The cut planes could be held in place by holding down the trigger button on the VR-controller to exactly sample nearby data values (see Fig. 3C).

Fig. 3. A–C. A) Volumetric particles, **B)** a movable large cut plane with a heatmap, and **C)** small cut planes to be fired from the user's hand.

The particle visualization was intended to give a quick overview of the pollution levels, while the cut planes were designed to facilitate a closer and more deliberate exploration. Several attributes concerning shape were investigated such as density and scale of the particles. However, we noted that the nature of a particle system in a realistic 3D-environment made it difficult to distinguish if changes in saliency were caused by color, size, or density. Thus, we chose in the end to only encode the scalar data into color. Tests to enhance visibility of particles were done throughout the development of the prototype, e.g., using different color scales, and encircling particles with a black outline [29].

10 different color scales were developed based on different principles, e.g., light to dark, or transparent to opaque. These were then tested in the model. Some scales were based on scales we developed in previous research projects. Others were, or were adapted from, already existing color scales. They covered different parts of the color circle and included different number of steps to facilitate investigation of how many steps visually worked in the visualization of particularly particles. The criteria for the choice of scales were that they should work in 3D and be easily distinguishable from the environment. Eventually four scales were chosen (see Table 1). Three of these scales (A, B, C) were designed for the most common color deficiencies.

To facilitate presence and enhanced understanding for the visualized data different perspectives were intended to be used in the model. To get a more general understanding of the air pollution data through an overview of the model the user could enter a birds-eye view; to facilitate experience of the model in more detail, and at human scale, a street-level view was included. Users could move freely between the different perspectives, choosing which to use.

Table 1. The four final scales implemented in the prototype version of the application.

	Scale	Reference
A	Diverging lightness scale ranging from blue to red	Adapted from ColorBrewer https://colorbrewer2.org/
B	Sequential scale from light to dark on the yellow-red part of the colour spectrum	Adapted from ColorBrewer https://colorbrewer2.org/
C	Diverging scale from blue to light yellow, adapted for common colour deficiencies	Adapted from Ware (2004)
D	Rainbow scale (Jet)	Scale from Matplotlib https://matplotlib.org/stable/tutorials/colors/colormaps.html

In addition, an investigation was done where specifically the distinguishability of visualized data against a detailed background was studied [29]. Through a Likert scale [41] 87 respondents evaluated the visibility and balance between the 3D-background and data visualization of particles. Screenshots were used with two background lightning conditions (light and dark), two separate color scales (see Table 1, scale B and D) as well as two camera perspectives (birds-eye view and street-level view) [29].

3.3 User Studies

Evaluations of the prototype were conducted during the whole development process, in which the visual visualization of data was one examined aspect. The evaluations were made as weekly reflections within the visualization team, reflections outside the team, in focus group discussions and workshops, lab tests, field tests and in a questionnaire study investigating the impact of background conditions on distinguishability (see Fig. 4).

- **Reflections within the visualization team.** In weekly meetings the team (researchers and developers) reflected together, as well as tested ideas in Unreal Engine. If a specific issue demanded more time, additional meetings or internal workshops were conducted to find a solution.
- **Reflections outside the visualization team.** Meetings with the expanded project network (the different partner organizations, comprising expertise's in science communication, environmental sciences, 3D-visualization, architecture, interaction design)

were conducted throughout the project with e.g., idea generation regarding different aspects of the development process, and problem solving of identified issues in the development.

- **External focus group discussions.** Meetings with stakeholder representatives (e.g., teachers and urban planners) were conducted throughout the project with needs inventory discussions and idea generation regarding various aspects of development. The groups varied in number of participants, from 3–16, and in areas of expertise according to development needs.
- **Questionnaire study.** A questionnaire study about distinguishability of data in virtual environments was conducted between March and June of 2021 with 87 participants from the public, through a master thesis affiliated with the project [29].
- **Lab tests.** Prototype testing was conducted through lab tests in June 2021 where 9 individual user tests were done remotely to identify and resolve errors in the prototype. Interviews were conducted with each user.
- **Field tests.** Field tests were conducted between November 2021 and March 2022 with approximately 190 students, aged 15–18, from 4 schools, as ethnographic studies to observe authentic learning activities with the prototype. Participants were informed about pollution before the sessions and had discussions afterwards. Observation notes were taken during the VR sessions, where participants were in groups of approximately 4. After each session, they were interviewed in groups ranging from 5–15 participants. Both discussions and interviews were recorded and later transcribed.
- **Analysis of results.** The data transcribed from the 15 group interviews and 14 discussion sessions constituted approximately 14 h of recorded material. An in-depth analysis of the transcribed material is currently ongoing. For this paper's purpose, an initial analysis was conducted using a Miro-board to categorize and identify qualitative material relevant to the present topics.

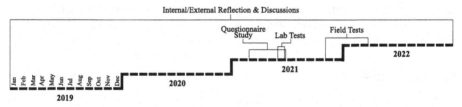

Fig. 4. Graph of user studies timeline from 2019 to 2022, describing data collection. This paper presents initial results. Additional analysis is currently ongoing.

4   Results

The result section is primarily based on qualitative user feedback and observations from the field tests with students connected to the focus for this paper. Results from the questionnaire study on distinguishability are presented briefly as well.

4.1 Interaction and Behavior

Based on observations of the interaction with the visualization tools during the field tests (volumetric particles, small cut planes, large cut planes), almost none of the participants used or showed an interest in using the large cut planes. This is reflected in comments made during the interviews, e.g., *"The wall [large cut plane] [...] wasn't really that useful [...]"* and *"It wasn't that clear on what we had to do to get it [the large cut planes] to work."*

Regarding color scales, few participants were observed to change from the default rainbow color scale of the application for the visualized data. This was noted in the interviews as well where participants, when asked if they tested the different color scales, typically answered with a simple "no."

Both perspectives (street-level and birds-eye view) were used by the participants. The function to switch between them was commonly used, with a few exceptions preferring to stay in one perspective. During the interviews and discussions, the participants elaborated on this behavior stating that, e.g., *"[The volumetric data] was helpful in birds-eye view [...] to see where the most [...] amount of pollution was, to get an overview [...] so that I could go down there [to street-level view] and explore more deeply."*

4.2 Understanding and Preferences

While there were both positive and negative comments relating to understanding of the visualization in VR the majority were positive about the overall experience and reported enjoying using it. Positive comments include that the application was "fun", "engaging" and "a novel approach to learning". Many of the given comments could be reflected in e.g.: *"You got a better idea than if you just [...] listened"*, and *"If it was just a type of computer program that you sit in front of [...], it wouldn't have been as interesting"*. Negative comments can be exemplified with remarks like *"It was fun playing in there, but I didn't learn anything"*. Other critical comments concerned difficulties understanding what the visualized data meant, and the complexity of VR as a medium. Some thought that VR helped their learning, and gave comments such as *"For me, who hadn't tested much in VR before, it was a fun way to learn new things. I probably would have learned less if I had done it in a different way,"* whilst others did not agree, e.g. *"I had probably learned more [...] with lessons [...] because I found it too complicated."*

Several comments regarding the visual design and the visualization concepts with particles, large cut planes, and small cut planes were also identified. Positive comments included that *"It [the particles] is good for showing [data] quickly and clearly."* The particles were not however considered to provide more in-depth information, exemplified with: *"[...] it is clear what [the particles] visualized, but what you only understood was that 'here is a little less' and 'here is a little more', no more than that"*. The functionality of the large cut plane was considered difficult to understand by many. Almost no one indicated a preference for the large planes. The smaller cut planes were generally considered easier to use and provided better understanding of the data because of the written values, e.g. *"I find the numbers [on the small cut planes] clearer, because with just colors it becomes a bit unclear."* Participants also liked how the colors changed as the cut planes moved through the data when fired, as they clarified the pollution levels throughout the environment. Furthermore, feedback indicated that using the small cut planes had a positive effect on their enjoyment and engagement, e.g. *"I think it [small cut planes] was the most fun! To be able to see how it [concentration of pollution] changes [...]. I liked the colors [...] and to see them change, and to be able to see how it [the numbers and colors] changed from one ppm [...] to another ppm, and then understanding that for example up there [above the buildings] there was less [concentrated pollution], and that is better [air quality]."*

Participants who used the default color scale (rainbow) commented that it was easy to understand, and that red meant high values and green and blue meant low values, e.g. *"I liked the colors [of the rainbow color scale]. [...] Blue is better [air quality] and [...] when it comes to red it is less good [air quality]"* and *"We [as people] think that greenish colors are better than red ones. That is the perception I have."* Many participants, however, commented that it was difficult to interpret the meaning of the colors alone, and that they would have liked them to be linked to a practical meaning and the impact on humans, e.g. *"What does the numbers and colors actually mean? What practical implications does it have on the human body?"* Another participant reported having a color deficiency and was unable to distinguish any data based on color.

During the interviews there were some statements regarding distinguishability, e.g., *"[...] when you were in [birds-eye view] it [the particles] was helpful. But, when you were on the ground [street-level view], it was quite difficult to distinguish."* These comments are in line with results from the investigation on distinguishability [29] where particles were noted by respondents as harder to distinguish in the street-level view compared to the birds-eye view. Neither of the two color scales (see Table 1, scale B and D) that were used was shown to be significantly more effective against the 3D-background (p = 0,677). However, both perspective (p = 0,002) and background light condition (p < 0,001) had a pronounced impact in this regard. Participants reported the particles as harder to see with a light background regardless of perspective, but on the other hand, in the dark conditions they reported difficulty seeing the urban environment instead [29].

5 Discussion

5.1 Visual Exploration of Data

While related research areas such as 3D-visualization and immersive analytics are advancing rapidly [8], many aspects of the conditions for such visualizations are changing quickly, e.g., with developing capabilities and applicability of technology, and much remains to be investigated. This paper informs continued research by presenting learnings, with potential solutions and initial assessments, connected to complex usage scenarios, specifically related to urban planning and communication with stakeholders. Further research, building on these starting points, is needed to draw more generalized conclusions.

Among participants, there was satisfaction with the solution using volumetric particles and the small cut planes. One reason for the preference towards the volumetric visualization could be that, as suggested by Gautier and Brédif [36], it provides a quick overview of the entire distribution of data in the model. The volumetric particles were visible as default and the small cut planes were created and manipulated easily, with one button press and direct hand movements. In contrast, few students used the large cut planes. This may be related to the more complicated interaction required for this feature; based on a menu that many participants missed, or because the users did not see the need for it, as they were satisfied with the combination of particles and small cut planes.

Color plays a key role in communicating information. Miscommunication can arise if a color scale is not used thoughtfully, as demonstrated in a study by Weninger [42], where noise pollution levels proved to be assessed higher if brighter colors were used. While most participants in our study were able to use the default rainbow scale to correctly associate which colors represented high and low pollution, it was repeatedly remarked that it was hard to associate colors to a specific pollution level and actual impact. One approach to address this, which was discussed during the project, was to use color scales based on limit values, such as air quality guidelines (AQG), relating to health impacts of air pollutants [43]. AQG are statistical constructs to protect the population from harmful air pollution levels. In reality, health effects do not suddenly appear at a certain pollution concentration. Rather, the health risk increases with exposure level in a way that can be complex. Consequently, it may be challenging to associate environmental risk with colors in a simple way.

When it comes to distinguishability, we did a lot of testing and self-evaluations to create better visual salience, such as giving the particles dark outlines. This improved visibility regardless of perspective, with the dark outline contrasting against bright backdrops and the lighter center against dark backdrops [29]. Another challenge was with how visual representations in a space occlude the space and data behind them [22, 32]. In our case, getting an overview of the particles was facilitated by wind simulation movement, ensuring that these did not obscure the same background space across time. This also avoided potential misinterpretation of random movement [22, 36], which was discussed as an option. To solve this problem, Christman et al. [22] suggests that a user could be given the option to restrict the area showing the air flow.

Concerning the two perspectives, the benefit of including both street-level view and birds-eye view was supported both by observed use and statements in interviews. It showed appreciation for having both options to use according to their preference and needs, and not only provide the traditional birds-eye view, as often in previous research [37, 38]. However, more data is needed to make reliable conclusions.

5.2   The Development and Result of the Prototype

We could see in the user tests that many participants liked to interact with the application and thought it was fun and looked nice. However, visual aesthetics is different from providing optimal understanding, and we realized that herein lays a conflict. While they liked the experience and aesthetics, they did not necessarily completely understand the content of it. Therefore, more work and further study is needed regarding e.g., guidance and selection of viewpoints.

The project had high ambitions regarding prototype and design development, where the aim was to incorporate many features and as many of the intended scenarios as possible. Many features were developed and implemented, but in the end, many features were not used by the participants. This, e.g., included a possibility to adjust color scale spans which few users explored. Perhaps an alternative development method could have been to work with a so-called vertical slice [44], an in-depth section of the application with only a selected few scenarios in focus.

User study comments and observations indicated that the resulting application was perhaps more suited for expert-users than for non-experts. The complexity of the application was a possible result of continuous influence from experts within the project team and from representatives of end-users involved throughout development. Earlier user testing and systematic communication with the actual end-users could have aided in balancing the complexity. This was however difficult to achieve due to the restrictions and safety precautions during the COVID-19 pandemic.

6 Conclusion

This paper explores complex air data visualization using interactive 3D and immersive VR in a transdisciplinary project on urban planning and stakeholder communication. Field test observations and interviews with students who used the implemented interaction methods and visualization techniques are presented, as well as a questionnaire study on distinguishability. One finding was the preference for a combination of volumetric particles and small cut planes over traditional large cut planes. Results indicate that students were satisfied with the quick overview provided by the particles, together with the easily manipulated small cut planes. Challenges were noted with color association and distinguishability. Particles were given dark outlines to improve visibility, and simulated wind movement helped alleviate issues with occlusion. In general, care must be taken that such movement does not misrepresent the data. The visualization included both street-level view and birds-eye view, with participants appreciating both options. However, several features were more suited to expert-users than non-experts, suggesting earlier and more systematic communication with end-users would have aided in balancing complexity.

In summary, this study provides insights, potential solutions, and initial assessments related to 3D-visualization and immersive analytics in urban planning and stakeholder communication. More research is needed to draw generalized conclusions and evaluate the potential of the implemented visualization concepts.

References

1. Garau, C., Pavan, V.M.: Evaluating urban quality: indicators and assessment tools for smart sustainable cities. Sustainability **10**, 575 (2018). https://doi.org/10.3390/su10030575
2. Senbel, M., Church, S.P.: Design empowerment: the limits of accessible visualization media in neighborhood densification. J. Plan. Educ. Res. **31**, 423–437 (2011)
3. Bailey, K., Blandford, B., Grossardt, T., Ripy, J.: Planning, technology, and legitimacy: structured public involvement in integrated transportation and land-use planning in the United States. Environ. Plann. B: Plann. Des. **38**, 447–467 (2011)
4. De Longueville, B.: Community-based geoportals: the next generation? Concepts and methods for the geospatial Web 2.0. Comput. Env. Urban Syst. **34**(4), 299–308 (2010). https://doi.org/10.1016/j.compenvurbsys.2010.04.004

5. Neuville, R., Pouliot, J., Poux, F., De Rudder, L., Billen, R.: A formalized 3D Geovisualization illustrated to selectivity purpose of virtual 3D city model. ISPRS Int. J. Geo-Inf. **7**, 194 (2018). https://doi.org/10.3390/ijgi7050194
6. Mathews, N.S., Chimalakonda, S., Jain, S.: AiR: an augmented reality application for visualizing air pollution. In: Proceedings – 2021 IEEE Visualization Conference – Short Papers, VIS 2021, pp. 146–150 (2021). https://doi.org/10.1109/VIS49827.2021.9623287
7. Moloney, J., Spehar, B., Globa, A., Wang, R.: The affordance of virtual reality to enable the sensory representation of multi-dimensional data for immersive analytics: from experience to insight. J. Big Data **5**(1), 1–19 (2018). https://doi.org/10.1186/s40537-018-0158-z
8. Ens, B., et al.: Grand challenges in immersive analytics. In: CHI Conference on Human Factors in Computing Systems Proceedings, CHI'21, ACM Association for Computing Machinery, The ACM CHI Conference on Human Factors in Computing Systems 2021, Virtual Conference, Japan (2021). https://doi.org/10.1145/3411764.3446866
9. Agius, T., Sabri, S., Kalantari, M.: Three-dimensional rule-based city modelling to support urban redevelopment process. ISPRS Int. J. Geo-Inf. **7**, 413 (2018). https://doi.org/10.3390/ijgi7100413
10. Billger, M., Thuvander, L., Stahre Wästberg, B.: In search of visualization challenges: the development and implementation of visualization tools for supporting dialogue in urban planning processes. Environ. Plann. B: Urban Anal. City Sci. **44**(6), 1012–1035 (2017). https://doi.org/10.1177/0265813516657341
11. José, R.S., Perez, J.L., González-Barras, R.M.: 3D Visualization of air quality data. In: Proceedings of the 11th International Conference "Reliability and Statistics in Transportation and Communication" (RelStat'11), pp. 1–9. Transport and Telecommunication Institute, Riga, Latvia (2011). ISBN 978-9984-818-46-7
12. Jerrett, M., et al.: Spatial analysis of air pollution and mortality in Los Angeles. Epidemiology **16**, 727–736 (2011). https://doi.org/10.1097/01.ede.0000181630.15826.7d
13. Veas, E., Grasset, R., Ferencik, I., Grünewald, T., Schmalstieg, D.: Mobile augmented reality for environmental monitoring. Pers. Ubiquit. Comput. **17**, 1515–1531 (2013). https://doi.org/10.1007/s00779-012-0597-z
14. Onyimbi, J.R., Koeva, M., Flacke, J.: Public participation using 3d web-based city models: opportunities for e-participation in Kisumu, Kenya. ISPRS Int. J. Geo-Inform. **7**(12), 454 (2018). https://doi.org/10.3390/ijgi7120454
15. Hajji, R., Yaagoubi, R., Meliana, I., Laafou, I., Gholabzouri, A.E.: Development of an integrated BIM-3D GIS approach for 3D Cadastre in Morocco. ISPRS Int. J. Geo Inf. **10**(5), 351 (2021). https://doi.org/10.3390/ijgi10050351
16. Judge, S., Harrie, L.: Visualizing a possible future: map guidelines for a 3D detailed development plan. J. Geovis. Spat. Anal. **4**(1), 1–21 (2020). https://doi.org/10.1007/s41651-020-00049-4
17. Papanastasiou, G., Drigas, A., Skianis, C., Lytras, M., Papanastasiou, E.: Virtual and augmented reality effects on K-12, higher and tertiary education students' twenty-first century skills. Virtual Reality **23**(4), 425–436 (2018). https://doi.org/10.1007/s10055-018-0363-2
18. Lamb, R., Etopio, E.A.: Virtual reality: a tool for preservice science teachers to put theory into practice. J. Sci. Educ. Technol. **29**(4), 573–585 (2020). https://doi.org/10.1007/s10956-020-09837-5
19. Szczepańska, A., Kaźmierczak, R., Myszkowska, M.: Virtual reality as a tool for public consultations in spatial planning and management. Energies **14**(19), 6046 (2021). https://doi.org/10.3390/en14196046
20. Glaas, E., Gammelgaard Ballantyne, A., Neset, T.-S., Linnér, B.-O.: Visualization for supporting individual climate change adaptation planning: assessment of a web-based tool. Landsc. Urban Plan. **158**, 1–11 (2017). https://doi.org/10.1016/j.landurbplan.2016.09.018

21. Dübel. S., Röhlig, M., Schumann, H., Trapp, M.: 2D and 3D presentation of spatial data: a systematic review. In: 2014 IEEE VIS International Workshop on 3DVis (3DVis), pp. 11–18 (2014). https://doi.org/10.1109/3DVis.2014.7160094

22. Christmann, O., et al.: Visualizing the invisible: user-centered design of a system for the visualization of flows and concentrations of particles in the air. Inf. Vis. **21**(3), 311–320 (2022)

23. De Klerk, R., Mendes Duarte, A., Pires Medeiros, D., Pinto Duarte, J., Jorge, J., Simões Lopes, D.: Usability studies on building early stage architectural models in virtual reality. Autom. Constr. **103**, 104–116 (2019). https://doi.org/10.1016/j.autcon.2019.03.009

24. Florea, C., et al.: Extending a user involvement tool with virtual and augmented reality. In: 2019 IEEE Conference on Virtual Reality and 3D User Interfaces (VR), pp. 925–926 (2019). https://doi.org/10.1109/VR.2019.8798299

25. Teles, B., Mariano, P., Santana, P.: Game-like 3D visualisation of air quality data. Multimodal Technol. Interact. **4**(3), 54 (2020). https://doi.org/10.3390/mti4030054

26. Isikdag, U., Sahin, K.: Web based 3d visualisation of time-varying air quality information. Int. Arch. Photogramm. Remote Sens. Spatial Inf. Sci. **XLII–4**, 267–274 (2018). https://doi.org/10.5194/isprs-archives-XLII-4-267-2018

27. Ujang, U., Anton, F., Rahman, A.: Unified data model of urban air pollution dispersion and 3d spatial city model: groundwork assessment towards sustainable urban development for Malaysia. J. Environ. Prot. **4**(7), 701–712 (2013). https://doi.org/10.4236/jep.2013.47081

28. Chen, P.: Visualization of real-time monitoring datagraphic of urban environmental quality. EURASIP J. Image Video Proc **2019**, 42 (2019). https://doi.org/10.1186/s13640-019-0443-6

29. Larsson, C.: Point of View: The Impact of Background Conditions on Distinguishability of Visualised Data in Detailed Virtual Environments (Dissertation) (2021). http://urn.kb.se/resolve?urn=urn:nbn:se:hv:diva-16751

30. Stahre Wästberg, B., Eriksson, T., Karlsson, G., Sunnerstam, M., Axelsson, M., Billger, M.: Design considerations for virtual laboratories: a comparative study of two virtual laboratories for learning about gas solubility and colour appearance. Educ. Inf. Technol. **24**(3), 2059–2080 (2019). https://doi.org/10.1007/s10639-018-09857-0

31. Stahre Wästberg, B; Billger, M., Forssén, J., Holmes, M., Jonsson, P., Sjölie, D., Wästberg, D.: Visualizing environmental data for pedestrian comfort analysis in urban planning processes. In: Proceedings for CUPUM 2017 – 15th International Conference on Computers in Urban Planning and Urban Management, Adelaide, Australia, 11–14 July 2017

32. Ware, C.: Information Visualization: Perception for Design. Morgan Kaufmann (2004)

33. Bláha, J.D., Štěrba, Z.: Colour contrast in cartographic works using the principles of Johannes Itten. Cartogr. J. World Mapp **51**, 203–213 (2014)

34. Borland, D., Taylor, R.M., II.: Rainbow color map (still) considered harmful. IEEE Comput. Graph. Appl. **27**, 14–17 (2007)

35. Grainger, S., Mao, F., Buytaert, W.: Environmental data visualization for non-scientific contexts: literature review and design framework. Environ. Model. Softw. **85**, 299–318 (2016)

36. Gautier, J., Christophe, S., Brédif, M.: Visualizing 3D climate data in urban 3D models. Int. Arch. Photogramm. Remote Sens. Spatial Inf. Sci. **XLIII-B4-2020**, 781–789 (2020). https://doi.org/10.5194/isprs-archives-XLIII-B4-2020-781-2020

37. Biljecki, F., Ito, K.: Street view imagery in urban analytics and GIS: a review. Landsc. Urban Plan. **215**, 104217 (2021)

38. Bartosh, A., Gu, R.: Immersive representation of urban data. In: SimAUD Conference Proceedings, SimAUD 2019, pp. 65–68. Atlanta, Georgia (2019)

39. Anderson, T., Shattuck, J.: Design-based research: a decade of progress in education research? Educ. Res. **41**(1), 16–25 (2012). https://doi.org/10.3102/0013189X11428813

40. Stahre Wästberg, B., Billger, M., Adelfio, M.: A user-based look at visualization tools for environmental data and suggestions for improvement—an inventory among city planners in Gothenburg. Sustainability **12**(7), 2882 (2020). https://doi.org/10.3390/su12072882

41. Robinson, J.: Likert scale. In: Michalos, A.C. (ed.) Encyclopedia of Quality of Life and Well-Being Research, pp. 3620–3621. Springer Netherlands, Dordrecht (2014). https://doi.org/10.1007/978-94-007-0753-5_1654

42. Weninger, B.: The effects of colour on the interpretation of traffic noise in strategic noise maps. In: 26th International Cartographic Conference Proceedings, ICC 2013, Dresden, Germany (2013)

43. WHO: WHO global air quality guidelines: Particulate matter (PM2.5 and PM10), ozone, nitrogen dioxide, sulfur dioxide and carbon monoxide. World Health Organization, Geneva (2021)

44. Ratner, I. M., Harvey, J.: Vertical slicing: smaller is better. In: 2011 Agile Conference, AGILE 2011, pp. 240–245. Salt Lake City, UT, USA (2011). https://doi.org/10.1109/AGILE.2011.46

Urban Design

Transforming Large-Scale Participation Data Through Topic Modelling in Urban Design Processes

Cem Ataman[1]([⊠]) [iD], Bige Tunçer[1] [iD], and Simon Perrault[2] [iD]

[1] Architecture and Sustainable Design, Singapore University of Technology and Design, Singapore, Singapore
cem_ataman@mymail.sutd.edu.sg

[2] Information Systems Technology and Design, Singapore University of Technology and Design, Singapore, Singapore

Abstract. The advancements in digital tools and data collection methods ensure the continuing growth of textual data obtained through large-scale participation processes in urban contexts. In order to extract the thematic content of such under-utilized textual datasets, topic modeling (TM) and content analysis have been deployed as promising AI-based Natural Language Processing (NLP) techniques. Yet, implementing such techniques has not been exploited in urban design domains due to the complexity of textual datasets and the lack of a systematic evaluation framework. In this paper, we addressed the challenges in the utilization of large textual data by using a real-world dataset collected via a digital participation platform in Madrid, Spain. Firstly, we identified prominent data structures and potential information embedded into the dataset by using a document-oriented NoSQL database. In this step, we systematically discussed data pre-processing steps to convert them into a series of structured data collections. Secondly, we evaluated three different TM algorithms, i.e. LDA, LSI, and HDP, according to a number of hyperparameters controlling the learning process. This step aimed to reveal the required number of topics to extract meaningful content through the algorithms. Lastly, we presented possible textual data visualization techniques to enable the use of textual information in digital participation processes. Consequently, this paper facilitates the use of large textual datasets by investigating data structures & processing, revealing the potentials of different TM algorithms, and eventually analyzing the results with the support of urban big data analytics and computational linguistic techniques for informed urban design processes.

Keywords: Digital Participation · Urban Big Data · Urban Design · Natural Language Processing · Topic Modelling · Hyperparameter Tuning

1 Introduction

The advancements in digital tools and data collection methods ensure the continuing growth of textual data obtained through large-scale participation processes in urban contexts. These multi-scale and multi-source datasets are valuable resources in informed

M. Turrin et al. (Eds.): CAAD Futures 2023, CCIS 1819, pp. 271–286, 2023.
https://doi.org/10.1007/978-3-031-37189-9_18

decision-making processes for inclusive and transparent urban designs, especially in the early design phases [3]. In urban practices, these digital tools aim at gathering information regarding urban issues through various participation modules, i.e. proposals, debates, and discussions, all expressed in natural languages in digital mediums [2]. Therefore, extracting the information and knowledge embedded into such textual data is essential to make informed decisions based on citizens' perceptions [15, 19].

In order to explore the content of digital participation data, AI-based computational linguistic methods, i.e. Natural Language Processing (NLP), content analysis, and Topic Modelling (TM), are deployed for the transition of collected data into urban designs and policies. Yet, the implementation of such techniques has not been exploited in urban design domains due to the complexity of textual datasets and the lack of a systematic evaluation framework. Accordingly, this paper explores the deployment of TM as an NLP technique by focusing on data preprocessing, algorithm selection, and data visualization.

2   Background

Citizen participation is essential to get insights from the inhabitants, and several studies in literature revealed the importance of qualitative data in urban design processes [7, 8, 14]. Nevertheless, analyzing participation data becomes challenging when participation is conducted in digital mediums due to the size of data and the complexity of information embedded into textual datasets. In order to achieve meaningful analysis by extracting the thematic content from texts and documents, TM has been deployed in many fields, including tourism [9], bioinformatics [13], political science [21], and eventually urban design [3].

The main challenge is to understand these models' accuracy and improve them to obtain meaningful results. These improvements are mainly conducted intuitively by creating the models first and changing parameters and hyperparameters (e.g., the number of topics and alpha and beta values) until a more accurate model is created. Furthermore, the existence of multiple TM algorithms with different features challenges making an informed decision as the benefits vary according to the type and size of datasets [5]. In most cases, an algorithm is selected at the beginning regardless of the dataset type or size, the topics are created without assessing their accuracy, and the results are analyzed based on the first two intuitive steps without an informed assessment.

In this study, we focus on the utilization of TM in digital participation for informed urban design processes. By concurrently evaluating three different TM algorithms while optimizing their hyperparameters (when applicable), we propose a framework to decide the proper algorithm for certain textual datasets and systematically improve their coherence and accuracy. Accordingly, this study aims at (a) exploring the prominent data structures and potential information embedded into big textual datasets by using a document-oriented NoSQL database, (b) understanding methodological issues in the analysis of such big datasets by using different TM algorithms with different hyperparameters, and (c) presenting possible data visualization techniques to enable the use of textual information collected through digital participation for urban design processes.

3   Methodology

This study focuses on the development of a framework for digital participation data analysis through a specific case study. However, it is important to acknowledge that different datasets may require different algorithms with optimized hyperparameters to effectively explore and analyze textual data. Thus, the framework and methods suggested in this study must be followed in every new dataset to determine the most suitable algorithm, topic numbers, and other hyperparameters. The following section will explain the deployed methods and the developed framework for the selected case study.

3.1   Case Study: Madrid

Madrid is chosen as a case study in this study because the Madrid City Council openly shares information on the participation process and datasets collected using the Decide Madrid digital participation tool, which includes various textual data types.

Participation Tool: Decide Madrid. Decide Madrid is a digital participation platform that deploys Consul Project as an open-source participation tool designed and launched by the Madrid City Council in 2015 [24]. The platform allows residents to engage with the local government and encourage citizen participation in developing ideas to improve city management. With their open-source and open-data policies, the Madrid City Council aims to conduct transparent decision-making processes on city-related issues.

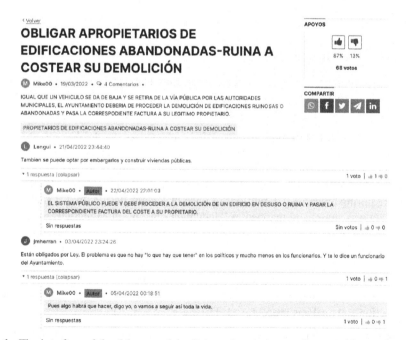

Fig. 1. The interface of the debate module. Debate description on the top with comments and replies below the debate topic.

Decide Madrid consists of several modules to enable different participation modes, i.e. debates, proposals, voting, and participatory budgets. Yet, the debate module is the main textual data source, as the primary purpose of this module is to shape government actions through discussions and ideas directly posted by the residents. The debate topics are generated by the admins or the residents with the features of comment, reply, and rate with the "I Agree" or "I Don't Agree" buttons (See: Fig. 1.).

Data Collection Process. This research explicitly focuses on the data obtained from the Debate module, in which people discussed issues related to the city, policies, and designs. Since its first launch in 2015, 470.000 residents have registered to participate via Decide Madrid. Each user has a username defined by themselves; therefore, the data collection process is completely asynchronous and anonymous. In the last seven years, 3838 debate topics have been created by 2390 different residents. Under the debate topics, 31359 comments and replies are posted with 404.830 votes by the registered users. The debate module has a hierarchical data structure, as users are able to comment on the debate descriptions or reply to previously posted comments and replies. In this structure, each debate topic has minor discussions about the main topic description.

The content of the debate topics varies from practical issues, e.g., the use of cycling lanes, the extension of public transportation hours at night, or the design of passages for people with wheelchairs; to policy-related discussions such as greener city strategies, waste management, tourism strategies, et cetera. Although each debate topic has a tagging system that helps connect the related discussions, many overlaps and repetitions exist in their content within the debate module. The participation platform is currently active and in use by Madrid residents.

3.2   Digital Participation Database

In order to overcome the complexity of the unstructured textual data coming from Decide Madrid, we deployed an open-source database management program to work on the data more efficiently, as explained in the following sections.

Document Oriented Database. MongoDB is selected as the database to store large volumes of textual data obtained from Decide Madrid. It is a document-oriented NoSQL database widely used for machine learning projects to filter, sort, and aggregate the data in the preparation and training processes. MongoDB stores data in JSON format, without requiring predefined definitions or upfront schema design. In this study, the big textual data was stored and mapped as "document objects" in MongoDB to ensure the accessibility of unstructured data during the analysis process. The data was segregated into clusters based on different participation modules, such as debates or proposals, and each cluster contained several collections of descriptions, comments, and sorted data. In the database, each comment and description was stored as a document with embedded information such as time, title, author, comment, and votes, which are later used for the data selection and analysis processes (1. Database" in Fig. 2.).

Data Selection. Digital participation generates substantial amounts of textual data, necessitating the selection of relevant information for meaningful analysis. To identify important data to obtain meaningful results, factors such as content, time, and the number of comments and supports are considered. To streamline the dataset, three steps were taken. Firstly, only topics with informative descriptions were included to ensure that the users conducted informed discussions. Secondly, we excluded the debate topics with only a few comments and replies (min_comment > 5), as they do not contribute to valuable insights and increase computational costs. Lastly, the debate topics that did not receive enough support by voting (min_support > 10) were removed, as they did not engage residents. Following these steps, the final dataset comprised 1,300 debate topics with 26,051 comments from 2015–2022.

3.3 Data Analysis

Data Pre-processing. Textual big data analysis requires a corpus, i.e. a large and structured set of texts. As raw data often contains irregularities and unnecessary texts such as URLs, user mentions (@), abbreviations, and advertisements, a thorough data cleaning process is essential to ensure the accuracy of NLP toolkits. In this research, we deployed functions from the "NLTK" Python library to clean the data by converting all texts to lowercase, tokenizing, removing punctuations, special characters, stopwords, and finally stemming and lemmatization.

After creating our list of words, we removed the ones that occurred less than a threshold (cut-off value = 10). We then formed bigrams and trigrams from words in the list that usually appeared together as phrases, such as "city council," "decision making," and "affordable housing," (min_count = 10, threshold = 50). This step was critical to understand the natural language patterns for the context in topic modeling.

Although we obtained a clean dataset after the previous steps, it still contained frequently occurring words that do not necessarily have significance for further analysis. Yet, these words force topic modeling algorithms to place them as a high priority due to their occurrence frequency in the entire corpus. In order to overcome this issue, we implemented the "Gensim" TF-IDF (Term Frequency – Inverse Document Frequency) algorithm, which removes all the ubiquitous words that do not add value to the clustering algorithms (low value = 0.03). At the end of the data cleaning and preparation, we obtained our final corpus for topic modeling (see: "2. Preprocessing" in Fig. 2.).

Topic Modelling (TM). We used TM algorithms to extract the topics from our corpus. TM is an unsupervised Machine Learning technique, as the labels are retrieved directly from the text without any prior labeling opposite to supervised models [5]. These algorithms are capable of determining the probability of a word or phrase belonging to a specific topic, i.e. a collection of words, based on their similarity or closeness. In this study, three different TM algorithms (see: "3. Data analysis" in Fig. 2.), namely Latent Semantic Indexing (LSI), Hierarchical Dirichlet Process (HDP), and Latent Dirichlet Allocation (LDA), were implemented and evaluated as explained in the following sections.

4 Topic Model Parameters

4.1 The Number of Topics

Despite the wide use of topic models in various domains, detecting the optimal number of topics remains a problem. The main challenge is the lack of an established metric to measure the quality of topics resulting from the topic model. In most TM algorithms, the topic number is an a priori parameter, which profoundly affects the capacity of the model and the legitimacy of the results resulting in an adequate number of topics [12]. In these models, a relatively smaller number of topics may generate broad and heterogeneous topics, while a large number of topics may lead to too specific topics, which would challenge their interpretability [17].

Nevertheless, it is not possible to suggest a general formula to decide the optimum topic number as it depends on the kind of the corpus (i.e. the terminologies, the field, the language), the size of the corpus (if it is small or big data), and the aim of the project based on the experience. Therefore, each participation dataset must be analyzed to identify the optimum topic number by examining several metrics and optimizing the hyperparameters, as explained below.

4.2 Evaluation Metrics

In the evaluation of topic models, the primary purpose is to investigate if the created topics serve the purpose of the project understandably and coherently. However, the evaluation becomes more complicated when the topic model is used for a qualitative task to explore the semantic themes as in digital participation. Specific evaluation metrics are necessary for these models to investigate the "semantic interpretability" of topics. These metrics particularly examine whether words in the topics represent a single coherent concept according to a person's subjective judgments.

Perplexity is one of the fundamental evaluation metrics widely used for language model evaluation. It measures the ability of a topic model to predict a new dataset after the training set by the exponential of the negative log-likelihood of the new dataset given the trained model [23]. In other words, the perplexity metric assesses what degree the model represents the statistics of the held-out data. A lower perplexity indicates better human interpretable topics as it is less uncertain or random in its predictions.

Although optimizing perplexity increases semantic interpretability, it does not directly correlate with human judgment. Therefore, other metrics, such as topic coherence, are combined with the perplexity measure. It also measures how interpretable the topics are to humans. Still, in this case, it calculates the top number of words with the highest probability of belonging to a particular topic [16]. Simplistically, the coherence score measures the co-occurrence of words in a topic to better correlate with human interpretation [12]. The results are dependent on the dataset that they are calculated from. For this reason, each model should be examined separately to trace the change in coherence score and the number of topics. In this study, we used the "u_mass" metric, as it is the fastest way of calculating the coherence score by checking how often two words appear together in the corpus.

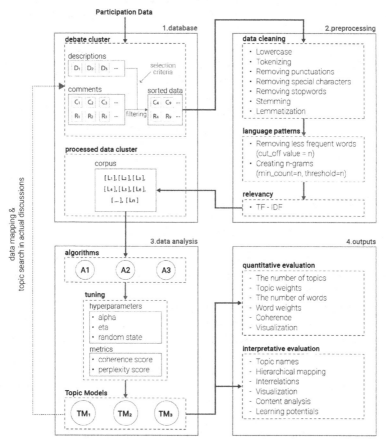

Fig. 2. An overall view of the research flow. "1. Database" presents the actions explained in Sect. 3.2, "2. Preprocessing" covers Sect. 3.3.1, "3. Data analysis" explains topics in Sect. 4, and "4. Outputs" refers to the discussion in this study (D: Description, C: Comment, R: Reply, L: List, A: Algorithm, TM: Topic Model).

4.3 Hyperparameters

Another critical step in the evaluation of Topic Models is hyperparameter tuning. It consists of finding a set of optimal hyperparameter values for a learning algorithm while applying it to a data set. However, topic models cannot change these parameters according to the researcher's insights. The optimal values depend on the algorithms and the dataset; therefore, it is essential to identify their values in advance to maximize the model's performance and produce better results with fewer data losses.

Topic modeling hyperparameters include the alpha (α), beta (β), gamma (γ), and random state. Alpha controls the distribution of topics over documents, while beta represents the distribution of words over topics [10]. Gamma is the concentration value in Dirichlet Processes, and random state is a performance measure used to increase efficiency [10]. In this study, the random state is fixed at 42 for all algorithms to ensure the reproducibility and comparability of results. Other hyperparameters are optimized, when possible, in algorithms.

It is important to emphasize that the optimal values of hyperparameters depend on the algorithms and datasets employed. To ensure data homogeneity and assess the learning potential of the algorithms, we adopted a widely accepted technique of 10-fold cross-validation in this study. This technique involves dividing the dataset into 10 non-overlapping folds, where each fold serves as the test set for the model trained on the remaining nine folds [22]. By applying this approach, we assessed the learning potential of our algorithms and ensured that our data was topologically homogeneous based on the optimized hyperparameters and selected topic numbers.

5   Topic Modelling Algorithms

5.1   Latent Dirichlet Allocation (LDA)

LDA is an unsupervised generative probabilistic method for modeling a corpus [6]. It is an intuitive approach that calculates the similarity between source data to reveal their respective distributions of each cluster over topics [11]. In LDA models, the documents are represented as random mixtures over latent topics characterized by a distribution of words and their probabilities. In other words, it is possible to interpret topics as a distribution of words in known vocabulary to extract thematic content from text-specific data [20]. The words with high probabilities in each topic reveal the topic context.

Although LDA topic models are widely used as powerful models, some limitations exist due to the mathematical methods embedded into their algorithms. Firstly, the number of topics must be fixed and identified before the formulation of the model. Additionally, LDA is a static and non-hierarchical model that does not allow for the sharing of data and topic evolution. Lastly, the LDA topic model creates uncorrelated topics, as Dirichlet topic distribution cannot capture correlations [4].

In this study, the LDA topic model is the first algorithm to classify texts in debates on a particular topic. We used the preprocessed dataset from our database and implemented the "Gensim" Python library for the LDA TM, which allows controlling different hyperparameters. After creating our base model, we started the sensitivity tests to determine the model hyperparameters, i.e. the number of topics, the Document-Topic Density (α), and the Word-Topic Density (β). We performed sequential tests, in which we kept one parameter constant while changing the others iteratively over the range of topics (10, 25), alpha (0.01, 1, 0.3), and beta (0.01, 1, 0.3) parameter values.

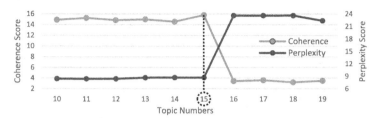

Fig. 3. Coherence and perplexity values of each topic number in LDA topic modeling.

To determine the optimal number of topics, we utilized coherence (u_mass) and perplexity scores with fixed values of α (= 0.01) and β (= 0.01). Our analysis revealed that the coherence score increased while the perplexity score decreased as the number of topics increased (Fig. 3). Based on the results, we selected 15 topics for our LDA model as it achieved the highest coherence score with one of the lowest perplexity scores.

After deciding the number of topics, we investigated the optimal α and β parameters. We selected the values that yielded minimum u_mass score for 15 topics, which resulted in α = 0.31 and β = 0.91. This adjustment led to a 7.4% improvement over the initial evaluation scores. By using the optimized hyperparameters based on coherence and perplexity scores, we created our LDA model, whose first five topics, with their names, keywords, and word weights, are presented below (Table 1).

Table 1. LDA model results of the first five topics with keywords and their weights.

Topic ID	1	2	3	4	5
Topic Name	Public Realm	Traffic	Dogs	Children	Public Transportation
Words	0.033 area 0.024 neighborhood 0.024 street 0.013 square 0.011 park 0.010 center	0.031 cars 0.018 sidewalks 0.016 traffic 0.015 opinion 0.014 vehicles 0.014 space	0.041 dogs 0.016 clear 0.015 seen 0.013 education 0.013 remainder 0.013 go	0.021 children 0.020 day 0.015 today 0.015 person 0.014 final 0.013 fines	0.038 m 0.031 proposals 0.023 taxi 0.019 public transport 0.016 house

5.2 Latent Semantic Indexing (LSI)

Latent Semantic Indexing (LSI) is based on Dimensionality Reduction algorithms, which aim to reduce classification dimensions to increase time efficiency [18]. It achieves this by combining words with similar meanings in similar text pieces. The LSI algorithm usually requires a large set of documents to receive accurate results; therefore, it is a convenient algorithm to be deployed in digital participation datasets. Its distinctive feature is that LSI ranks the topics by itself, which means it outputs them in a ranked order. Yet, it also requires a pre-identified topic number intuitively, and the algorithm does not allow conducting hyperparameter tuning, unlike LDA models.

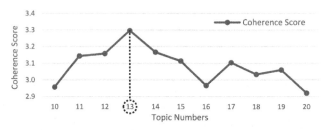

Fig. 4. Coherence values on each topic number in LSI topic modeling.

To formulate our LSI model, we implemented the "Gensim" python library for our final dataset as for the LDA model. We traced the change in coherence scores to identify the optimal number of topics as the only control parameter in LSI algorithms. The coherence score had its maximum when the topic number was 13 (Fig. 4). Accordingly, we formulated the LSI topic model with 13 topics and presented the results below (Table 2).

Table 2. LSI model results of the first five topics with keywords and their weights.

Topic ID	1	2	3	4	5
Topic Name	Administration	People	Neighborhood	Cycling	City
Words	0.676 city hall 0.320 do 0.267 person 0.213 people 0.162 zone 0.151 go 0.124 I think	0.455 do 0.383 person 0.318 I think 0.313 zone 0.279 also 0.253 city hall 0.184 proposal	0.540 zone 0.329 neighborhood 0.292 I think 0.180 do 0.167 city hall 0.089 neighbors	0.610 city 0.302 proposal 0.282 go 0.271 people 0.263 person 0.118 bike 0.106 do	0.526 city 0.441 also 0.410 neighborhood 0.235 zone 0.215 people 0.143 city hall

5.3   Hierarchical Dirichlet Process (HDP)

Another topic model designed to help summarize and organize large archives of texts is the HDP model. It is a mixed-membership model for unsupervised analysis of grouped data, incorporating a vector of subject-specific weights rather than forcing each subject to belong to a single (topic) cluster [1]. HDP algorithm is an extension of LDA and addresses the topic number issue (i.e. the number of topics is identified a priori). It also uses a Dirichlet process to capture the uncertainty in the topic numbers. Still, unlike LDA and LSI, HDP infers the number of topics from the data as a random variable.

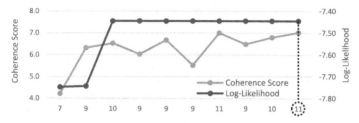

Fig. 5. Coherence and log-likelihood values for each topic number in HDP topic modeling.

In this study, we implemented the HDP model by deploying the "Tomotopy" Python library to initiate a model object with parameters like weight tokens, token frequency, alpha, and gamma values. For our dataset, we used words appearing in at least ten documents (comments or replies) and excluded the words that appeared in more than 15. We set the concentration parameters as $\alpha = 2$ and $\gamma = 1$ to keep the topic number like in previous models to allow the comparison. Ultimately, we initiated the model training with 15517 documents (total number of words = 192450, total number of vocabularies = 3212). The HDP algorithm identified 11 topics with the highest log-likelihood value based on our dataset (Fig. 5.) with the topic keywords and their weights (Table 3).

Table 3. HDP model results of the first five topics with keywords and their weights.

Topic ID	1	2	3	4	5
Topic Name	Pollution in Aluche	Learning	Childcare	Pedestrian Movement	Historical Center
Words	0.0255 neighbor 0.0255 aluche 0.0255 crystals 0.0255 problem dirt neglect 0.0255 next full bottles 0.0191 stucked cases	0.0666 go 0.0489 can 0.0355 leaving impromptu scribbles 0.0356 first blackboard 0.0355 expression brainstorm	0.0429 responsibility owner 0.0368 children knowledge 0.0368 child play 0.0368 child walkers 0.0368 free toy	0.0391 qued pedestrian circule 0.0391 left 0.0168 work 0.0168 circle 0.0168 spread 0.0168 close 0.0168 part 0.0168 executed	0.0539 museum 0.0216 grounds 0.0216 madrid 0.0144 historic garages 0.0144 cooperative 0.0108 historical

5.4 The Comparison of Topic Models

This study revealed disparate outcomes across the algorithms despite utilizing the same participation data for model formulation. By optimizing hyperparameters, the optimal topic numbers were determined as 15, 13, and 11 for the LDA, LSI, and HDP models, respectively. Subsequently, we subjected these three topic models to a 10-fold cross-validation, and employed the mean coherence scores of the implemented topic models to compare their learning potential and computed their standard deviations (SD) and coefficient of variation (CV) to establish the topological homogeneity of the dataset.

The mean coherence scores, along with their SD and CV values, were computed for the LDA, LSI, and HDP models, resulting in scores of 4.75 (SD = 0.2, CV = 4.21), 3.11 (SD = 0.14, CV = 4.5), and 13.37 (SD = 1.12, CV = 8.37), respectively. The mean coherence scores obtained from 10-fold cross-validation were lower for LDA and LSI algorithms than the coherence scores used to determine topic numbers, suggesting that the results are meaningful for the given dataset, but may not perform well when presented with new data. From the findings, we can infer that algorithms with greater control over hyperparameters exhibit poor responsiveness to new datasets. Conversely, the HDP algorithm demonstrated a higher average coherence score than that of the entire dataset, indicating its effectiveness in dealing with new datasets, as it determines the optimal topic number on its own. In terms of the dataset topology, the CV values for all three algorithms were below the generally accepted value of 10%, establishing homogeneity in random parts of the dataset. Table 4 presents a comprehensive overview of the hyperparameters, metrics, and results of TM and 10-fold validation for each algorithm.

When we compared the algorithms, LDA has its advantages in the hyperparameter tuning to reveal the optimal topic number with more control and accuracy. Yet, the tuning process necessitates a longer computing time as all the values in a particular range with defined intervals need to be evaluated. In contrast, the LSI model provides faster results in a ranked order since the only control variable is the topic number in the algorithm. In the HDP model, the number of topics is a random variable deriving from the data rather than specified in advance. Although the computing time is relatively longer than LSI models, the HDP model is still more efficient compared to LDA models without the necessity of defining a control variable or hyperparameter tuning.

The models' results for our dataset also vary in semantic interpretability. The LSI algorithm results in a model with many overlapping keywords in different topics. The obtained topics, such as neighborhood and city, do not provide specific inputs for designers and policymakers and challenge the interpretation of the topics. Compared to the LSI model, HDP and LDA models provide more accurate results with more specific keywords for our dataset. Both models' abstraction levels are similar, and the topic contents are clear enough to reveal the semantic content. As both algorithms provide comparable outputs, the HDP model is preferable for our dataset, considering the affordable computing cost and time efficiency.

Although all three algorithms yield similar outputs in terms of topics, keywords, and word weights, it is crucial to visualize the results for decision-makers to utilize them effectively. In this study, we investigated several visualization techniques commonly used in topic modeling studies across different fields (Fig. 6). To begin, we utilized word

Table 4. The inputs and outputs of topic models with LDA, LSI, and HDP algorithms.

Algorithm	LDA	LSI	HDP
Hyperparameters	α = 0.31 β = 0.91 Random state = 42	Random state = 42	α = 2 γ = 1 Random state = 42
Coherence Score	15.8	3.3	7.03
Perplexity	8.7	NA	NA
Topic Number	15	13	11
10-Fold Cross-Validation Results	Mean = 4.75 SD = 0.2	Mean = 3.11 SD = 0.14	Mean = 13.37 SD = 1.12
Topic Names (First 5)	- Public Realm - Traffic - Dogs - Children - Public Transportation	- Administration - People - Neighborhood - Cycling - City	- Pollution in Aluche - Learning - Childcare - Pedestrian Movement - Historical Center

clouds to display the frequency of terms in our dataset for the HDP model. Next, we employed t-SNE (t-distributed stochastic neighbor embedding) as a statistical method to plot the LSI topics on a 2D map, indicating the proximity of topics and the associated discussions. These two visualization techniques were applied to all three models to facilitate their interpretation. For the LDA topic model, a specific visualization method is available in the "Gensim" library, which is the intertopic distance map. This technique plots the topics in a 2D space, with the size of the topic circles proportional to the number of words associated with each topic in our dataset. This approach enables the identification of the most relevant terms for a given topic and the dataset.

Designers can use the presented visualization techniques to help decision makers understand participation data and its insights. Word clouds can provide a quick overview of the most frequent terms and themes, while the t-SNE method can reveal relationships between different topics, enabling identification of potential areas for collaboration or integration. The intertopic distance map can identify relevant terms and be used to develop targeted communication strategies. By leveraging these visualizations, designers can support effective decision-making and positive outcomes. However, it is acknowledged that domain-specific visualization techniques are necessary for decision-makers in urban design domains to effectively communicate and analyze the data. To interpret the data in a meaningful way for domain-specific purposes, specific visualization techniques should be designed to meet the needs of designers and policymakers based on predefined learning points and expectations.

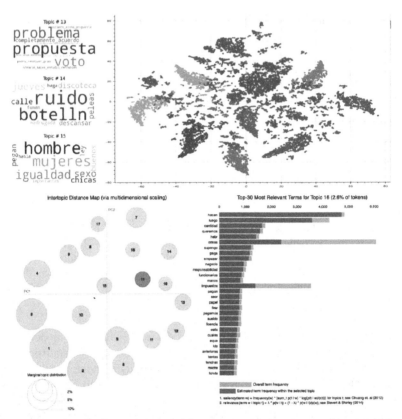

Fig. 6. Different visualization techniques of the topic models based on Decide Madrid data. Top Left: word clouds for the HDP model, Top Right: t-SNE for the LSI model, Bottom Left: intertopic distance map & Bottom Right: Relevant Terms for the LDA model.

6 Conclusion

Based on the understanding that different participation processes require varying approaches in urban design practices, designers and policymakers must exercise more discretion when deploying analysis and evaluation methods for different datasets. In light of this, our study delved into large textual data structures using a preliminary evaluation framework to analyze digital participation data. Our focus was on topic models as an NLP technique, exploring different algorithms with varying hyperparameters and outputs. We developed practical means of optimizing hyperparameters to increase topic model coherence while ensuring the correct number of topics. Our goal was to simplify this framework to enable urban designers and policymakers to use qualitative evaluation metrics in their practices.

Nevertheless, there are several limitations to our study. Firstly, we only explored one segment of the participation data, namely the debate module. Integrating proposals and polls into this evaluation framework would provide a more comprehensive assessment of the participation data. Secondly, our dataset is city-scale, making it difficult to trace location-based data. We suggest studying neighborhood-scale datasets with the same evaluation framework and integrating topic models into GIS data. Lastly, while our study concurrently examines three TM algorithms, we evaluated each algorithm individually. More automated systems are necessary to increase the efficiency of the evaluation process for designers and policymakers.

Further research is necessary to investigate interpretive evaluation metrics that would enable the use of big textual data in urban practices. While our study presents a preliminary framework for evaluating the quantitative aspects of topic models, we recommend further studies to explicitly discuss the potential applications of these models in the urban context to develop domain-specific visualizations and insights that facilitate effective communication and data analysis. This would enable designers and policymakers to define learning points and implement them in new design contexts.

In conclusion, our study presents the potential applications of big textual data, focusing on data structures, processing, visualization, and analysis, with the support of cutting-edge urban big data analytics and computational linguistic techniques. We hope this research will inform urban design processes and empower designers and policymakers to make more informed decisions.

Acknowledgment. This research is supported by "Designing mobile-friendly cartograms for visualising geospatial data" Grant, from the Ministry of Education, Singapore, under its Academic Research Fund Tier 2 programme (award number MOE-T2EP20221-0007) and by Singapore International Graduate Award (SINGA).

References

1. Airoldi, E., Blei, D., Xing, E., Fienberg, S.: A latent mixed membership model for relational data. In: Proceedings of the 3rd international workshop on Link discovery – LinkKDD'05, pp. 82–89 (2005)
2. Ataman, C., Tuncer, B.: Urban interventions and participation tools in urban design processes: a systematic review and thematic analysis (1995–2021). Sustain. Cities Soc. **76**, 103462 (2022)
3. Ataman, C., Tunçer, B., Perrault, S.T.: Asynchronous digital participation in urban design processes: qualitative data exploration and analysis with natural language processing. In: POST-CARBON – Proceedings of the 27th CAADRIA Conference, pp. 383–392 (2022)
4. Vyankatrao Barde, B., Madhavrao Bainwad, A.: An overview of topic modeling methods and tools. In: 2017 International Conference on Intelligent Computing and Control Systems (ICICCS), pp. 745–750 (2017)
5. Blei, D.M.: Probabilistic topic models. Commun. ACM **55**(4), 77–84 (2012)
6. Blei, D.M., Ng, A.Y., Jordan, M.I.: Latent dirichlet allocation. J. Mach. Learn. Res. **3**, 993–1022 (2003)
7. Dembski, F., Wössner, U., Letzgus, M., Ruddat, M., Yamu, C.: Urban digital twins for smart cities and citizens: the case study of Herrenberg, Germany. Sustainability **12**(6), 2307 (2020)
8. Dunne, C., Skelton, C., Diamond, S., Meirelles, I., Martino, M.: Quantitative, Qualitative, and Historical Urban Data Visualization Tools for Professionals and Stakeholders, pp. 405–416 (2016)
9. Egger, R. (ed.): Applied Data Science in Tourism: Interdisciplinary Approaches, Methodologies, and Applications. Springer International Publishing, Cham (2022)

10. Jacobi, C., van Atteveldt, W., Welbers, K.: Quantitative analysis of large amounts of journalistic texts using topic modelling. Digit. J. **4**(1), 89–106 (2015)

11. Jelodar, H., et al.: Latent Dirichlet allocation (LDA) and topic modeling: models, applications, a survey. Multimed. Tools Appl. **78**(11), 15169–15211 (2018). https://doi.org/10.1007/s11 042-018-6894-4

12. Krasnov, F., Sen, A.: The number of topics optimization: clustering approach. Mach. Learn. Knowl. Extr. **1**(1), 416–426 (2019)

13. Liu, L., Tang, L., Dong, W., Yao, S., Zhou, W.: An overview of topic modeling and its current applications in bioinformatics. Springerplus **5**(1), 1–22 (2016). https://doi.org/10.1186/s40 064-016-3252-8

14. Mueller, J., Hangxin, L., Chirkin, A., Klein, B., Schmitt, G.: Citizen design science: a strategy for crowd-creative urban design. Cities **72**, 181–188 (2018)

15. Mazhar Rathore, M., Paul, A., Hong, W.-H., Seo, H., Awan, I., Saeed, S.: Exploiting IoT and big data analytics: defining smart digital city using real-time urban data. Sustain. Cities Soc. **40**, 600–610 (2018)

16. Röder, M., Both, A., Hinneburg, A.: Exploring the space of topic coherence measures. In:Proceedings of the Eighth ACM International Conference on Web Search and Data Mining, pp. 399–408. https://doi.org/10.1145/2684822.2685324 (2015)

17. Sbalchiero, S., Eder, M.: Topic modeling, long texts and the best number of topics. Some problems and solutions. Qual. Quant. **54**(4), 1095–1108 (2020). https://doi.org/10.1007/s11 135-020-00976-w

18. Nareshkumar Singh, K.S.H., Dickeeta Devi, S., Mamata Devi, H., Mahanta, A.K.: A novel approach for dimension reduction using word embedding: an enhanced text classification approach. Int. J. Inform. Manag. Data Insights **2**(1), 100061 (2022)

19. Tekler, Z.D., Low, R., Choo, K.T.W., Blessing, L.: User Perceptions and adoption of plug load management systems in the workplace. In: Extended Abstracts of the CHI Conference on Human Factors in Computing Systems, pp. 1–6 (2021)

20. Wang, Y., Taylor, J.E.: Urban crisis detection technique: a spatial and data driven approach based on latent Dirichlet Allocation (LDA) topic modeling. In:Construction Research Congress 2018, pp. 250–259 (2018). https://doi.org/10.1061/9780784481271.025

21. Wilkerson, J., Casas, A.: Large-scale computerized text analysis in political science: opportunities and challenges. Annu. Rev. Polit. Sci. **20**(1), 529–544 (2017)

22. Wong, T.-T., Yeh, P.-Y.: Reliable accuracy estimates from k-fold cross validation. IEEE Trans. Knowl. Data Eng. **32**(8), 1586–1594 (2020)

23. Zhao, W., et al.: A heuristic approach to determine an appropriate number of topics in topic modeling. BMC Bioinformatics **16**(S13), S8 (2015)

24. Decide Madrid. https://decide.madrid.es

AI-Assisted Exploration of the Spirit of Place in Chinese Gardens from the Perspective of Spatial Sequences

Zihao Wang$^{(\boxtimes)}$ and Xiaofei Zhang$^{(\boxtimes)}$

Ecole nationale supérieure d'architecture Paris-Malaquais, Paris, France
xiaofei.zhang2021@outlook.com

Abstract. The Chinese garden style is not only a decorative aesthetic but also a composition of spaces. Chinese gardening is about creating narrative spatial sequences, and the composition of the views in each space is also important. We propose to simulate these approaches with AI. Firstly, we used the pix2pix algorithm to generate a plan layout and determine the paths that connect the spatial sequences. secondly, we will enter the first-person view. Whereas traditional architects design by plans and axonometric drawings, our work focuses on the 'spirit of place'. We used Isovist to simulate the spatial elements that one sees as he or she moves through space over time and encoded these spatial elements, then we chose RNN (LSTM) to learn this spatial sequence composition of the 'spirit of place', and to generate a practical project with the 'spirit of place' in Chinese gardens. The RNN (LSTM) model can handle the information of the sequence well, that is, the input before and the input after are related, and in a spatial sequence, the spatial nodes before and the spatial nodes after will also jointly form a sense of spatial rhythm. Finally we tried to materialize the model with CycleGAN to refine the workflow of the auxiliary architectural design. We tested this methodology in a renovation program for a Chinese village to observe the 'spirit of place' created by AI(Artificial Intelligence). This approach could potentially be applied to other user experience-oriented aided designs. Our objective is to establish a workflow that enables architects to collaborate better with AI.

Keywords: Artificial Intelligence · Spatial Sequences · Chinese Gardens · Spirit of Place · Case Study

1 Introduction

1.1 Context

The first digital turn in architecture changed our way of making, then, in the 1990s, the advent of parametric design changed the way we design, and at the

---

Z. Wang and X. Zhang—These authors contributed equally to this work.

© The Author(s), under exclusive license to Springer Nature Switzerland AG 2023
M. Turrin et al. (Eds.): CAAD Futures 2023, CCIS 1819, pp. 287–301, 2023.
https://doi.org/10.1007/978-3-031-37189-9_19

same time, the advent of BIM helped to optimize the workflow. [1] The major breakthrough in AI in recent years has been achieved through statistical methods based on big data samples. So in this new era, how is AI affecting the discipline of architecture?

As Stanislas Chaillou says, machine learning has shown great potential in recent years, and is a quantum leap from the world of heuristics to the world of statistics. His ArchiGAN project in 2019 has sparked much discussion and reflection within the architectural community. In the past, the multi-objective optimization used in computer-aided design used the logic of a genetic algorithm, which is a heuristic process. Today, machine learning is an inductive black box which defines the internal relationships of the sample through a large amount of data [2]. Hao Zheng and Weixin Huang first explored the application of the GAN algorithm on architectural planning [3]. Stanislas Chaillou has further developed ArchiGAN based on it, while a UCL graduate's work GlitchGAN's [4] analogy on different spatial types also inspired us. According to the characteristics of Chinese gardens, we tried to use the RNN algorithm.

Previously, generative design in the field of architecture was designed to optimize objective standards. Different from traditional deductive algorithms, AI is an inductive method. The concept of genius loci (spirit of place) [5] was first proposed by the Norwegian architectural theorist Norberg-Schulz, whose theoretical research applied the methods of architectural phenomenology. The place involves not only the material form, but also the spiritual meaning. The theory starts from a more microscopic point of view, i.e. the human sensation. It perceives the atmosphere from the collection of things and envisions the shape of the designed building from the atmosphere. This design method first feels the spirit of the place, that is, the atmosphere, and then creates the atmosphere. Everyone's memory is unique, and the role of architecture and landscape is to awaken previous memories stored in the mind. In recent years, the development of AI and big data has opened up a new idea for us, that is, the use of machine learning to understand the possibility of user experience. Isovist is a method often used in architecture to study the range of human vision in space, and we believe it is appropriate for our study. It allows us to encode the information in the viewable field of the model into a format that machine learning can understand.

1.2   Learning from Chinese Gardens

The Industrial Revolution gave birth to the homogenization of big cities. The French anthropologist Marc Augé named these spaces "Non-place": "If a place can be defined as identity, relational and historical, a space that can be defined neither as identity, nor as relational, nor as historical defines a non-place" [6]. In 1995, Koolhaas defined this phenomenon in his article "The Generic City" [7]. As urbanization spreads to the countryside, architecture in rural China is also becoming generic. As inequality between urban and rural areas and the gap between the rich and the poor widened, large numbers of Chinese lost faith in the culture of the past and turned to modernity.

Accompanying the densification of cities and the dramatic decline of the rural population is the growing awareness of uneven urban and rural development. Some intellectuals have proposed another construction model, that is, using the local cultural industry to revitalize the countryside, rather than blindly developing and constructing.

Shanshui and Tianyuan, two art movements, have played a crucial role in Chinese culture and the history of Chinese architecture. The Chinese garden originated in the Song Dynasty and reached its peak in the Yuan, Ming and Qing dynasties. It is regarded as a comprehensive art form. After that, the architectural style was affected by globalization. The Chinese garden is a key subject of architectural education in China, which enables Chinese architects to understand their own culture. Therefore, we decided to include the gardens of Yuan, Ming and Qing dynasties as the research object because the sample size of them is large enough, which makes our research method feasible. With the development of AI in the field of statistics, we take the real Chinese gardens as the database, summarize knowledge about how to build Chinese aesthetics, identity and atmosphere, and apply it to a contemporary Chinese construction project.

Narrative Approach. Many prose texts and poems often describe the experience of spatial changes. For example, the most famous of them is by Tao Yuanming, which is entitled 'Peach Blossom Source', is set in the Jin Dynasty and describes a utopia: "He noticed with surprise that the grove had a magic effect, so singularly free from the usual mingling of brushwood, while the beautifully grassy ground was covered with its rose petals. He went further to explore, and when he came to the end of the grove, he saw a spring which came from a cave in the hill. Having noticed that there seemed to be a weak light in the cave, he tied up his boat and decided to go in and explore. At first the opening was very narrow, barely wide enough for one person to go in. After a dozen steps, it opened into a flood of light. He saw before his eyes a wide, level valley, with houses and fields and farms." The spatial experience described here, from the narrowest to the widest space, is often compared to the design of the Lingering Garden in Suzhou. Moreover, in Chinese gardens, every pavilion or house, every element of the space, has its own name, which usually comes from a poem. When designing Chinese gardens, the study of Shanshui painting occupies a crucial position. Shanshui painting is an important branch of Chinese painting, which uses the scattered, flat and distant perspective like "walking on the mountain", looking at walking, the focus constantly changing, so this kind of painting can paint in the format of very long scrolls, including thousands of kilometers of rivers and mountains [8]. Based on these points, modern analysis of Chinese gardens tends to focus on spatial sequences, which is the reason why we choose them as the starting point of our research.

Visual Composition. As the spirit of place is the identity, character and atmosphere of a place, it is a human perception of place, i.e. determined by the sense data of place. This sense data is derived from the physical properties of space, such as the shape of the space, light, materials, etc. The percentage of external information received by the human brain through each of the five senses is 1% for taste, 1.5% for touch, 3.5% for smell, 11% for hearing and 83% for sight [9]. Accordingly, two related visual approaches, narrative spatial sequence, and visual composition, are selected to study the "spirit of place" of Chinese gardens.

Visitor experience plays a key role in Chinese gardens. In the basic approach to Chinese garden design there is a method called "relative landscape", i.e. in the garden or in the building, or on the pavilion or on the platform, you can see the hall, the mountain, the bridge, the trees. Conversely, in the corridor, on the bridge, in the hall and other places you can see the pavilion, the platform, the building. This method of designing from viewpoint A to viewpoint B, and from viewpoint B to viewpoint A, is called "relative landscape".

Other garden design techniques are "landscape borrowing" and "landscape framing". That is, a beautiful visual composition is formed when we look at distant elements through objects that hinder the foreground view (through doors, windows, columns, trees, stones, etc.).

Furthermore, the conceptual approach of the Chinese garden is characterized by the imitation of nature. And the border between architecture and nature is not clear, which is called "gray space" by Kurokawa. These approaches to Chinese garden design influence people's visual perception of space, which we consider to be a visual composition.

1.3 Case Study

We located the experimental case in Guyuan Village, Nanping City, a village in the north of Fujian Province. Fortunately, the government is trying to reactivate the village by developing new sectors of the economy. Inspired by the Yue Opera season in Japan and the Bishan project in China, the Jvkou County Government in Fujian Province decided to create a rural art festival to revitalize the countryside. Various programs have been developed around this project: artist-in-residence, exhibition, local handicraft workshop, heritage workshop, residential workshop, farm, restaurant, theater, bookstore, etc.

Nowadays, the pursuit of modern Chinese rural construction has created a large number of "non-places" without the spirit of place. On the contrary, the Rural Art Festival aims to find another architectural model to awaken the countryside with cultural industries and develop in a sustainable and slow manner. Therefore, we will preserve the existing architectural heritage and restore it by borrowing the spirit of Chinese gardens, plan it as a whole, endow it with value, and integrate it with the environment.

The Chinese garden itself is a space for visiting and walking. Since the Song Dynasty, some gardens have been used as urban parks open to the public, for example: Jinming Pond Garden is a royal garden open to the public. In the gardens, corridors and implicit paths connect the buildings in a rhythmic sequence of spaces, while the artificial elements: pond, stone and mounts are all works created by artists. These artificial landscape elements have different properties.

The properties of water basins are: Viewable; Non-accessible; Extended; Open; Empty; Quiet; the properties of stones are: Visual focus; Foreground, the properties of the artificial mounds are: Participatory; Accessible; Closed and Secret; the properties of the corridors are: Walkable; Semi-open; Zig zag; Continuous assembly. We observed an analogy between scenography and Chinese gardening. The idea was to use the "spirit of place" of the Chinese gardens to organize the programs around the festival, renovating the existing buildings on the site and giving them new functions, linking them with new corridors and paths, and proposing the artists' arrangements and properties for their works.

We believe that AI can create a new atmosphere by combining the "spirit of place" of Chinese gardens with contemporary art festivals. As for why we choose Chinese gardens as a case, first of all, the analysis of Chinese gardens in the past is subjective, and we try to use machine learning to verify these design techniques. Second, the festival, a land art festival in an ancient village, itself represents a visiting experience. One will appreciate a series of installations, which we believe are similar to the narrative spatial sequence of the garden, where the spirit of the landscape and the site are well integrated. Finally, the spirit of place of Chinese gardens is the yearning of Chinese scholars for pastoral and romanticism since ancient times, but this poetic pursuit has been ignored under the impact of globalization today, so we should devote ourselves to restoring this lost natural aesthetic.

2   Methods

The research mainly focuses on two aspects: how to generate the spatial sequence with the spirit of Chinese garden place and how to apply this spatial sequence to rural construction. Different from classical architectural research, we prefer to enable machine learning to participate in the creative process and quantify the real world in the digital world. Therefore, we plan the detailed method according to the logic of AI. In addition, we explore how AI cooperates with human beings to complete creation in the design stage. In our workflow, the interaction between architects and AI is in chronological order, but it is not absolutely linear; architects still play a decisive role.

Before introducing the methodology, we will consider the tool we are going to use, which allows us to determine the detailed methodology. In fact, the proliferation of many programming tools has pushed architects to implement AI projects more simply. This project adopts PyTorch, an open source machine learning library developed based on Torch.

In recent years, there are a variety of algorithms in the field of machine learning, so it is necessary to choose the most suitable algorithm for our architectural goals according to our analysis of the "spirit of place" in Chinese gardens. As we pointed out at the beginning, narrative spatial sequence and visual composition are the most important principles of the "place spirit" of Chinese gardens (landscapes), so we will also follow these two principles to define AI models.

In order to generate spatial sequences, two neural network models, pix2pix and RNN (LSTM), are prepared for simulation. Spatial sequence is a series of spaces, so it is very important to learn the path of spatial sequence of Chinese gardens. First, we use pix2pix to generate a path and an approximate spatial layout, and then we use RNN to generate a specific spatial sequence. The traditional artificial neural network is generally composed of a series of layers, and the input of each layer comes from the output of the previous layer. Each layer is made up of independent neurons. However, the spatial sequence of Chinese gardens is composed of multiple spaces, which are compared and changed. Therefore, if each space is regarded as an incoming neuron, it is reasonable to arrange the space according to the order of the visit. In other words, the former space will also affect the experience of the latter space, so the neural network is required to remember the data of the previous space. That's why we chose RNNs, which are particularly suitable for time series analysis, which is often used to generate music or poetry. Like the spatial sequence of Chinese gardens, rhythm is needed between every part of music or poetry.

The methodology of this work is actually the designer's visual experience in space, so the immersive perspective of first-person users is crucial to us. This suggests that the atmosphere of watching a single frame during a tour is also very important, so we use CycleGAN for a style transfer, which can attribute local styles to the spatial sequences generated.

2.1   Generate Plane Informations for Spatial Sequences

Plans are a kind of abstract understanding of three-dimensional space by architects, which is accumulated through knowledge. In this step, the dataset we use are the existing Chinese garden plans, which are selected, mapped and compiled in an atlas by professional architects. On the basis of this atlas, according to the experience of architectural learning, we selected 48 Chinese gardens that are similar to our site scale and thought to be worth studying, traced them, vectorized them in Rhino, including the boundary, buildings, entrances' location, corridors, ponds, stones, rockery, plants, paths implied in the garden, then we augment the dataset to 521 plans.

Then, these data are imported into the neural network pix2pix for training, hoping that it can learn the position of people walking on the plane and generate viewable location points. Since the site we chose, Guyuan village, has many old, abandoned buildings, we decided to preserve and renovate them for festival-related programs to give them new value. Therefore, in this step, we import our site boundary, existing buildings, and entrance locations into the model. Then the model will generate the corridor, ponds, stones, artificial mounds, plants and implied paths. The ponds, stones, artificial mounts are the original man-made landscape in the gardens, and we propose that artists use these locations and create in their place works of similar dimensions in connection with the site; the corridors, where there are often windows to gardens, can be used both for walking and for showing the landscape as a continuous montage, so that the corridors generated in our project are not only walkable spaces but also galleries for visitors. These galleries are used for the exhibition of paintings and photographs. The paths generated in this step are prepared for the subsequent generation of spatial sequences (Fig. 1).

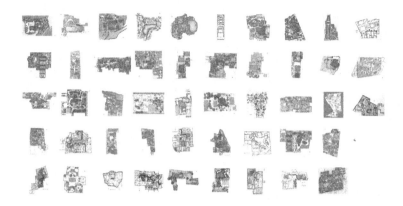

Fig. 1. Atlas for preparing the datasets

2.2   Generate Spatial Sequences

When we are in different spaces, what gives us different feelings? When we are in a space, in addition to the area of the space, the façade is more important from a human point of view, whether the space is open or closed, etc.

After we get the track of people walking in the garden by learning the plan, we will enter the first perspective of people in this step. Traditional architects do design by drawing plans, axonometric drawings, etc., while our work focuses on the spirit of the place, that is, from the perspective of user experience. In the preface to the "An architecture towards Shanshui", Wang Shu holds that the garden is designed from the perspective of peoples' experience, which he compares with the description in Alain Robbe-Grillet's novel "Jealousy", in which Grillet also uses a groundbreaking immersive perspective rather than the omniscient perspective of the traditional writer. The tour's path is the process of people moving in space with the passage of time, and the spatial sequence is the order and change of the space through which this process is studied.

We modeled the vector floor plan described in the previous step in Rhino from the photo. Due to the limited data, we finally selected 10 gardens as the base dataset. Compared with pixel data, vector data uses less data to record information, so it is regarded as a critical task to convert building spatial information into matrix codes that computers can understand. We summarize a set of codes for translating building space, and use data such as spatial coordinates, height, size, wall opening and closing degree to encode the building space. The input condition is the planar information, 108 digits, and the output is the 3D information containing the height and the opacity possibilities of the walls, 256

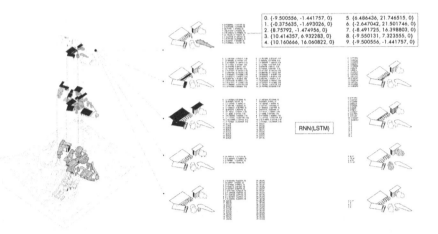

Fig. 2. RNN encoding

digits. For example, if we rotate the cube on a plane, the x and y coordinates of the four points will change, but it will still be the same cube. In other words, it is easy for humans to understand that this is the same cube, but it is a completely different code for machine learning. In order to augment the data, the models are also rotated, ending up with a total of 45 models. Several itineraries are set up in each model to cover elements of the entire garden, and the turning points of the paths are where the user stops to observe during the visit. In these positions, in order to analyze its area of sight, we use the plug-in Isovist in Grasshopper to draw a circle with a radius of 20 m centered on these positions, while the walls or columns of buildings, the columns of corridors, the rockeries, and the overhigh trees block the sight line, resulting in a closed polyline range in which objects are defined as viewable. For each itinerary, the information about the elements seen at each turning point is considered a cell in RNN, arranged by the chronological order of the visit. We use Grasshopper to encode these objects. The location and modulus of the building and corridor have been determined, and the RNN algorithm will predict whether the columns are open or closed to each other, with the opening encoded as 0 and the closure encoded as 1. We thought it might be possible to output a number of 0–1 to determine the degree of closure of the wall. Besides, the RNN algorithm will also predict the elements in the original garden (rockery, stone height), which will be used as a size reference for artists in our scheme. The height of the plant will also be predicted, which will allow us to choose the plant species later. After using Grasshopper encoding, the file is exported as a CSV table file, and then used for RNN training. When the RNN model was trained well enough, the loss values declined significantly, we tried to generate a result and evaluate it with architectural knowledge. We input the information of the site model to RNN in order to generate these prediction results, which are also in CSV table files, then they are imported into Grasshopper and decoded with the same logic. Because there are multiple itineraries within a site, the results generated by these itineraries may overlap. For example, for opacity between the same two columns, the probability that there is a wall between it from one angle is 5%, but the probability that there is a wall between it from another angle is 80%, in which case we decide to take the average, 42.5%. The prediction results of machine learning (as a branch of AI in statistics) are always expressed in the form of percentage. We therefore define opacity in terms of these probabilities. We have customized a rule based on the characteristics of Chinese gardens: when the probability is less than 40%, we defined it as empty, when the probability is more than 60%, we defined there to be a wall, and when the probability is between 40% and 60%, it means that the wall may be hidden from time to time and so we think it is uncertain and will be represented by a semi-transparent wall, for example all kinds of leaky windows in Chinese gardens. Finally, we need to get the integration of these results to form a complete model (Figs. 2, 3, 4 and 5).

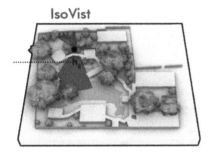

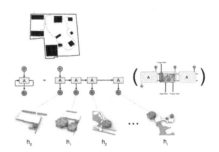

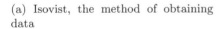

(a) Isovist, the method of obtaining data

(b) Transform the elements seen in each trip into RNN data

Fig. 3. Converting 3D model information into RNN data

After obtaining the results of RNN, our 3D model of the site has been generated, and the properties of the space are roughly determined, partially open and partially closed. Then we will enter the stage of architect's design, and the programs will be arranged according to the properties of the space.

2.3 Materialization

With regard to visual composition, our aim is to restore the 'spirit of place' of Chinese gardens and to integrate Chinese aesthetics into modern functional architecture. In fact, the problem is how to integrate the local Chinese aesthetics into the architectural project. In this regard, we want to use a CycleGAN algorithm.

As previously discussed, the immersive view of the user is very important to us, and in the previous step we defined the places where people will stop and

Fig. 4. The sequence seen by a person on a route, the 3D information of these architectural elements and the frames seen by the person, using the Retreat & Reflection Garden as an example

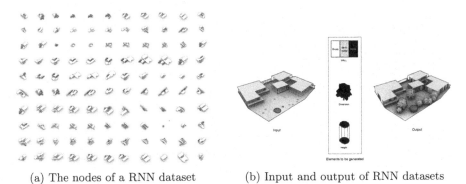

(a) The nodes of a RNN dataset   (b) Input and output of RNN datasets

Fig. 5. RNN datasets

look, and in these places we use the plugin "Vray for Grasshopper" to export images that the person sees as a series of montages following the itineraries, but at this stage the materials are represented simply by colors and are not yet determined. We use CycleGAN, and the datasets are exports of Vray perspective views and photographs of existing local buildings (Figs. 6 and 7).

3   Results and Reflexions

We test this method of AI-aided design with a real case. pix2pix is already a mature algorithm, and we can see that the plan layout generated has the characteristics of Chinese gardens, and the paths are generated which are needed for RNN input information. When the RNN model is trained well enough, we use it to generate 3D information for the case study. Although we have 12,000

(a) The frames exported by Vray for Grasshopper, and style transferred by CycleGAN

(b) Screenshot of the walkthrough animation

Fig. 6. Generation of CycleGAN

Fig. 7. Plan and perspective drawings by architects inspired by the results of the three steps

iterations, to prevent the problem of model overfitting, we stop iterating before the model fully converges to the training data set. Since the intention of our training is to learn a very subjective element like the spirit of place, the level of loss value cannot correctly express the accuracy of the model training. Therefore, we compare the test results with the rules summarized by some Chinese garden experts in practice, and use our professional background as architects and landscape architects to judge the accuracy of the model. For example, we referred to the content in Analysis of Classical Chinese Gardens [10], which says that in Chinese gardens it is important to focus on the relationship between hiding and revealing and the layering of space. In the final model we generated, we can easily feel the rich layering in the space in the first picture, with the foreground/midground/background interpenetrating. In the second picture we can see the relationship between hiding and revealing, so the judgment of the result using our discipline expertise is more convincing than loss value. This also proves the usefulness of the previous knowledge of architectural landscape in the age of AI (Figs. 8, 9, 10 and 11).

Fig. 8. Result of pix2pix, Generation of plane information for the Guyuan village

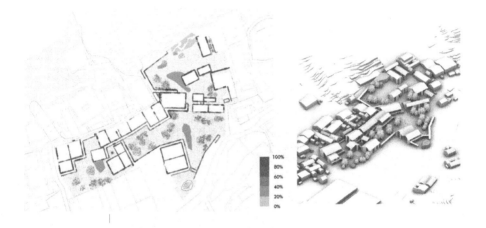

Fig. 9. The opacities of the façade, and the heights of artworks, predicted by trained RNN

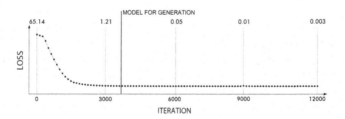

Fig. 10. Loss value of RNN training

(a) Result of RNN, Generation of 3D informa-   (b) Ke Garden in Dongguan b.
tion for the Guyuan village   Suzhou Classical Garden

Fig. 11. Comparison of the generated results and a real Chinese garden

4   Conclusion

As well as seeing the drop in loss values, as architects we should also use our expertise to assess the results generated. We pick out one of the routes to show, and we can see the dynamic relationship of the space in the motion picture, where we can see that there is a rich variation of real and imaginary changes in size and rhythm, with effects such as landscape framing and landscape borrowing. One of the views actually worked for us: the AI generated a towering sculpture above the roof in a small courtyard so that people outside the courtyard could also see a corner of the sculpture showing. These surprises can provide good reference meaning for the architect.

In this thesis we use three algorithms. From the first pix2pix algorithm, we can see that AI has a great prospect of development in design assistance, and is already being applied in a mature way on drawing plans. The second RNN algorithm is the focus of this research and is what we propose based on this feature of spatial sequences. As the dataset is vector data, we found it difficult to converge the training results with insufficient data, and we kept experimenting with the coding approach, as well as augmenting data. Eventually, it has acceptable applicability. The third algorithm is CycleGAN, which was only used in this stage as an experiment to preview the application of local materials. Although not all three parts of our workflow are implemented or sufficiently automated yet, we are optimistic about AI-assisted design. As the technology develops, with the advent of the big data era, statistics-based machine learning can encapsulate the intelligences of learned architects and inspire the architects of today.

The purpose of our project is to explore for the first time the application of RNN in spatial sequences and to integrate it into the architect's workflow. In the RNN phase, we selected 10 gardens, which each have several itineraries, rotated them to augment the data, and then encoded their 3D information as our dataset. This is due to the limited available 3d information, more Chinese gardens are yet to be digitized by peers, and the training results will be enhanced by increasing the amount of data. Because machine learning still relies on a large number of samples.

Regarding the generality of this method, in this experiment, the three steps are separated, but it is viable to integrate the trained models into a plug-in, and other architects can train another space type by replacing the datasets themselves, such as commercial spaces, attractive city streets, similar user experience-oriented space design, etc., but the datasets should be chosen carefully. Machine learning still relies on a large number of samples. To extend our approach to industry, a large amount of labeled data is needed, and then using common machine learning parametric tuning methods to improve the technical details will give better training results (Fig. 12).

The fear of AI comes from a lack of understanding. In fact, in our workflow, architects and AI interact; although it is chronological, it is not absolutely linear, as architects always have a decisive involvement in the AI part and can always review and adjust. AI-assisted design requires not only the manipulation of code, but especially knowledge of architecture. The AI tool that assists in the

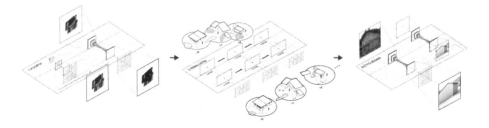

Fig. 12. The workflow shows how AI assists architects with these three steps

design of buildings can therefore only be trained by the architect; it is difficult for the programmer to apply AI directly to architecture, and the architect will probably turn into a meta-architect. When this AI tool has the ability to perceive and judge, we think it already has some creativity, maybe even beyond some architects. And it is too early to worry about AI replacing human jobs, because humans are still far from a strong AI and the theory of everything. A simple example: a computer can only read numbers and it learns datasets which can be existing experiences, but human architects can also think and find new inspiration in life, in society. AI is not meant to replace humans, but rather to benefit ordinary people when they can master it, and decentralized individualistic technology is liberating.

References

1. Carpo, M.: The Second Digital Turn: Design Beyond Intelligence. MIT Press, Cambridge (2017)
2. Chaillou, S.: The Advent of Architectural AI (2019). https://towardsdatascience. com/the-advent-of-architectural-ai-706046960140
3. Huang, W. Zheng, H.: Architectural Drawings recognition and generation through machine learning. In: ACADIA 2018: Re/Calibration: On Imprecision and Infidelity (2018)
4. Luo, Y., Lyu, Y., Yang, X., Zhao, H.: Glitch. Arch. (2020). https:// bproautumn2020.bartlettarchucl.com/rc18/glitch-arch
5. Norberg-Schulz, C.: Genius Loci: Towards a Phenomenology of Architecture. Academy Editions Ltd., New York (1980)
6. Augé, M.: Non-places introduction to an anthropology of supermodernity. Verso (1995)
7. Koolhaas, R.: The Generic City. Sikkens Foundation, Sassenheim (1995)
8. Wang, X.: An Architecture Towards Shanshui. Tongji University Press, Shanghai (2015)
9. Rosenblum, L.D.: See What I'm Saying: The Extraordinary Powers of Our Five Senses. W. W. Norton & Company, New York (2011)
10. Peng, Y.: Analysis of the Chinese Classical Garden (2005)
11. Bachelard, G.: La poétique de l'espace, 11 edn. PRESSES UNIVERSITAIRES DE FRANCE - PUF, Paris (2012)

12. Isola, P., Zhu, J.-Y., Zhou, T., Efros, A.A.: Image-to-Image Translation with Conditional Adversarial Networks (2018). arXiv:1611.07004
13. Karoji, G., Hotta, K., Hotta, A., Ikeda, Y.: Pedestrian dynamic behaviour modeling (2019)
14. Karras, T., Laine, S., Aila, T.: A Style-Based Generator Architecture for Generative Adversarial Networks (2019). arXiv:1812.04948
15. Koh, I., Nagy, D., Lau, D., Locke, J., Stoddart, J., Villaggi, L.: The augmented museum 10 (2020)
16. Negroponte, N.: The Architecture Machine. The MIT Press, Cambridge (1972)
17. Rhee, J., Veloso, P., Krishnamurti, R.: Integrating building footprint prediction and building massing (2020)
18. Wang, R., Zhao, D., Benjamin, D.: Project Discover: An Application of Generative Design for Architectural Space Planning (2017). https://doi.org/10.22360/simaud. 2017.simaud.007
19. Wang, X.: Arcadia Volume I Painting & Garden. Tongji University Press, Shanghai (2014)

A Parametric Tool for Outdoor Shade Design: Harnessing Quantitative Indices and Visual Feedback for Effective and Efficient Climatic Design of Streets

Or Aleksandrowicz[(✉)] ⓘ and Ezra Ozery

BDAR Lab, Technion – Israel Institute of Technology, Haifa, Israel
oraleks@technion.ac.il

Abstract. To date, there is a lack of orderly and data-based methods for quantifying, evaluating, and benchmarking street-level outdoor shade in streets and urban public spaces. The lack of such methods impedes the effective design of walkable and liveable outdoors in locations where shading is essential for significantly mitigating outdoor heat stress. To address this shortcoming, we have developed **Kikayon**, a relatively simple parametric tool that allows planners and designers to easily compare the effect of design alternatives on outdoor shade provision based on building geometry and tree canopy morphologies while taking into account the variance of exposure to solar radiation at different times. The tool calculates several shade and tree indices, some we have originally developed, for each street design, giving users quick and straightforward feedback and enabling them to quantitatively compare design alternatives. Our tool is implemented as a **Grasshopper** code that harnesses several components of the **Ladybug Tools** suite.

Keywords: Urban Microclimate · Shade Maps · Parametric Urban Design · Heat Stress · Outdoor Shade

1 Introduction

The most common indicator of urban overheating is higher air temperatures, usually recorded at street level. An increase in air temperatures can have diverse negative effects, including exacerbated outdoor heat stress (during daytime and nighttime alike), deteriorated air quality, and even increase in mortality rates [1–5]. However, previous research has consistently demonstrated that during the hot season, direct and diffuse solar radiation play a significant role in the generation of excessive daytime heat stress in multiple geographic locations, while outdoor shade provision has shown to be able to significantly mitigate it [6–13]. Shade also reduces health risks caused by exposure to UV radiation [14] while also having the potential to reduce street-level air temperatures and cooling loads in buildings because it decreases the insolation, and thereby the heat absorption, of man-made surfaces in our cities [15].

M. Turrin et al. (Eds.): CAAD Futures 2023, CCIS 1819, pp. 302–316, 2023.
https://doi.org/10.1007/978-3-031-37189-9_20

The potential of shade provision to substantially reduce heat stress is implicit in the most widely used outdoor thermal comfort models, such as, for example, the Physiologically Equivalent Temperature (PET), the Universal Thermal Climate Index (UTCI), and the Index of Thermal Stress (ITS) [16–20]. Such models consider the effect of exposure to shortwave and longwave radiation through inputs like the mean radiant flux intensity or its derivative, mean radiant temperature (MRT) [15]. Exposure to direct and diffuse solar radiation has a decisive effect on increase in MRT values and thus also on the perceived heat stress according to each of these models. In practice, the capacity of urban design to control other factors considered by common comfort models, such as air temperature, relative humidity, and wind speed, is rather limited [21]. Shade, on the other hand, is almost entirely determined by how urban streets and open spaces are designed and maintained, even at a rather local level. Therefore, in many locations, one can argue that tackling inadequate outdoor shading may be the single most important climatic task an urban planner can engage in for reducing urban heat stress.

Cities that already promote climatic urban planning and design policies are usually faced with difficulties in translating them into evidence-based concrete actions [22–24]. This applies also to outdoor shade: while the target of intensifying outdoor shade is widely recognized in climatic action plans around the world [25–30], the detailed planning of its implementation does not yet follow clear, systematic, and replicable methodologies. Besides a handful of initial attempts to suggest quantitative criteria for street shade evaluation [31, 32], the design of outdoor shade almost always relies on basic rules of thumb that at times can be misleading or result in inefficient allocation of resources (for example, in excessive street tree planting in streets that are well shaded as a result of street and building geometries and street orientation). Moreover, assuming that quantitative standards for outdoor shade provisions can be developed and adopted by planning authorities, designers must be provided with simple and design-oriented tools that can enable them to quickly create, evaluate, and improve design alternatives in light of their effect on outdoor shade and the investments required to provide it. This article describes such an attempt that combines quantitative indices for outdoor shade evaluation previously developed by the first author [32, 33] with the development of a parametric tool that applies these indices to assist designers in evaluating shading strategies for streets.

2 Scientific Background

The design tool we have developed enables users to evaluate how effective their street designs are in terms of outdoor shade provision by calculating two indices developed by the first author: a Shade Index and a Shade Availability Index. In addition, and since trees are considered to be one of the most effective street elements for enhancing outdoor shade, the tool also implements a commonly used Tree Canopy Cover Index for quantitatively evaluating the cover provided by tree canopies in a street, which may be indicative to their contribution to street-level shade. One advantage of these indices lies in the way they describe the quality of outdoor shade provision using a unitless scale, thus making them ideal for straightforward evaluation of certain aspects of climatic design even by designers with little knowledge in urban climatology and its complex measurement methodologies. The following sections describe these indices in detail.

2.1 Street Shade Index

A street Shade Index (SI) describes on a scale of 0 to 1 the ratio between the blocked insolation at ground level at a certain location and the maximum insolation of an unobstructed horizontal surface at the same time and location. The higher the value, the higher the shading. This indicator considers shade produced by all elements in an urban environment: buildings, trees, and other shade-giving elements. It can be formulated as follows:

$$SI_p = 1 - \left(\frac{Insolation_p}{Insolation_r} \right) \tag{1}$$

where SI_p is the SI at a certain point, $Insolation_p$ is the insolation at that point, and $Insolation_r$ is the insolation at an unobstructed reference point during the same period. When applied to a street segment or a specific part of a street, SI is calculated as an average of all sampled point SI values contained in that area (the sampling density depends on user preference, though a sampling rate higher than 1 m may overlook fine differences in spatial shade distribution). The SI depends on the date and time of calculation: different dates and times will produce different SI values for the same location and urban morphology. While it is more effective to calculate SI values for mid-summer, when daytime air temperatures are at their peak and heat stress is at its highest level, it is possible to use other dates as reference dates for shade evaluation (for example, during spring and autumn). To evaluate the overall shading effect of street and building geometry during daytime hours, it is better to use the cumulative exposure of ground level during a time range that represents all or most of daytime hours. Nevertheless, it is also possible to calculate SI values for a certain hour, or for a short time range of a couple of hours.

2.2 Sidewalk Shade Availability Index

While SI provides an accurate description of the amount of blocked global radiation at street level, it may not reflect well the spatial distribution of shade in the available space for pedestrian traffic. Here, a different index is suggested to quantify the spatial distribution of shade on sidewalks. The Shade Availability Index (SAI) describes on a scale of 0 to 1 the ratio of daytime hours during a specific time range in which at least 50% of a sidewalk area is shaded (kept unexposed from direct shortwave radiation). The calculation of the SAI is therefore time-dependent: as with the SI, the same street configuration can result in different SAI values not only on different dates, but also when calculating shade availability in different periods. It is advisable to calculate the SAI for the same date and time as the SI, as a complementary index that focuses not on the degree of heat stress caused by exposure to global radiation but rather on the existence of a viable choice for a pedestrian to continuously walk in the shade along a sidewalk.

2.3 Street Tree Canopy Cover

The third index calculated by the tool is that of a Street Tree Canopy Cover (TCC). It describes, on a scale of 0 to 1, the ratio between the projection area on a horizontal surface of all tree canopies located within a street and the total area of the same street. The higher the street TCC value, the higher the tree canopy cover of the street. TCC values can indicate the likelihood of a street to enjoy high levels of street-level shading cast by wide-canopied trees. Nevertheless, since street-level shade depends also on the shade cast by buildings and other physical objects, TCC alone may not describe well the overall shade conditions in a certain street segment, especially where TCC values are low. Therefore, while street TCC can provide complementary information on the dependence of certain street configurations on trees for street-level shade provision, evaluation of shade provision should be done based on the SI and SAI described above.

3 Kikayon: Tool Description

The tool we have developed, named **Kikayon**, is a parametric tool implemented as a **Grasshopper** code while harnessing several components of the **Ladybug Tools** suite [34] to calculate SI and SAI values. It allows planners and designers to easily compare the effect of design alternatives of a street configuration on outdoor shade provision based on building geometry and tree canopy morphologies while taking into account the variance of exposure to solar radiation at different times. Based on this geometry, the tool automatically computes the indices described in Sect. 2 for each sidewalk or the entire street. In addition, the tool provides visual feedback, presenting the typical pattern of shade distribution across the street, which helps to interactively tweak the design based on the quantity and quality of shade provision. Designers are expected to use the tool by following an ordered sequence of actions described below.

3.1 Inputs

User inputs are separated into several input types: street geometry, number and types of shading elements, calculation setup variables, and different display options. Street geometry assumes a street which consists of a road for vehicular traffic and a sidewalk flanking each of its sides. **Street geometry** is thus parametrically created by defining numerical values for the following elements:

1. **Length and orientation** of the modelled street section.
2. Width of each of the **sidewalks**, calculated as the cumulative width of the following street elements: walking strip, street furniture strip, bicycle lane, and a planting strip. Providing width for the walking strip is mandatory while all other components are optional.
3. **Road** width, calculated as the cumulative width of a traffic strip and an optional parking strip (next to one of the two sidewalks or next to both).
4. The width of an optional central **walking boulevard**.
5. **Building** geometry for each of the street sides separately, created by defining the following components: number of buildings, building depth, front and lateral building line, entrance level height, number of typical floors above entrance level, height of a typical floor, front and lateral setbacks of a typical floor (optional), roof level height, and front and lateral setbacks of the roof level (optional). While building design on each of the street sides is determined independently (which means also that one side

of the street can have no buildings at all), the current version of the tool assumes all buildings along a sidewalk are similar and evenly distributed along the sidewalk.

We recommend that the initial calculation of SI and SAI values will be performed without the inclusion of additional shading elements, to evaluate outdoor shade levels resulting from the basic elements of the street: its height-to-width ratio, its building geometry, and its orientation. After performing such an initial calculation, it is easier to consider and explore different shading strategies by adding the three types of shading elements included in the tool: trees, awnings protruding from building facades, and colonnades. It is possible to use each of the three shading element types separately or simultaneously.

The inclusion of trees in the model is done by choosing from a predefined list of tree types representing different tree geometries or by creating custom tree types. **Tree type** geometry is determined by providing numerical values for the following variables: tree trunk radius, tree trunk height, tree canopy radius, and tree canopy height. It is assumed that trees will be planted on one or two rows in each of the sidewalks and the optional central walking boulevard. The type and number of trees in each of these rows, as well as their distance from the curbs, are separately controlled. Users can therefore create an intricate design of tree planting, combining different tree geometries and positions in different planting rows. It is important to note that for keeping radiation calculation times low, tree canopies are represented in the model as uniformly enclosed geometric volumes. This means that unlike real tree canopies, through which some solar radiation is transmitted (usually between 10% to 15% of incident radiation), the modelled canopies do not transmit radiation. It is worth noting that a different approach to light transmissivity modelling, which is based on assigning a light transmissivity value to a solid object, is currently not applicable when using Ladybug's Incident Radiation component.

The two other external shading elements are dependent on the predefined building geometries of each of the street sides. Users can set the depth of a **protruding awning** and its height above ground level or the depth of a **colonnade** running along the front façade of each building (a colonnade's height is identical to the height of the ground floor). As with trees, the inclusion of each of these elements is optional.

Before initializing the calculation procedure of the output indices, users are required to define some of the calculation settings. First, to ensure correct calculation of street-level exposure to incoming shortwave radiation, users need to select a standardized weather file reflecting the geographic location of the modelled street from the online EPW repository at climate.onebuilding.org. Next, users should define the date and hours of calculation, depending on the seasonal effect of shade that is being considered and the simulated geographic location. For example, to calculate outdoor shade conditions during the height of summer in Tel Aviv-Yafo, we set the calculation date to 6 August and the calculation time between 07:00 and 16:00 (standard time, totalling 10 h). Since EPW weather files contain a mixture of clear sky and overcast conditions, a graphical representation of incoming solar radiation levels during the analysis period informs users whether the selected analysis period reflects clear sky conditions. The analysis of outdoor shade conditions should take into account the maximum heating effect of solar radiation, and therefore the selected analysis period should not include hours of significantly low incoming direct solar radiation.

Other variables that can be defined by the users include some display and results-saving options since results for each run are displayed as a combination of visual and textual outputs (see below). Users can set a destination folder to automatically save screenshots that consist of a top view of the design, including a "heat map" of ground level irradiance levels, a compass sign indicating street orientation, a reference top view of a sky dome showing incoming direct radiation levels during the analysis period, and a textual representation of the analysis results.

3.2 Calculation and Outputs

Using the input geometry and weather data, the tool applies the Incident Radiation component of the **Ladybug Tools** suit to calculate the cumulative incidence of global solar radiation (direct and diffuse) on all ground surfaces in the street model. For calculating SI values, additional calculation of the solar incidence on an unshaded horizontal reference cell outside the modelled street is simultaneously executed. The **Ladybug** component performs separate calculation for each grid cell of the radiation-receiving surface. Calculation grid size can be controlled by the user: we recommend starting preliminary calculations at a cell size of 4 m but performing detailed calculations at a cell size of 1 m or less. Calculation time of a 200 m long street with a cell size of 1 m may take about 45 s.

It is important to note that the Incident Radiation component does not take into account reflected radiation from any of the model's surfaces, and therefore underestimates the levels of solar irradiance at ground level. This means that the calculated SI values are to a certain degree lower than what would have resulted from a calculation method that takes into account reflected solar radiation within the simulated street canyon. Yet, since in most realistic cases surface albedo levels within a street canyon are expected to be relatively low (0.10 to 0.20) for vertical and horizontal surfaces alike, we believe that Ladybug's lack of consideration of reflected solar radiation cannot significantly distort the insights derived from comparing SI values of different street designs.

To determine the SI and SAI values, the raw output of the Incident Radiation calculation is further analysed by the tool based on additional code. Output SI values are calculated for each sidewalk separately, for the central boulevard (if existing), for the road, and for the entire right of way (sidewalks, boulevard, and road) of the street. Output SAI values are calculated for each sidewalk and for the central boulevard (if existing) separately.

Calculation of TCC values for the entire area of the street is based on the geometric properties of the trees and the street. Other calculated outputs relating to trees are the number of trees used in the entire model, and the number of trees per area unit (in our case, a dunam, which equals 1000 sqm) in the entire model, which represents the tree density of the street. When evaluated with respect to the SI and SAI values, these three quantitative outputs can give indications of the added value of increased number of trees in terms of their effect on outdoor shade.

4 Use Scenario Example

4.1 Baseline Geometry

The developed tool is designed to provide relatively quick and straightforward feedback on employing different design strategies for outdoor shade provision in a way that can easily support design decisions based on quantitative outputs. In the following example, set in the Mediterranean city of Tel Aviv-Yafo, we begin with a predefined 200 m long street with a central pedestrian boulevard 15 m wide flanked by two roads 7.5 m wide and two sidewalks 8 m wide (totalling in a right of way 46 m wide). The street is positioned on a north-south axis so that the main building facades on both sides of the streets face east and west respectively. The buildings on each of the street sides have different geometric features, as described in Table 1. A 3D rendering of the baseline street geometry, as it is presented to the users, appears in Fig. 1.

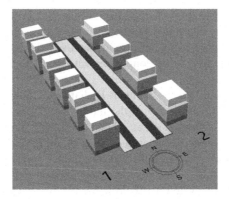

Fig. 1. A 3D representation of the baseline model, before adding shading elements.

Shade values for this street design were calculated for 6 August between 07:00 and 16:00 (default calculation time). The results screenshot (Fig. 2) details the resulting SI and SAI values, as well as a top-view representation of the radiation exposure levels in kwh per sqm across the entire area of the street, a compass symbol showing the street's orientation, a circle representing the corresponding sky dome indicating the incoming radiation during the calculated hours, and details on the calculated period, location, and source weather file.

While the results of the baseline calculation show relatively low SI values in both sidewalks (0.27), they also indicate that for several hours during that day each of the sidewalks provide good shading conditions (represented by an SAI value of 0.4). This is a result of the combined effect of street orientation and building geometries. While the central boulevard also benefits from some shade cast by the buildings, its SI and SAI values are lower than the corresponding values of each of the sidewalks, because of the relative openness of the central section of the street to the sky.

Table 1. Street geometry properties of the design example.

Street side	Western (side 1)	Eastern (side 2)
Number of buildings	6	4
Building frontage width [m]	17.3	30
Front building line [m]	0	0
Lateral building line [m]	8	10
Building depth [m]	22	22
Entrance level height [m]	5	5
Number of typical floors	6	4
Typical floor height [m]	3	3
Front setback of a typical floor [m]	0	0
Lateral setback of a typical floor [m]	0	0
Roof level height [m]	8	8
Front setback of a typical floor [m]	3	3
Lateral setback of a typical floor [m]	3	3

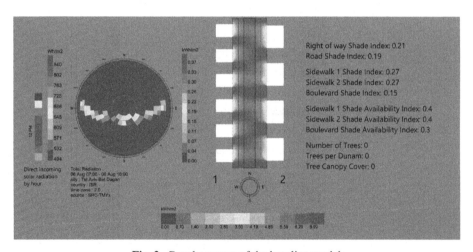

Fig. 2. Results screen of the baseline model.

4.2 Design Scenarios for Increased Shade

Assuming that street and building geometries must be kept unchanged, the baseline results mean that improvement of the shade conditions in the sidewalks and the boulevard has to rely on additional shading elements. The following design steps thus explored several shading strategies: adding one row of trees to the sidewalks or two rows of trees to the boulevard and redesigning the buildings to include a colonnade on each of the street sides (without changing their number, width, or height). Details on the additional shading element in each of the design scenarios (six in total) appear in Table 2, while 3D renderings of each design scenario appear in Fig. 3.

Table 2. Shading elements description for each design scenario.

	Design Scenario	A	B	C	D	E	F
Western sidewalk (one row of identical trees)	Number of trees	8	0	0	0	0	0
	Distance between tree trunks [m]	25	–	–	–	–	–
	Tree canopy height [m]	6.4	–	–	–	–	–
	Tree canopy radius [m]	4.5	–	–	–	–	–
	Tree trunk height [m]	1.9	–	–	–	–	–
	Tree trunk radius [m]	0.3	–	–	–	–	–
	Colonnade depth [m]	–	–	–	5	5	5
Eastern sidewalk (one row of identical trees)	Number of trees	16	0	0	0	0	0
	Distance between tree trunks [m]	12.5	–	–	–	–	–
	Tree canopy height [m]	6.8	–	–	–	–	–
	Tree canopy radius [m]	2.5	–	–	–	–	–
	Tree trunk height [m]	9	–	–	–	–	–
	Tree trunk radius [m]	0.3	–	–	–	–	–
	Colonnade depth [m]	–	–	–	5	5	5
Boulevard (two rows of identical trees)	Number of trees	0	10	15	15	15	12
	Distance between tree trunks [m]	–	20	13.3	13.3	13.3	16.6
	Tree canopy height [m]	–	6.4	6.4	6.4	4.0	6.4
	Tree canopy radius [m]	–	4.5	4.5	4.5	2.5	4.5
	Tree trunk height [m]	–	1.9	1.9	1.9	3.0	1.9
	Tree trunk radius [m]	–	0.3	0.3	0.3	0.3	0.3

4.3   Results

Comparison of the calculation results of the design scenarios and the baseline case (Table 3) can assist in understanding the contribution, as well as the limits, of applying different shading strategies. While it is always possible to integrate numerous street trees with wide and lush canopies in a street design to secure high levels of outdoor shade, realistic and effective shade tree planting must recognize the inherent difficulties of providing adequate underground space for achieving a developed tree-root system without which canopies will remain undeveloped and ineffective in terms of shading. This means that when wide canopied trees are part of the design, this must go hand in hand with careful planning of underground technical infrastructures in the trees' vicinity. The bigger the tree canopy, the larger the underground soil volume it requires [35]. The different design scenarios reflect this limitation: they try to balance between the inclusion of a reasonable number of wide-canopied trees and other shading strategies that do not rely on trees.

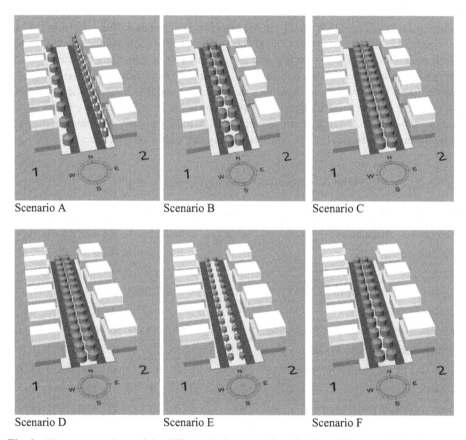

Scenario A   Scenario B   Scenario C

Scenario D   Scenario E   Scenario F

Fig. 3. 3D representations of the different design scenarios of adding different shading elements.

The first design scenario (scenario A) used two different tree types planted along each of the sidewalks (12 trees in each row). Wide-canopied trees were positioned on the western sidewalk, and trees of narrow and high tree canopies (which require less underground soil volume) were positioned on the eastern sidewalk. This difference was reflected in the resulting SI and SAI values of each of the sidewalks, although the eastern sidewalk's SAI remained the same as in the baseline design, marking lower shading efficacy of the narrow-canopied trees used in that sidewalk. Adding two rows of wide-canopied trees to the central boulevard (scenario B) instead of the sidewalk trees significantly improved the boulevard's SI and SAI values, but at the same time had only a small contributing effect to the shading of the western and eastern sidewalks.

Increasing the number of boulevard trees by 50% (from 20 to 30, scenario C) had a significant effect on shade quality in the boulevard (increasing SI values from 0.53 to 0.71 and SAI values from 0.50 to 1.00) but will probably require much higher initial investment because of the doubling of wide-canopied trees. Changing the tree types of the boulevard trees to smaller, and thus less expensive, trees (scenario E) significantly reduced their effectiveness in providing more than reasonable shade in the boulevard. An effective compromise between the number of trees and their shading effect in the boulevard was reflected in scenario F, which was based on slightly reducing the number of boulevard trees from 30 to 24 while still using the wide-canopied trees.

Table 3. Calculation results for all design scenarios.

Design Scenario	Baseline	A	B	C	D	E	F
Right of way Shade Index	0.21	0.32	0.39	0.48	0.56	0.39	0.51
Road Shade Index	0.19	0.30	0.36	0.44	0.50	0.34	0.45
Western sidewalk Shade Index	0.27	0.49	0.30	0.31	0.46	0.43	0.46
Eastern sidewalk Shade Index	0.27	0.40	0.30	0.31	0.48	0.45	0.47
Boulevard Shade Index	0.15	0.20	0.53	0.71	0.73	0.40	0.63
Western sidewalk Shade Availability Index	0.40	0.60	0.50	0.50	0.60	0.60	0.60
Eastern sidewalk Shade Availability Index	0.40	0.40	0.40	0.40	0.70	0.60	0.70
Boulevard Shade Availability Index	0.30	0.30	0.50	1.00	1.00	0.50	1.00
Number of trees	0	24	20	30	30	30	24
Trees per dunam [1000 sqm]	0	2.61	2.17	3.26	3.26	3.26	2.61
Street Tree Canopy Cover	0	0.09	0.14	0.21	0.21	0.06	0.17

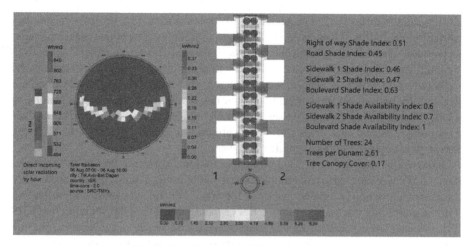

Fig. 4. Results screen of Scenario F model.

As for the shading of the sidewalks, scenarios D, E, and F integrated colonnades into the building design on both sidewalks instead of the trees. The colonnades provided slightly better SI and SAI values in both sidewalks, making them an effective alternative to tree planting with all its inherent limitations and difficulties. Scenario F (Fig. 4) thus reflected a combined shading strategy that while using the same number of trees as in scenario A, significantly increased the shading conditions in all parts of the street.

5   Discussion and Conclusion

The above use example demonstrates the flexibility of the developed tool in analysing outdoor shading conditions under a variety of street and building geometries, as well as the employment of different shading strategies. By providing quantitative and relatively quick feedback on the shading qualities of different design combinations, the tool is expected to enhance user engagement with climatic design and improve the shading efficacy of a variety of design options. While the tool can be used for design optimization, it is more important to use it for considering the advantages and disadvantages of a wide variety of shading alternatives.

Since shade trees are expensive elements to implement and maintain, the tool can help in achieving a reasonable balance of increasing shading conditions while keeping shading costs low. Future versions may also include components that would quantify the financial cost of tree planting based on required tree canopy dimensions and the resulting underground soil volume for the tree's root system. At the same time, the tool also enables the users to quickly understand to what extent the avoidance of tree planting, which is currently perceived as one of the most effective urban design tools for urban cooling [36], may become detrimental to summer thermal comfort of pedestrians and other road users. The visual presentation of street-level shade distribution assists in figuring out better shading options and their possible effectiveness.

While the current level of scientific knowledge in urban climatology can support effective climatic urban design, the implementation of proactive climatic urban design is still generally limited [22, 23]. Arguably, this can be attributed to professional hurdles restricting the ability of design professionals to thoroughly understand and implement the products of scientific research in routine design tasks. The developed tool attempts to streamline the integration of climatic considerations in urban design while bridging the gap between scientific knowledge and design considerations. Users of **Kikayon** do not need to fully understand the intricate and complex nature of microclimatic phenomena in cities. Instead, they can rely on several simple indices that reflect the effect of their design on pedestrian heat stress while concentrating on familiar design tasks as giving shape to a street, its buildings, and its vegetation.

Kikayon is designed as a design-assisting tool and therefore prefers the ease and speed of use over the creation of complex street and building geometries. It substantially differs from common computational tools for urban climate simulation and evaluation of urban morphologies, as, for example, ENVI-met [37] and the UMEP plugin for QGIS [38], in the relative speed of model creation, ease and speed of calculation, and the presentation of the climatic effect of design through several unitless indices that can be easily communicated to professionals with little knowledge in urban climatic processes. This, however, comes with a price, not only because the tool limits itself to shade quantification alone, but also because the street geometries that can be calculated by the tool have to conform to certain design conventions and assumptions. The latter limitation can be addressed in future versions of the tool, which may enable users to perform calculations on street and building geometries that cannot be properly represented using the existing version of the tool. However, user control of a wide range of design features in the current version means that the tool is already useful for analysing numerous instances of real-life urban design scenarios.

Acknowledgement. The development of the tool was supported by the Israel.100 initiative. The authors would like to thank Arch. Shamay Assif for his ongoing support and Arch. Omri Ben-Chetrit for his valuable comments.

References

1. Arnfield, A.J.: Two decades of urban climate research: a review of turbulence, exchanges of energy and water, and the urban heat island. Int. J. Climatol. **23**, 1–26 (2003). https://doi.org/10.1002/joc.859
2. Johansson, E., Emmanuel, R.: The influence of urban design on outdoor thermal comfort in the hot, humid city of Colombo. Sri Lanka. Int. J. Biometeorol. **51**, 119–133 (2014). https://doi.org/10.1007/s00484-006-0047-6
3. Kleerekoper, L., van Esch, M., Salcedo, T.B.: How to make a city climate-proof, addressing the urban heat island effect. Resour. Conserv. Recycl. **64**, 30–38 (2012). https://doi.org/10.1016/j.resconrec.2011.06.004
4. Nikolopoulou, M., Baker, N., Steemers, K.: Thermal comfort in outdoor urban spaces: understanding the Human parameter. Sol. Energy. **70**, 227–235 (2001). https://doi.org/10.1016/S0038-092X(00)00093-1
5. Oke, T.R.: Canyon geometry and the nocturnal urban heat island: comparison of scale model and field observations. J. Climatol. **1**, 237–254 (1981). https://doi.org/10.1002/joc.3370010304

6. Colter, K.R., Middel, A., Martin, C.A.: Effects of natural and artificial shade on human thermal comfort in residential neighborhood parks of Phoenix, Arizona, USA. Urban For. Urban Green. **44**, 126429 (2019). https://doi.org/10.1016/j.ufug.2019.126429

7. Coutts, A.M., White, E.C., Tapper, N.J., Beringer, J., Livesley, S.J.: Temperature and human thermal comfort effects of street trees across three contrasting street canyon environments. Theoret. Appl. Climatol. **124**, 55–68 (2015). https://doi.org/10.1007/s00704-015-1409-y

8. Middel, A., Krayenhoff, E.S.: Micrometeorological determinants of pedestrian thermal exposure during record-breaking heat in Tempe, Arizona: introducing the MaRTy observational platform. Sci. Total Environ. **687**, 137–151 (2019). https://doi.org/10.1016/j.scitotenv.2019.06.085

9. Middel, A., AlKhaled, S., Schneider, F.A., Hagen, B., Coseo, P.: 50 grades of shade. Bull. Am. Meteorol. Soc. **102**, 1–35 (2021). https://doi.org/10.1175/BAMS-D-20-0193.1

10. Aleksandrowicz, O., Pearlmutter, D.: The significance of shade provision in reducing street-level summer heat stress in a hot Mediterranean climate. Landsc. Urban Plan. **229**, 104588 (2023). https://doi.org/10.1016/j.landurbplan.2022.104588

11. Lee, H., Holst, J., Mayer, H.: Modification of human-biometeorologically significant radiant flux densities by shading as local method to mitigate heat stress in summer within urban street canyons. Adv. Meteorol. **2013**, 1–13 (2013). https://doi.org/10.1155/2013/312572

12. Lee, I., Voogt, J.A., Gillespie, T.: Analysis and comparison of shading strategies to increase human thermal comfort in urban areas. Atmosphere (Basel). **9**, 91 (2018). https://doi.org/10.3390/atmos9030091

13. Shashua-Bar, L., Pearlmutter, D., Erell, E.: The influence of trees and grass on outdoor thermal comfort in a hot-arid environment. Int. J. Climatol. **31**, 1498–1506 (2011). https://doi.org/10.1002/joc.2177

14. Gandini, S., Autier, P., Boniol, M.: Reviews on sun exposure and artificial light and melanoma. Prog. Biophys. Mol. Biol. **107**, 362–366 (2011). https://doi.org/10.1016/j.pbiomolbio.2011.09.011

15. Erell, E., Pearlmutter, D., Williamson, T.J.: Urban microclimate: Designing the Spaces Between Buildings. Earthscan, London and Washington, DC (2011)

16. Bröde, P., et al.: The universal thermal climate index UTCI compared to ergonomics standards for assessing the thermal environment. Ind. Health. **51**, 16–24 (2013). https://doi.org/10.2486/indhealth.2012-0098

17. Matzarakis, A., Mayer, H., Iziomon, M.G.: Applications of a universal thermal index: physiological equivalent temperature. Int. J. Biometeorol. **43**, 76–84 (1999). https://doi.org/10.1007/s004840050119

18. Matzarakis, A., Muthers, S., Rutz, F.: Application and comparison of UTCI and PET in temperate climate conditions. Finisterra. **49** (2014). https://doi.org/10.18055/Finis6453

19. Givoni, B.: Man, Climate and Architecture. Elsevier, Amsterdam (1969)

20. Pearlmutter, D., Berliner, P., Shaviv, E.: Integrated modeling of pedestrian energy exchange and thermal comfort in urban street canyons. Build. Environ. **42**, 2396–2409 (2007). https://doi.org/10.1016/j.buildenv.2006.06.006

21. Aleksandrowicz, O., Vuckovic, M., Kiesel, K., Mahdavi, A.: Current trends in urban heat island mitigation research: observations based on a comprehensive research repository. Urban Clim. **21**, 1–26 (2017)

22. Hebbert, M., Mackillop, F.: Urban climatology applied to urban planning: a postwar knowledge circulation failure. Int. J. Urban Reg. Res. **37**, 1542–1558 (2013). https://doi.org/10.1111/1468-2427.12046

23. Mills, G.: Urban climatology: history, status and prospects. Urban Clim. **10**, 479–489 (2014). https://doi.org/10.1016/j.uclim.2014.06.004

24. Erell, E.: The application of urban climate research in the design of cities. Adv. Build. Energy Res. **2**, 95–121 (2008). https://doi.org/10.3763/aber.2008.0204

25. Shorris, A.: Cool Neighborhoods NYC: A Comprehensive Approach to Keep Communities Safe in Extreme Heat. New Yotk City's Mayor's Office of Recovery and Resiliency, New York (2017)
26. Osmond, P., Sharifi, E.: Guide to Cooling Strategies, pp. 1–72 (2017)
27. Brandenburg, C., Damyanovic, D., Reinwald, F., Allex, B., Gantner, B., Czachs, C.: Urban Heat Island Strategy: City of Vienna. Vienna Environmental Protection Department (MA22), Vienna (2018)
28. Francis, J., Hall, G., Murphy, S., Rayner, J.: Growing Green Guide: A Guide to Green Roofs, Walls and Facades in Melbourne and Victoria, Australia. Department of Environment and Primary Industries, State of Victoria, Melbourne (2014)
29. Ruefenacht, L., Acero, J.A.: Strategies for Cooling Singapore. Singapore ETH Centre, Singapore (2017)
30. Tel Aviv-Yafo Municipality: Climate Adaptation Action Plan (2020)
31. Peeters, A., et al.: A decision support tool for calculating effective shading in urban streets. Urban Clim. **34**, 100672 (2020). https://doi.org/10.1016/j.uclim.2020.100672
32. Aleksandrowicz, O., Zur, S., Lebendiger, Y., Lerman, Y.: Shade maps for prioritizing municipal microclimatic action in hot climates: learning from Tel Aviv-Yafo. Sustain. Cities Soc. **53**, 101931 (2020). https://doi.org/10.1016/j.scs.2019.101931
33. Aleksandrowicz, O.: Mapping and management of urban shade assets: a novel approach for promoting climatic urban action. In: Khan, A., Akbari, H., Fiorito, F., Mithun, S., and Niyogi, D. (eds.) Global Urban Heat Island Mitigation, pp. 1–27. Elsevier, Amsterdam, Netherlands ; Kidlington, Oxford, England ; Cambridge, Massachusetts (2022)
34. Sadeghipour Roudsari, M., Pak, M.: Ladybug: a parametric environmental plugin for Grasshopper to help designers create an environmentally-conscious design. In: BS2013: 13th International IBPSA Conference, Chambéry, France (2013)
35. Urban, J.: Two different approaches to improve growing conditions for trees comparing Silva cells and structural soil. Am. Soc. Consult. Arborists. **46**, 5–14 (2013)
36. Aleksandrowicz, O., Vuckovic, M., Kiesel, K., Mahdavi, A.: Current trends in urban heat island mitigation research: observations based on a comprehensive research repository. Urban Clim. **21**, 1–26 (2017). https://doi.org/10.1016/j.uclim.2017.04.002
37. ENVI-met v. 4.3.2. http://www.envi-met.com (2018)
38. Lindberg, F., et al.: Urban Multi-scale Environmental Predictor (UMEP): an integrated tool for city-based climate services. Environ. Model. Softw. **99**, 70–87 (2018). https://doi.org/10.1016/j.envsoft.2017.09.020

Urban Shaderade. Building Space Analysis Method for Energy and Sunlight Consideration in Urban Environments

Francesco De Luca[1]([⊠]) [iD] and Abel Sepúlveda[1,2] [iD]

[1] Department of Civil Engineering and Architecture, Academy of Architecture and Urban Studies, Tallinn University of Technology, Ehitajate Tee 5, 19086 Tallinn, Estonia
francesco.deluca@taltech.ee
[2] Institute of Design and Civil Engineering, Architecture and Intelligent Living, Karlsruhe Institute of Technology, Englerstraße, 7, 76131 Karlsruhe, Germany
abel.luque@kit.edu

Abstract. The built environment contributes significantly to climate change being responsible of a large portion of global energy use and CO_2 emissions. In the scientific community it become evident that designing urban environments and buildings which do less harm or have a neutral impact on the environment is not enough anymore to realize sustainable cities. This paper presents a method to help design buildings with a positive impact on the context, filling vacant lots in dense urban environments. The method defines optimal building boundaries to reduce energy use of existing surrounding premises, while guaranteeing them adequate solar access. A two-step computational workflow was developed. In the first step building space positive, negative and neutral effect on cooling, heating and electric lighting energy use is analyzed in consideration of shading factors and beam solar energy. In the second, the effect of the positive and neutral building space on sunlight exposure of neighboring premises is considered, generating energy and sunlight optimized conceptual building massing. Initial results of the method application in several urban conditions, different building use and scenarios and in different cities are presented and discussed.

Keywords: Climate change · Resilient urban design · Building energy use · Sunlight exposure · Environmental simulations · Computational design

1 Introduction

Nowadays sustainable development is in the agenda of most countries and international organizations. Among the United Nations' seventeen Sustainable Development Goals, several concern the development of cities [1]. Goal 11 Sustainable Cities and Communities sets as targets for 2030 the resource efficiency and mitigation and adaptation to climate change of cities. Also in 2020 the building sector was one of the main responsible for resource depletion and global warming due to its 36% and 37% share of global energy use and energy-related CO_2 emissions, respectively [2].

© The Author(s), under exclusive license to Springer Nature Switzerland AG 2023
M. Turrin et al. (Eds.): CAAD Futures 2023, CCIS 1819, pp. 317–332, 2023.
https://doi.org/10.1007/978-3-031-37189-9_21

The loss of undeveloped and pristine lands on the outskirts of cities due to built environment expansion reduces and harms natural habitats and ecosystem, suggesting a more efficient use of brownfields and urban voids. In this regard, environment policies were developed as the Roadmap to a Resource Efficient Europe with the goal of "no net land take" by 2050 [3]. Since early 1990s urban intensification and the compact city have been investigated for the potential positive effects on sustainability due to reduced energy consumption and greenhouse gas emissions associated with reduced travel distances, and on social liveliness and economic activity [4].

Urban form and density, buildings mass and distance significantly influence the solar radiation received by building facades which is one of the main factors of indoor thermal and visual comfort, and energy use [5, 6]. Window features, properties and size can reduce solar gains, thus cooling energy use, during the warm season and take advantage of direct solar radiation during the cold season, thus reducing heating energy use [7, 8], while also guaranteeing adequate daylight during the entire year [9, 10].

The availability of natural light is one of the most appreciated characteristics of building interiors improving surfaces contrast and color rendering [11]. Daylight has a positive effect on the psychological and physiological wellbeing of building occupants, improving also workers' satisfaction and productivity [12]. Since humans spend the majority of time indoor, the yearly and daily variability of sunlight entering the building interiors help significantly to entrain occupants' circadian rhythm [13].

Beyond the sustainability approach, in recent times is getting relevance the research question about the possibility to realize buildings not only less harmful or with a neutral effect on the environment, but with a positive impact on the existing urban environment and natural systems [14]. This paper is a contribution to answer the research question proposing a method to use in sustainable development of compact cities, for the design of buildings with a positive impact on the urban environment.

2 Background

Defining the building massing which guarantees access to solar radiation of existing premises is a critical step during the early stages of the design process for the design of sustainable buildings and neighborhoods. The solar envelope method allows the designer to determine the maximum volume a new building cannot exceed to guarantee direct sunlight on surrounding facades [15]. The inputs for the generation are: the building plot; the shadow fences, i.e., the lines on the surrounding facades above which sunlight must be guaranteed; and the sun azimuth and altitude angles of the hours between the required start and end of the analysis period. Since three decades, computer tools have been developed to generate solar envelopes using the inputs of the conventional method [16]. The solar envelope resembles an irregular pyramid.

An advancement of the solar envelope method and computer tools was introduced to tackle the limitation of the start-and-end hour period selection [17]. This input makes the conventional method inefficient when solar access must be guaranteed for a daily number of hours or a percentage of actual sunlight hours as prescribed by existing regulations. The advanced method firstly sorts the available sun rays for specific days and windows according to different criteria, and secondly selects the required quantity. Thus, the designer can generate solar envelopes in consideration of a qualitative approach, e.g., using hours with higher solar radiation entering the building that is beneficial for occupants' health, or of a quantitative approach, e.g., using hours with higher sun altitude to realize larger buildable volumes, or through trade-offs.

The Reverse Solar Envelope (RSE) is a new method [18]. Differently than the previous methods and computer tools generating solar envelopes translating vertically points of a grid and then using the points to generate the envelope Mesh, RSE subtracts cells from the maximum buildable volume if they obstruct the surrounding windows during select hours. Additionally, the new method expands further the sun ray selection criteria of the advanced method. The RSE method is particularly efficient in dense urban environments generating several massing variations of a larger volume comparing previous methods, and allowing initial performance studies on the building envelope and interiors. Both the advanced and new RSE method were included in the solar envelope tools of Solar Toolbox, a plug-in for Grasshopper [19].

A number of studies investigated the potential of the solar envelope method at block scale to allow solar radiation on building facades maximizing passive solar gains, to reduce energy need during the cold season [20, 21]. The solar envelope was used to determine buildings massing and distance to analyze energy uses of a typical development in different US cities and relative densities as indicator of walkability or transit service [22]. The existing studies using solar envelopes for energy consideration are based on the limited time selection method start-and-end hour. The shading envelope is a method developed for hot climates, to investigate urban massing strategies for limiting direct solar radiation to improve outdoor thermal comfort [23]. Further investigations developed methods for generating shading envelopes in consideration of the new building energy use reduction and of self shading criteria [24, 25].

As the solar and shading envelope methods which allow and control direct solar radiation at the urban scale, window shading devices have been largely investigated for the same scope at the building scale [26]. Two shading device design methods are relevant for the present work. The Cellular shading method subdivides the shading device surface in theoretical cells, and calculates the shading importance of each cell in blocking or allowing solar radiation to the interior [27]. The shape of the device is determined through accumulation of cells hourly shading importance during the analysis period, according to predicted hourly heating and cooling energy use of the room. The Shaderade method, also based on the subdivision of the shading device or three-dimensional shading volume, added consideration of the undesirable fraction of solar beam energy causing cooling load for every time step of the zone thermal simulation [28]. Thus, the cell transmittance or its inclusion is considered for the determination of the translucent or opaque energy optimized shading device, respectively.

This paper presents the current step of a research for the development of a method to help investigating new building forms with the potential of reducing energy use while guaranteeing healthiness of existing premises in urban environments [29]. Building on the new generation methods for solar envelopes and shading devices with energy considerations, the scientific novelty of the proposed method is to consider the influence of discretized elements of the potential building volume on energy use and on required sunlight exposure of surrounding premises, to help realizing context energy and solar access optimized conceptual building forms. The steps of a computational workflow, method's initial application and results are presented.

3 Methods

To develop the method for analyzing the building space influence on energy use and solar access of existing premises, finalized to help realizing conceptual building forms with a positive impact on the existing urban environment, a computational workflow was developed (Fig. 1). In this study, we refer to building space as the maximum buildable volume subdivided in cells, each with an associated positive, negative or neutral impact on energy use and positive or negative on solar access. We also refer to solar access and sunlight exposure as interchangeable terms.

The steps of the computational workflow were: 1) three-dimensional modeling of urban environments; 2) parametric modeling of test room and clustering of building façade samples for rooms location; 3) parametric modeling of urban scenarios, thermal modeling, energy and solar access simulation in the existing situation; 4) calculation of building space cell effect on energy use of surrounding premises; and 5) generation of context energy and solar access optimized conceptual building forms.

The computational workflow was realized in the Rhinoceros and Grasshopper environment [30]. Three-dimensional and parametric modeling was done using standard tools. Solar radiation and energy simulations were performed using the validated simulation software Radiance [31], and EnergyPlus [32], respectively, through the climatic, daylight and energy modeling tools of ClimateStudio [33], a plug-in for Grasshopper. Sunlight hour calculations were performed using Ladybug Tools [34], an environmental design plug-in for Grasshopper. For the development of the computational workflow, several custom components and automation functions were developed by the authors.

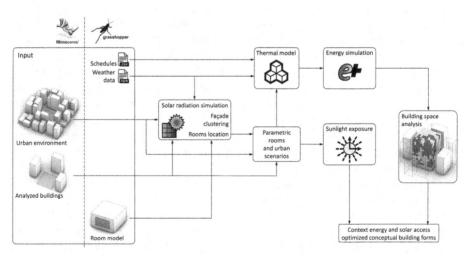

Fig. 1. The computational workflow realized for the development of the method.

3.1 Urban Environments

Three hypothetical urban environments were realized in three cities located at very different latitudes to analyze the developed method in different urban and climatic conditions. The three urban environments present the same plots and buildings used for energy and solar access analysis, but different surrounding buildings layout and height to obtain three urban densities: low; medium; and high (Fig. 2).

Eight 35 × 35 m plots constituted the existing urban environment. A central square plot of the same size, was used as the new building volume location. The distance between opposite plots was 10 m. The three existing buildings used for analysis were located opposite the central plot toward north, east and west, and present heights of 35 m, 42 m and 49 m, in all the three urban scenarios. In the eight plots surrounding the central one, different building layouts were used in the three urban environments different for density. Consequently, the building footprint polylines were extruded to generate solids with heights between 8.3 and 28.5 m, 10.5 and 49 m, and 31.3 and 74 m for the low, medium and high density urban environment, respectively, corresponding to 11.5 m^3/m^2, 17.6 m^3/m^2, and 29.1 m^3/m^2, respectively.

The cities were: Tallinn, Estonia (59.41°N, 24.83°E); Paris, France (49.02°N, 2.53°E); and Athens, Greece (37.92°N, 23.93°E). The Köppen–Geiger climate classification are, respectively: Cold, no dry season, warm summer (Dfb); Temperate, no dry season, warm summer (Cfb); and Temperate, dry hot summer (Csa).

Fig. 2. The urban environments with low, medium, and high density (left to right). The buildings used for energy and solar access analysis (red). The new building plot in the center (grey).

3.2 Façade Samples Clustering

For energy and solar access analysis, the facades of the existing buildings facing the central plot and oriented south, west and east, were selected and subdivided in samples 3.5 m in size, representing window locations, with sensor points at 1.75 m for higher simulation accuracy. To shorten long energy simulation time needed for all the rooms of the façades taking into account the effect of the building space, three representative rooms per façade were selected following efficient and validated methods [35]. Being solar beam allowed or blocked by the new building massing the main factor for energy use and solar access variation, the representative rooms were located performing annual direct solar radiation simulations on the façade samples (Fig. 3), and through k-means clustering, using the Grasshopper plug-in LunchBox [36]. The simulation results were divided in three clusters per façade and the façade sample with the solar radiation closer to the cluster centroid, i.e., the average solar radiation value, was selected. The clustering was performed for the three urban environments and three cities, producing nine groups of nine representative rooms, three per façade.

The room was 5 m wide, 5 m deep, and 3 m high, which represent the size of a residential living room with kitchen, and of a multi-desk office room. A section of the computational workflow inserted automatically the test room in the location of the selected façade samples on the existing analyzed buildings (Fig. 3). For the direct solar radiation simulations, a reflectance of 36.9% was used for all urban surfaces.

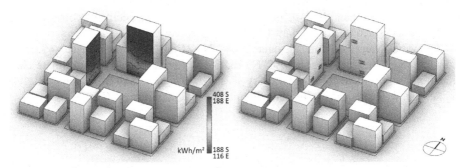

Fig. 3. Solar radiation analysis for Tallinn low density urban environment, façade sample clusters and centroids (left). S = south facade, E = east facade. The test rooms inserted (right).

3.3 Urban Scenarios and Performance Parameters

To obtain diverse urban scenarios and a large number of variations, in addition to the three urban environments the following building uses combinations and room parameters were used. Building use were office and residential. The scope was to investigate the potential of the method for buildings differently influenced by solar energy, being office cooling dominated, and residential heating dominated.

Three building use scenarios were used: mixed; only office; and only residential. For the mixed cases, one building was assigned one use and the other two the other use, for six combinations. For the only office and only residential cases, either one, two or all three buildings were used, for seven combinations each use. When not all three buildings were used, the others were considered surrounding buildings. The rooms of a building had the same use. Three window-to-wall-ratios were used: 40% (WWR 40); 60% (WWR 60); and 80% (WWR 80). The algorithm performed energy simulations and sunlight hour calculations for 54 combinations of the mixed building use and 63 combinations of each building use, office or residential, for every city.

Cooling, heating, electric lighting and transmitted beam solar radiation energy simulations were performed for all the urban environments, WWRs, building use scenarios combinations, and cities, for the existing situation, i.e., without the new conceptual building form. The main envelope and thermal zone parameters are presented in Table 1. The glazing systems select did not have solar protection to test rooms more subject to energy variations due to allowed or blocked solar beam energy.

Table 1. Parameters used. EW = ext. Wall, W = win., U_{tot} = total therm. Transm., T_{vis} = visible transm., SHGC = Solar Heat Gain Coefficient, int. Walls, floor and ceil. = adiabatic, PD-ED-LD = people-equipment-lighting power density, IT-ID = illuminance threshold-dimming, HS-CS = heating-cooling setpoint, MV = mech. Vent., NV S-WOA = nat. Vent. Setpoint-win. Oper. Area.

Envelope Parameters	Off	Res		Off	Res
EW U_{tot} (W/m^{2}K)	0.25	0.25	W T_{vis} (%)	0.71	0.8
W U_{tot} (W/m^{2}K)	0.84	1.85	W SHGC (-)	0.6	0.71
Thermal Parameters					
PD (p/m^2)	0.16	0.08	HS-CS (°C)	21–25	21–27
ED (W/m^2)	8	4	MV (L/s/m^2)	2	0.5
LD (W/m^2)	4	6	NV S-WOA (°C - %)	-	22–0.1
IT (lux)	500	300	ID	Cont	On/Off
Occupancy Schedules					
Off. – Monday-Friday - 18–6 0 7 0.2 8–9 0.6 10–11 0.7 12 0.4 13 0.6 14–15 0.7 16 0.6 17 0.2					
Res. – Monday-Sunday – 22–5 1 6–8 0.5 9–12 0.1 13–15 0.2 16–18 0.5 19–21 0.8					

Solar access assessment was performed according to the standard EN 17037:2018 [37]. The standard recommends a minimum of 1.5 h of sunlight exposure during a day between February 21st and March 21st. For the study, the latter day was selected. Standard's assessment principles were followed, including the use of a minimum solar altitude. The recommended angles are: 8° for Tallinn; 13° for Paris; and 20° for Athens. The results of energy use and sunlight exposure simulation of neighboring premises for the exiting situation, were then compared with the results when the new conceptual building forms were used.

3.4 Building Space Analysis

For building space analysis relative to energy use of surrounding premises, the central square plot measuring 35 m in size, was extruded for a height of 52.5 m to obtain the maximum buildable volume. This was higher than the analyzed buildings to have influence on the entire façade. Consequently, the volume was subdivided in cubic cells of 3.5 m in size representing a floor to floor distance, for a total of 1500 cells.

At select hourly time steps of the energy use and beam solar energy annual simulations performed for the existing situation, i.e., without new building volume obstruction, the algorithm developed for the study assessed the potential influence of each cell of the building space on the energy uses of each surrounding room. The main assumption was that a cell blocking beam solar radiation during an hour would shade the window reducing cooling energy use when present, or would prevent solar gains increasing heating energy use when present, and would increase electric lighting use when present. A cell (absence) allowing beam solar energy would have the opposite effect on energy uses, increasing cooling and reducing heating and electric lighting.

The algorithm used sun vectors to determine at which hour of the year sun was potentially visible from each room, considering orientation and obstructions, and selected energy results for the relevant hours. Consequently, the algorithm calculated shading factors, i.e., the ratio of the size of the potential shadow of each cell on each window to the window size, and beam factors remapping to the range [0–1] the intensity of solar beam energy for each room, for every relevant hour. The potential effect of a cell in reducing energy uses was calculated according to Eq. 1:

$$CE_{i,w} = \sum_{h=1}^{n}(sf \: x \: bf) \: x \: (C - H - L) \tag{1}$$

where $CE_{i,w}$ (-) is the effect of cell i on window w, h are the relevant hours used for energy and solar beam simulations for each window (room), n are all the relevant hours, sf and bf are the shading and beam factors (0–1), respectively, C, H and L are the simulated cooling, heating and electric lighting energy use (kWh), respectively.

The effect of a cell in blocking solar beam energy was: a) positive if cooling energy was larger than the sum of heating and electric lighting and higher if the shading and solar beam factors were larger; b) negative in the opposite case and worse if shading and beam factors were larger; and c) neutral if there was no energy use, no window obstruction or no beam energy (overcast sky). The algorithm generated one building space summing for each cell the effects on all the windows, for every combination of urban environment, WWR, building use scenario and city (Fig. 4). We use then the term voxel as the cell with an associated positive, negative or neutral value.

3.5 Energy-Solar Envelope

In the last step of the computational workflow, the building space constituted only by the voxels with positive and neutral effect on energy use of the existing surrounding premises was selected, tested against solar access requirements and refined if necessary, and finally was used to investigate its effects on variations of energy use and to validate its potential in guaranteeing sunlight exposure of the considered premises, for all the combinations of urban environment, WWR, building use scenario and city.

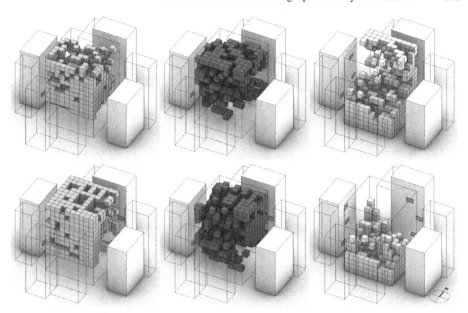

Fig. 4. Building space with positive, negative and neutral (left to right) effect on energy use of existing rooms in the medium density urban environment, for residential buildings facing west and east, WWR 60, in Paris (top), and for mixed use buildings, WWR 80, in Athens (bottom).

The algorithm firstly assessed if the analyzed windows were receiving the required minimum sunlight exposure in the existing situation, secondly calculated the sunlight hours when the building space voxels used were obstructing the windows, and finally eliminated select voxels to allow required sunlight to the compliant windows. To eliminate the least amount and the voxels less beneficial for energy, a component realized for the study firstly selected sun vectors (sunlight hours) from outside the maximum volume. If these were not enough, then selected additional sun vectors using the following order: 1) not intersecting any voxel; 2) intersecting only voxels with a neutral effect; and lastly 3) intersecting also or only voxels with positive effect.

The voxels intersecting the selected sun vectors were then excluded (Fig. 5). The resulting building space with a positive and neutral effect on energy use and solar access of existing premises was used as an energy-solar envelope (ESE), i.e., maximum volume and conceptual building form. After culling isolated voxels, the algorithm merged the remaining in one or more polysurfaces used for simulations (Fig. 5).

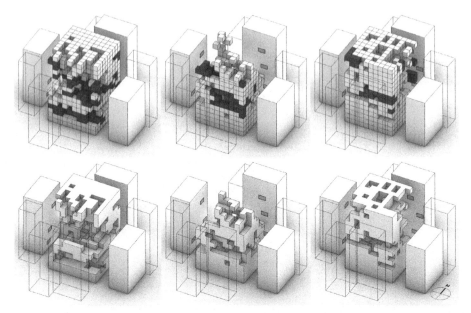

Fig. 5. Building space voxels used (white) and excluded (grey) for conceptual building massing generation (bottom) in the medium density urban environment and for mixed use buildings, with WWR 60 in Tallinn (left), and with WWR 80 in Paris (center) and Athens (right).

4 Results

This section presents comparisons of results of energy use simulations and solar access calculations for the considered existing premises in the existing situation and when ESEs were used. The algorithm simulated iteratively all the 54 combinations for the mixed use building scenarios, and the 63 combinations for each single use building scenario, for a total of 180 combinations for each city. The results are presented for the sum of energy uses and for cooling, heating and electric lighting for all the rooms analyzed together, to investigate the influence of the optimized conceptual building forms on the existing urban environment as a whole.

4.1 Building and Energy Use

In this section, aggregated and per energy use results per building use scenario are presented for the three cities. In the city of Tallinn, using the optimized conceptual building forms, the sum of energy uses decreased of from 0.6 to 30% among the mixed use building combinations, varied from an increase of 0.4 to a decrease of 17.1% among the office building scenarios, and decreased of from 1.2 to 40.3% among the residential building scenarios (Fig. 6). For Paris, the sum of cooling, heating and electric lighting decreased of up to 37.3% for the mixed use scenarios, and of from 1.3 to 22.3% for the office building combinations, and varied from an increase of 3.5% to a decrease of 43.7% for the residential building scenarios (Fig. 6).

In the city of Athens, whole energy use decreased for all the building use scenarios and combinations. Energy use decreased of from 14.3 to 58.1% for the mixed use buildings, of from 10.5 to 44.6% for the office buildings, and of from 19.4 to 64.6% for the residential buildings (Fig. 6). These initial results show that the ESE method helped to decrease energy use in almost all the analyzed cases.

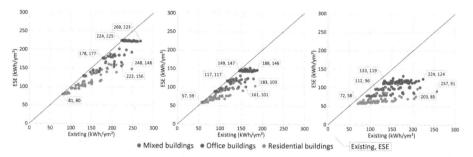

Fig. 6. Energy variations per building use scenario in Tallinn, Paris and Athens (left to right).

Consequently, the different energy use variations per building use scenario are presented and briefly discussed. In Tallinn, when the energy-solar envelope was used cooling energy use was reduced of 10 to 48, of 10 to 82, and of 3 to 20 kWh/m^{2}y for the mixed use, office and residential building scenarios, respectively. While cooling energy use always decreased, heating energy use varied from a decrease of 45 to an increase of 13 kWh/m^{2}y for the mixed use building combinations, increased of 7 to 32 kWh/m^{2}y for office buildings, and varied from a decrease of 91 to an increase of 1 kWh/m^{2}y for residential buildings. As expected, electric lighting energy increased for all combinations, of 3 to 6, of 2 to 4, and of 1 to 8 kWh/m^{2}y for the mixed use, office and residential building scenarios, respectively (Fig. 7).

In Paris, cooling energy was reduced of 9 to 46, of 12 to 74, and of 5 to 21 kWh/m^{2}y for mixed use, office and residential buildings, respectively. Heating energy varied from a reduction of 34 to an increase of 10 kWh/m^{2}y for mixed use buildings, increased of 7 to 29 kWh/m^{2}y for office buildings, and varied from a decrease of 68 to an increase of 3 kWh/m^{2}y for residential buildings. Electric lighting increased from 1 to 6, from 1 to 4, and up to 8 kWh/m^{2}y for mixed use, office and residential buildings, respectively (Fig. 7). In Athens, energy use variation followed similar patterns as in Tallinn and Paris, though with different ranges. Cooling energy was reduced of 23 to 98, of 27 to 129, and of 19 to 63 kWh/m^{2}y, for mixed use, office and residential building scenarios, respectively. Heating energy use varied from a reduction of 50 and 127 to an increase of 12 and 1 kWh/m^{2}y for mixed use and residential buildings, respectively, and increased of 6 to 25 kWh/m^{2}y for office buildings. Electric lighting increased of 3 to 6, of 2 to 4, and of 2 to 8 kWh/m^{2}y for the mixed use, office and residential building scenarios, respectively (Fig. 7).

The results show that the algorithm succeeded in generating ESEs which reduced significantly cooling energy use of existing premises for all the analyzed cases, mostly reduced but also increased heating energy and always increased electric lighting energy

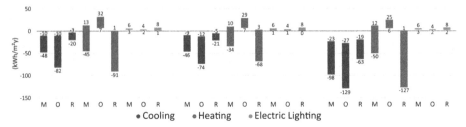

Fig. 7. Energy use variations between the existing situation and when ESEs were used for the cities of Tallinn, Paris and Athens (left to right). M = mixed use, O = office, R = residential.

use. The authors argue that the reason for heating energy use increases of office buildings was being these cooling dominated for internal loads, the algorithm failed to assign the right weight to beam energy. Thus the algorithm used a number of voxels with an actual negative effect, which reduced passive solar gains.

As a matter of fact, heating energy use was almost always reduced for residential buildings. Being these heating dominated the algorithm in these cases succeeded to assign the right weight to beam solar energy thus eliminated the right voxels to guarantee passive solar gains. Electric lighting energy use was always increased due to its small weight on the total energy use and due to the shading nature of the optimized conceptual building forms. This is also the reason for the better performance of ESEs in the hotter climate of Athens.

4.2 Urban Environment and WWR

This section presents the influence of urban environment density and WWR on total energy use variations (Fig. 8). In Tallinn, the smallest and largest energy reductions were always for office and residential buildings, respectively. Reductions ranged from 1.3 to 14.6%, from 0.4 to 12.6% and from -0.4 (increase) to 8.6%, for the low, medium and high density environments, respectively, using WWR 40, from 5.5 to 29.1%, from 3.4 to 23.6%, and from 1.3 to 23.5%, for the low, medium and high urban density, respectively, using WWR 60, and from 12.1 to 40.3%, from 7.1 to 29.2%, and from 3 to 16.1%, for the three urban densities, respectively, using WWR 80.

In Paris, using WWR 40, both smallest and largest reductions were recorded for residential buildings, ranging from 4.8 to 18.1%, from 1.7 to 13% and from -3.5 (increase) to 7.6%, in the low, medium and high density environment, respectively. When WWR 60 was used the smallest and largest reductions were recorded for office and residential buildings, respectively, ranging from 12.3 to 38.7%, and from 9.5 to 25% for low and medium urban density, respectively. In the high density urban environment, both smallest and largest reductions were recorded for residential buildings, from 4.3 to 11.1%. Using WWR 80, the smallest and largest reductions were recorded for office and residential buildings, respectively, ranging from 18.9 to 43.7%, and from 8.3 to 15.3%, in the low and high density urban environment, respectively. In the medium density environment, reductions ranged from 14.9 to 29.9%, recorded for office and mixed use buildings, respectively.

In Athens, as in Tallinn, the smallest energy use reductions were recorded always for office buildings and the largest for residential buildings. Considering the low, medium and high density urban environments, reductions ranged from 20.8 to 39.8%, from 15.4 to 33.7%, and from 10.5 to 21.6%, respectively, when WWR 40 was used, from 32.1 to 56.6%, from 27.2 to 48.1%, and from 17.6 to 34.4%, respectively, using WWR 60, and from 40.8 to 65%, from 33.9 to 56.6%, and from 25.4 to 42.3%, respectively, when WWR 80 was used.

As expected, the largest energy use variations were recorded for low density environment and high WWR, in all cities. However, the method showed significant potential of reducing energy use of existing premises also in medium and high density environments, and for smaller WWRs, making it applicable in different of urban settings.

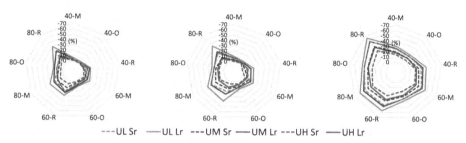

Fig. 8. Total energy use variations per urban environment density and WWR for Tallinn, Paris, and Athens (left to right). 40, 60, 80 = WWR (%), M = mixed use, O = office, R = residential, UL, UM, UH = urban density low, medium, high, Sr, Lr = smallest, largest reduction.

4.3 Sunlight Exposure

Two sunlight exposure results are presented considering the building use and WWR variations for every urban environment, in the three cities (Fig. 9). The first is about the sunlight allowed by ESEs to the window of every variation in relation to the standard EN 17037:2018 requirement. The second is about reduction of annual sunlight hours on all the windows of every variation comparing the existing situation.

Results show that the algorithm succeeded in generating conceptual building forms which guaranteed the required 1.5 h of sunlight exposure to every window for all variations and urban environments, in the three cities. The annual sunlight exposure was inevitably reduced (Fig. 9). Most of the largest minimum and maximum reduction were recorded for high density urban environments. This outcome shows that the use of ESEs worsens sunlight exposure mostly in urban situations where it is already scarce. Further, the minimum reductions larger in Athens than in Tallinn and Paris show that the shading effect of ESEs can be higher at low latitudes. The reason is that at high latitudes, considering all the year, sunlight is blocked by the urban environment more than at low latitudes, thus the use of ESEs reduces less sunlight exposure.

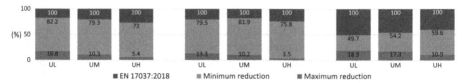

Fig. 9. Standard compliance and sunlight reduction for Tallinn, Paris and Athens (left to right).

5 Conclusion

The paper introduces a method to help design new buildings with a positive impact on existing urban environments. Building on recent advancements on solar envelopes and shading devices generation with energy consideration, a computational workflow was realized which: a) subdivided the buildable volume; b) analyzed the effect of building space cells on energy use and on sunlight exposure of surrounding existing premises; and c) generated conceptual building forms optimized to reduce energy use while guaranteeing adequate solar access. The method was applied to a large number of cases, different for urban density, room WWR, and building use, in three cities.

Results showed that the proposed method succeeded to generate forms capable of reducing energy use of existing premises. In Tallinn, and Paris, in few cases energy use was not affected, whereas for most cases it decreased, for a maximum of 40.3 and 43.7%, respectively. In Athens, energy use reduction was more significant, ranging from a minimum of 10.5 to 65%. Sunlight exposure of the existing premises was always guaranteed by the conceptual building forms, according to the standard used. In the paper, detailed results are presented and briefly discussed. The whole forms generated can be used to study the optimal location of the new building façades and roofs, and to outline floor plans "slicing" the conceptual building mass.

The main limitations of the presented method are three. Building space voxels mutual obstruction, which would influence solar radiation on surrounding rooms, was not considered to reduce long computation time. The conceptual forms are fragmented, thus hardly usable for real building design. The select rooms represent efficiently the energy use of the entire building facade, but not its sunlight exposure. They were used for consistency of performance assessments in the initial development of the method.

Being promising, the method will be developed further. The algorithm will be improved to correctly consider the cooling demand of office rooms, to compute voxels mutual obstruction, and to optimize energy simulation times to consider all the façade rooms. The usability of the generated building massing will be improved through the use of larger cells and the possibility to analyze trade-offs between existing premises energy use reduction and new building massing uniformity. Additionally, the method will be implemented as a design tool for Grasshopper, which will consider also energy and sunlight exposure of the new building massing in the form-finding process.

Acknowledgments. The research was supported by the grant Smart City Center of Excellence (AR20013).

References

1. United Nations: Transforming our World: The 2030 Agenda for Sustainable Development. UN, New York (2015)
2. United Nations Environment Programme: 2021 Global Status Report for Buildings and Construction. UN, Nairobi (2021)
3. European Commission: Roadmap to a Resource Efficient Europe. EU, Brussels (2011)
4. Jenks, M.: The acceptability of urban intensification. In: Williams, K., Burton, E., Jenks, M. (eds.) Achieving Sustainable Urban Form, pp. 242-250. E & FN Spon, London (2000)
5. De Luca, F., Naboni, E., Lobaccaro, G.: Tall buildings cluster form rationalization in a Nordic climate by factoring in indoor-outdoor comfort and energy. Energy Build. **238**, 110831 (2021)
6. Strømann-Andersen, J., Sattrup, P.A.: The urban canyon and building energy use: urban density versus daylight and passive solar gains. En. Build. **43**(8), 2011–2020 (2011)
7. De Luca, F., Voll, H., Thalfeldt, M.: Comparison of static and dynamic shading systems for office buildings energy consumption and cooling load assessment. Manage. Environ. Quality: An Int. J. **29**(5), 978–998 (2018)
8. Yu, X., Su, Y.: Daylight availability assessment and its potential energy saving estimation - A literature review. Renewable Sustainable Energy Rev. **52**, 494–503 (2015)
9. De Luca, F., Sepúlveda, A., Varjas, T.: Multi-performance optimization of static shading devices for glare, daylight, view and energy consideration. Build. Environ. **217**, 109110 (2022)
10. De Luca, F., Dogan, T., Kurnitski, J.: Methodology for determining fenestration ranges for daylight and energy efficiency in Estonia. In: Simulation Series **50**(7), pp. 47–54, 9th SimAUD. SCS, San Diego, USA (2018)
11. Reinhart, C.F.: Daylighting Handbook I: Fundamentals Designing with the Sun. Building Technology Press, Cambridge, USA (2014)
12. Andersen, M., Mardaljevic, J., Lockley, S.M.: A framework for predicting the non-visual effects of daylight – Part I. Light. Res. Technol. **44**(1), 37–53 (2012)
13. Lockley, S.W.: Circadian rhythms: influence of light in humans. In: Encyclopedia of Neuroscience, pp. 971–988. Academic Press, Cambridge, USA (2009)
14. Reed, B.: Shifting from 'sustainability' to regeneration. Buil. Res. Inf. **35**, 674–680 (2007)
15. Knowles, R.L.: The solar envelope: its meaning for energy and buildings. Energy Build. **35**(1), 15–25 (2003)
16. Alkadri, M.F., De Luca, F., Turrin, M., Sariyildiz, S.: Understanding computational methods for solar envelopes based on design parameters, tools, and case studies: a review. Energies **13**(13), 3302 (2020)
17. De Luca, F., Dogan, T.: A novel solar envelope method based on solar ordinances for urban planning. Build. Simulation: An Int. J. **12**(5), 817–834 (2019)
18. De Luca, F., Dogan, T., Sepúlveda, A.: Reverse solar envelope method. A new building form-finding method that can take regulatory frameworks into account. Automation in Construction **123**, 103518 (2021)
19. Solar Toolbox. https://www.food4rhino.com/en/app/solar-toolbox. Accessed 3 Mar 2023
20. Okeil, A.: A holistic approach to energy efficient building forms. Energy Build. **42**, 1437–1444 (2010)
21. Vartholomaios, A.: The residential solar block envelope: a method for enabling the development of compact urban blocks with high passive solar potential. Energy Build. **99**, 303–312 (2015)
22. Niemasz, J., Sargent, J., Reinhart, C.F.: Solar zoning and energy in detached dwellings. Environ. Plann. B. Plann. Des. **40**, 801–813 (2013)
23. Emmanuel, R.: A hypothetical 'shadow umbrella' for thermal comfort enhancement in the equatorial urban outdoors. Archit. Sci. Rev. **36**(4), 173–184 (1993)

24. Capeluto, I.G.: Energy performance of the self-shading building envelope. Energy Build. **35**, 327–336 (2003)
25. Alkadri, M.F., De Luca, F., Turrin, M., Sariyildiz, S.: A computational workflow for generating a voxel-based design approach based on subtractive shading envelopes and attribute information of point cloud data. Remote Sensing **12**(16), 2561 (2020)
26. Lechner, N.: Heating, Cooling, Lighting. Sustainable Design Methods for Architects. 4th edn. John Wiley & Sons, Hoboken (NJ), USA (2015)
27. Kaftan, E., Marsh, A.: Integrating the cellular method for shading design with a thermal simulation. In: Proceedings of 1st International Conference on Passive and Low Energy Cooling for the Built Environment (Palenc 2005), pp. 965–970. Santorini, Greece (2005)
28. Sargent, J.A., Niemasz, J., Reinhart, C.F.: Shaderade: combining rhinoceros and energyplus for the design of static exterior shading devices. In: Proceedings of Building Simulation 2011, pp. 310–317. IBPSA, Sydney (2011)
29. Sepúlveda, A., De Luca, F.: A novel multi-criteria method for building massing based on energy performance and solar access. The mixed solar envelope (MSE) method. In: Proceedings of 40th eCAADe Conference, vol. 1, pp 649–658. eCAADe, Ghent (2022)
30. Rhinoceros and Grasshopper. https://www.rhino3d.com/. Accessed 3 Mar 2023
31. Ward, G.J.: The RADIANCE lighting simulation and rendering system. In: Proceedings of SIGGRAPH'94 Conference, pp. 459–472. Orlando, USA (1994)
32. EnergyPlus. https://energyplus.net/. Accessed 3 Mar 2023
33. ClimateStudio. https://www.solemma.com/climatestudio. Accessed 3 Mar 2023
34. Ladybug Tools. https://www.ladybug.tools/. Accessed 3 Mar 2023
35. Dogan, T., Reinhart, C.F.: Shoeboxer: an algorithm for abstracted rapid multi-zone urban-building energy model generation and simulation. En. and Build. **140**, 140–153 (2017)
36. LunchBox. https://apps.provingground.io/lunchbox/. Accessed 3 Mar 2023
37. European Commission: EN 17037:2018 Daylight in buildings. EU, Bruxelles (2018)

Design Space Recommendation: Assisting Users to Manage Complexity in Urban Design Optimisation

JingZhi Tay[(✉)] [iD], F. Peter Ortner[iD], Peng Song[iD], Anna Claudia Yenardi[iD], and Zebin Chen[iD]

Singapore University of Technology and Design, 8 Somapah Road, Singapore 487372, Singapore
jingzhi_tay@mymail.sutd.edu.sg

Abstract. In the context of developing a generalizable and user-friendly computational urban design tool, this study proposes and test a method of sensitivity analysis and propose a visualization technique to 1) improve user understanding of interactions between model parameters and objectives; 2) improve speed and accuracy of optimization through intelligent reduction of the ranges in design space. Sensitivity analysis of optimisation results from morphology-based, non-linear urban models can be ineffective due to computational costs limiting the number of samples possible, and large ranges of search due to users' inexperience with the model parameters. In response to these challenges, this paper puts forward a method of identifying well-performing parameter ranges, and tests three parameter-clustering experiments to improve optimization efficiency and quality of outputs in comparison with a baseline NSGA-III optimization. These methods are applied to results from an urban design optimization tool implemented within the context of Singaporean urbanism. The proposed method shows improvement in optimization convergence, especially when tighter parameter clustering is implemented. A visualization technique to share insights from the proposed parameter clustering method to an eventual user of the design tool is explored in the final section which emphasizes on informing users to define better search boundaries.

Keywords: Urban Design Optimisation · Design Space Exploration · Machine Learning · Tool Development · Clustering Algorithm · Complexity

1 Introduction

With a view to creating a user-friendly urban design tool that can generate quick, accurate design simulations for a variety of urban sites in South-East and East Asia, this paper presents methods of simplifying and automating design space exploration. Computational urban design models are growing more extensive and intricate as city planners seek to address complex and interlinked economic, social and sustainability goals. These models are employed to support data-driven, evidence-based urban design approaches, with increasing demand from both academic and commercial sectors in recent years

© The Author(s), under exclusive license to Springer Nature Switzerland AG 2023
M. Turrin et al. (Eds.): CAAD Futures 2023, CCIS 1819, pp. 333–344, 2023.
https://doi.org/10.1007/978-3-031-37189-9_22

(Calixto et al. 2021; Wortmann, 2017). Multi-objective optimisation (MOO) for computational urban design exploration has been successfully demonstrated with several notable limitations: although it is possible to identify better performing results by searching more effectively using optimisation methods or by searching more extensively with more computing power, the results generated present growing complexity with each added parameter or objective (Koenig et al. 2020).

To improve the efficiency of multi-objective optimization for complex urban design models, this study tests methods of sensitivity analysis to 1) improve understanding of interactions between objectives and model parameters by proposing a visualization technique; and 2) reduce the span of each design space dimension by modifying parameter ranges for further optimisation search based on clustering analysis of initial samples.

Sensitivity analysis (SA) is widely used to better understand uncertainty between model inputs and outputs in fields as diverse as environmental simulation, fintech and epidemiology. (Borgonovo and Plischke, 2016). Prior to conducting SA, a sample set of the model is obtained either via random sampling or a more systematic method such as Saltelli sampling (Saltelli, 2002). Subsequently, various analysis methods can be applied to gather insights from the sample set, such as Fourier Amplitude Sensitivity Test, Method of Morris, and Linear Regression (Cukier et al. 1973; Morris, 1991). However, when analyzing morphology-based urban models for the purposes of architecture and urban design, the effectiveness of global sensitivity analysis is poorer due to the non-linear nature these models and relatively low sample sizes possible due to high computational costs of certain simulations. While carrying out optimisation without SA is possible, it may be ineffective, as user behavior with incomplete understanding of the model typically includes defining larger ranges for each parameter in hopes of including their ideal solution.

In this paper, we propose a method to automatically refine the range of search using clustering analysis on a sample set, with the goal of improving the effectiveness of subsequent optimisation search. We integrate the Density-Based Spatial Clustering of Applications with Noise (DBSCAN) algorithm with the Non-dominated Sorting Genetic Algorithm III (NSGA-III) to achieve a novel design search workflow (Blank et al. 2019; Ester et al. 1996). The proposed methodology is tested on an urban design model based on development patterns in Singapore. The model generates urban fabric, including road network, land parcels, and building geometry simultaneously based on a set of variable parameters. According to changing needs of urban design models, designers can extend the use of this demonstrated models by including new design parameters and objectives. To test the proposed methods, optimization results are presented and compared (Sect. 3). Final discussion identifies potential application to improve user understanding of parameter effects on optimization outcomes within the context of improving the user-interface for an urban design optimisation tool (Sect. 4).

2 Methodology

This section describes the model setup (Sect. 2.1) and the proposed method and design of clustering experiments (Sect. 2.2) in greater detail. Our model is described in 3 steps: the Urban Fabric Model used to subdivide the boundary into smaller parcels, the

Building Geometry Model used to populate buildings at each parcel and the Optimisation Model used for design search. Our targeted search method is elaborated in 3 phases: the sampling phase, the clustering phase, and the optimisation phase. At the end of this section, we describe the design of experiments to evaluate the impact of two parameters of the clustering algorithm on the outcome of optimisation.

2.1 Model Setup

An initial parametric design model is created to closely match the urban fabric of private residential development in Singapore. The parametric model firstly generates the 'urban fabric': road network and land parcels by subdividing a larger development area. Subsequently, the model generates 'building geometry': oriented 3D geometries based on design logic and local development guidelines and constraints. Our parametric model was built in the Rhino and Grasshopper environment, using two open-source datasets: 1) Singapore's latest masterplan for 2019 and 2) OpenStreetMap. The parameters and fitness functions used are summarized in Tables 1 and 2, respectively. Although some objectives are described as maximization in Table 2 due to clearer semantics, all seven objectives are formulated as minimization problems in our simulation model. Secondly, an optimisation model, built in Python, iterates the parametric model to automatically search for best performing solutions.

Urban Fabric Model. The core algorithm for this model is based on an algorithm for 2D placement of streamlines from the Computational Geometry Algorithms Library (Mebarki, 2023; Yang et al. 2013). We replaced each vector in the vector field with a pair of orthogonal vectors, where one vector is either parallel or perpendicular to the nearest boundary edge. This modification ensures that the resulting streamline curve reflects the tangents of a given boundary shape, allowing it to be applicable to any urban context. Figure 1 summarizes the steps taken by the model to result in subdivided parcels and roads, using the first two parameters and involving the first four fitness functions described in Table 1 and Table 2. Road widths and parcel boundaries are generated automatically via offsets of the initial streamlines. The urban fabric is generated within the context of a 2-ha study area, with newly generated street center lines linking into the existing network of road center lines.

Building Geometry Model. This model simulates the development of residential blocks, constrained by local guidelines and populated with an edge-following logic. The model constraints include site setback of each parcel and maximum building height allowed. These constraints are determined, following Singapore development regulations, by the category of adjacent roads to the parcel, the height of the building to be placed in the site and the zoning of the site/ location within Singapore. Building footprint is selected, using parameters 3 and 4, from a library of existing building footprints in Singapore arranged in order of increasing width and depth. The model calculates the number of buildings to be placed on site using parameter 7, and tests possible orientations on a grid of points generated on site using parameters 5 and 6. The edge-following logic used in our algorithm systematically attempts to orient the building footprint on each point on the grid, facing the closest parcel edge and starting with the points closest to the

Table 1. Parameters used to generate solutions in parametric model.

Parameter	Description	Type / Range	Scale
Cross Point U	U parameter for selecting pair of streamlines	Float / 0 - 1	Precinct
Cross Point V	V parameter for selecting pair of streamlines	Float / 0 - 1	Precinct
Bkey X Scale	Select width of block key based on normalized widths of all block keys	Float / 0 - 1	Building
Bkey Y Scale	Select depth of block key based on normalized depths of all block keys	Float / 0 - 1	Building
Grid Angle	Angle to rotate the plane used to generate initial grid of points	Float / 0 - 180	Parcel
Grid Spacing	Spacing between each point in the grid	Float / 20 - 30	Parcel
Parcel Storey Scale	Parameter to govern trade-off between number of blocks and height of each block	Float / 0 - 1	Parcel

Table 2. Fitness functions used to evaluate solutions in parametric model.

Fitness Function	Description	Scale
Total Road Length	Minimize road coverage	Precinct
Mean Area Deviation	Minimize variance in area of subdivided parcels	Precinct
Average Betweeness Index	Maximize network betweeness of generated roads within the local context	Precinct
Av. Straightness Index	Maximize straightness of generated roads within the local context	Precinct
Av. GPR Efficiency	Maximize the total floor area allowable for the parcel	Parcel
Av. NS to EW Aspect Ratio	Minimize façade exposure to east-west orientation for reducing heating due to direct sun exposure	Building
Av. View Obstruction	Minimize view obstruction from units by penalizing any building placed too close to another building	Building

boundary of the parcel, until all buildings are in place without intersections. The steps for building generation are summarized in Fig. 2, and the final output from the combined parametric model is shown in Fig. 3, for selected objectives. A consistent color gradient ranging from green to yellow to red is applied for both objectives visualized in Fig. 3, signifying best to worst performance.

Optimization Model. Multi-objective optimization of the model is set up using a customized NSGA-III algorithm, developed within a Python framework, which is used in all MOO runs in this paper (Blank and Deb 2020). NSGA-III is a genetic algorithm that

Fig. 1. Progressive steps to generate urban fabric: 1) Site selection (red dashed line) and computation of best-fit orientation plane (grey plane); 2) Generation of all streamlines (teal); 3) Selection of streamline pair to subdivide plot.

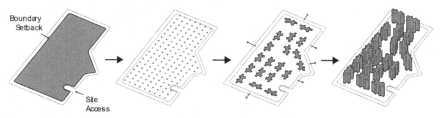

Fig. 2. Building geometry generation; a) setback and access; b) grid implementation; c) edge-oriented key plan layout; d) volumetric geometry definition.

can be used for mixed-integer programming for many objectives using reference directions to direct its search. These reference directions are generated with the Das-Dennis method, to uniformly sample all 7 parameter dimensions (Das and Dennis, 1998). We implement a genetic algorithm, with the survival selection following NSGA-III methods. The mating process, however, is modified such that the crossover and mutation process will repeat until the parameters of the new child solution is within specific ranges.

2.2 Targeted Search Method: NSGA-III + DBSCAN

Computational cost of simulations used in urban design model is not trivial, and with the possibility of failed solutions due to any discontinuous functions resulting in many wasted simulations. Accordingly, our strategy aims to avoid failed simulation runs by testing for regions of each parameter that are prone to failed solutions.

We propose a method employing DBSCAN to automatically identify parameter ranges that will result in more efficient, subsequent optimization. In contrast to weighted-sum scalarization approach, this method does not seek to reduce the number of dimensions, as the interactions between complex fitness functions cannot be fully established with a small sample set. In the first step, we generate a sample set by running 5 generations with 50 individuals each using the NSGA-III algorithm.

In the second step, we extracted a subset of the random sample to be used for clustering. This subset is extracted by firstly, removing samples that failed to give a complete solution, and secondly, by removing samples that fall within a minimum percentile for any fitness function. For example, with a fitness percentile (FP) value of 0.9, we only

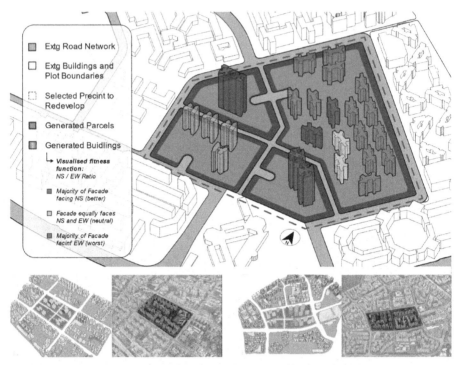

Legend text within figure:

- Extg Road Network
- Extg Buildings and Plot Boundaries
- Selected Precinct to Redevelop
- Generated Parcels
- Generated Buidlings
  - ↳ *Visualised fitness function: NS / EW Ratio*
  - Majority of Facade facing NS (better)
  - Facade equally faces NS and EW (neutral)
  - Majority of Facade facinf EW (worst)

Fig. 3. Visualization of parametric model outputs for three residential sites in Singapore: Tampines, Jurong and Punggol (Clockwise from top). NS to EW Aspect Ratio objective values show for Tampines and Jurong, and View Obstruction for Punggol, with corresponding satellite imagery (Google) for Jurong and Punggol shown for reference.

include solutions that perform in the top 90 percentile and remove solutions that perform in the bottom 10 percentile for any objective. FP is one of the two clustering parameters tested in our methodology.

In the third step, we clustered the normalized data points for each parameter using DBSCAN algorithm. The algorithm generates clusters by firstly, grouping points that are within a fixed epsilon distance (EPS) from each other and secondly, ensuring that each group has at least a fixed minimum number of samples. For our tests, we fixed the minimum number of samples to be 3 and assigned EPS as the second clustering parameter to be tested. The final output from this step returns a set of parameter ranges derived from the clusters identified from the algorithm.

In the fourth step, we conducted optimization using the new set of parameter ranges. Using the modification as described in Sect. 2.1.3, the mating process between each generation is repeated until success at a negligible computing cost relative to the total time taken for optimisation.

Finally, we repeat steps 2 to 4, for a total of 3 experiments. The baseline test results are created using the same model setup and optimized with a default NSGA-III. We will further discuss the results in the next section.

3   Results and Discussion

To improve our understanding of how model parameter clustering affects design search, we create three clustering options using different sets of clustering parameters. The 3 options differ in terms of how tightly the algorithm constrains the ranges. The goal of the first clustering option is to act as a baseline test for this method, using parameter values that loosely constrain the clustering. The goal of the second clustering option is to achieve fewer ranges per parameter, using only relatively better performing solutions from the samples during clustering, using a lower value of 0.7 for fitness percentile. The goal of the third clustering option is to identify more specific ranges per parameter using a lower value of 0.03 for EPS.

3.1   Results from Clustering Parameter Ranges in Search Space

The results generated by clustering parameter ranges with the three separate settings are visualized in Fig. 4. Each clustering option is represented in a single plot with seven rows, representing the seven parameters used in the model. The refined search range for each experiment is shown in grey. The identified ranges from our clustering algorithm are visualized as a colored hatch.

The first two rows in Fig. 4 show that the ranges for parameters "Cross Point U" and "Cross Point V" should be reduced as values lying on the extreme ends are consistently not included in any of the 3 clustering options. These parameters govern the point to select our pair of streamline curves to be used for site subdivision. Selecting streamline curves with the intersection point near the edges of our site boundary results in poor subdivision of the site that greatly increases the likelihood of failed building solutions and are therefore excluded. For parameters "Bkey X Scale" and "Bkey Y Scale", our results show that it is possible to populate a site with footprints of varying aspect ratios, as ranges identified span nearly the entire domain. However, well performing solutions tend to cluster at a specific dimension for depth, as seen in the clearly separated clusters for "Bkey Y Scale" in clustering option 2. For the last two parameters, "Grid Spacing" and "Parcel Storey Scale", we can conclude that while the entire range can result in possible solutions, better performing solutions cluster around the upper ranges. This conclusion is not unexpected as these two parameters affect the density of building footprints generated, directly conflicting with the density-based objectives of "Av. GPR Efficiency" and "Av. View Obstruction".

3.2   Performance Comparison of Clustering Method in the Solution Space

To compare the performance for each clustering experiment versus the baseline, we plotted for each objective the minimum, average and maximum score attained in each generation (Fig. 5). A total of 50 individuals evolving over 30 generations (1,500 samples) for each experiment is considered for this analysis. Each of the seven objective is plotted separately. These charts show that complexity in the results arising from conflicting relationships between multiple objectives would not allow us to evaluate the results one objective at a time. Instead, summary indices are required to improve our

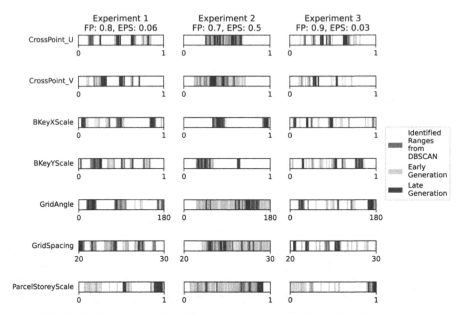

Fig. 4. Design space and specific parameter ranges for three clustering options.

interpretation of the results from these charts; we use the normalized average score and hypervolume indicators (Figs. 5, 6).

The chart in the middle column of the bottom row of Fig. 5 charts a normalized average score for each experiment and the baseline. The score is firstly, normalized within the global minimum and maximum scores of each objective and secondly, averaged across all seven normalized objective scores. As no experiment reaches a score of 0, it can be concluded that at least 1 pair of objectives conflict. The normalized average score is visualized using a histogram format in Fig. 6 to show the distribution of the score achieved by each experiment. On the right of Fig. 6, we see that solutions with the best normalized average score are from experiments 1 and 3. Furthermore, results from experiment 3 show a positive skew as compared to the baseline, indicating that a larger percentage of results from experiment 3 perform better than the baseline. On the left of Fig. 6, we chart the hypervolume achieved for each experiment per generation (cite). Signs of plateauing in the chart indicate diminishing improvements made to the pareto front and imply convergence of results from the algorithm. From Fig. 6 we see that experiments 1 and 3 are the fastest to converge as their respective hypervolume indicator starts above the baseline, and crosses over to below the baseline around generation 12. As experiments 1 and 3 are shown to converge more quickly and to produce solutions of quality similar to the baseline, the study results suggest that smaller EPS values, which produce tighter clusters, are providing the greatest benefit for target search in the model.

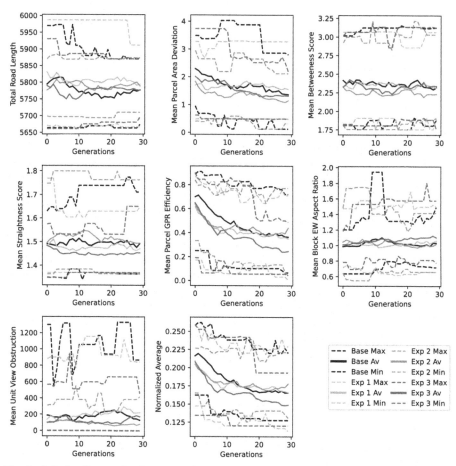

Fig. 5. Solution Space – Performance of solutions over each generation for all experiments in all objectives and with normalized average score.

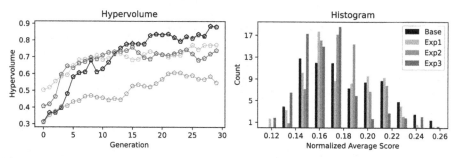

Fig. 6. Hypervolume per generation for each experiment (left) and histogram of solution count versus normalized average score for each experiment (right).

4   From Parameter Clustering to Design Space Recommendation

The parameter clustering method presented above is developed with the goal of supporting an urban design optimization tool capable of generating results for many possible sites and meeting the requirements of a variety of users. In this context the proposed method shows value as a means of automatically limiting design space for quicker optimization and of providing information to the end-user on well-performing parameter ranges.

The relative speed of convergence and quality of results demonstrated by experiments 1 and 3 (Fig. 6) suggests that the proposed targeted search method represents a viable method of automatically creating an initial or default optimization run – allowing a user to quickly identify a satisficing result. With a generalizable urban design model, there is demand to continuously add new model parameters and evaluation methods – making automated tuning of parameters a valuable tool.

To support the end-user to identify a result that best meets their requirements, the parameter clustering method can be employed not to automatically limit optimization search space, but instead to inform the user on parameter ranges that result in well performing results. An example of how this design space recommendation could be achieved is shown in Fig. 7, which shows a user interface (UI) for a web-based version of the urban design optimization tool presented in this paper. In this UI, the performance of a parameter value is visualized with a color gradient inserted below a two-handled range slider. As the user adjusts the range sliders to define their design requirements

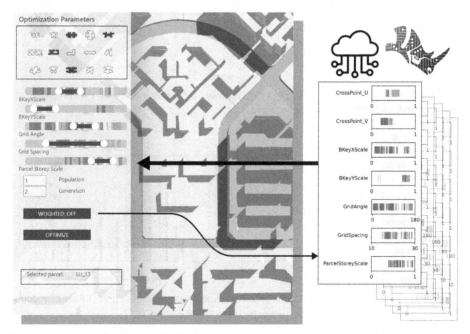

Fig. 7. Prototype of web-based tool. Ranges for each parameter are progressively updated as random samples are evaluated on a cloud server and analyzed with user preferred weightings.

for a given parameter, the performance gradient affords awareness of well-performing parameter ranges within or adjacent to their design preferences. This design space recommendation system can be further personalized by adjusting the gradient according to the objective weightings assigned by the user. The gradients can be remapped to reflect which parameter ranges result in best performance in highly weighted objectives. This method would allow the user to choose parameter ranges that reflect not only their design preferences but also deliver the objective values they prioritize.

5 Conclusion

This paper presents a method of automatically clustering parameter ranges using DBSCAN based on the position of well-performing results in a random sample. After testing three proposed clustering options in comparison with a baseline optimization, the proposed method is shown to achieve faster convergence to a solution. In the context of a generalized urban design optimization tool, the proposed method presents value in providing quick or default optimization settings for a frequently changing model. Additionally, the method shows promise in supporting user understanding of well-performing areas in the design space.

Future investigation may be able to maintain the quicker convergence demonstrated for this method, while improving its closeness to the global optimal result. The concept of dynamic parameter control can be integrated into this method such that the EPS and FP parameters are continuously being refined instead of being a static variable. Additionally, the research team plans to conduct more extensive user testing, which will provide insight on the ability of the parameter clustering visualization to inform more satisfactory outcomes for different users and on a variety of urban sites. If the method is able to assist users to quickly identify results that satisfactorily meet their design preferences, then lengthy optimization in search of a global optimum can be shown to be unnecessary.

References

Blank, J., Deb, K., Roy, P.C.: Investigating the normalization procedure of NSGA-III. In: Deb, K., et al. (eds.) EMO 2019. LNCS, vol. 11411, pp. 229–240. Springer, Cham (2019). https://doi.org/10.1007/978-3-030-12598-1_19

Blank, J., Deb, K.: Pymoo: multi-objective optimization in python. IEEE Access. **8**, 89497–89509 (2020)

Borgonovo, E., Plischke, E.: Sensitivity analysis: a review of recent advances. Eur. J. Oper. Res. **248**(3), 869–887 (2016)

Calixto, V.: A layered approach for the data-driven design of smart cities. In: Globa, A., van Ameijde, J., Fingrut, A., Kim, N., Lo, T.T.S. (eds.), PROJECTIONS - Proceedings of the 26th CAADRIA Conference - Volume 2, The Chinese University of Hong Kong and Online, Hong Kong, 29 March - 1 April 2021, pp. 739–748. CUMINCAD (2021)

Cukier, R.I., et al.: Study of the sensitivity of coupled reaction systems to uncertainties in rate coefficients. I Theory. J. Chem. Phys. **59**(8), 3873–3878 (1973)

Das, I., Dennis, J.E.: Normal-boundary intersection: a new method for generating the pareto surface in nonlinear multicriteria optimization problems. SIAM J. Optim. **8**(3), 631–657 (1998)

Ester, M., et al.: A Density-Based Algorithm for Discovering Clusters in Large Spatial Databases with Noise (1996)

Koenig, R., et al.: Integrating urban analysis, generative design, and evolutionary optimization for solving urban design problems. Environ. Planning B: Urban Analytics City Sci. **47**(6), 997–1013 (2020)

Mebarki, A.: 2D placement of streamlines. In: CGAL User and Reference Manual. CGAL Editorial Board, 5.5.2 edition, (2023)

Morris, M.D.: Factorial sampling plans for preliminary computational experiments. Technometrics **33**(2), 161–174 (1991)

Saltelli, A.: Making best use of model evaluations to compute sensitivity indices. Comput. Phys. Commun. **145**(2), 280–297 (2002)

Wortmann, T.: Opossum - introducing and evaluating a model-based optimization tool for grasshopper. In: Janssen, P., Loh, P., Raonic, A., Schnabel, M.A. (eds.), Protocols, Flows, and Glitches - Proceedings of the 22nd CAADRIA Conference, Xi'an Jiaotong-Liverpool University, Suzhou, China, 5–8 April 2017, pp. 283–292. CUMINCAD (2017)

Yang, Y.-L., et al.: Urban pattern: layout design by hierarchical domain splitting. ACM Trans. Graph. **32**(6), 1–12 (2013)

Digital Design, Materials
and Fabrication

Augmented Performative Design: A Workflow Utilizing Augmented Reality for Structurally Stable Masonry Design

Yang Song[(✉)], Asterios Agkathid, and Richard Koeck

School of Architecture, University of Liverpool, Liverpool, UK
`yang.song@liverpool.ac.uk`

Abstract. This paper presents an experimental performative design workflow for masonry structures, incorporating augmented reality (AR) technology and real-time stability simulation. We aim to resolve the lack of structural stability simulation in currently existing immersive design workflows. Our workflow consists of two phases: a) an AR design phase, in which interactive inputs, such as gestures, screen-based inputs, or marker recognitions, are translated to a design algorithm for the masonry structure generation and preview as holograms; b) an AR-assisted structural stability simulation phase, in which the designed structure is simulated on its structural performance attributes (without mortar or other adhesives) and shown as an AR holographic result animation on-site with the structurally optimised feedback. Our research findings highlight the advantages of the proposed workflow, which allows users modification suggestions and design implementation control for the assembly of structurally stable masonry structures, as well as the challenges arising regarding design diversity.

Keywords: Augmented Reality (AR) · Immersive Design · Structural Stability Simulation

1 Introduction

With the emergence of *AutoCAD* in 1942 and the rise of computer-aided design (CAD) in 1950, the way of architectural design and performance gradually changed. Architects turn handmade models and sketches into more accurate computer graphics to show their design ideas and drafts (Qi et al. 2021). Although the 2D-based drawing method can accurately express the design draft, it still has limitations and cannot meet the needs of architects to present their designs (Robertson and Radcliffe, 2009). In the era of the rapid development of information technology, the 3D modelling method provides architects with a better visual effect of space for the design scheme. The use of 3D digital models in architecture enabled by the constant advances in computer graphics is becoming increasingly accurate (Moural et al. 2013). However, virtual models seen on the screens lack the building scale, which is a disadvantage to the real experimentation of space and affects the expression of the designer's creative thinking (Dorta et al. 2016).

© The Author(s), under exclusive license to Springer Nature Switzerland AG 2023
M. Turrin et al. (Eds.): CAAD Futures 2023, CCIS 1819, pp. 347–360, 2023.
https://doi.org/10.1007/978-3-031-37189-9_23

In the last decade, the new immersive technologies of mixed reality (MR), have become available and usable, together with tools facilitating easy design and preview of 3D-4D creations (Barczik, 2018). Since then, MR has emerged as an increasingly valuable tool for architectural visualisation (Morse and Soulos, 2019). The MR tool for the early architectural design phases makes it possible to annotate, extend, and label real working models in a virtual environment in real-time (Schubert et al. 2012). Especially for the relatively mature virtual reality (VR) technology, its immersion, real-time rendering, interaction and natural user interface, have played an essential role in the architectural immersive design process (Achten and Vries, 2000). VR immersive technologies integrated with gesture and haptic devices have been used extensively in architectural design and construction projects (Kontovourkis et al. 2021). For example, Arnowitz et al. showed in their research project, VR provides spatial awareness and a sense of immersion that gives a unique opportunity for architects to perceive and evaluate the space of their design models. They created an application, *vSpline*, which uses an *HTC Vive* VR headset and controllers to provide tactile feedback for traditional sketched and small-size physical models. This application enables designers and architects to create, modify, understand and explore complex geometric relationships in VR immersive design process (Arnowitz et al. 2019). However, since VR ultimately exists in the virtual space, this immersive design environment lacking fundamental physical world properties will lead to blurred users' perception of space and distorted control of the design scale (Lin and Muslimin, 2019).

In contrast to VR, which takes place in purely virtual scenes, augmented reality (AR) allows us to overlay virtual objects with our real-world environment that can be viewed and interacted with in real time (Weissenbock, 2021). The character of AR technology, which combines virtual holograms with reality and provides data interactivity, gives the potential to the architectural design process for an immersive space experience to feel and perfect the design before the structure is built (Song et al. 2021). In an AR environment, design drafts can become more lively, convenient, and intelligent. Plus, with AR, the virtual products or graphic technology are not only for simulation but also for obtaining practical higher values (Choo et al. 2009). For example, Sumitomo et al. show in their system that the designers can use their gestures to participate in AR immersive design. The system will transform the physical gesture inputs into digital parameters for models developed in the AR environment. They prototype digitally augmented craftsmanship and present a hand gesture-based AR-assisted design method to challenge the traditional role of the artisan (Hahm et al. 2019). However, the current limitations in the calibration techniques and display and tracking technologies are the biggest obstacles preventing AR from being realistic in practical uses (Tang et al. 2003). Although AR has been used in some preliminary design methods, because of its virtual features, the design freedom is exceptionally high, resulting in the lack of corresponding physical attribute simulation. Therefore, many immersive designs still remain in the virtual environment or preliminary design discussions, and few physical structures have been constructed successfully through the AR immersive design method.

Architectural modelling radically evolved throughout its history, and the current integration of MR, especially AR, in the corresponding design tasks is limited chiefly to enhancing visualisation. Little to none of these tools attempt to tackle the challenge of modelling within immersive environments, which calls for new input modalities in order to move away from the traditional mouse and keyboard combination (Coopens et al. 2018). Moreover, architecture design, in fact, relying on 2D screen-based devices for 3D-4D manipulations, does not seem to be effective as it does not offer the same degrees of freedom (Dorta et al. 2016).

This paper presents an experimental performative design workflow for masonry structures, incorporating AR and real-time structural stability simulation. The developed augmented-performative design workflow using digital outcomes to explore and validate how AR changes the current immersive design methods for post-construction in the early design stages.

2 Methodology

The development and validation of the *Augmented Performative Design* research project consist of two phases: a) an AR immersive design phase, in which interactive inputs are translated to a design algorithm directly in AR through gestures, screen-based inputs, or markers recognition on AR devices (e.g., smartphones or *Microsoft HoloLens 1*); b) an AR-assisted structural stability simulation phase, in which the designed structure is simulated on its structural performance attributes and shown as an AR holographic animation on-site with the structurally optimised feedback. This research aims to propose an AR immersive design with a physical simulation workflow to meet the rationality of the post-construction and keep masonry designs structurally stable. Moreover, the validation experiment of this research uses a comparative method to explore the differences in the construction of the preliminary AR immersive design results in compliance with and without structural stability feedback.

The employed performative design workflow is driven by an instant connection between 3D-modelling software (*Rhinoceros7*), parametric design environment (*Grasshopper*), AR holographic immersion plugin (*Fologram*), and structural stability simulation plugin (*PhysX.GH*) (see. Figure 1). *Fologram* is a third-party application programming interface (API) developed by architects for architects, which could extract human gestures, screen-based interactions, device locations, and markers information. The *Fologram* user interface (UI) provides a bridge to interact and modify the related parameter sliders in the parametric design tool from *Grasshopper* through AR. *Fologram* works with its integrated graphical algorithm editor *Grasshopper*, ubiquitous tools in architectural design, and can easily be integrated into established immersive design workflows. *PhysX.GH* is an open-source rigid body simulation tool for *Grasshopper*, which provides the animation analysis of stability for stacked rigid bodies based on gravity and physical friction factors. The results of the immersive design will be transmitted to *PhysX.GH* for calculation and presented to the designer in the form of AR holograms, so as to give feedback on the structural stability of the preliminary design draft.

In the *Augmented Performative Design* workflow, the hardware includes a handheld device – *iPhone 11*, and a head-mounted display (HMD) – *Microsoft HoloLens 1*(for the experiments in this article). We also use a laptop for back-end running and debugging. These devices are connected to a WIFI router in the same IP address network environment for transforming the data from different stages and live streaming comments on design

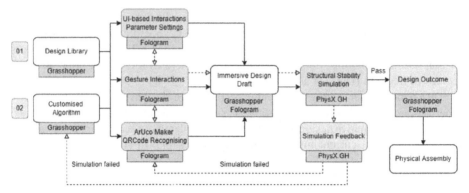

Fig. 1. This is the flowchart of the *Augmented Performative Design* research. It described each essential step of the AR immersive design and AR-assisted structural stability simulation (in green), the related plugin and software (in red), as well as the key outputs from immersive design and structural simulation (in blue).

software and plugins to visualise and output response ports. We will virtually evaluate the immersive design phase by setting the design algorithm and interacting with AR inputs for masonry outcomes. We will evaluate the structural stability simulation results by using the robotic arm to physically assemble design drafts, both with and without stimulation and modification feedback, to explore the necessity and accuracy of the structural stability simulation results. To achieve that, the hardware also includes a *UR 10e* robotic arm with *Robotiq 2F-140* gripper for the physical validations.

3 Experiments and Findings

3.1 Phase A: AR Immersive Design

The purpose of the AR immersive design is to upgrade the conventional design and preview media from screen-based to complete immersion by using AR UI, gestures, makers recognition, etc., to transform the interactive inputs from the physical world into the virtual design language, as well as to map the 1:1 scale design drafts in the physical space as holograms to realise the immersive design drafts preview.

In the AR immersive design phase, the shape generation algorithms have already been pre-set in the design library, representing different brick combinations for the fundamental masonry structures, such as brick columns and walls (see Fig. 2). All the interactions from the AR environment, including UI, markers, and gesture inputs, will be recorded as design inputs and linked to the corresponding parameters in the design file through the *Fologram* plugin. These interactive inputs affect how the bricks are arranged and combined, the shape and height, design constraints, etc., to provide a variety of design results (see Fig. 3). For complex shapes, architects can customise the design algorithms and interventions according to their needs and extract interactive parameters in AR immersive design phase through *Grasshopper*.

For example, we use the customised design of a masonry pavilion to validate the AR immersive design phase. The design algorithm for shape generation was pre-set

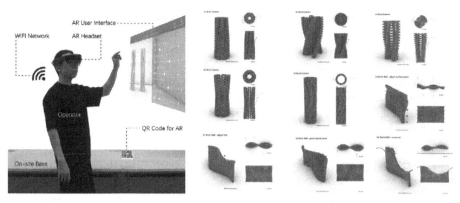

Fig. 2. The user chooses the masonry prototype from the pre-designed library through the AR UI before continuing the AR immersive design phase.

Fig. 3. These are the process screenshots of the AR immersive design steps. The designer uses hand gestures and UI interactions in AR to change the parameters, including the brick size, number of bricks per layer, column radius, column height, etc., and preview the pre-construction outcome on-site in a 1:1 scale as holograms.

in *Grasshopper*. In this case, we specify that the masonry pavilion is generated from reference points and based on adjustable curves. Therefore, we have formulated the design algorithm framework and extracted the key parameters, including setting reference points, controlling the curve shape, defining the combination of bricks, etc., according to the requirements of the interactive inputs. First, the user places multiple QR codes on the design site to define the base shape of the pavilion and converts them into reference points in *Grasshopper* using the marker recognition method in AR through *Fologram*. The user can adjust the QR code position and preview the base shape line in

AR in real time. Second, after the reference points are determined, the edge line of the base shape will be extruded as an adjustable curve in *Grasshopper*. There are some interactive control points attached to the curve virtually. Designers can use hand gestures to grab and move the control points in AR to adjust and modify the mesh shape and preview the result drafts holographically in real time on-site. Last, the specific combination and generation method of bricks will be developed in *Grasshopper*, as well as the related adjustable parameters will be extracted into AR UI for dynamic editing and holographic preview in a 1:1 scale (see Fig. 4).

Fig. 4. The process screenshots of a customised algorithm for masonry pavilion immersive design. The user can set and scan QR codes for edge line design, extrude and design the shape control curves by hand gestures, and modify the brick connection details in AR UI.

Our findings highlight that the AR immersive design experiment does fulfil our predetermined assumptions, which upgrades the conventional design and preview media from a 2D screen to complete 3D immersion in AR for design drifts. However, this AR virtual design environment is separated from the physical factors of the real world. Although the design algorithms in *Grasshopper* have corresponding restrictions specified according to the physical common sense, the obtained design results cannot guarantee structural stability. Therefore, the structural stability simulation needs to be added to the augmented performative design workflow to meet the structural rationality, so as to facilitate the subsequent physical construction process. Phase B was thus born to provide reliable physical simulations supporting the design drafts from Phase A.

3.2 Phase B: AR-Assisted Structural Stability Simulation

The AR-assisted structural stability simulation aims to augment the AR immersive design by adding physical property constraints, and the structural simulation feedback, to provide reasonable structural performance predictions and design modification suggestions immersively before the design is built. This structural stability feedback will reduce the drawbacks caused by the overly input freedom design phase (Phase A).

The simulation algorithms have already been pre-set in the *Grasshopper* through *PhysX.GH* plugin, representing different physical simulation input data, such as material properties, friction coefficients, rigid simulation samples, etc. All the input data will be transformed into the core calculator in *PhysX.GH* for the structural stability simulations. The result animations will be shown on-site as 3D holograms in AR through *Fologram* to predict the physical performances, including stability, shaking, and collapsing states. This physical simulation can be applied to the individual brick unit of object stacking in non-adhesive systems, as well as the overall masonry structure in adhesive systems.

For example, we use the AR immersive design outcome from the previous phase (Phase A) to verify the AR-assisted stability simulation feedback process. Due to the limitation of the robotic arm manipulation radius in the following physical verification step in the lab, we divided the pavilion design draft into 16 parts for the simulation instead of the whole structure. First, the user is required to set the corresponding physical simulation parameters according to the on-site environmental properties in the '*PX Material*' component. Second, the simulation requires an analogue plan which will be the on-site plane geometry, as well as the physical action samples which will be the brick unit geometries from the masonry structure. Last, all the above pre-set data and geometries will be connected to the '*PX Simulate*' component for the physical calculation. To preview the physical performance as AR holographic animation on-site, we need to connect the calculation results from the '*PX Simulate*' to the *Fologram*, so that users can preview the simulation results in AR (see Fig. 5).

As a result, in this case, this part of the pavilion did not pass the stability simulation. According to the simulated holographic animation, above the middle and upper part, the structure becomes unstable and starts to collapse, which means this part of the pavilion can not stand by itself in the real physical world during the assembly process. In order to verify the reliability and the accuracy of this simulation, and to compare the difference between the calculated results and the physical performance, we used a robotic arm to carry out the physical construction of this part to record the structural stability performance (see Fig. 6). Evidently, this part of the pavilion did collapse just above the structural centre. Although it was not the actual falling brick obtained in the simulation, the physical representation shows similar results around that falling brick. To a certain extent, the structural stability simulation can indeed predict the physical performance and mechanical stability during the construction process. After this preliminary experiment, we repeated the corresponding steps to simulate and assemble 15 other parts of the pavilion structure that will or will not collapse during the simulation. The physical results are roughly similar to the simulations.

Our findings highlight the validity of the simulation, which does predict the stability performance of the structure. Due to the safety restrictions of the robotic lab and the clay brick wall collapse safety analysis, we have to use foam bricks instead of clay bricks to

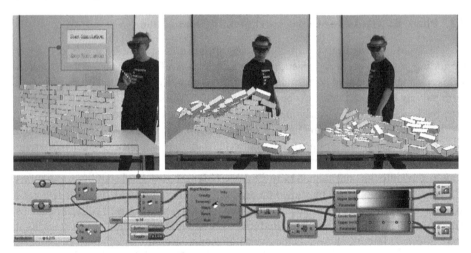

Fig. 5. The process screenshots of the structural stability simulation for a part of the masonry pavilion design. The user activates and previews the structural performance simulation through AR UI (above). The related *Grasshopper* workflow file shows the simulation logic (below).

Fig. 6. The structural stability simulation of the pavilion part (left) and the physical performance during the robotic assembly process (right).

keep both operators and devices safe. The pavilion designed in Phase A is divided into 16 parts for the stability simulation verifications. All simulation results are consistent with physical performance. In the 16 parts of validation, the tolerance between the number of bricks that collapsed in the simulation and the number of bricks that collapsed was no more than five, and their collapsed positions were similar. It can be summarized that at the preliminary design draft stage, this physical simulation can give approximately structural stability suggestions. However, the foam bricks used in this experiment were prefabricated in the factory, so there is almost no individual tolerance. The actual bricks might have tiny imperfections that make them behave differently from the ideal brick, so the clay bricks' stability needs to be verified in future research or by using methods such as average error. In addition, the physical performance animation generated by *PhysX.GH* is instantaneous, and the user does not know which part of the design drift needs modification. Thus, the stability simulation desperately needs feedback to guide the designer in modifying the results from the AR immersive design.

For the simulation feedback, in this case, we brought the '*dynamics*' outputs from the '*PX Simulate*' results into the feedback algorithm. First, the algorithm will extract the barycentric coordinate position of each brick element before and after the simulation, and find the absolute value of the corresponding movements in the X, Y, and Z three-axis directions. Second, the above value will be quantitatively defined as a 'stable' state from 0 to a quarter of the distance of the shortest brick unit; a 'shaking' state from a quarter to less than half the distance; as well as a 'collapse' state for more than half. The simulation feedback will be displayed as brick stability states in different colors through AR to alert the designers, as green represents 'stable', yellow represents 'shaking', and red represents 'collapse' (see Fig. 7). Next, the part concentrated mostly on 'red' means that they will collapse for the physical assembly process. The simulated design draft must be carefully modified in AR immersive design (Phase A). Last, the revised design draft will be re-entered into the simulation system until the red area no longer appears in the result. To validate the positive effect of the feedback system on structural stability, we brought the modified structure into the corresponding robotic assembly process. As a result, the final physical structure performed the same as the simulated results in AR, showing structural stability, which means that the feedback has played an essential role in assisting designers in modifying the design shapes and improving structural stability. This simulation feedback can be applied not only to the parts but also to the overall revision of the pavilion (see Fig. 8).

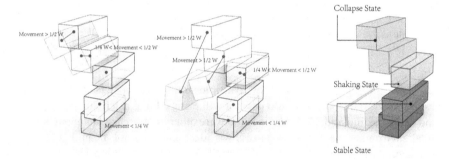

Fig. 7. The calculation (left) and preview (right) logic of the stability simulation feedback system.

Our findings highlight that the AR-assisted structural stability simulation and the modification feedback do fulfil our pre-determined assumptions, which assist the AR immersive design with physical simulations and feedback for further modifications. The simulation provides reasonable structural performance prediction before the physical structures are built. Similarly, the feedback augments the meaning of simulations, providing suggestions for revising unreasonable structural design drafts. This simulation and feedback reduce the drawbacks of AR immersive design (Phase A) away from lacking physical constraints. The proposed structural modification feedback was validated by the robotic assembly of a pavilion design part, before and after the modifications following the feedback (See Fig. 9). The simulation feedback can roughly but not exactly guide the design shape for improving physical performance.

Fig. 8. The process screenshots of the structure modification and the physical re-simulation for a part of the masonry pavilion design. After the simulation feedback (color state definition) (a), the user modified the shape mesh in AR by hand gestures to modify structural stability (b). The designer returns the structure to the simulation (c) and previews the result (d). The user verifies the accuracy of the simulation through the physical performance during the robotic assembly (e,f).

However, the detailed quality of the real-time simulation depends on the number of bricks designed. If the number is too large, it will bring a lot of calculations, resulting in lag and delay in simulation results. Since there are other interference factors in the actual physical environment, the simulation result is ideal and does not mean that the actual performance will completely follow the simulation. The simulation feedback is also the approximate interval information. Users are only guided on whether the modified structure can pass the stability simulation or not, but they do not know how to modify it according to the feedback, which takes them multiple modification attempts. Phase B is aimed at architects, only for now, with structural experience and basic structural knowledge.

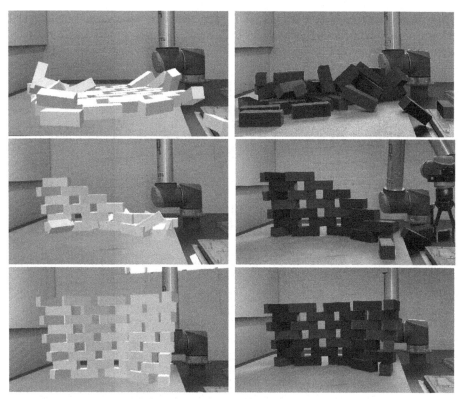

Fig. 9. These are the user tests to validate the proposed AR-assisted structural stability feedback. The user modifies the design in AR following the structural modification feedback (on the left from top to bottom). The simulation results (on the left) can only provide a rough reference for the physical structure performance (on the right). After following the structural modification feedback, the designed structure improved the physical performance (on the right from top to bottom).

4   Conclusion

The *Augmented Performative Design* research project developed and verified a real-time and interactive workflow from design to simulation in the early architectural design stage for masonry structures. Closely practising the AR immersive design as well as AR-assisted structural stability simulation, it can be concluded that the proposed workflow does fulfil our pre-determined assumptions and offers a new way to design, modify, and simulate the drafts on-site through AR in real time. Architects are able to preview their digital designs out of the 2D-based media, as well as interact and communicate with the related on-site physical environment with a 3D-4D immersive perception in AR. Moreover, to reduce the exceptionally high design freedom and the lack of corresponding physical attribute simulation, architects can also simulate the structural stability and get related feedback for further modification in AR before it is built physically. Using AR technology, our performative design workflow enables the users to improve their cognition and understanding of space, triggers reflections and remodelling of the architectural design process, cultivates their design creativity and outcome variety, and simulates and gets structural modification feedback for the assembly of structurally stable masonry structures.

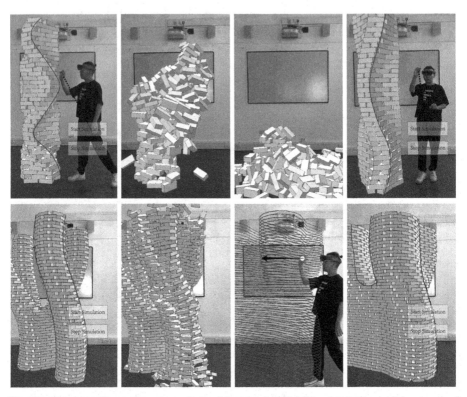

Fig. 10. These are the process screenshots of the physical stability simulation and the structural modification for more masonry structures, such as a column (above) and a pavilion (below).

We designed, simulated, and modified different masonry proposals in the same way through the performative design workflow, including walls, columns, facades, etc. (see Fig. 10). The obtained virtual results also verified and confirmed the findings from the experiment of this paper. However, there are limitations and space for further improvement. The AR immersive design experiments only occur in initial discrete brick-based masonry structures. Whether this design method is suitable for other more complex architectural scale designs remains to be verified by future experiments. Moreover, according to the structural stability simulation, the user is only guided on whether or not the modified structure can pass the simulation. However, the user does not know how to modify it according to the feedback. In future research, the simulation part or part that accompanies the immersive design with simulation can be transferred to the design software system such as *Unity* with a real-time physical calculator; or through machine learning in Artificial Intelligence (AI) to generate the structural modification suggestions after extensive summarisations. Additionally, there are some issues based on the AR device hardware, such as hologram drift or inaccurate motion capture, etc., which need to wait for the device to update or add extra sensors.

In conclusion, our *Augmented Performative Design* research project provides designers with an immersive design method, unprecedented design constraints based on the real physical world, simulation, and related structural stability feedback for the preliminary design drafts through AR. We will systematically develop this workflow for more different architectural structure forms or materials in different proposals, and combine other software or AI tools to improve this system smoothly and reasonably for the initial architectural design drafts.

References

Qi, X.N., Lo, T.T., Han, Y.S.: Designing for human-VR interaction - how VR interaction can be designed to bring better design participation. In: Proceedings of the 26th International Conference of the Association for Computer-Aided Architectural Design Research in Asia (CAADRIA) 2021, **2**, pp. 163–172. CAADRIA, Hong Kong (2021)

Robertson, B.F., Radcliffe, D.F.: Impact of CAD tools on creative problem-solving in engineering design. Computer-Aided Design **41**, 136–46. Butterworth-Heinemann, Newton, MA (2009)

Moural, A., Dias, M.S., Eloy, S., Pedro, T.: How space experimentation can inform design - immersive virtual reality as a design tool. In: Proceedings of the International Conference of the Ibero-American Society of Digital Graphics (SiGraDi), pp. 182–185 (2013)

Barczik, G.: From body movement to sculpture to space – employing immersive technologies to design with the whole body. In: Proceedings of the 36th Education and research in Computer Aided Architectural Design in Europe (eCAADe) 2018, **2**, pp. 781–788 (2018)

Morse, C., Soulos, F.: Interactive façade detail design reviews with the VR scope box. Proceedings of the Association for Computer Aided Design in Architecture (ACADIA) **2019**, 422–429 (2019)

Schubert, G., Artinger, E., Yanev, V., Petzold, F., Klinker, G.: 3D virtuality sketching – interactive 3D sketching based on real models in a virtual scene. Proceedings of the Association for Computer Aided Design in Architecture (ACADIA) **2012**, 409–418 (2012)

Achten, H., Vries, B.: DDDOOLZ – a virtual reality sketch tool for early design. In: Proceedings of the 5th Conference on Computer Aided Architectural Design Research in Asia (CAADRIA), Singapore, pp. 451–460 (2000)

Kontovourkis, O., Konstantinou, A., Kytizi, N., Tziourrou, P.: Built-in immersive vr technology for decision-making in design and simulation of a flexible shading device. In: Proceedings of the Arab Society for Computer Aided Architectural Design (ASCAAD), pp. 190–200 (2021)

Lin, X.H., Muslimin, R.: Reshape – rapid forming and simulation system using unmanned aerial vehicles for architectural representation. In: Proceedings of the 24th International Conference of the Association for Computer-Aided Architectural Design Research in Asia (CAADRIA) 2019, Hong Kong, **1**, pp. 413–422 (2019)

Arnowitz, E., Morse, C., Greenberg, D.P.: vSpline – physical design and the perception of scale in virtual reality. Proceedings of the Association for Computer Aided Design in Architecture (ACADIA) **2017**, 110–117 (2019)

Weissenbock, R.: Augmented quarantine – an experiment in online teaching using augmented reality for customized design interventions. In: Proceedings of the 39th Education and research in Computer Aided Architectural Design in Europe (eCAADe) 2021, **2**, pp. 95–104 (2021)

Song, Y., Koeck, R., Luo, S.: Review and analysis of augmented reality (AR) literature for digital fabrication in architecture. Autom. Constr. **128**, 103762 (2021)

Choo, S.Y., Heo, K.S., Seo, J.H., Kang, M.S.: Augmented reality – effective assistance for interior design. In: Proceedings of the 27th Education and research in Computer Aided Architectural Design in Europe (eCAADe), Session **18**, pp. 649–656 (2009)

Tang, A., Owen, C., Biocca, F., Mou, W.M.: Comparative effectiveness of augmented reality in object assembly. new techniques for presenting instructions and transcripts. CHI 2003, April 5–10, 2003, Ft. Lauderdale, Florida, USA. **5**(1), pp. 73–80 (2003)

Hahm, S., Maciel, A., Sumitiomo, E., Rodriguez A.L.: FLOWMORPH: Exploring the human-material interaction in digitally augmented craftsmanship. In: Proceedings of the Conference on Computer Aided Architectural Design Research in Asia (CAADRIA), Hong Kong, pp. 553–562 (2019)

Coopens, A., Mens, T., Gallas, M.: Parametric modelling within immersive environments – building a bridge between existing tools and virtual reality headsets. In: Proceedings of the 36th Education and research in Computer Aided Architectural Design in Europe (eCAADe) 2018, **2**, pp. 711–716 (2018)

Dorta, T., Kinayoglu, G., Hoffmann: Hyve3D and the 3D cursor: architectural co-design with freedom in virtual reality. Int. J. Architectural Comput. **14**(2), 87–102 (2016)

A Design-to-Fabrication Workflow for Free-Form Timber Structures Using Offcuts

Dominik Reisach[1]([✉]) [iD], Stephan Schütz[2] [iD], Jan Willmann[3] [iD], and Sven Schneider[2] [iD]

[1] Department of Architecture, ETH Zurich, Zurich, Switzerland
reisach@arch.ethz.ch
[2] Faculty of Architecture and Urbanism, Bauhaus University Weimar, Weimar, Germany
[3] Faculty of Art and Design, Bauhaus University Weimar, Weimar, Germany
{sven.schneider,jan.willmann,stephan.schuetz}@uni-weimar.de

Abstract. Currently, the timber industry in the European Union incinerates up to 80% of its waste wood, releasing its embodied CO_2 into the atmosphere while producing energy. This practice also affects offcuts, a waste material from structural timber production, mostly because of aesthetic imperfections. However, there are potential architectural use cases for this material that extend its lifetime beyond downcycling. Therefore, we propose to employ these offcuts as load-bearing elements of free-form timber structures and present an integral design-to-fabrication workflow suitable for this task. In this paper, we discuss the underlying method in detail, specifically (1) the computational design process to optimally place timber offcuts and to compute wood joints, (2) the transfer of design data into a robotic fabrication process, and (3) the integration of these findings into a unifying design-to-fabrication workflow and its architectural implications. This process minimizes material waste and facilitates the design and buildup of offcuts into structural configurations, including their dis- and reassembly. The resulting timber morphologies consist of non-standard material aggregated under digital guidance, giving them a distinct aesthetic expression. A series of digital experiments demonstrated the capabilities of the conceived method. Finally, we prove the feasibility of the proposed workflow with the design and robotic fabrication of a full-scale Offcut Demonstrator under real-world conditions.

Keywords: Waste Wood · Upcycling · Circularity · Computational Design · Design-for-Disassembly · Timber Structures · Robotic Fabrication

1 Introduction

1.1 Background

Recently, the European Union (EU) proposed an "Action Plan" for the sustainable development of its industries that also affects the architecture, engineering,

---

Pursued as thesis project and demonstration study at Bauhaus University Weimar.

© The Author(s), under exclusive license to Springer Nature Switzerland AG 2023
M. Turrin et al. (Eds.): CAAD Futures 2023, CCIS 1819, pp. 361–375, 2023.
https://doi.org/10.1007/978-3-031-37189-9_24

and construction sector. This sector is responsible for an estimated 38% of the waste production and an estimated 36% of the CO_2 emissions in the EU [1]. The proposed strategies to achieve these goals include (1) a transition to a circular economy and (2) an increasing use of materials with a low or negative carbon footprint [2], e.g., timber, which could turn buildings into global carbon sinks [3]. This paper investigates the combination of both strategies and demonstrates that the implementation of digital tools creates new horizons for addressing the sustainability challenges in the field of architecture and construction.

Timber Industry Practices. In this research, we examined the supply chains, waste management, and manufacturing practices of the timber industry in central Germany. Of particular interest are the amount and types of waste wood generated. In general, the local manufacturers successfully use a so-called "Cascading Utilization" to extend the lifespan of wood products—as it is standard in the timber industry elsewhere in the EU [4]. According to this principle, wood is converted at the end of its product life and downcycled to, for example, particle-based products, wood insulation, bark mulch, or to produce pulp and paper. However, most of the waste wood—up to 80%—is directly incinerated to generate energy [5] and, as such, it is highly questionable if this is the best use of this valuable resource.

More specifically, there are different types of waste wood generated in the process of cutting raw logs into engineered timber beams and planks, including, for example, sawdust, woodchips, and bark, but also offcuts.[1] These offcuts are produced in large quantities (see Fig. 1) and, with careful selection, they are still usable for architectural purposes beyond downcycling and incineration.

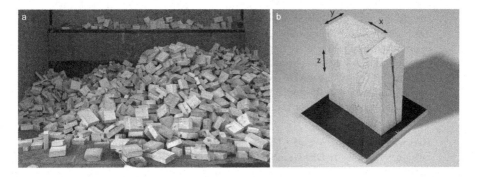

Fig. 1. a) The offcut collection space of a manufacturer with an annual production of approximately 12.000 m³ of offcuts. b) Offcut with dimensions, where x and y determine the cross-section and z the length.

---

[1] The offcuts covered in this work exclusively consist of wood from Norway spruce (*Picea abies*), a softwood with typical anisotropic characteristics [5]. It is the most commonly available tree species to timber manufacturers in the mentioned region.

1.2 State of the Art

The timber industry started implementing digital technologies from the mid-1980s onward. These technologies include large-scale CNC machinery and a range of scanning tools with the goal of optimizing the production of standardized timber products [6]. This development paralleled academic research, which fostered non-standard approaches of timber processing by deploying novel digital tools that range from CNC machines with three axes to create complex structures [7,8] to multi-axis fabrication machinery, i.e., five-axis CNC machines and six-axis industrial robotic arms. Combined with a variety of tools and end-effectors, different manufacturing processes and fabrication routines became feasible in the academic context [9,10].

The first academic research projects using robotics in timber architecture explored assembly task applications. In particular, stationary [11], gantry-mounted [12], and mobile robots [13] were used to layer timber elements. Additionally, collaborative efforts between a robot and humans [14], multiple robots [15], or a combination of both, i.e., multiple robots and humans [16,17], demonstrated the assembly of more complex structures. More recent research illustrated how wood joints can be assembled with industrial robotic arms using specialized end-effectors [18,19] in conjunction with machine learning algorithms [20] and by using collaborative robots [21].

Concurrent with these advances in robotic non-standard timber construction, novel computational planning tools promoted the development of respective free-form timber structures [22]. This development enabled the construction of timber plate shell structures with bespoke cassette modules [23] and recycled cross-laminated timber (CLT) cutouts [24] as well as timber grid shell structures [14, 25]. Further optimization of the design, planning, and fabrication processes led to increased form freedom of curve-based glulam structures [26]. Additionally, these novel computational tools facilitated experimentation with the recycling of waste wood for load-bearing structures, such as timber walls and ceilings made from CLT waste [27,28], truss structures [29], timber plate structures [30], reciprocal frame structures [31], and complex reconfigurable structures [32].

1.3 Research Motivation and Objective

The previous section showed that research in digital fabrication and non-standard timber construction is constantly progressing and that powerful computational design and engineering tools are becoming increasingly available. However, a large number of approaches are still centered around using standardized engineered timber and conventional planning and building routines, which are heavily constrained in terms of circularity and waste-free construction. Therefore, in this paper, we investigate the potential of non-standard, circular, and low-waste design concepts and planning methods for upcycling timber offcuts. For this purpose, we developed new algorithmic design tools and combined them with digital fabrication processes to handle complex geometries, resulting in a novel computational method for designing low-engineered free-form timber structures.

This method transfers these design concepts into 1:1 applications, thus facilitating interconnections across multiple scales, while demonstrating the overall feasibility of the approach. It integrates several steps that together constitute the design-to-fabrication workflow (see Fig. 2):

a) Digitizing the offcuts to create a database that facilitates the handling of their non-standard cuboid geometries.
b) Designing and optimizing curve-based structures.
c) Combining the design proposals with the offcut data to simulate the optimized positioning and orientation of every offcut, including their joinery.
d) Generating fabrication data and respective robotic control code.
e) Fabricating and assembling the physical offcuts as described by the digital blueprint.

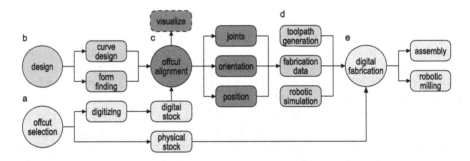

Fig. 2. The proposed design-to-fabrication workflow: a) Offcut selection and database, b) architectural design and form finding, c) offcut alignment and joinery, d) fabrication data and robotic control code, e) digital fabrication including manufacturing and assembly.

2   Methodology

2.1   Design Concept of Offcut Structures

The concept for the design of offcut structures is grounded on curve-based representations of structural configurations. These curve-based structures can be optimized with various form-finding tools, with examples ranging from simple arch structures to complex double-curved free-form shapes. On each of these optimized curves, the cuboid offcuts are placed in an iterative process, starting at one end of a curve and finishing at the other. Thereby, the intersection between a curve and both ends of an offcut occurs at the same place, e.g., the center of the surfaces. That allows the elements to follow the course of a curve perfectly, while the wood fibers are aligned with the principal forces. However, this also results in gaps and overlays between the offcuts. Therefore, their position is adjusted—i.e., they are pushed together—to create larger overlays. Then, the median of an overlay determines how every element is trimmed, resulting in angular cuts on both ends of every offcut. With this procedure, the alignment of offcuts on curves is completed successfully. However, the offcuts' cuboid geometries lead to an irregular discretization of the curves. For this reason, their placement order is optimized, positioning shorter offcuts at stronger curvatures and longer offcuts on more straight sections (see Fig. 3).

Once an alignment is completed, loose wood joints connect all elements, which—besides their ecological and economic upsides—have the benefit that structures can be disassembled or even recycled waste-free, while simultaneously retaining the maximum length of every offcut. Two types of joints are used in this work: (1) Loose tenon joints for compressive stresses and (2) loose spline joints that also resist tensile stresses. These loose joints are milled out of smaller pieces of waste wood. Finally, intersections between assembled components of a structural configuration are solved by cross-lap joints.

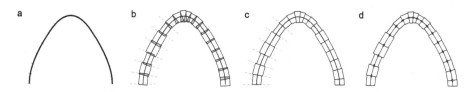

Fig. 3. Design concept of offcut structures: a) Curve design, b) offcuts on curve with overlaps, c) trimmed offcuts, d) loose joints connecting the offcuts.

2.2 Algorithms and Computational Implementation

The complexity of the design, planning, and fabrication of offcut structures—which stems primarily from the large number of offcuts, their spatial relationship, and the unique geometry of each element—requires the implementation of appropriate algorithmic design tools. In this context, this research presents a multi-component plugin that is implemented in McNeel's *Rhino/Grasshopper* software framework as a *.NET* library written in *C#*, with the open-source development data available at GitHub [33]. The plugin facilitates the simplicity and consistency of all steps of the proposed method, beginning with the extraction of data from databases to compute instances of a unique Class called *Offcut* used in all components. In particular, it provides the ability to align, place, and adjust Offcut instances on curves, compute wood joints, and efficiently export fabrication-relevant data.

Offcut Database. The algorithm works with digital representations of offcuts and, therefore, requires their digitization. These representations are mere abstractions of the real objects since they are conceived as perfect cuboids. This approach simplifies subsequent computational operations as well as the digitization itself, as only the main dimensions are necessary and, therefore, measured.[2]

---

[2] In this work, the three main dimensions of each offcut were measured manually.

Here, the x and y dimensions define the offcut's cross-section, while the z dimension represents the length in the direction of its wood fibers. The data is then transferred to a database that is of a CSV or XLS file type. In this process, every Offcut receives a numerical ID that is marked on the equivalent physical object as well.

In *Grasshopper*, a specific component can access these databases and compute basic Offcut instances, containing each entry's ID, its x, y, and z dimensions, as well as its volume, which is calculated based on these dimensions. The Offcut Class possesses additional properties, allowing it to store further values—e.g., geometrical Boundary Representations (BReps) and Planes—that are computed in subsequent steps of generating offcut structures. Additionally, it is possible to deconstruct Offcut instances to access and manipulate their data and reassemble them again. That enables an efficient and informed planning process with less waste and reduced fabrication times since relevant information, e.g., how much material is lost, is available at all times.

Offcut Alignment. In this context, an "Alignment" relates to the positioning and orientation of offcuts on curves in a way that they follow the curvature precisely without gaps or overlays. To achieve this, an algorithm has been developed that positions the offcuts in an iterative placement process (see Fig. 4). It is compiled in a *Grasshopper* component and requires curve and offcut data to perform this operation. The algorithm starts with setting the curve domain to an interval between 0 and 1 (reparametrization), with the initial iteration beginning at 0. Already here, it chooses what offcut to place, which, in this case, is the first item of the database. Then, to compute the offcut geometry at the desired location, there are three planes required. The first plane is computed as a perpendicular frame at the start of the curve. Its origin serves as the center point of a sphere with a radius that is equivalent to the z dimension of the offcut to place, multiplied by an adjustment value that is below 1. This variable value ensures that the offcuts placed overlap without gaps. The algorithm then computes another perpendicular frame—i.e., the second plane—at the intersection of the curve and the sphere.[3] The available two planes are now used to compute a linearly interpolated "average" plane that determines the orientation of the final offcut. This plane is then moved in the negative direction of its z-axis by half the length of the current offcut. In the next step, the algorithm uses the average plane to compute a rectangle with an adjustable base point that is then extruded by the length of the current offcut along the plane's z-axis. The resulting geometry is the Offcut at its final position. Finally, to create the correct angle at its ends, it is trimmed with the first and the second plane. Afterward, the open faces are closed, a clean-up is performed, and the placed Offcut instance is removed from the list to avoid its repeated use, ending the iteration. Now, the current second plane becomes the new first plane in the next iteration.

---

[3] If the sphere is not close to the start or the end of the curve, there will be two intersection points, of which only the second one is considered.

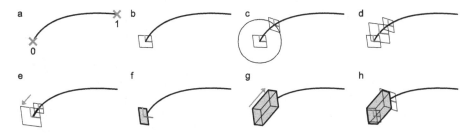

Fig. 4. Computational placement process: a) Curve reparametrization, b) first plane, c) second plane, d) average plane, e) moved average plane, f) rectangle placement, g) extrusion to cuboid, h) trimming the cuboid with the first and second plane.

While approaching the end of the curve, the algorithm checks the distance between the origin of the current second plane and the endpoint of the curve at every iteration. If this distance is smaller than the Offcut with the greatest length still available, it places this specific element as the last item of the alignment and performs the trimming operation. As soon as the algorithm completed the alignment process on all curves, it returns the Offcut instances used, though updated with BRep, plane, and volume data.

Furthermore, the alignments are adjustable through torsion and rotation, which is useful to create optimal intersections between multiple components. For this purpose, the alignment component computes an interval out of two angles (start and end) provided and remaps it onto an interval between 0 and 1, which describes the curve domain. Every point on the curve is now related to a specific angle, determining the rotation of the planes at every intersection. Different angles at the start and end result in a gradual rotation of offcuts and, thereby, in torsion, whereas the same angles result in a complete rotation.

As of now, the way the list is sorted determines the placement sequence of offcuts. This might result in a coarse discretization—specifically if the curve has a strong curvature—and, therefore, in increased material waste, reduced structural performance, and less aesthetic quality. Therefore, the algorithm uses an optimized placement strategy that matches the z dimensions of the Offcut instances to the curve's curvature. The minimum and maximum values of the curvature determine its domain, which is linked to the domain of the minimum and maximum z dimensions. Now, every point on the curve corresponds to a certain Offcut instance, meaning that at every iteration, it is determined which item to place to achieve an optimized alignment.

The algorithm uses a different optimization logic if it performs the placement on a linear curve. Here, the trimming procedure is redundant—and, therefore, excluded—since the Offcut instances are aligned perfectly. Rather, it places the longest Offcut instance left on the list on every iteration, thereby reducing the number of elements and joints to a minimum, resulting in reduced material waste and fabrication times.

Wood Joints. There are two components—one for tenon and one for spline joints—to compute the loose wood joints in iterative processes between the Offcut instances on the curves (see Fig. 5). This procedure requires the first and second plane of each instance that determine the orientation of the joints. The algorithm computes the base points of the joints—after their number is specified—in consideration of the x and y dimensions of the adjacent Offcut instances. Then, it creates their specific shapes, fillets the corners according to the radius of the milling tool, extrudes the shapes, and performs a clean-up on the final BRep geometry. This procedure also accounts for the fact that the first and last instances only have one neighboring Offcut.

Finally, the algorithm solves the intersection between the BRep geometries of the joints and the Offcut instances with a *Boolean Difference* operation, resulting in the negative joint shape. To reduce computation times, this method is executed only once per Offcut instance on an array of the relevant BReps. Additionally, every iteration is performed in a multithreaded loop—in this case, in a data-parallel operation—, meaning that the algorithm performs multiple iterations concurrently. After the *Boolean Difference* succeeded, the BRep faces are merged, split, and returned. Then, the data of every Offcut instance is updated, including the BRep geometry and volumes, and now contains the joint geometry and volume.

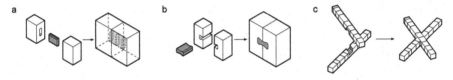

Fig. 5. Offcut wood joints: a) Loose tenon joint, b) loose spline joint, c) cross-lap joint.

The intersections between multiple components are solved with cross-lap joints, which facilitate an almost ideal overlap depending on the components' orientations. To the detect these intersections, the *RhinoCommon* Class *CurveIntersections* is used on the curves. It returns the amount and the types—i.e., end-to-end, cross-over, or T-intersection—of intersections occurring on the curves and marks them with points. Then, the algorithm detects the three offcuts of every component that are closest to the intersection point. Their BRep data and planes are used to compute a BRep geometry with the correct orientation to cut the Offcuts. After the *Boolean Difference* operation succeeded, the component updates and returns the data of every Offcut affected.

2.3 Robotic Setup and Control

The robotic setup to mill the offcuts consisted of a *Universal Robots UR10e* being mounted on a custom fabrication table. The table in front of the robot had a wooden plate with a machine vise to fixate the offcuts. A custom milling end-effector was used for subtractive manipulation, consisting of a tailored aluminium mount and a *BIAX RE 2860* spindle.

The toolpaths were generated in *Autodesk Fusion 360*, which imports *Rhino3d*-native files (see Fig. 6). For this purpose, a component of the plugin facilitates the placement of offcuts, moving them to the fabrication table in the digital environment and computing a stock model of the original offcut that encloses the fabricated shape. Subsequently, the two geometries are exported. After the data is imported, the final toolpaths are generated and then exported as coordinates in a CSV file. This CSV file is read in *Grasshopper*, where another component extracts the coordinates and computes points. Finally, the correct planes for each milling operation can be assigned to these points. To simulate the robotic process and to generate the machine code in the *URScript* language, the *Grasshopper* plugin *HAL Robotics* was used.

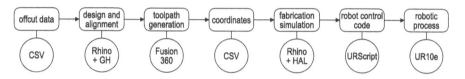

Fig. 6. Data flow diagram, proposing a seamless transfer between different software environments at all stages of the overall workflow.

3   Results

3.1   Design Space Exploration

The developed software tool functions independently of any structural system since it simply places offcuts on curves, regardless of their shape. The design space exploration examines the potentials and limitations with several examples of offcut structures in 2D and 3D.

In 2D, simple frame structures give insights into the connections and transitions of alignments. If the curve transition is sharp rather than smooth, the algorithm either produces a coarse alignment (see Fig. 7a) or fails to solve the instance completely, requiring the curves to be split (see Fig. 7c). Filleting the corners fixes this issue and results in the optimization algorithm placing the shortest offcuts there (see Fig. 7b and d).

The 3D examples give insights into the spatial relationships between the offcuts and components. The algorithm creates arch and vault structures without issues (see Fig. 7e). Creating a mass-timber structure is possible as well, though it is required to standardize the thickness of the material to avoid gaps (see Fig. 7f). Additionally, example *g* demonstrates that the software tool is capable of computing an alignment with an exaggerated torsion and six joints perfectly (see Fig. 7g). In contrast, the main challenge is the intersection of multiple components (see Fig. 7h). Here, form-finding tools could optimize the curves and minimize the issues. In the end, the software tool provides the means to design an offcut structure that can be fabricated.

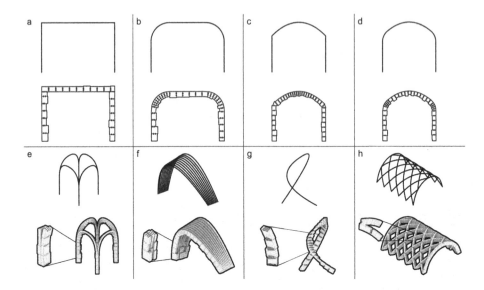

Fig. 7. Design space exploration: a)–d) 2D examples that gave insights into connections and transitions. e)–h) 3D examples that examined the components' spatial relationships, including torsion and intersection.

3.2   Offcut Demonstrator

The design, robotic fabrication, and assembly of a demonstrator served as proof of concept and validated the feasibility of the design-to-fabrication workflow and, as such, the general applicability of the proposed method by making use of all features. This Offcut Demonstrator consisted of three arches and had an equilateral triangular footprint with a side length of three meters, which fit perfectly into a circle with a diameter of 3.5 m. Each of these arches had a different height, i.e., 1.25 m, 1.5 m, and 1.75 m, and was primarily designed as a compression structure. The algorithm placed the offcuts on these arc-shaped curves and rotated them by 30° so that they met at the base points with a perfect fit (see Fig. 8). Finally, the loose joints were computed, consisting exclusively of two spline joints which demonstrated sufficient stiffness and tensile strength in preliminary studies. Here, a tool radius of eight millimeters was used to produce a smooth fillet, which guaranteed a strong fit. In the end, the complete composition consisted of 81 individual offcuts and 162 loose spline joints, with the first arch having 24 offcuts, the second 25, and the third 32.

The conceptualization of the design was followed by the toolpath generation, the simulation of the robotic milling process, and the material preparation. This meant sorting, marking, and pre-cutting the offcuts chosen by the algorithm. Afterward, the robot milled every offcut as designated in the blueprint

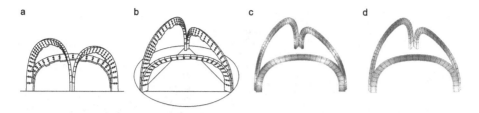

Fig. 8. Final design of the Offcut Demonstrator: a) Front view, b) axonometric drawing, c) & d) structural assessment: axial stresses & displacement, simulated with Karamba3D [34].

(see Fig. 9). Contrary to this bespoke fabrication process, the loose spline joints were produced serially. Here, an extra 0.1 mm was milled off to enable their insertion while simultaneously maintaining a strong fit. In total, this resulted in approximately 35 h of milling and 0.074 m^3 of waste produced during fabrication, whereas the complete Offcut Demonstrator had a volume of 0.23 m^3 (0.22 m^3 for the offcuts and 0.01 m^3 for the joints).

Finally, the arcs were assembled as determined by the blueprint of the design by joining the offcuts on ground. Then, they were lifted and carried to their designated position in the ensemble. Here, the feet of the adjoining arcs, i.e., the offcuts touching the ground, were screwed together to ensure gap-free connections. The assembled Offcut Demonstrator is depicted in Fig. 10.

Fig. 9. Robotic fabrication and assembly of the Offcut Demonstrator: a) & b) Milling of the loose spline joints, with the mortise being milled from two sides, c) offcuts and joints for one arch, d) assembled arch.

Fig. 10. The Offcut Demonstrator proves that the proposed method facilitates the design and fabrication of low-engineered free-form offcut structures. Photographs by Michael Braun.

4   Discussion and Conclusion

This paper demonstrates the potential of non-standard, circular, and low-waste design concepts and planning methods for upcycling locally available timber off-cuts. It specifically builds on previous work on free-form glulam structures and several circular waste timber projects. Combining both topics, in this research, we propose a method that proved effective in the upcycling and utilization of offcuts for architectural applications, fostering novel digital-material design techniques that facilitate the design of low-engineered free-form timber structures. This non-trivial task was solved by developing computational design tools that serve the entire workflow from design to fabrication. These tools select and place offcuts optimally to significantly reduce material waste during fabrication. They also compute wood joints, enabling whole structures to be dis- and reassembled and interconnecting different phases of the material life-cycle and the architectural design process. Furthermore, the proposed method demonstrates the feasibility of transferring design data into a robotic fabrication process, which enables transitioning into 1:1 applications and creates interconnections across multiple scales.

Notwithstanding these results, future work could further optimize and expand the proposed method. In particular, improvements to the computational design tool could streamline the data transfer and the toolpath generation, combining them in a single software framework. The implementation of scanning methods, specifically at the factory level, could automate the digitization of offcuts. Furthermore, material modifications, e.g., external joints made of hardwood, could enhance the stiffness and tension strength of offcut structures, whereas strength tests of whole components could bridge the gap to real use cases beyond academic experimentation. Here, the structural, dimensional, and procedural scalability of the method remains a key challenge because building with a large number of relatively small bespoke elements is still labor-intensive and complex. A database containing countless offcuts might increase the complexity even more. Therefore, future research should address this challenge by exploiting offcut imperfections and by creating an adaptive feedback loop between the design and available parts. However exploratory, this research suggests a potential use-case for waste wood, i.e., offcuts, in architecture beyond downcycling and contributes to the discourse on resource consumption and sustainable construction.

Acknowledgements. This research was conducted as a master's thesis in the framework of the M.Sc. Architecture program at the BAUHAUS UNIVERSITY WEIMAR, Germany. It was supported with a *Bauhaus Degree Completion Scholarship* granted by the university. Offcuts were sponsored by RETTENMEIER HOLZINDUSTRIE HIRSCHBERG GMBH. The authors especially thank Ringo Gunkel, Christian Hanke, Matthias Henkelmann, and Christian Möhwald for their support in the university's workshops. Special thanks go to Michael Braun, who played an essential role in the installation and testing of the robot system.

References

1. Observatory, E.C.S.: Analytical report: Improving energy and resource efficiency. Technical report, European Commission, Brussels (2018)
2. Commission, E.: Directorate-General for Environment: A new circular economy action plan: For a cleaner and more competitive europe. Technical report, European Commission, Brussels (2020)
3. Churkina, G., et al.: Buildings as a global carbon sink. Nat. Sustainabil. **3**, 269–276 (2020). https://doi.org/10.1038/s41893-019-0462-4
4. Mair, C., Stern, T.: Cascading utilization of wood: a matter of circular economy? Curr. Forest. Rep. **3**(4), 281–295 (2017). https://doi.org/10.1007/s40725-017-0067-y
5. Dinwoodie, J.M.: Timber: Its Nature and Behaviour. E & FN SPON Online Taylor & Francis, London (2000)
6. Willmann, J.: Digitale Revolution im Holzbau: Roboter, Narration, Entwurf. In: Rinke, M., Krammer, M. (eds.) Architektur fertigen: Konstruktiver Holzelementbau, pp. 137–142. Triest Verlag, Zurich (2020)
7. Wójcik, M., Strumiłło, J.: Behaviour-based Wood Connection as a Base for New Tectonics. In: Keitsch, M. (ed.) Proceedings of the 20th Annual International Sustainable Development Research Conference, pp. 170–184. Resilience: the New Research Frontier, Norwegian University of Science and Technology, Trondheim (2014). https://doi.org/10.21427/D71R50
8. Liu, Y., Lu, Y., Akbarzadeh, M.: Kerf bending and zipper in spatial timber tectonics: a polyhedral timber space frame system manufacturable by 3-axis CNC milling machine. In: ACADIA 2021: Realignments (2021)

9. Takabayashi, H., Keita, K., Hirasawa, G.: Versatile robotic wood processing based on analysis of parts processing of Japanese traditional wooden buildings. In: Willmann, J., Block, P., Hutter, M., Byrne, K., Schork, T. (eds.) Robotic Fabrication in Architecture, Art and Design 2018. pp. 221–231. Springer, Zurich (2018). https://doi.org/10.1007/978-3-319-92294-2_17

10. Vercruysse, E., Mollica, Z., Devadass, P.: Altered behaviour: the performative nature of manufacture. In: Willmann, J., Block, P., Hutter, M., Byrne, K., Schork, T. (eds.) Robotic Fabrication in Architecture, Art and Design 2018, pp. 309–319. Springer, Zurich (2018). https://doi.org/10.1007/978-3-319-92294-2_24

11. Willmann, J., Knauss, M., Bonwetsch, T., Apolinarska, A.A., Gramazio, F., Kohler, M.: Robotic timber construction: expanding additive fabrication to new dimensions. Autom. Constr. **61**, 16–23 (2016). https://doi.org/10.1016/j.autcon.2015.09.011

12. Apolinarska, A.: Complex Timber Structures From Simple Elements: Computational design of novel bar structures for robotic fabrication and assembly. Ph.D. thesis, Swiss Federal Institute of Technology Zurich (ETHZ), Zurich (2018). https://doi.org/10.3929/ethz-b-000266723

13. Helm, V., Ercan, S., Gramazio, F., Kohler, M.: Mobile robotic fabrication on construction sites: dimRob*. In: 2012 IEEE/RSJ International Conference on Intelligent Robots and Systems, Vilamoura, pp. 4335–4341 (2012). https://doi.org/10.1109/IROS.2012.6385617

14. Eversmann, P., Gramazio, F., Kohler, M.: Robotic prefabrication of timber structures: towards automated large-scale spatial assembly. Constr. Rob. **1**, 49–60 (2017). https://doi.org/10.1007/s41693-017-0006-2

15. Wagner, H., et al.: Flexible and transportable robotic timber construction platform - TIM. Autom. Constr. **120**, 16–23 (2020). https://doi.org/10.1016/j.autcon.2020.103400

16. Thoma, A., Adel, A., Helmreich, M., Wehrle, T., Gramazio, F., Kohler, M.: Robotic fabrication of bespoke timber frame modules. In: Willmann, J., Block, P., Hutter, M., Byrne, K., Schork, T. (eds.) Robotic Fabrication in Architecture, Art and Design 2018, pp. 447–458. Springer, Zurich (2018). https://doi.org/10.1007/978-3-319-92294-2_34

17. Thoma, A., Jenny, D., Helmreich, M., Gandia, A., Gramazio, F., Kohler, M.: Cooperative robotic fabrication of timber dowel assemblies. In: Leopold, C. (ed.) Research Culture in Architecture: Cross-Disciplinary Collaboration, pp. 77–87. Birkhäuser, Basel (2020). https://doi.org/10.1515/9783035620238-008

18. Robeller, C., Weinand, Y., Helm, V., Thoma, A., Gramazio, F., Kohler, M.: Robotic integral attachment. In: Menges, A., Sheil, B., Glynn, R., Skavara, M. (eds.) Fabricate 2017: Rethinking Design and Construction, pp. 92–97. UCL Press, London (2017). https://doi.org/10.2307/j.ctt1n7qkg7.16

19. Leung, P., Apolinarska, A.A., Tanadini, D., Gramazio, F., Kohler, M.: Automatic assembly of jointed timber structure using distributed robotic clamps. In: Globa, A., van Amoijde, J., Fingrut, A., Kim, N., Lo, T. (eds.) Projections, vol. 1, pp. 583–592. The Chinese University of Hong Kong, Hong Kong (2021). https://doi.org/10.52842/conf.caadria.2021.1.583

20. Apolinarska, A.A., Pacher, M., Cote, H.L.N., Pastrana, R., Gramazio, F., Kohler, M.: Robotic assembly of timber joints using reinforcement learning. Autom. Constr. **125**, 103569 (2021). https://doi.org/10.1016/j.autcon.2021.103569

21. Kramberger, A., Kunic, A., Iturrate, I., Sloth, C., Naboni, R., Schlette, C.: Robotic assembly of timber structures in a human-robot collaborative setup. Front. Rob. AI **8** (2022). https://doi.org/10.3389/frobt.2021.768038

22. Stehling, H., Scheurer, F., Roulier, J.: Bridging the gap from CAD to CAM: concepts, caveats and a new grasshopper plug-in. In: Gramazio, F., Kohler, M., Langenberg, S. (eds.) Fabricate 2014: Negotiating Design and Making, pp. 52–59. UCL Press, London (2017). https://doi.org/10.2307/j.ctt1tp3c5w.10

23. Alvarez, M., et al.: The BUGA wood pavilion: integrative interdisciplinary advancements of digital timber architecture. In: ACADIA 2019: Ubiquity and Autonomy. pp. 490–499. The University of Texas at Austin School of Architecture, Austin (2019). https://doi.org/10.52842/conf.acadia.2019.490

24. Robeller, C., von Haaren, N.: Recycleshell: wood-only shell structures made from cross-laminated timber (CLT) production waste. J. Int. Assoc. Shell Spat. Struct. **61**, 125–139 (2020). https://doi.org/10.20898/j.iass.2020.204.045

25. Yuan, P., Chai, H., Yan, C., Zhou, J.: Robotic fabrication of structural performance-based timber gridshell in large-scale building scenario. In: ACADIA 2016: Posthuman Frontiers, pp. 196–205. University of Michigan Taubman College, Ann Arbor (2016). https://doi.org/10.52842/conf.acadia.2016.196

26. Svilans, T.: Integrated material practice in free-form timber structures. Ph.D. thesis, The Royal Danish Academy of Fine Arts, Copenhagen (2020). https://doi.org/10.5281/zenodo.4545124

27. Poteschkin, V., Graf, J., Krötsch, S., Shi, W.: Recycling of cross-laminated timber production waste. In: Leopold, C. (ed.) Research Culture in Architecture: Cross-Disciplinary Collaboration, Birkhäuser, Basel, pp. 100–112 (2020). https://doi.org/10.1515/9783035620238-010

28. Mangliár, L., Hudert, M.: Re:Shuffle. In: ICSA2022: Critical Practices, pp. 50–51. Aalborg University, Aalborg (2022)

29. Salas, J., et al.: Sourced Wood (2018). https://www.iaacblog.com/programs/sourced-wood/

30. Najari, A., et al.: Good Wood: Robotic Upcycling (2018). https://www.iaacblog.com/programs/good-wood-robotic-upcycling/

31. Castriotto, C., Tavares, F., Celani, G., Larsen, O.P., Browne, X.: Clamp links: a novel type of reciprocal frame connection. Int. J. Arch. Comput. **20**(2), 378–399 (2022). https://doi.org/10.1177/14780771211054169

32. Sunshine, G.: Inventory: CAD for medium resolution materials. In: ACADIA 2022: Hybrids & Haecceities. University of Pennsylvania, Weitzman School of Design, Philadelphia (2022)

33. Reisach, D.: Spruce Beetle (2022). https://doi.org/10.5281/zenodo.7157251

34. Preisinger, C.: Linking structure and parametric geometry. Arch. Des. **83**(2), 110–113 (2013). https://doi.org/10.1002/ad.1564

Towards an AI-Based Framework for Autonomous Design and Construction: Learning from Reinforcement Learning Success in RTS Games

Ahmed Elmaraghy[1] [ID], Jacopo Montali[2(✉)] [ID], Marcello Restelli[3] [ID],
Francesco Causone[3] [ID], and Pierpaolo Ruttico[3] [ID]

[1] ETH Zurich, Zurich, Switzerland
[2] Algorixon, Parma, Italy
jacopo@algorixon.com
[3] Politecnico di Milano, Milan, Italy

Abstract. The present study summarizes the state-of-the-art research in deep reinforcement learning (DRL) techniques in the architecture, engineering and construction industry and it formulates a general framework for autonomous design and construction. The framework is inspired by the noticeable success of DRL and imitation learning algorithms in real time strategy (RTS) games, which normally require efficient resource planning and long-term strategic coordination. The objective of the proposed framework is to reduce data segregation and loss of project information. The prevention of data leakage is achieved by replacing the linear process with an iterative one where the consequences of design decisions on the construction process (and vice versa) are understood in a virtual environment simultaneously. The proposed framework also exploits recent advances in simulated physics-based environments, like game engines. Designers and builders can therefore simulate on-site scenarios and exchange views on the required design and construction goals early in the project. The multi-objective optimization problem is then passed to artificial agents. These agents train on achieving the project goals under the supervision of a team of humans. The tacit knowledge transferred to the brain of the agents can later be deployed on-site through execution robots. The proposed approach is demonstrated by a proof-of-concept software application, showcasing a brick pavilion project. Design and construction constraints are first imposed by the user. Agents are then trained using a DRL algorithm.

Keywords: Artificial Intelligence · Deep Reinforcement learning · Architecture Agency · Construction

1 Introduction

The architecture, engineering, and construction (AEC) industry is the largest consumer of energy and raw materials worldwide. It consumes over 40% of global resources and energy and generates waste amounting to 50% of material waste worldwide [1]. These

M. Turrin et al. (Eds.): CAAD Futures 2023, CCIS 1819, pp. 376–392, 2023.
https://doi.org/10.1007/978-3-031-37189-9_25

issues are exacerbated by other problems like low productivity rates and on-site hazardous tasks [2]. Despite the active promotion of building information modeling (BIM) as a solution for better collaboration and data integration, current BIM tools still struggle to break down information silos among teams and do not automatically incorporate expert knowledge into projects, given its mere information management nature. To bridge this gap, advances in artificial intelligence (AI) can provide a comprehensive view of the data flow process, transforming the current linear process into a circular feedback loop. This paradigm can allow designers and builders to address conflicting constraints before fabrication begins while integrating digital fabrication in computational design workflow [3].

1.1 Challenges of AI Application in the AEC Industry

The proper implementation of AI in AEC industry is primarily hindered by lack of innovation caused by inadequate investments in research and development [4], as well as risk aversion. Furthermore, the AEC industry lags behind other industries in terms of digital transformation due to the absence of innovative procurement routes and the fragmentation of project stakeholders [3]. This creates challenges for data integration, making it difficult to incorporate AI into the process without adding complexity and potentially leading to unreliable results [5].

Interoperability issues also arise due to the diverse discipline-specific software and development tools adopted, hindering the scalability of AI applications. The absence of any mature AI application on the market that integrates different tools and technologies like robotics and real-time sensing data into one single framework further supports this issue [6]. To overcome these challenges, several researchers suggest the development of a holistic framework that provides an interoperable platform for integrating different technologies [5–8].

This framework should use the heuristic knowledge and intuition of field experts to guide the process of algorithm selection and results interpretation [5]. The interaction in such an environment could adopt game-play techniques that are user-friendly to make it easier for human users to quickly grasp the game mechanics and interact with the environment without much knowledge of the "black-box" behind [8]. Accordingly, a potential solution to creating such a framework is building upon the success of AI applications in several sophisticated domain [7], such as real-time strategy games.

1.2 AI Applications in RTS Games

Real-time strategy (RTS) games are a sub-genre of strategy games in which a player needs to build a virtual economic system by gathering resources, building and training units, researching technologies, and sometimes also by developing military power to defeat opponents [9]. In RTS games, players must attempt to multitask by working on all aspects of the game simultaneously. The challenges imposed in RTS games makes it a fertile ground for AI research [10]. These challenges include aspects like resource and time management, decision making under uncertainty, spatial and temporal reasoning, real-time planning, and use of domain knowledge.

Despite these demanding challenges, Google DeepMind's AlphaStar, an AI agent, managed to defeat top professional players of "StarCraft II". This accomplishment is considered a significant achievement, as "StarCraft II" is one of the most complicated RTS games ever created [11]. DeepMind regarded this victory as evidence of the effectiveness of general-purpose machine learning techniques used in training AlphaStar, specifically self-play deep reinforcement learning and imitation learning. The AI company believes that these techniques can be applied to other complex domains [12]. Therefore, in this paper, the application of AI strategies that promote long term planning like DRL in RTS games is taken as an inspiration for the development of a framework for autonomous design and construction.

2 Literature Review

Researchers have been using DRL algorithms to solve several problems in the AEC industry such as design and construction optimization, urban planning, energy efficiency, and assembly of building elements. In preliminary space planning, some researchers have used DRL techniques to solve game-like schemes representing architectural concepts. One example is an abstract grid-world testbed developed by Kotov and Vukorep [13] that involved coloring or occupying grids to achieve specific goals like generating different zoning areas based on architects' requirements. DRL algorithms have also been applied to solve early-stage urban planning problems such as space allocation. Saha et al. [14] used Deep Q-Learning (DQN) algorithm to assign geometric shapes and functions to each urban space, aiming to produce building layouts that meet the designer's goals and constraints.

On the other hand, DRL has contributed to building design in several aspects like generative design, circular economy workflows, assembly configuration, and optimization of building elements. One example of generative design is Wang and Snooks [15] use of multi-agent systems to encode global design intentions in a pavilion project. Agents in the form of mesh graphs took actions by moving each mesh vertex in a cuboid space, with rewards given for achieving global design intentions such as spatial coverage and topological performance.

Regarding circular economy, DRL algorithms have been utilized to promote material reuse and cradle-to-cradle concepts in the AEC. In [16], DRL algorithms, like DQN, was used to optimize the reuse of steel and timber elements for the generation of truss structures. Huang [17] tested the applicability of game engines like Unity to design new shell structures from an available repository of wasted materials to prevent energy-intensive recycling processes.

Several researchers have also worked on the design of reconfigurable structures that can learn, through DRL, to generate intelligent assemblies that fulfills structural and disassembly requirements [18, 19]. Similarly, DRL has been utilized in designing specific building elements to achieve certain performance criteria. Examples include optimizing the movement of kinetic facades shading panels for controlling solar radiation [20], or generating floor plans that satisfies specific architectonic goals [21].

Moreover, DRL techniques have also gained significant attention in construction tasks and planning including the automated assembly of building elements. For instance,

Apolinarska et al. [22] optimized robotic assembly of timber lap joints using an adapted Ape-X DDPG algorithm. The DRL agent was trained entirely in a simulation environment and successfully deployed in the real world, with a noticeable ability to generalize situations not experienced during training.

Accordingly, this paper integrates discrete efforts in academia and practice towards an AI-centered framework for autonomous design and construction. The framework tackles the currently discretized process that is motivated by human intuition to address complexity through skills specialization and generation of in-depth experience after working on specific challenges. A proof-of-concept application is introduced based on the framework. The results are then discussed along with future considerations.

3   Framework

In this section, the proposed framework is introduced, where AI, construction robotics and digital fabrication are situated at the core. It is based on 3 distinct teams: (1) a supervision team of humans that possess design and construction knowledge, and (2) a team of AI agents that progressively learn in a simulated environment the tacit knowledge needed for producing adequate design and construction methods. Finally, (3) an execution team of construction robots that are controlled by the agents and that transforms the agents' orders into real world actions. The execution robots translate on-site perceived data into situations like what encountered in the simulated environment. The framework is composed of 5 phases: design intent, site data acquisition, design and construction legislation, training of agents, and robotic execution (Fig. 2).

3.1   Design Intent

It is the phase in which a designer starts to transform the project requirements into initial sketches and drawings or preliminary models. The AI contribution in design inspiration can be categorized into 3 different cases:

- *No AI Contribution:* The Designer fully contributes towards the initial design ideas with no AI support.
- *Partial Contribution*: The AI agent acts as a catalyst for inspiration. This could be in the form of an AI assistant like text-to-image deep learning models that produce unique inspirational images from a user input of descriptive words (the likes of Midjourney [23] and Dall-E-2 [24]). Different forms of semi-supervised learning that are trained on a certain style or a blend of different styles could also be an option to support the designer. This solution is thus considered one of the forms of human – machine collaboration in the early phases of design where the AI agent act as the designer assistant.
- *Full Contribution*: The preliminary design is entirely produced by the agent with no or limited human intervention. In other words, the design intent can be implicitly formulated in the succeeding "legislation" phase, where the builder/architect formulates a group of design and construction goals for the agent to train on. An example could be a building in which its essential functionality is heavily prioritized over following a certain style or a complex form. However, advances in AI could offer fully automated preliminary design solutions in the future that can produce complex forms as well.

3.2 Site Data Acquisition:

This phase guarantees the representation of real-world elements and parameters inside the simulated environment. The accuracy of capturing on-site data is crucial to ensure proper training of the agents on the actual site conditions. With the presence of modern technologies like laser scanning, drones, ground penetrating radars and sensors for capturing climate, wind and other necessary conditions, the task of recreating a digital twin that mimics on-site real-world conditions could be achieved.

3.3 Design and Construction Legislation

This phase is the principal divergence from the traditional workflows currently adopted in the AEC industry. The team of designers and builders are the major contributors during this phase. The team meets at an early stage of the project with the aim of translating the design and construction objectives into goals that the agents are entitled to realize. These goals shall guide, at a later stage, the training process of the agent(s). To achieve this, the team first checks the conformity of the simulation environment with real world parameters. Afterwards, the team discusses how to represent the simulated environment in an understandable form for the AI agents. This could be done by formulating the problem as a Markov decision process (MDP) (Fig. 1).

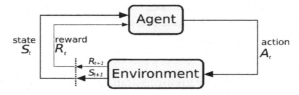

Fig. 1. Markov Decision Process (MDP)

For a MDP, the following 4 main elements should be defined:

- The observation parameters that the agent should record in each state of the game.
- The range of possible actions that could be selected by the agent in each state.
- The reward function that encourages the agent when a certain milestone in the project is achieved or penalizes the agent upon failure to complete the required task.
- The agent's "brain" that represents the algorithm used to search for an optimal policy for solving the formulated problem. Each algorithm is characterized by a set of hyperparameters that controls its behavior and should be fine-tuned to optimize its performance and help it reach an optimum policy.

Transforming the design and construction problem into a MDP is a quite complex iterative process that requires intensive planning by the human team. For this reason, planning considerations from the human team could include:

- Hierarchical division of the problem into sub-problems to facilitate better learning process for the agents.

- Combination of more than one algorithm to solve discrete parts of the problem.
- Selection of the type of agents involved in the process (single vs multi-agent environment, competitive vs collaborative agents...).

3.4 Design and Construction Learning

In this phase, the main contribution is provided by the agents, which learn to perform the tasks allocated by the human team. The human team becomes the supervisor that monitors the performance and progress of the agents, and it intervenes when necessary. For example, the human team could alter some algorithm parameters or hyperparameters to speed up the training process or to modify the MDP formulation. Therefore, both phases of design and construction legislation and learning are considered as an iterative process of continuous modification and fine-tuning of parameters.

3.5 Robotic Execution

After the learning process is finalized with satisfactory results, the executing robots are then equipped with the "agents' brains". The robot sensors capture real world observation that are translated afterwards into values understood by the agents in the "simulated environment". The agents feed the right action to each robot based on the approximation of the observed input data to the nearest similar observation in the simulated environment. The robot would finally formulate the perceived order from the agent's brain into an actual step or an action in the real world.

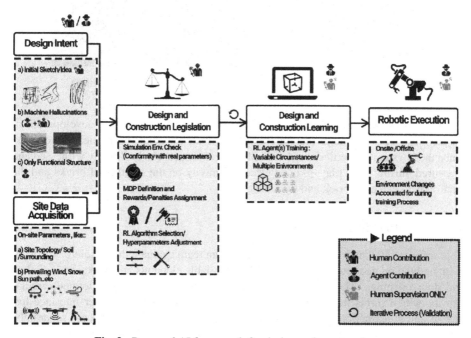

Fig. 2. Proposed AI framework for design and construction.

4 Proof-of-Concept: RoBuilDer

Robotic Builder and Designer (RoBuilDer) is developed in 2 versions, namely: "human version" (UV) and "robot version" (RV). UV enables designers to experiment with and evaluate different game parameters. RV provides an appropriate environment for the agents to acquire the tacit knowledge needed by exploring different paths and learning. The scope of the software application is focused on the "design and construction legislation" phase and the "learning of agents" phase of the framework. The "robotic execution" step is regarded out of this research scope, since literature is already available, both for on- and off-site situations.

The project proposed is a brick pavilion. It is inspired from a functional design perspective by the Pantheon's Rotunda and from the execution side by "the Endless wall" project of Gramazio Kohler Research [25]. The pavilion inherits the brick-like nature of "the Endless Wall" project, while design goals are mainly inherited from the Pantheon's Rotunda by realizing an open-ended dome. The presence of the top opening increases the stability of the dome and enables the use of passive strategies such as natural sunlight (Fig. 3) and stack ventilation.

Fig. 3. Imaginary visualization inside and outside of the brick pavilion

RoBuilDer was developed in Unity. This game engine allows the development of physics-based environments that would help simulate the actual properties of the brick. Once the physical properties are identified, several aspects can be directly witnessed and measured during game play, like the effect of gravity on the stability of bricks and the friction between each brick and the other.

4.1 RoBuilDer: Human Version

This software application is built to let the human team evaluate the simulated environment and the related rewarding scheme. The available elements are bricks, and window and door openings. These elements can only be placed along a pre-defined circular canvas, which forms the outline of the pavilion's foundation. The user interface has design and construction indicators that show the progress of each process separately. This allows the visualization of the impact of a series of pre-defined actions on each of design and construction goals instantly by displaying two separate progress bars (Table 1). The interface also includes a counter for accumulated scores of rewards and penalties. Figure 4 and 5 show some of the main functionalities of the HV, such as how rewards and penalties work and how a design or a construction progress bar works.

Table 1. Examples of tasks that contribute to making progress in the relevant design or construction progress bars.

Progress bar	Associated tasks
Design	Stack Ventilation, Cross Ventilation, Windows having adequate Sill height, Adequate Placement of Door Opening, Consistent Brick Perforations
Construction	Each completion of a full level of bricks

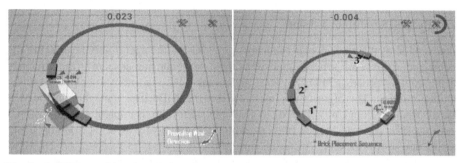

Fig. 4. (left) Two positive rewards for window placement at adequate sill height and efficient stack ventilation (through air circulation between the placed window and the dome). (right) Penalty for placement of non-consecutive brick placement

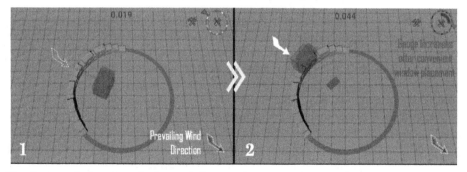

Fig. 5. (1 - > 2) Snapshot of before and after completing a design task and monitoring the increase of progress on the design progress bar.

Table 2 provides a summary of probable game-over scenarios within the game.

Table 2. Summary of probable game-over scenarios

Type of Failure	Causes of Failure
Design	**For Openings** (Doors and Windows): No Opening Placement, Inadequate Door Placement, Inadequate Sill Height for Windows, Poor Cross Ventilation, Poor Stack Ventilation **For Bricks**: Uneven large gaps in the design
Construction	Poor Brick Placement Technique*
	Brick Collapsing

* Placement of bricks in such a way that consumes more "energy" by the executing robot or may increase the obstacles in the robot paths caused by the placed bricks themselves

4.2   RoBuilDer: Robot Version

This version of the game covers the learning process of the agents and the effect of the parameters and rewards scheme imposed by the supervision human team on the efficiency of the training process. PPO [26] was used as the DRL algorithm to help achieve the optimal solution. This algorithm is integrated into the RoBuilDer game through the ML-Agents framework in Unity. The ML-Agents PPO algorithm is implemented in PyTorch and runs in a separate Python process, which communicates with the running Unity application over a socket [27].

To speed up the learning process, concurrent Unity instances were used for training. This parallel learning approach enables agents to encounter a more diverse range of game circumstances during training and adapt to them better than an agent trained in a single environment. The game's logic was modeled as an MDP with appropriate states, actions, and rewards to simulate the design and construction of the pavilion.

Game Actions: The main action in the game is to place a brick or a door or window opening. From the agent's perspective, the output is a numerical value that is translated as an exact position to place a certain unit. Both discrete and continuous actions were experimented in RoBuilDer. For continuous actions, an action vector of size 3 was indicating the planar coordinates on the canvas, and the height. The discrete actions, however, were composed of a 4-step process to place a certain unit in the game. The first step determines whether to insert a brick on the lowest incomplete level of bricks or to place it onto any of the higher levels available. Then, step 2 decides an integer value that represents a segment out of 12 segments dividing the circular grid. Afterwards in step 3, the agent selects a certain sub-segment in the segment previously selected in step 2. In step 4, the agent selects one of the insertion points available in the sub-segment. Once the 4-step process is executed, a unit will actually be placed and observed at the final location chosen at step 4 (Fig. 6).

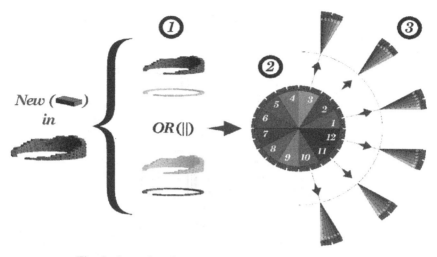

Fig. 6. General Action Strategy Scheme for placing a brick.

Observations: The agent's "brain" receives two types of observations: fixed-size and variable-size vectors. Fixed-size vectors represent all the environment parameters existing from the start to the end of each episode, such as counters for placed units or wind direction. Variable-size vectors, also known as "buffer sensors" in ML-agents, report a variable number of observable objects in the scene, with each object having a fixed number of observations to report. For instance, a buffer sensor is used to report each newly placed brick's information, including its position, tilting vector, and relationships with other units (Table 3).

To reduce the number of observables, the concept of "naked bricks" was adopted. These bricks are the most "active" in the game because they allow the placement of a new brick directly above them (Fig. 7). Among these bricks are the most unstable and less supported ones. Observing these bricks provides some insights to the neural network on the how bricks instability evolves.

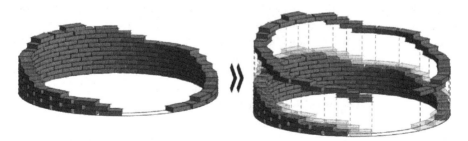

Fig. 7. Naked bricks demonstration

Goals and Rewards: The HV provides the designer with the chance to experiment different rewards and goals in order to learn how to set up the training process for the

Table 3. Examples of observation variables used in the buffer sensor for each naked brick

Obs. Variables:	Explanation	Notes
Relative Height	Relative Height of the observable brick with respect to the starting height of the current phase	Equation 1 (float)
Normalized Tilting Value (Dot product Between brick local Z-Axis and world Z-Axis)	Measure of how much a brick is "Tilting". It is a brick property. If the "tilting value" decreases, it means that the instability of the brick increases, therefore the observation value reported to the neural network is negative in case of tilting	Equation 2 (float)
Is Last Brick?	Checks whether this observable brick is the last brick placed in the game or not	Bool (int)
Notes	Reward/Penalty Calculation	

$$penalty = \frac{(\text{Current Brick Height})_{abs} - (\text{Height at phase start})_{abs}}{(\text{Max possible Height during current phase})_{relative\ to\ phase\ start}}$$

Case Tilting Value < Threshold: penalty = -1 *
Case Tilting Value > Threshold: penalty =

$$\left\{ (-1) \left[\frac{1}{1 + e^{\frac{1 - Tilting\ Value}{1 - Threshold}}} \right] \right\}^{**}$$

* Tilting Threshold is a factor with a value less than one. If the tilting dot product gets lower than this value, the game is terminated
** The reason for using a "sigmoid function" is to increase the negative value fed to the neural network once a signal of slight tilting is captured

AI agent. Several small incentives were experimented to encourage the agent to exhibit a particular behavior. An example of these incentives is a small value (+0.01) that was given to the agent each time it placed a brick in a consecutive order. Working with a sparse environment was also explored. In such an environment, the agent is not given any reward until it completes the required structure within the required standards. Similarly, the agent is not penalized while learning until a game termination scenario is reached. At that point, a relatively large negative penalty is given to the agent for failure in the episode. The game termination scenarios in the game are mainly classified under 3 main categories: goal accomplished, brick collapsed, and fatal design error (Fig. 8).

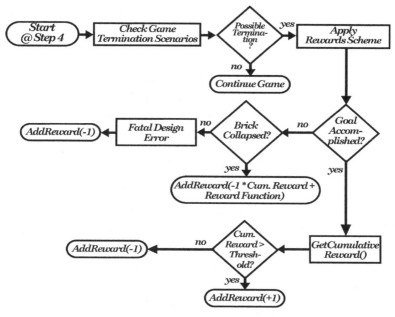

Fig. 8. Diagram for game termination scenarios

Table 4 summarizes some examples of the different rewards and penalties that were tested during the training of agents:

5 Discussion

The RV has witnessed several iterations of trial and error to establish a reward scheme and select optimal hyperparameters for an effective learning process. However, the challenge of reducing project complexity emerged and was addressed using various methods. First, the entire design and building process was divided into separate phases, with the brick placement process divided hierarchically into three phases during training. Furthermore, discrete action vectors eventually replaced continuous ones, since the agent was failing to converge to an optimum solution in most of the cases.

For instance, when using discrete action vectors, around 50 million training steps were sufficient to achieve a satisfactory level of proficiency in the most complicated brick placement phase, the Dome Creation. By the end of training, the agent could build up to 25 levels of bricks in the dome structure while maintaining a stable brick structure and respecting design requirements. In contrast, using continuous actions would consume the same number of steps without accomplishing even six brick levels stacked properly within the dome structure.

In addition, within each phase, curriculum learning was introduced. Curriculum learning is a way of decomposing a difficult task into subtasks starting from a simple level and increasing difficulty till reaching exactly what is required in the main task. This idea is well-established, as it mimics how humans typically learn [27]. Moreover, the

Table 4. Examples of in Game Agent Reward Functions

Trigger Event	Reward Value	
Consecutive Brick-Placement	Reward = Continuous Brick Reward *Count of Consecutive Bricks	
High Brick Placement	Case Brick height relative to current lowest incomplete level > Threshold: Penalty = $\left(\dfrac{Relative\ Height\ Difference}{Height\ of\ 1\ brick}\right)^2$ *High Brick Penalty * 2	
Windows Instance Evaluation	Dot Product of $\left(\begin{array}{l} Window\ Forward\ unit\ Vector \\ \quad and\ Wind\ Unit\ Vector \end{array}\right)$ Reward = (dotProduct) * Stack Ventilation Max Reward	
Windows Interim Evaluation (Evaluation of Cross Ventilation After the total number of Windows Required is Achieved.)	Step1: Pick window with maximum fulfilment of stack Ventilation Step2: Pick the furthest window from that selected window Step3: Calculate Dot Product between Both Windows Step4: Reward or Penalty= Max Dot Product in List * Dot Product of 2 Windows * Cross Ventilation Max Reward	

agent was trained in both sparse versus abundant reward-environment. When rewards were sparse, each agent could get a + 1 point if managed to place all the necessary bricks properly and reach the required final height of that phase and -1 otherwise. During learning, the agent discovered a quick way to reach the goal by stacking the bricks exactly over each other, thus finishing one column of bricks and moving to the other.

While this behavior was not witnessed in the abundant rewards scenario because of the presence of incentives for placing bricks adjacent and consecutive to each other on closer levels, the agent was, however, able to surpass this placement scenario in the sparse reward environment. Eventually, a significant outcome was the ability of the agent to devise unexpected strategies, while not violating any game rules and succeeding in reaching the anticipated goals.

It has to be noted that, during training in sparse environments, curiosity reward was added to the hyperparameters of the PPO algorithm, since no rewards were introduced during game play. This module works best when the agent receives rare or infrequent rewards. The curiosity reward was expected to encourage the agent to explore states that weren't visited often during gameplay [28]. The training time was, however, longer than the time taken for the agent to determine how to handle and place bricks within an extrinsic rewards scheme that is defined external of the DRL algorithm.

The motivation for experimenting with a sparse environment was to assess the level of self-sufficiency that the agents could achieve without excessive intervention from human supervisors. Another challenge that the supervisors could focus on is determining whether the agents' solutions are feasible or if game mechanics need to be updated to prevent those strategies in the future. The primary goal of the iterative process in creating the reward scheme was to provide small incentives to the agent in specific situations where convergence to a solution is not achieved, or when the human supervisors deem that the final goal is not satisfactorily met.

During training, a design issue arose in the creation of the pavilion's dome, where gaps between bricks became increasingly wide with increasing height, making it impossible to reconnect bricks on top due to missing supports. While the agent could have still met the required final height, this was aesthetically unacceptable. To address this issue, a "gap check" was introduced in the game mechanics to detect and report gaps as an observation to the agent. A penalty was added whenever gaps were detected during the final execution phase, resulting in a more coherent dome with bricks placed at reasonable distances from one another (Fig. 9).

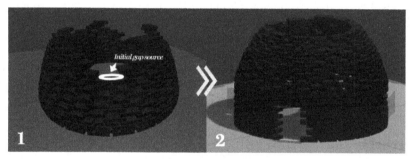

Fig. 9. (1) Gap propagation leading to failure in complying with design requirements and then (2) coherent brick placement by the agents after the introduction of gap penalty

On the other hand, in HV, there was an opportunity to experiment and test various methods and reward schemes that could be applied to RV. This was made possible by providing the "supervisor" a set of parameters that could be adjusted and then observing their impact on gameplay. These parameters included values related to the reward scheme and game mechanics-related variables such as the tilting limit of bricks before they were deemed unstable, the material and physical properties of bricks affecting their behavior during placement, and more.

Once a change was agreed upon, it was integrated into the agent's training canvas and evaluated against performance metrics to track progress. The whole process involved a meta-learning challenge for the human team to transfer tacit knowledge and experiences into tangible incentives and goals to guide the agent. The back-and-forth between testing and implementation phases allowed for instant feedback. Interruptions to the training process were recovered by reusing the neural network's learnt weights as initialization parameters when training resumed.

6 Conclusion and Future Work

This paper introduced a new framework for autonomous design and construction of buildings. The framework combines the success of DRL algorithms in complex domains like RTS games and the latest advancements in simulation environments supported by game engines, along with modern sensing technologies and construction robotics. The framework established the relationship between AI and humans in the AEC industry, where the highest level of autonomy is achieved when agents find solutions within the constraints set by designers and builders. In this new workflow, the human team serves as a supervisor to the agents, directing them towards areas of knowledge that may lead to solutions not previously explored by humans.

While the paper proposed a generalized framework, there are still several aspects that need to be addressed in the future, especially regarding project-specific workflows. As each project in the AEC industry is unique, detailed pipelines starting from scope identification, selection of optimization algorithms, simulated environments, and execution robots should be addressed and shared among the community. Further research can also be directed towards the adaptability of trained algorithms to be used on future projects with no or limited additional training. The paper also provided a proof of concept, named RoBuilDer, that covers part of the proposed framework by introducing a human and robot version. Although the robotic execution part was not the focus of the paper, it aligns with the framework's insights regarding the combination of one or more algorithms to execute the final project. When executing the learnt behavior, the use of traditional or DRL-supported path planning algorithms can help transfer the simplified points or actions taken in the simulation environment into real world actions.

References

1. Dean, B., Dulac, J., Petrichenko, K., Graham, P.: Global status report 2016. Global Alliance for Buildings and Construction (2016)
2. Felbrich, B., Schork, T., Menges, A.: Autonomous robotic additive manufacturing through distributed model-free deep reinforcement learning in computational design environments. Constr. Robot. **6**, 15–37 (2022)
3. Remes, J., Krishnan, J.M.: Solving the productivity puzzle: the role of demand and the promise of digitization. Int. Prod. Monitor **35**, 28–51 (2018)
4. Fulford, R., Standing, C.: Construction industry productivity and the potential for collaborative practice. Int. J. Project Manage. **32**, 315–326 (2014)
5. Rafsanjani, H.N., Nabizadeh, A.H.: Towards digital architecture, engineering, and construction (AEC) industry through virtual design and construction (VDC) and digital twin. Energy Built Environ. **4**, 169–178 (2023)
6. Zabin, A., González, V.A., Zou, Y., Amor, R.: Applications of machine learning to BIM: a systematic literature review. Adv. Eng. Inform. **51**, 101474 (2022)
7. Darko, A., Chan, A.P.C., Adabre, M.A., Edwards, D.J., Hosseini, M.R., Ameyaw, E.E.: Artificial intelligence in the AEC industry: scientometric analysis and visualization of research activities. Autom. Constr. **112**, 103081 (2020)
8. An, Y., Li, H., Su, T., Wang, Y.: Determining uncertainties in AI applications in AEC sector and their corresponding mitigation strategies. Autom. Constr. **131**, 103883 (2021)

9. Ontañón, S., Synnaeve, G., Uriarte, A., Richoux, F., Churchill, D., Preuss, M.: A survey of real-time strategy game AI research and competition in starcraft. IEEE Transactions On Computational Intelligence And AI In Games **5**, 293 (2013)

10. Buro, M.: Real-Time Strategy Gaines: A New AI Research Challenge (2003)

11. Burgar, C.: The 10 Hardest RTS Games Ever Made, Ranked. https://gamerant.com/real-time-strategy-rts-games-hardest-ranked/. Accessed 24 Sep 2022

12. AlphaStar: Grandmaster level in StarCraft II using multi-agent reinforcement learning. https://deepmind.com/blog/article/AlphaStar-Grandmaster-level-in-StarCraft-II-using-multi-agent-reinforcement-learning. Accessed 10 Mar 2022

13. Kotov, A.V.: Gridworld architecture testbed. In: Proceedings of the 39th eCAADe Conference - Volume 1, University of Novi Sad, Serbia, pp. 37–44 (2021)

14. Saha, N.H.: Space allocation techniques (SAT). In: Proceedings of the 40th Annual Conference of the Association of Computer Aided Design in Architecture (ACADIA), pp. 248–257, CUMINCAD, Online (2020)

15. Wang, D.S.: Intuitive behavior - the operation of reinforcement learning in generative design processes. In: Proceedings of the 26th CAADRIA Conference, pp. 101–110, CUMINCAD, The Chinese University of Hong Kong and Online, Hong Kong (2021)

16. Apellániz, D., Pettersson, B., Gengnagel, C.: A flexible reinforcement learning framework to implement cradle-to-cradle in early design stages. Towards Radical Regeneration. In: Gengnagel, C., Baverel, O., Betti, G., Popescu, M., Thomsen, M.R., Wurm, J. (eds.), pp. 3–12. Springer International Publishing, Cham (2023)

17. Huang, C.: Reinforcement learning for architectural design-build - opportunity of machine learning in a material-informed circular design strategy. In: Proceedings of the 26th CAADRIA Conference, pp. 171–180, CUMINCAD, The Chinese University of Hong Kong and Online, Hong Kong (2021)

18. Hosmer, T.T.: Spatial assembly with self-play reinforcement learning. In: Proceedings of the 40th Annual Conference of the Association of Computer Aided Design in Architecture (ACADIA), pp. 382–393, CUMINCAD, Online and Global (2020)

19. Wibranek, B.: Reinforcement learning for sequential assembly of sl-blocks - self-interlocking combinatorial design based on machine learning. In: Proceedings of the 39th eCAADe Conference, pp. 27–36, CUMINCAD, University of Novi Sad, Serbia (2021)

20. Dai, S.: Reinforcement learning-based generative design methodology for kinetic facade. In: van Ameijde, J., Gardner, N., Hyun, K.H., Luo, D., Sheth, U.(eds.), POST-CARBON - Proceedings of the 27th CAADRIA Conference, Sydney, 9–15 April 2022, pp. 151–160 CUMINCAD (2022)

21. Veloso, P.K.: An academy of spatial agents - generating spatial configurations with deep reinforcement learning. In: Werner, L., Koering, D.(eds.), Anthropologic: Architecture and Fabrication in the Cognitive Age - Proceedings of the 38th eCAADe Conference, TU Berlin, Berlin, Germany, pp. 191–200 (2020)

22. Apolinarska, A.A., et al.: Robotic assembly of timber joints using reinforcement learning. Automation in Construction **125** (2021)

23. Midjourney. https://www.midjourney.com/home/. Accessed 18 Oct 2022

24. DALL·E 2. https://openai.com/dall-e-2/. Accessed 18 Oct 2022

25. Gramazio Kohler Research. https://gramaziokohler.arch.ethz.ch/web/e/forschung/216.html. Accessed 18 Mar 2022

26. Proximal Policy Optimization — Spinning Up documentation. https://spinningup.openai.com/en/latest/algorithms/ppo.html#quick-facts. Accessed 31 Mar 2022

27. Unity ML-Agents Toolkit (2022). https://github.com/Unity-Technologies/ml-agents
28. Pathak, D., Agrawal, P., Efros, A.A., Darrell, T.: Curiosity-Driven Exploration by Self-Supervised Prediction (2017)

A System for Truss Manipulation with Relative Robots: Designing and Prototyping HookBot

Burak Delikanlı[(✉)] [iD] and Leman Figen Gül [iD]

Istanbul Technical University, Istanbul, Turkey
{burak.delikanli,fgul}@itu.edu.tr

Abstract. This paper presents design details and proof-of-concept prototypes of a relative robot, so called 'HookBot', customized for autonomous discrete assembly of round-shape truss system elements. Initially, it examines robotic assembly systems for truss structures and highlights the similar and different features of these studies. The novel design characteristics of HookBot that differ from similar studies are the locomotion kinematics and the adhesion mechanism that improves the degree of freedom (DoF). Thus, two versions of prototype were developed, focusing on locomotion kinematics. The final version of the prototype has successfully executed the functions that enable translational, rotational, and helical locomotion on a truss strut. Besides, problems during prototyping, further improvements, successes, and failures were examined. Further, a system for truss manipulation with relative robots defines the place of HookBot within autonomous discrete assembly systems and coordinates with subsystems. Consequently, the paper discusses an autonomous system whose principal actor is relative robots and its potential contribution to future architectural services and construction systems. Also, the paper concludes that autonomous discrete assembly can be implemented with further prototypes and the examined design details.

Keywords: Automation in Construction · Robotics in Architecture · Relative Robots · Assembly Robots · Truss Systems

1 Introduction

Mobile fabrication [1–5] and assembly [6–13] robots, which are novel and emerging technologies, are gaining the interest of architects and designers [14, 15]. Architectural robots were adopted mainly from the manufacturing industry to automate the construction process in the late twentieth century [16]. Although these robots successfully perform repetitive tasks, they might have trouble adapting to flexible designs and processes [17, 18]. Today, researchers have turned to manipulating, developing, and redesigning robots for alternative conditions, environments, and designs, usually specialized to operate in controlled environments such as dedicated manufacturing facilities [19–21]. Articulated robots offer an excellent opportunity for architecture with their software libraries,

Supplementary Information The online version contains supplementary material available at https://doi.org/10.1007/978-3-031-37189-9_26.

M. Turrin et al. (Eds.): CAAD Futures 2023, CCIS 1819, pp. 393–409, 2023.
https://doi.org/10.1007/978-3-031-37189-9_26

user-friendly interfaces, and universal natures allowing manipulation [20]. Neverthe-less, mobilizing large-scale robots on site can be challenging as they are designed for factory conditions [1–3]. Thus, relatively small-scale robots emerge as an alternative by successfully performing custom fabrication and assembly tasks using simpler behaviors [4–13]. These robotic studies can be categorized into four groups regarding manipula-tion methods and robotic morphologies [1]: stationary robots (Fig. 1a), mobile robots with an extended manipulator (Fig. 1b) [2], mobile robots with a standard manipulator (Fig. 1c) [3], and relative robots (Fig. 1d) [4–13]. Among these four approaches, rela-tive robots offer more flexible and efficient solutions concerning the degree of freedom (DoF), lightweight, and speed [22].

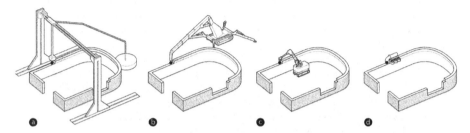

Fig. 1. A categorization for robotic approaches in architecture [1].

Therefore, this paper focuses on the possibilities and opportunities for using rel-ative robots in architecture. Relative robots [22], whose applications in architecture have recently emerged, are derived from climbing robots [23], which are generally used for inspection and maintenance in engineering. These robots, which locomote rel-ative to the structures they produce, can fabricate with 3D printable concrete [4] and fibrous materials [5]. Besides, they can collaboratively erect large structures from small components by carrying, climbing, stacking, and assembling modules [6, 7]. Further, unmanned aerial vehicles (UAVs), suitable for carrying light and small building materi-als, can autonomously stack [8] and assemble [9] structural components. Also, rod-like components are appropriate materials for small robots. Thus, relative robots can effi-ciently locomote truss structures by carrying, reconfiguring, and assembling [10–13]. Experimental studies about mobile robotics in architecture that can locomote structures emphasize innovative building materials, and new robot designs [4–13, 15].

Consequently, this research proposes a novel robotic truss manipulation system that increases the degree of freedom by combining the reconfigurable adaptation mechanisms of pipe-climbing robots [24, 25] and exemplifies a customized distributed assembly system [26]. Besides, we contribute to the automation research of truss assembly with a novel collaborative robotic system design and proof-of-concept prototypes for round-shape truss system elements. The system highlights integrated autonomous processes in truss structures and presents a fully autonomous case from design to production with the coordination, collaboration, and combination of so-called 'HookBot' robots. Therefore, the paper, within ongoing research, introduces the details of a relative robot 'HookBot', the possibilities and opportunities of the assembly system, and the first observations and implications of physical prototypes.

2   Background

Truss structures are usually used to achieve long spans, minimize the weight, and support heavy loads. They also consist of struts and joints arranged to form one or more triangles since the joints are fixed and cannot transmit moments. Truss structures have different carrying mechanisms according to their forms. Significantly, space trusses vary from plane trusses in their geometric possibilities. Even if the strut and joint dimensions differ, assembly techniques have no substantial variance. The repetitive assembly process of truss structures is a task suitable for robotic automation. Therefore, automation solutions using robots at different scales have been proposed. Two leading solutions differ for increasing automation in truss assembly: one of them is stationary systems [27], which generally work with an articulated robot that can move in X and Y directions through a frame larger than the structure, and the other is mobile systems [10–13, 28] that can easily move and locomote on or around the structure. Technological advancements in mobile climbing robots are pushing robotic assembly to become increasingly smaller, lighter, accessible, and customizable. Thus, mobile robots climbing structures are promising research due to their degree of freedom and size advantages (Fig. 2) [14, 15].

Fig. 2. Various types of structure manipulation robots [Top: 10, 12, 13; Bottom: 14–16].

Early research on the automated robotic assembly of space truss structures was aimed at the construction and configuration of space stations in the nineties. These systems suggested automated assembly with robotic arms [27] and re/configuration with mobile climbing robots teleoperated by human operators [28]. Nevertheless, the applications of these robots were limited as automation technologies were not mature enough. Besides, climbing robots that can reconfigure structures were studied for more general use, such as maintenance and inspection, for tasks that could be dangerous and difficult for human workers after the second millennium. Since these studies focus on the kinematic design of robots, the specific tasks describing cases are superficial [10, 11]. Studies focusing on automated architectural and structural possibilities in the last five years include system designs as well as robots [12, 13]. These studies highlight the significance of mobile climbing robots for automated architecture research. Nevertheless, different characteristics such as the degree of freedom (DoF), dimensions, and mechanisms should be researched for different cases and materials.

Therefore, this paper presents the case of HookBot as an example of an autonomous manipulation (re/configuration and dis/re/assembly) system for space truss structures with round-shape struts. HookBot, a system for truss manipulation with relative robots, presents a strategy customized for autonomous locomotion such as climbing, traversing, manipulating, and reconfiguring in/on truss structures. Besides, HookBot aims to use an integrated and autonomous assembly method in truss structures for design and construction tasks such as semi-open spaces, temporary or permanent pavilions, and logistic bridges. The system would provide benefits such as time, material, labor, cost savings, site safety, accident prevention, mass customization, design democratization, and automation in organizational process management.

3 Designing HookBot: A Relative Robot

Climbing Robots are specialized mobile robotic systems that can move over walls, ceilings, roofs, and geometric structures and perform various tasks. They can perform hazardous maintenance tasks such as cleaning high-rise buildings, inspecting nuclear power plants and petrochemical facilities, and welding ships or pipelines. Besides, they differ from other mobile robots with their characteristics, such as the lightness of weight, high mobility, and reliable grasping mechanisms. Also, they can locomote with legged, wheeled, tracked, translational, and cable-driven mechanisms and adhere with suction, magnetic, gripping, rail-guiding, and biomimetic components to various structures [23].

Relative robots, considered as climbing robots with different classes according to locomotion and adhesion mechanisms, can locomote on structures they build or assemble [22]. There are many experimental studies on fabrication [3–5] and assembly [6–12] of structures with different materials by these robots, which emphasize building material design and robotic concepts [14, 15]. Unlike these examples, Leder et al. [13] proposed a system that stands out by lightening, simplifying, and minimizing features. Nevertheless, the single-axis design of the assembly robots in that system reduces the degree of freedom and creates new complexities. This single-axis design, which limits its mobility, can be overcome with a reconfiguration mechanism applied in pipe-climbing robots [24, 25]. The proposed assembly system consists of multiple collaborative HookBot robots designed to use the truss system for its locomotion. Moreover, HookBot, a climbing robot that can dis/re/assemble in truss systems, differs from the literature regarding its reconfiguration system and the degree of freedom [26].

3.1 Initial Design

Many morphologies are investigated for the initial design of HookBot. Robots with legged locomotion and gripping adhesion are preferred by many researchers [7, 23].

This general morphology is quite successful for orthogonal traversing in truss systems. However, adapting to space truss designs having variable strut angles might be challenging in terms of inverse kinematics. Another outstanding design is "Shady3D" [10] or "Hinge Robot" [11, 12], with tracked locomotion and gripping mechanism. This design, which is promising for orthogonal and diagonal transitions and reconfiguration of trusses, does not offer convincing suggestions for transferring new struts into the system. On the other hand, the "Single-axis Robotic Assembly System" [13] offers a highly simplified morphology. The system aims to increase the degree of freedom by articulating the simplified single-axis mechanism of the robot. Although this strategy is promising with a simple hierarchy, it might have payload limitations. Therefore, the reconfiguration mechanism of pipe-climbing robots [24, 25] that allows rotational locomotion can benefit assembly systems. These robots can track pipes and reconfigure their direction and rotation. Thus, this strategy increases the degree of freedom of single-axis robots. The design of HookBot utilized a combination of these strategies as explained below.

3.2 Adhesion Mechanism

HookBot can grab and release struts thanks to the mechanical gripper. The gripping mechanism of the robot consists of two-wheeled fingers, which provide adhesion to the struts (Fig. 3a). Pipe-climbing robots [24, 25] utilize a similar gripping mechanism. Two-wheeled fingers connect with linear and twisting joints to the robot (Fig. 3b). Before grasping, linear joints move in the Z_0 direction (Fig. 3c). Then, the linear joints move in the $-Z_0$ direction and grasp the robot after placing it on the strut (Fig. 3d). After grasping, the twisting joints reconfigure the θ_1 angle of the two-wheeled fingers according to the locomotion direction (Fig. 3e). Linear joints offer active adhesion that adapts to different angles by retracting, as twisting the two-wheeled mechanism changes the strut grasp angle (Fig. 3f).

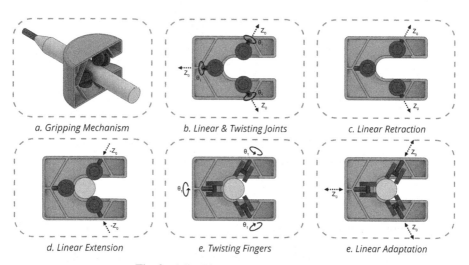

a. Gripping Mechanism   b. Linear & Twisting Joints   c. Linear Retraction

d. Linear Extension   e. Twisting Fingers   e. Linear Adaptation

Fig. 3. Adhesion mechanism of HookBot

3.3 Locomotion Kinematics

The HookBot has translational, rotational, and helical locomotion by tracking struts in the truss system thanks to the reconfigurable mechanical gripper (Fig. 4a). The two identical halves of the robot are connected with a twisting joint. Each side has the same mechanism (Fig. 4b). One half grasps a strut, while the other half manipulates another strut (Fig. 4c). Two robots would have a higher degree of freedom (DoF), connecting each other with a strut as a passive module, as in the "Three-dimensional Truss climbing Robots" study [10]. The locomotion subsystem consisting of two robots and a passive module can thus traverse from one strut to the next (Fig. 5).

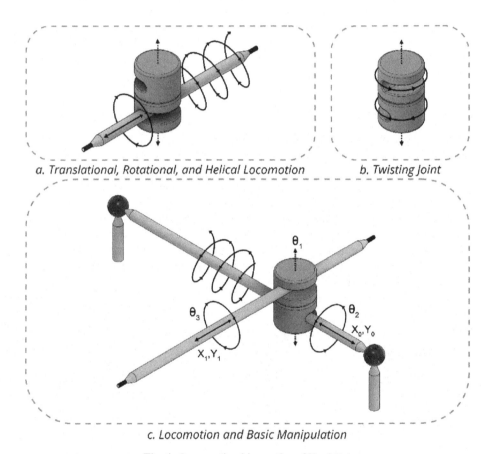

a. Translational, Rotational, and Helical Locomotion   b. Twisting Joint

c. Locomotion and Basic Manipulation

Fig. 4. Locomotion kinematics of HookBot

3.4 Providing Material

The HookBot collaboration transfers the struts, which are a passive part of the locomotion mechanism, to the truss system and moves them to the exact assembly positions. Besides,

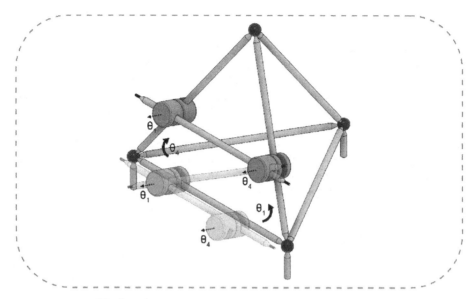

Fig. 5. The locomotion subsystem traversing between struts

the material-providing strategy for the "Distributed Robotic Timber Assembly" study [13] utilizes a similar logic. The subsystem providing material requires the collaboration of at least three HookBot. Therefore, the robot grasping the locomotion subsystem carries a strut (Fig. 6). Firstly, robots assemble the rail-track system on the construction site. HookBot on the rail-track system collaborate to provide material to the structure (Fig. 6). When struts are transferred to the truss structure, they are moved to the exact assembly positions.

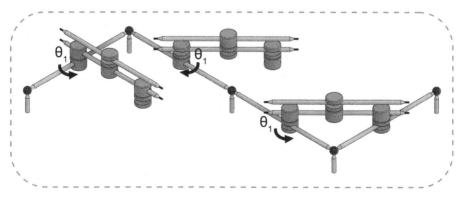

Fig. 6. The material-providing subsystem carrying a strut on the rail-tracks

3.5 Self-positioning

One of the most characteristic issues of relative robots is self-positioning. Two position-ing methods come to the fore: the first is to receive motion feedback from the motors, and the second is to perceive the environment instantly. Nigl et al. [11] showed that motion feedback would not be consistent due to slippage in climbing robots. Therefore, they developed bidirectional gears to solve the slippage problem. Nevertheless, this strategy requires replacing all struts with bidirectional gears and might cause excessive com-plexity. The other method is to collect location data with sensors instantly. Integrated gyro-inclinometer sensors and OptiTrack or IR LED tracking systems can instantly scan and determine the position of each HookBot. In this way, the trajectory planning of robots whose positions are updated instantly can be calculated.

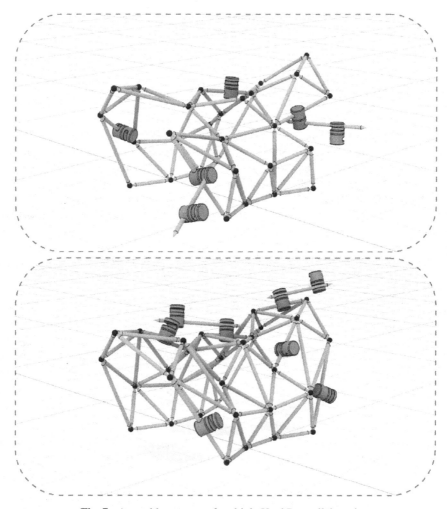

Fig. 7. Assembly process of multiple HookBot collaboration

3.6 Assembly Details

The traditional truss connections consist of bolts and nuts. Struts are assembled to nodes when at the correct position. Extending bolts are used in these details because the strut mechanics are not flexible. Struts and nodes design of "Structure-reconfiguring Robots" study [11] has male-female connection detail that does not require stretching. Hook-Bot can utilize a similar connection detail design, which is suitable for simple robotic assembly. Consequently, the material-providing subsystem rotates and assembles the struts into the correct position. Also, in different scenarios, robots can disassemble or reassemble the system (Fig. 7).

4 Prototyping HookBot: A Proof-of-Concept Chassis

To demonstrate locomotion kinematics in a physical robotic system and the proof-of-concept, hardware prototypes were developed (Figs. 8 and 10). These prototypes focus on the chassis, emphasizing the details rather than the shell. Besides, paper struts were used instead of metal struts as a structural component in the experiments due to the lightness in the test phase.

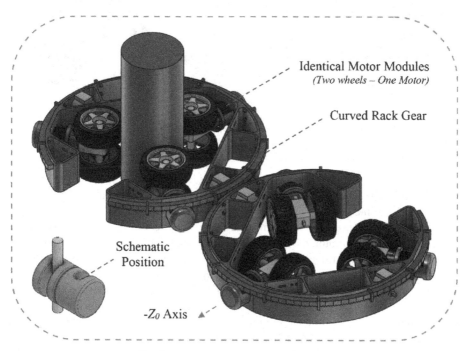

Fig. 8. CAD drawings of Prototype V.01

4.1   Prototype V.01: A Chassis with Curved Rack Geared DC Motors

The adhesion mechanism of HookBot is one of the complications of the system. Parallel retraction in the $-Z_0$ direction of the motor modules is desired for HookBot to reconfigure its directions and grasp the struts. The gripping mechanism providing adhesion offers an active adaptation to diameter changes. This mechanism allows HookBot to grasp, rotate, and carry the struts. In addition, this function would increase the adhesion by keeping the friction force within the desired range. However, developing this mechanism with locomotion kinematics will likely cause new complications. Therefore, Prototype V.01 focused on locomotion kinematics and preferred a passive adhesion mechanism that provides adhesion to diameter changes from 77 to 82 mm. This mechanism consists of elements that apply pulling force to the identical motor modules towards the paper strut. Thanks to the pull springs increase the friction force of the wheels and provide passive adhesion (Fig. 9).

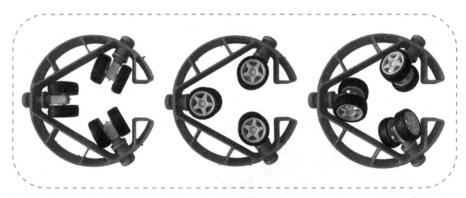

Fig. 9. The physical Prototype V.01 (Left to Right: Translational, rotational, and helical configuration)

In addition, the gripping mechanism has three identical wheel modules. Low-budget "DC Gearbox TT Motors" were tested for Prototype V.01. The Gearbox reduces DC motor speed and increases torque by 1:48 ratio. Although these DC motors offer easy driving, they have inaccurate positioning. Since the two wheels in each motor module are connected to the same motor, they drive in the same direction. Thus, another mechanism is needed to change the direction of the wheels. As all three motor modules are driven simultaneously, another motor can coordinate wheel directions. In Prototype V.01, this mechanism was solved with a curved rack gear rotating around two robot modules due to avoid complications. This mechanism is controlled by a stepper motor that provides more accurate positioning. Each motor module is connected to the curved rack with Z-axis free geared cases, so its movement is not affected during direction change. In addition, the drive control of the stepper motor provides a frictional force against slipping as it consumes power in the stop position.

The designed chassis was produced with a 3D printer using Polylactic Acid (PLA) material. The flexible nature of the material is a predictable disadvantage. Additionally, mass-produced motor driver circuits can control the motors. These driver circuits can be easily integrated with an Arduino, a programmable circuit board. Since the motor signalization is identical, relatively larger driver circuits that offer to drive more motors can use instead of different drivers. Thus, Prototype V.01 offers control with the low-budget L298N drivers. Also, an infrared (IR) sensor is attached for the remote control. The signalization of the receiver, which offers a much simpler control than Bluetooth or Wireless connections, is more disadvantageous. Nevertheless, it offers easy control for simple commands. Besides, two identical robot modules are connected with stepper motors, and prototypes have been tested with external power supplies rather than rechargeable batteries.

4.2   The Prototype V.02: A Chassis with Independent Stepper Motors

Prototype V.02 is significantly similar in form to Prototype V.01. Since the diameter of the paper strut used does not change, the positions of the wheels are the same. The major problem of Prototype V.01 was the flexing of the chassis in the opposite direction to the pulling force of the motor modules in the strut direction. Material selection and the frameless design of motor connections are the two identified causes of this problem. In Prototype V.02, the parts where the motor modules are connected to the chassis have been strengthened. Also, a blended material that is more resistant to flexing was preferred. These changes largely solved the flexing problem of the chassis. Another issue of Prototype V.01 is that the motor modules are controlled by a separate motor. The friction problems of the curved rack gear are an obstacle to achieving the desired efficiency. For this reason, the motor modules were redesigned in Prototype V.02. Stepper motors are preferred instead of DC motors. Two wheels and two motors have replaced the combination of two wheels and one motor. Thus, the connections of the modules have been made more independent. Accordingly, when the wheels rotate in the same direction, the robot moves translationally; when they rotate in the opposite direction, each module rotates in each Z_0 axis. Besides, silicone wheels with high adhesion were preferred in Prototype V.02. In addition, connection details of mass-produced rollerblade wheels and motors were adapted with rapid prototyping methods (Fig. 11).

In addition, the pulling springs partially prevent independent rotation of the motor modules in Prototype V.01. Also, the roughness of connections increases friction. Therefore, connections have been replaced with a mechanism containing rail shafts and linear bearings. The spring around the rail shaft works as a suspension and provides passive adhesion. Nevertheless, stepper motors caused new control limitations. Stepper motors, which are more complex to control than DC motors, can be controlled with Pulse-width Modulation (PWM) signals sent from Arduino to the ULN2003 driver circuit. The speed, direction, and position of stepper motors are controlled with sequential currents from the driver board. Even if the motors have equal control, this driver circuit draws too much current so motor signaling will be disrupted. To overcome this problem, individual signals or signal multiplier circuits can be used to control each motor. Therefore, Prototype V.02 drives each motor simultaneously with separate signals with Arduino Mega. Four PWM signals go to each of six motors in three motor modules and the motor that connects the robot modules. Consequently, the motors are controlled with 28 signals for stop, forward, backward, rotation, reverse rotation, and twisting functions. Furthermore, how HookBot robots would collaborate, and the assembly system would coordinate with

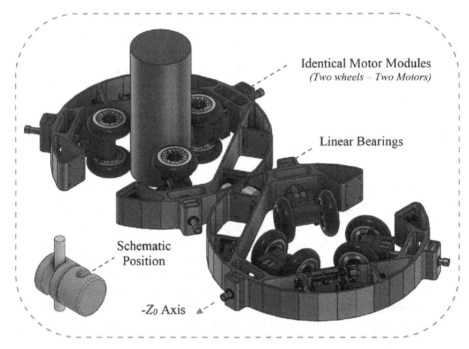

Fig. 10. CAD drawings of Prototype V.02

other subsystems needs to be studied. Thus, the following section presents a hypothetical case for an autonomous system.

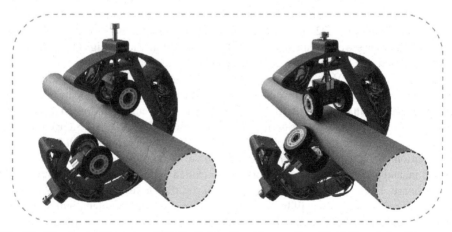

Fig. 11. The physical Prototype V.02 (Left: Rotational locomotion; Right: Translational locomotion)

5 Discussion

5.1 A System for Autonomous Truss Manipulation

A system whose principal actor is mobile assembly robots can explain and exemplify the working mechanisms and functioning of autonomous processes. Besides, a discussion conceptualizing a general construction system might present the potential contribution of relative robots [26]. In an autonomous architectural system, the client should contact the system for their needs. The client can order the design task to the system using a mobile application and sign a smart contract. Subsequently, mobile robots can transmit site surveying data such as wind, insolation, and 3D scans to the system with various sensors. With the analysis of all data, the design stage can start. Experimental studies were conducted for this stage with HumanUI, a Grasshopper 3D interface. In the application interface, the client can start with basic geometries, select struts and joints properties, and get guidance from the system. In addition, since the process has an integration, structural analyses can instantly conduct with cloud computing. Also, using interactive tools such as Augmented Reality (AR) at this stage will provide a more immersive experience for the client. Thus, tools offering direct Rhinoceros 3D integration, such as Fologram, are promising, but delays and interface problems are open to improvement. After the inspection process, the client can approve or renovate the project for the next stage of the system.

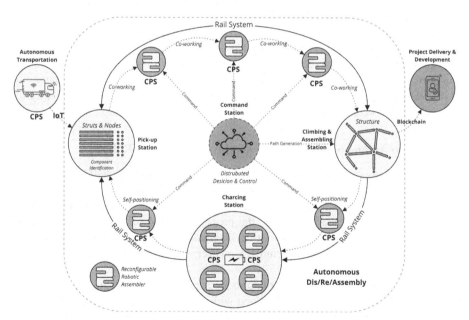

Fig. 12. A conceptual representation of the autonomous dis/re/assembly process [26]

Consequently, approved construction documents would send to an autonomous dedicated manufacturing facility. The production of structural components and the site preparation process is carried out simultaneously. Autonomous demolition robots can remove existing structures at the construction site, and autonomous excavators can provide earth-moving. Besides, this system promises to achieve the four characteristics of Industry 4.0: horizontal networking of similar production facilities in the same sector, vertical integration of the production network in the supply chain, through-engineering of the entire building lifecycle, and increasingly efficient processes with exponential technologies [29]. After the prefabrication and manufacturing process, autonomous vehicles can transport the robotic system and structural components to the construction site. With the preparation of the construction site and the construction of the substructure, HookBots become the principal actor of the system.

Initially, the "Charging Station" (Fig. 12) containing HookBots and the "Pick-up Station" with marked components such as struts and nodes are dropped off at the construction site. Each HookBot in this system works as a cyber-physical system (CPS) with distributed decision and control. Distributed operations, widely used in autonomous control systems in the aviation industry, provide collaborative swarm control to the system [30]. Thus, the "Command Station" assigns tasks to HookBots in standby mode at the charging station. Afterward, they began to assemble a specialized rail system that enabled locomotion between stations. HookBots move collaboratively on the rail system and transfer the components of the truss structure to the "Climbing and Assembling Station". At this station, they erect the structure by climbing and assembling it. After a while, HookBots with low batteries head to the charging station and are replaced by charged ones. After the structure's assembly, HookBots gather and are picked up from the construction site at the charging station. After the assembly, the client can receive or renovate the project. This stage can include a detailed feedback process, and the client can reactivate the system in case of possible changes in the project, such as the re/dis/assembly process for recycling or reusing. Consequently, the conceptualized system provides a hypothetical case for autonomous discrete assembly. The implementation of the system depends on transdisciplinary studies and the development of integration technologies between innovations.

5.2 Road to Autonomous Discrete Assembly with Relative Robots

Achieving an autonomous discrete assembly with relative robots is challenging in many respects. Three groups stand out: mechanical, hardware, and control/integration issues. HookBot prototypes have a successful mechanical design, focusing on locomotion kinematics and proof-of-concept. Besides, the proposed design offers a higher degree of freedom (DoF) and advanced locomotion kinematics compared to examined relative robots that can move on truss structures. However, the design of HookBot is also one of the limitations that complicate the system. Since the gripping mechanism that offers active adhesion requires compact solutions, experiments such as grasping, carrying, and placing the paper struts could be done with external assistance. Completing these tests autonomously and prototyping the other issues examined in the third chapter will improve the mechanical design.

Overcoming the hardware problems depends on testing the improved prototypes with appropriate battery, motor, and weight combinations. We used an external power supply when performing the experiments, but mobile robots should be tested with internal batteries. However, the passive weight of internal batteries in relative robots can be a limitation. Thus, optimized battery, motor, and weight possibilities need to be tested, and wireless charging or robots charging each other should be considered. Another limitation of these experiments is that the payload calculations are not included in this study. Nevertheless, these prototypes, produced with a low budget, have efficiency beyond expectations with simple designs.

Overcoming control/integration problems depends on digital robot simulations. We used a programmable circuit board to drive all the motor modules in coordination. Also, we were able to prototype the integration of simple modules, such as IR, within the scope of the paper. The integration of the IR module has been able to control the functions that provide translational, rotational, and helical locomotion. Reducing uncertainties about these issues and making simulations will strengthen the applicability of the system. Nevertheless, this study collects and presents promising data on relative robot design, prototyping, and system research.

6 Conclusion and Future Remarks

As a system for autonomous truss manipulation, HookBots describe the potential contribution of relative robots to future architectural services and construction systems. In addition, HookBot proposes an integrated approach that differs from studies focused on robot design by offering a design and build strategy as an order-to-delivery system. Further, the developed prototype offers a new mechanical design, which can successfully fulfill the locomotion kinematics in the third chapter. Nevertheless, prototyping all the issues presented in the third chapter would support interoperability. Unique truss structures, which can be notably complex for human workers, can be quickly built with these robots. Therefore, the research on autonomous discrete assembly for architectural services is promising. However, designing and implementing a fully autonomous system requires transdisciplinary studies.

In addition, using these robots in building structures blurs the boundaries of the worker-material or active-passive sides of the traditional construction industry. Robots moving on the structures become a part of the structure while reconfiguring, manipulating, and building. Thus, this novel discrete assembly strategy offers an alternative to bulky construction robots and describes a different future.

For future studies, it is planned to control and simulate robots with many parameters with the Robot Operating System (ROS). Simulations will be useful for constructing a system algorithm. Besides, future prototypes should focus on other design details in the discussion. Also, this paper does not focus on the essential payload calculations for relative robots but could be studied physical prototypes for further research. Ultimately, a roadmap can be achieved for autonomous truss manipulation system with discrete assembly that will work with coordination, collaboration, and combination of multiple HookBots.

References

1. Dörfler, K., et al.: Mobile robotic fabrication beyond factory conditions: case study Mesh Mould wall of the DFAB HOUSE. Constr. Robot. **3**(1–4), 53–67 (2019). https://doi.org/10.1007/s41693-019-00020-w

2. Keating, S.J., Leland, J.C., Cai, L., Oxman, N.: Toward site-specific and self-sufficient robotic fabrication on architectural scales. Sci. Robot. **2**(5) (2017). https://doi.org/10.1126/scirob otics.aam8986

3. Giftthaler, M., et al.: Mobile robotic fabrication at 1:1 scale: the In situ fabricator. Constr. Robot. **1**(1–4), 3–14 (2017). https://doi.org/10.1007/s41693-017-0003-5

4. Minibuilders: Minibuilders. The Institute for Advanced Architecture of Catalonia (2016). https://iaac.net/project/minibuilders/. Accessed 1 July 2022

5. Kayser, M., et al.: FIBERBOTS: an autonomous swarm-based robotic system for digital fabrication of fiber-based composites. Constr. Robot. **2**(1–4), 67–79 (2018). https://doi.org/10.1007/s41693-018-0013-y

6. Werfel, J., Petersen, K., Nagpal, R.: Designing collective behavior in a termite-inspired robot construction team. Science **343**(6172), 754–758 (2014). https://doi.org/10.1126/science.124 5842

7. Jenett, B., Abdel-Rahman, A., Cheung, K., Gershenfeld, N.: Material-robot system for assembly of discrete cellular structures. IEEE Robot. Autom. Lett. **4**(4), 4019–4026 (2019). https://doi.org/10.1109/lra.2019.2930486

8. Augugliaro, F., et al.: The flight assembled architecture installation: cooperative construction with flying machines. IEEE Control. Syst. **34**(4), 46–64 (2014). https://doi.org/10.1109/mcs.2014.2320359

9. Lindsey, Q., Mellinger, D., Kumar, V.: Construction with quadrotor teams. Auton. Robot. **33**(3), 323–336 (2012). https://doi.org/10.1007/s10514-012-9305-0

10. Yun, S., Rus, D.: Self-assembly of modular manipulators with active and passive modules. In: 2008 IEEE International Conference on Robotics and Automation (2008). https://doi.org/10.1109/robot.2008.4543410

11. Nigl, F., Li, S., Blum, J.E., Lipson, H.: Structure-reconfiguring robots: autonomous truss reconfiguration and manipulation. IEEE Robot. Autom. Mag. **20**(3), 60–71 (2013). https://doi.org/10.1109/mra.2012.2201579

12. Melenbrink, N., Kassabian, P., Menges, A., Werfel, J.: Towards force-aware robot collectives for on-site construction 382–391 (2017). https://doi.org/10.52842/conf.acadia.2017.382

13. Leder, S., Weber, R., Wood, D., Bucklin, O., Menges, A.: Distributed robotic timber construction. In: Proceedings of the 39th Annual Conference of the Association for Computer Aided Design in Architecture, pp. 510–519 (2019). https://doi.org/10.52842/conf.acadia.2019.510

14. Ardiny, H., Witwicki, S., Mondada, F.: Construction automation with autonomous mobile robots: a review. In: 2015 3rd RSI International Conference on Robotics and Mechatronics (ICROM), pp. 418–424 (2015). https://doi.org/10.1109/icrom.2015.7367821

15. Petersen, K.H., Napp, N., Stuart-Smith, R., Rus, D., Kovac, M.: A review of collective robotic construction. Sci. Robot. **4**(28) (2019). https://doi.org/10.1126/scirobotics.aau8479

16. Bock, T.: Construction robotics. Auton. Robot. **22**(3), 201–209 (2007). https://doi.org/10.1007/s10514-006-9008-5

17. Gharbia, M., Chang-Richards, A., Lu, Y., Zhong, R.Y., Li, H.: Robotic technologies for on-site building construction: a systematic review. J. Build. Eng. **32**, 101584 (2020). https://doi.org/10.1016/j.jobe.2020.101584

18. Melenbrink, N., Werfel, J., Menges, A.: On-site autonomous construction robots: towards unsupervised building. Autom. Constr. **119**, 103312 (2020). https://doi.org/10.1016/j.autcon.2020.103312

19. Figliola, A., Battisti, A.: Post-industrial Robotics. Springer, Singapore (2020). https://doi.org/10.1007/978-981-15-5278-6

20. Brell-Cokcan, S., Braumann, J.: Rob|Arch 2012. Springer, Vienna (2013). https://doi.org/10.1007/978-3-7091-1465-0

21. Bonwetsch, T.: Robotic assembly processes as a driver in architectural design. Nexus Netw. J. **14**(3), 483–494 (2012). https://doi.org/10.1007/s00004-012-0119-3

22. Carney, M., Jenett, B.: Relative robots: scaling automated assembly of discrete cellular lattices. In: Volume 2: Materials; Biomanufacturing; Properties, Applications and Systems; Sustainable Manufacturing, June 2016. https://doi.org/10.1115/msec2016-8837
23. Chu, B., Jung, K., Han, C.S., Hong, D.: A survey of climbing robots: locomotion and adhesion. Int. J. Precis. Eng. Manuf. 633–647 (2010). https://doi.org/10.1007/s12541-010-0075-3
24. Sattar, T., et al.: Mobile wall climbing and swimming robots to inspect aircraft, storage tanks, pressure vessels and large infrastructure. In: 24th ISPE International Conference on CAD/CAM, Robotics and Factories of the Future, Koriyama, Japan, June 2008. https://openresearch.lsbu.ac.uk/item/87q84
25. Lee, S.H.: Design of the out-pipe type pipe climbing robot. Int. J. Precis. Eng. Manuf. 1559–1563 (2013). https://doi.org/10.1007/s12541-013-0210-z
26. Delikanlı, B., Gül, L.F.: Towards to the Hyperautomation - an integrated framework for Construction 4.0: a case of Hookbot as a distributed reconfigurable robotic assembly system. In: Proceedings of the 40th International Conference on Education and Research in Computer Aided Architectural Design in Europe (eCAADe) [Volume 2] (2022). https://doi.org/10.52842/conf.ecaade.2022.2.389
27. Rhodes, M.D., Will, R.W., Quach, C.C.: Verification tests of automated robotic assembly of space truss structures. J. Spacecr. Rocket. 32(4), 686–696 (1995). https://doi.org/10.2514/3.26670
28. Nechyba, M.C., Xu, Y.: SM2 for new space station structure: autonomous locomotion and teleoperation control. In: Proceedings of the 1994 IEEE International Conference on Robotics and Automation (1994). https://doi.org/10.1109/robot.1994.351337
29. Deloitte: Industry 4.0. Deloitte (2015). https://www2.deloitte.com/ch/en/pages/manufacturing/articles/manufacturing-study-industry-4.html. Accessed 11 July 2022
30. Chen, H., et al.: From automation system to autonomous system: an architecture perspective. J. Mar. Sci. Eng. 9(6), 645 (2021). https://doi.org/10.3390/jmse9060645

Material-in-the-Loop Fabrication: A Vision-Based Adaptive Clay 3D Printing Workflow on Indeterminate Sand Surfaces

Özgüç Bertuğ Çapunaman⑩, Emily K. Iseman⑩, and Benay Gürsoy(✉)⑩

The Pennsylvania State University, University Park 16802, PA, USA
{okc5048,eki5054,bug61}@psu.edu

Abstract. The limitations of the robotic fabrication workflows in seeing, sensing, and responding to the changes in the work envelope typically impose a "unidirectional information flow." To overcome these challenges, we previously explored ways to augment the robotic fabrication
by using vision-based reconstruction frameworks to digitally capture the
work envelope and inform toolpath generation accordingly [9]. Building
upon this framework, we present an adaptive clay 3D printing workflow that employs computer vision and computer graphics algorithms
to scan, register, and accurately reconstruct the robot's work envelope.
This allows the robotic clay 3D printing operations to be carried out on
indeterminate (unknown) and complex surfaces and enables accounting
for emergent material behaviors in consequent robotic tasks. We explore
the applicability of this vision-based framework through a case study in
which we fabricated interlocking components by robotically 3D printing
clay on re-configurable sand formworks. Within this workflow, the moldable sand mixture is shaped by the robot into reusable and temporary
formwork, which acts as a supporting structure during the fabrication
and firing of the clay panels. Based on our early findings, We postulate
that this approach can help facilitate the integration of such temporary
formworks in robotic fabrication tasks as a more sustainable and customizable alternative.

Keywords: Digital Fabrication · Additive Manufacturing · Adaptive
Fabrication · Computer Vision · Robotic Clay 3D Printing

1 Introduction

Although robotic arms provide precision and control in the fabrication process,
their limitations in seeing, sensing, and responding to the changes in the work
envelope in real-time impose a "unidirectional information flow" from the digital model to the robotic manipulator [4]. This unidirectionality poses significant
challenges as to what the designers can conceptualize and fabricate. It is often
necessary to work with digitally predetermined work envelopes and to assume
that they do not change during the fabrication process. To overcome these challenges, researchers have been exploring ways to augment robotic fabrication and

© The Author(s), under exclusive license to Springer Nature Switzerland AG 2023
M. Turrin et al. (Eds.): CAAD Futures 2023, CCIS 1819, pp. 410–424, 2023.
https://doi.org/10.1007/978-3-031-37189-9_27

interaction with various computational feedback loops and Machine Learning algorithms.

In this paper, we present an adaptive clay 3D printing workflow that uses the vision-based reconstruction framework we presented in a previous publication [9]. This vision-based reconstruction framework employs various computer vision and computer graphics algorithms to scan, register, and accurately reconstruct robot's work envelope within its coordinate system. This allows us to carry out various robotic tooling operations with high levels of precision on indeterminate (unknown) surfaces. By enabling vision-based sensing feedback, we build upon Brooks' perspective to "sense [the world] appropriately and often enough" [6] to account for material indeterminacy during the robotic tooling operations. In contrast to contemporary offline programming approaches, this prevents relying on assumptions or simulations that have limited applicability due to the lack of accuracy and/or computational inefficiency.

Within the scope of this paper, we present a case study on conformal clay 3D printing on non-planar sand surfaces to fabricate interlocking clay panels, which we call material-in-the-loop fabrication. We present a novel moldable sand mixture that can be shaped by the robot into reusable and reconfigurable temporary formworks that provide support during the fabrication, transportation, and firing of clay panels. Through this workflow, we aim to (1) eliminate formwork waste that was previously the outcome of fabricating geometrically complex and unique, additively manufacture clay components and (2) enable a feedback loop between the digital representations and the fabricated outcomes by scanning and digitally reconstructing the emergent material features in sand and clay. By eliminating waste and enabling a feedback loop between digital representations and physical outcomes, we speculate that this process has the potential to significantly improve the efficiency and sustainability of 3D printing complex non-planar geometries.

2   Background

2.1   Clay 3D Printing and the Use of Sand as Reusable Formworks for Fabricating Complex Geometries

Clay is a material that has found extensive use in architecture to fabricate architectural components throughout history. As a highly malleable material, clay can be manipulated using a wide range of tools and techniques and be recycled and reused indefinitely prior to firing in high-temperature kilns. This vitrification process physically alters the characteristics of the clay body, increasing in density and strength, and becoming impervious to water in the process. Despite such advantages, manipulating clay bodies has traditionally required a high level of skill and mastery.

In recent decades, the open design and maker culture enabled the free flow of knowledge and resources in many areas, including how clay can be 3D printed [21]. This enabled various designers, architects, artists, and makers who did not have any previous experience with the material to experiment and work with clay by eliminating the need for hands-on crafting skills. Today, different approaches in additive manufacturing using clay span various techniques [11], with numerous commercial or open-source fabrication tools available for end users. However, the material extrusion approach, which was also employed in this study, is considered one of the most prominent techniques due to its ease of implementation and extensive documentation.

Clay 3D printing, however, has its constraints and affordances that are causally linked to the various material and tool parameters employed in fabrication. Within the body of work exploring this fabrication technique, discovering these causal relationships for emergent material features (i.e., textures and surface effects) has been a recent topic of interest [19,28,31]. Additionally, some researchers explored material deposition beyond conventional planar layers using parametrically generated machine code [1,13,16].

Furthermore, there have been various efforts in the literature to design component-based architectural elements using clay 3D printing. Among these efforts are the "Woven Clay" project by Friedman et al. [20], "Seed Stitches" and "G-Code Clay" projects by Emerging Objects [16,17], "Clay Non-Wovens" project by Rosenwasser et al. [33], and "RoboSense 2.0" project by Bilotti et al. [5]. More relevantly, In the "Informed Ceramic" project, Ko et al. propose robotic 3D clay printing on double-curved formwork surfaces to fabricate a component-based self-standing structure [25]. Although they demonstrate successful application of clay 3d printing on arbitrary surfaces, the foam formwork waste is not easily recyclable. Moreover, the proposed workflow relies on carefully coordinated formwork milling and clay 3d printing tasks to ensure that the fabricated formwork is true to its digital representation. This paper presents a case study in which a novel moldable sand mixture is used as temporary formworks for 3D printed clay components. A similar approach of using sand and gravel as temporary formworks for casting double-curved concrete components was tested in the Philips Pavilion by Le Corbusier and Xenakis back in 1958 [34]. In an earlier attempt to computationally manipulate sand formworks, researchers also experimented with parametrically designed toolpaths to explore resulting emergent geometries [23]. More recently, researchers have developed the "print-cast" technique for robotic additive manufacturing of concrete components using conformal toolpaths [2,3,36]. Using this technique, concrete is robotically 3D printed "over a subtractive shaped substructure of CNC tooled compacted green sand." and reduces the formwork waste for complex double-curved geometries. However, the robotic fabrication workflow still requires working within a digitally predetermined work envelope. We postulate that the proposed vision-based adaptive workflow can help bridge the gap between existing bodies of work in clay 3D printing and temporary formworks.

2.2 Vision-Based Sensing and Reconstruction in Robotic Additive Manufacturing

Over the past decade, there has been a growing interest in exploring ways to augment robotic fabrication workflows by equipping robotic arms with various sensor feedback loops and, more recently, Machine Learning protocols to address complex design-fabrication challenges. Menges argues that these advancements will lead to the emergence of "new cyber-physical production systems" that "have advanced sensing and learning capabilities," "are increasingly self-aware", and "can self-predict, self-organize and self-configure" [27]. One line of research towards this goal focuses on vision-based sensing and reconstruction systems that can inform robotic tooling operations by adapting the digital representations based on the data collected from robot's work envelope [7,15,18]. Researchers also started to explore how these sensor feedback loops can initiate interactive conversations between the designer and the robot in creative design scenarios [8,26,35].

More relevantly, a research effort to explore the fabrication of clay lattice structures [1,22] successfully demonstrated how a closed-loop feedback system can inform consequent tooling operations using a non-contact measurement method in spatial clay 3D printing applications using an industrial robot arm. Nicholas et al. [30] integrated real-time scanning and Neural network prediction for generating conformal toolpaths to robotically 3D print PETG plastic on unknown and arbitrarily shaped 3D surfaces. They conclude that their scanning and reconstruction workflows offer "great possibilities for smart environment-aware iterative robotic fabrication." Another recent study by Naboni et al. [29] explores conformal concrete 3D printing on indeterminate gravel surfaces with the integration of vision-based sensing and reconstruction systems. These studies are closely related to the work presented in this paper. Together, they illustrate the emerging possibilities of integrating vision-based sensor feedback loops into robotic fabrication workflows toward a "cyber-physical" fabrication paradigm.

3   Material-in-the-Loop Fabrication Workflow

Building upon our previous work [9], our goal is to inform tooling operations to respond to emergent material behavior. Within the scope of the proposed workflow, we employ various robotic manipulation tasks, and computational frameworks to (1) formatively shape sand formworks, (2) scan the resulting form to (3) inform conformal toolpath generation, and (4) additively fabricate clay panels. These tasks were planned and simulated using Robots plugin [32] in Grasshopper for Rhinoceros 7 and executed using an ABB IRB 2400 industrial robotic arm using various custom End-of-Arm-Tools (EoAT).

Figure 1 depicts an exemplary cycle of fabrication with the proposed material-in-the-loop workflow, in which information flows between digital and physical environments and constitutes the foundation upon which we can negotiate between design intention and material indeterminacy. For example, in formative fabrication of sand formworks, the design intention is enacted by the robotic arm using a simple compacting motion with discrete point targets. However, the resulting form is inherently indeterminate, since the outcome is not only a consequence of the forces acting upon it but also the material's affordance to be shaped [14]. Within the proposed workflow, such emergent material behaviors

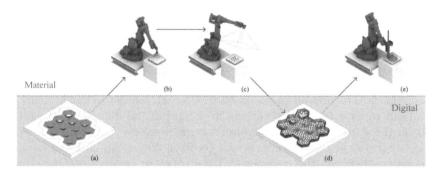

Fig. 1. Workflow diagram of the proposed workflow: (a) motion planning for sand compaction task, (b) formative shaping of the formwork, (c) 3D capture and reconstruction of the resulting form, (d) conformal toolpath generation, (e) additive fabrication of the panels.

are geometrically captured and reconstructed in the digital environment using an RGBD camera. This, in turn, informs the toolpath generation and enables the robot to act upon the work object using conformal material extrusion. Simply, this information flow from physical to digital and back allows us to work with the material and respond to material emergence on a component level, while still maintaining control over the global design.

3.1   Materials

Complex panel geometries require supporting structures in transportation and post-processing. Previously, we tested our approach with commercial kinetic sand as a proof-of-concept [9]; however, additives in this material are not suitable for use in ceramics kilns. In this study, we sought to create a moldable support solution that can stay in place through all stages a moldable sand, specifically suited to our needs, considering moldability, support, and use in a ceramics kiln.

In the moldable sand recipe presented here (Table 1), the granular components, namely silica sand and grog, are highly refractory and, therefore, resistant to the heat of the kiln firing. The variation in particle size helps the granules pack and stabilize, enabling it to retain its shape during the forming process and to offer support to the clay 3D printed atop. The additives include a material used to make the sand moldable, as well as a material used to disperse the molding element. CMC gum powder (carboxymethylcellulose), which is a cellulose gum, is used as a binding agent together with water. Additionally, household dish soap is added to break the surface tension of the water, reducing the amount of water needed in the mixture. For our application, we found that minimal water is ideal, as it takes up volume in the sand, causing the support structure to shrink as it evaporates, which can potentially disturb the panel before vitrification. These additives burn out during firing, leaving the refractory materials ready for reuse in our process.

Table 1. Moldable Sand Mixture Recipe.

Ingredients	Weight per 100 g
Silica Sand (50 mesh)	46 g
Grog (35–47 mesh)	34.25 g
Dish Soap	1.25 g
Water	11 g
CMC Gum	5.5 g

In addition to the moldable sand mixture, we used SIO-2 ANETO 3D Clay in testing the proposed workflow and the fabrication of the artifact. This was done to minimize inconsistencies in material behavior during extrusion and firing. This material has a recommended bisque temperature of 1000 °C and is recommended to be fired around 1230–1270 °C for 0.0% porosity (water absorption). It is also rated for dry shrinkage of $\sim$ 6.0% and firing shrinkage of 9.8%.

3.2   Formative Fabrication of Sand

In shaping the sand formwork, the compacting EoAT (Fig. 2-c) is pressed into the sand formwork sequentially based on the global design parameters ($d_{compact}$) as shown in Fig. 2-a. Together with the additives detailed in the earlier section, this compaction allows the sand formwork to hold the shape of the imprinted EoAT geometry. In addition to compacting the formwork at contact regions, this motion displaces material within the sand volume. In our tests, the sand mixture is framed by a CNC-milled mold to contain the material displacement within the component footprint and minimize cracks that can occur due to compressive stresses applied. This results in protrusions in the sand formwork in undisturbed regions (Fig. 2-a). Although some computational techniques like Finite-Element Analysis (FEA) and Discrete Element Method (DEM) can be used to estimate such emergent deformations, this would require tedious characterization of the material behavior. Even then, such approaches can be computationally intensive

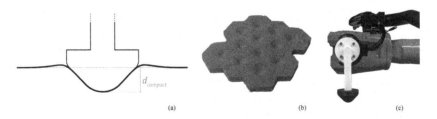

(a)   (b)   (c)

Fig. 2. Formative fabrication approach used in shaping the sand formwork; (a) simplified diagram of the compacting motion, (b) resulting form of the compacted sand formwork (c) custom shaping End-of-Arm-Tool.

and potentially inaccurate due to inconsistencies in sand-mixture homogeneity, moisture content, and other unforeseen environmental factors.

3.3 Computer Vision Framework

The vision-based reconstruction framework presented in this paper relies on a Microsoft Azure Kinect Developer Kit RGBD Camera and an NVIDIA Jetson Nano microprocessor (Fig. 3-b) to capture synchronized RGBD images at known six degrees-of-freedom robot poses. Using the calibrated robot pose information together with the captured RGBD images, the Truncated Signed Distance Function (TSDF) integration algorithm is used to produce a highly accurate mesh representation of the scanned region within the robot coordinate system. Our previous publication [9], provides further details and validation of these vision-based hardware and software components and examines the impact of operational parameters such as surface typology and camera pose configurations on digital reconstruction accuracy, and briefly demonstrates fabrication feasibility.

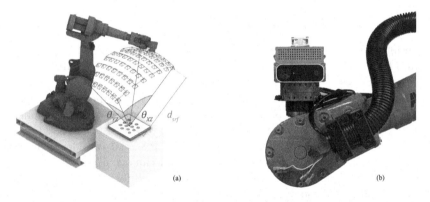

(a)   (b)

Fig. 3. Computer vision framework employed in the proposed workflow: (a) Hemispherical camera trajectory with $d_{srf} = 1000\,\text{mm}$, $\theta_{XZ} = 90$, $\theta_{YZ} = 60$, and $n_{pose} = 64$, (b) the hardware assembly used in scanning the work object.

Within the scope of this paper, we utilized an 8×8 camera pose configuration ($n_{pose} = 64$) and a voxel size (V_{size}) of $1/512\,\text{m}$ ($\approx 1.9\,\text{mm}$). As discussed in our earlier publication, we found these configurations to be highly accurate for reconstructing regions up to 600 mm by 600 mm. Additionally, we used operational parameters such as camera distance to surface (d_{srf}), angle spanned in XZ (θ_{XZ}), and YZ (θ_{YZ}) planes by the hemispherical trajectory (Fig. 3-a). These parameters were set at 1000 mm, 90°, and 60°, respectively, in light of our earlier validation study.

3.4   Conformal Toolpath Generation

In contrast to planar slicing approaches common in conventional 3D print-ing workflows, the generation of conformal toolpaths plays an important role in ensuring evenly spaced material deposition across nonplanar build surfaces. While planar toolpaths can be projected onto a non-planar surface, this often results in significant deviations in geodesic distance between toolpath traces and consequent undesirable gaps or overlaps of extruded material.

To address this challenge, we have implemented a Differential Growth Curves (DGC) algorithm that generates a meandering conformal pattern on the recon-structed mesh geometry (Fig. 4). Alternatively, Geodesic Heat Method [12] and the Fast Marching Method [24] can be used to calculate geodesic distances on complex surfaces. However, DGC approach results in a continuous path which is favorable due to the limitation in our clay extruder to stop material flow. In our implementation, the bounding curve for the intended component was used as the seed curve and gradually grown to fill the given surface. The discretization of this seed curve was based on the hexagonal lattice circle packing theorem [10] with which the number of discrete nodes can be approximated as follows;

$$n_{node} \approx \frac{A_{srf}}{\pi(r_{nozzle})^2} \times \frac{\pi\sqrt{3}}{6} = \frac{\sqrt{3}A_{srf}}{6(r_{nozzle})^2} \tag{1}$$

This approach ensures the DGC toolpaths are computed efficiently and allow dynamic discretization of the seed curve even when additional geometrical fea-tures, such as holes, are introduced on the mesh geometry.

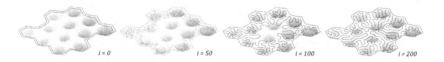

$i = 0$ $i = 50$ $i = 100$ $i = 200$

Fig. 4. Generation of Differential Growth Curves across iterations for a single layer of material deposition.

Within the scope of this implementation, we also explored control over the geodesic distance between toolpath traces ($d_{geodesic}$) to experiment with differ-ent levels of porosity that can be achieved using this technique (Fig. 5). When $d_{geodesic} \leq R_{nozzle}$, the generated toolpath results in a watertight surface, while $d_{geodesic} > R_{nozzle}$ results in a strand of extruded clay across the surface (Fig. 5). Since the proposed toolpath generation inherently provides evenly-spaced spaced traces, the DGC approach remains relevant even when individual layers aren't designed for complete coverage of material deposition across non-planar surfaces. Moreover, this characteristic presents a design variable that can be parametri-cally controlled to achieve visual or functional such as increased surface area and controlled light and air porosity.

Fig. 5. Examples of resulting toolpaths when the geodesic distance between toolpath traces are varied.

In order to ensure structural stability and adequate material strength, multiple layers of toolpaths are generated using this approach. In between these layers, the growth patterns were also randomly shuffled to achieve an irregular lattice of overlapping materials. Although the random shuffling of the growth pattern is beneficial regardless of the $d_{geodesic}$ parameter, the structural benefits of this approach are most evident in layers with $d_{geodesic}$ value greater than R_{nozzle}. By overlapping different strands of deposited clay, the shuffled growth patterns greatly help minimize weak points prone to cracking.

3.5 Additive Manufacturing of Panels

In the additive fabrication of the panels, we used an open-source clay extruder appropriated for use with an industrial robot arm [19] (Fig. 6-b). This clay extruder assembly provides direct control over the material extrusion rate using a digital output group that encodes the value in 7 bits with $2^7 = 128$ discrete extrusion rate values. Additionally, we used a nozzle with a diameter of 7 mm and a layer height of 3.5 mm.

The robot target locations were based on the DGC pattern generated for each formwork. These points were evaluated against the reconstructed mesh for the surface normal vector. However, due to the size constraints of the clay extruder used in this workflow, we observed collisions with the formwork, as well as the other linkages of the robot. To circumvent this issue, we implemented a simple robot pose optimization using a weighted average between the surface normal vector ($\vec{v}_{normal}$) and the world-Z vector ($\vec{v}_z$) as follows (Fig. 6-c):

$$\vec{v}_{optimized} = \frac{(c \times \vec{v}_{normal}) + ((1 - c) \times \vec{v}_z)}{2} \tag{2}$$

In the example panels presented in this paper, we used a constant weight of $c = 0.5$, and were able to carry out the clay extrusion task with no collisions. For more complex geometries, this weight can be adjusted accordingly.

3.6 Post-processing of Panels and Reclamation of the Sand Mixture

Following the additive fabrication of the panel, the moldable sand framework is used as support in transportation and firing process and is backed by a concrete board for additional rigidity. After approximately 50% of the moisture has

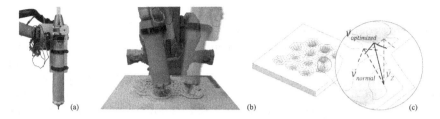

Fig. 6. Robotic clay 3D printing: (a) clay extruder End-of-Arm-Tool (b) additive manufacturing of panels, (c) robot pose optimization based on surface normal ($\vec{v}_{normal}$) and world-Z ($\vec{v}_z$) vectors.

evaporated from the clay in the open air, the panel can be transferred into the kiln together with the sand formwork and the concrete board. This approach minimizes the risk of cracking the panel before it is fired. The panel requires no specific firing schedule, as the sand mixture and concrete board bring negligible additional moisture despite their thickness. The panel can in fact, withstand a relatively fast firing schedule because of the open nature of its structure.

Following the firing process, the refractory ingredients can easily be reclaimed, which simply involves re-adding the non-refractory additives per recipe. The sand can be used indefinitely after every firing and reclamation cycle.

4   Design Experiment

As a proof-of-concept of how the proposed workflow can be employed in the conceptualization and fabrication of architectural elements, we have carried out a design experiment for a parametrically designed wall. Using the fabrication methods, computational frameworks, and material systems detailed in the earlier sections, this design experiment exemplifies a scenario in which the vision-based augmentation of the robotic fabrication workflows allows precise material deposition on indeterminate sand formworks. Since this design experiment is solely a demonstration of the proposed workflow, design parameters were not based on any specific performance criteria or site conditions.

4.1   Global Design

For fabrication feasibility, we adopted a modular wall design strategy which allows us to divide the global design into smaller units that can be easily fabricated and fired with the available extruder and kilns, respectively. This also allows us to reclaim and reuse sand mixture used in the formwork in between fabrication of different components. Based on the physical restrictions of the kiln, each unit was set to contain a total of 12 grid points and fits within a circular area with a diameter of 400 mm. The outlines of the components were based on the underlying hexagonal grid, which is also intended to work as an interlocking connection detail.

Within this parametric system, spacing parameters between toolpath traces ($d_{geodesic}$) were calculated per module basis, using the average noise value obtained within a tile (Fig. 7-d). For modules with the highest average compaction, the DGC pattern was generated using $d_{geodesic} = R_{nozzle}$, which results in a closed surface. In the generation of the conformal toolpaths for modules with the least compaction, on the other hand, we used $d_{geodesic} = 2R_{nozzle}$ to achieve a sparse material deposition for a see-through effect, as illustrated in Fig. 7-b.

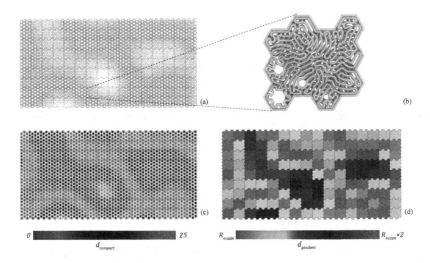

Fig. 7. Global design parameters; (a) geometric design, (b) simulated material deposition and porosity of a panel, (c) $d_{compact}$ parameter visualization, (d) $d_{geodesic}$ parameter visualization.

4.2 Fabricated Modules

In this study, we have chosen a module for fabrication primarily due to the variation in $d_{compact}$ values we obserced across the module, which best showcases the parametric openings introduced for compaction points when $d_{compact} \geq 15\,\text{mm}$.

In our testing of the formative fabrication of the underlying formwork, we observed the successful shaping of the moldable sand mixture using the series of compacting motions discussed in the earlier sections. Similarly, 3D scanning and geometrical reconstruction stages also yielded promising results, providing a high level of accuracy in digitally representing the state of the formwork. During the additive manufacturing of the clay panel, however, we observed minor deviations in extrusion width. Although this error is partially due to the composite error of the computer vision framework, we believe that the main limitation is due to the size and limited control of the clay extruder (Fig. 8).

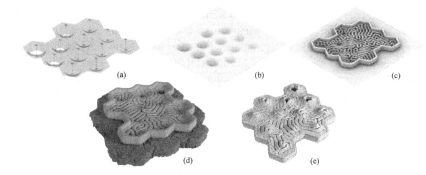

Fig. 8. A fabricated clay panel throughout different stages of the proposed workflow: (a) compaction motion planning, (b) scanned and digitally reconstructed sand formwork, (c) generated conformal toolpath, and kiln-fired panel (d) before and (e) after formwork removal.

5   Discussion

The material-in-the-loop fabrication workflow proposed in this paper presents new opportunities for working with complex and indeterminate material systems. It also allows computational systems to work with and adapt to emergent material behavior. Although the design experiment carried out is not based on real-world design and performance criteria, it showcases what can be achieved using this workflow. One potential application of the design language made available by this research is that the natural cooling performance of bisque ceramic can be exploited for natural ventilation. This workflow can yield parametrically controlled passive cooling with a large surface area achieved using DGC patterns. Similarly, since the DGC patterns can also be used to control light, we speculate the design and fabrication of parametric shading façade elements as another application. However, more research is needed to explore the full potential of this approach with more complex global geometries and component typologies, and develop a design space that takes advantage of these newly added capabilities.

As a study exploring a novel fabrication workflow, many tasks, material and computational systems presented in this paper have inherent implementation limitations. One of the major limitations we observed was the clay extruder EoAT. First, the extruder assembly is greatly limited in allowing the fine-tuning of extrusion rates. Since the extruder primarily relies on a single piston to extrude clay, changes in pressure build-up can cause an uneven extrusion rate. Additionally, the clay extruder EoAT's size limits robot poses due to crashes with the work object or the robot linkages. This results in limited applicability on geometries with steep curvature changes. In our experiments, we observed less than desirable first layer height at large curvature changes since the nozzle could not be oriented perpendicular to the sand formwork due to this limitation.

Although we demonstrated that the proposed workflow can manipulate and adapt to emergent forms, establishing control over the formative shaping of the sand formworks remains mostly unexplored. We believe that gaining better control over the formative fabrication task can enable more complex global geometries and expand the design space. Furthermore, especially in larger-scale fabrication scenarios, we acknowledge material limitations of clay bodies such as shrinkage and warpage, as a major challenge. More research is needed on how such behaviors can be actively mitigated or compensated for using computational approaches. We speculate that the vision-based reconstruction framework can also be a valuable tool in both informing the formative fabrication task and analyzing material behavior across the fabrication process.

6 Conclusion

In this paper, we present a novel vision-based adaptive and assistive fabrication workflow for the additive manufacturing of clay panels on indeterminate surfaces, that we call material-in-the-loop fabrication. Within a growing body of knowledge, the proposed workflow exemplifies how material indeterminacy can be captured and used to inform fabrication operations. The design experiment showcased here only captures one of many design approaches we believe possible with the proposed workflow. We believe that the flexibility of the vision-based adaptive and assistive workflows presented exceeds far beyond the geometric complexities and material systems employed. Our approach can easily be appropriated for large-scale applications and may be beneficial in scenarios where environmental variables and material properties pose significant challenges for robotic automation.

One such large-scale application scenario can be the robotic additive manufacturing of concrete on sand or gravel surfaces that have larger granules. Some examples dealing with sand or gravel as a formwork already exist [18–20, 30]; however, we believe that the workflow presented in this paper can be used to bridge the gap between the formative fabrication of such formworks and the adaptive additive manufacturing of architectural components. In addition to the augmentation of robotic fabrication, the moldable sand mixture detailed in this paper presents a novel and sustainable alternative for rigid formworks and can eliminate formwork waste when 3D printing complex clay components. As discussed earlier, most of the ingredients in this sand mixture can be reclaimed and reused repeatedly. Furthermore, this approach eliminates any additional support material in additive manufacturing and prevents cracking during the transportation, drying, and firing of the 3D printed artifacts.

Finally, as a future line of inquiry, we aim to explore the integration of continuous workflows to oversee and monitor the robotic fabrication processes in real time. Synchronous monitoring of the robot's work envelope is the next advancement towards a "cyber-physical" production paradigm that can allow co-creative interactions with the robots and open new avenues of creative inquiry. step towards a "cyber-physical" production paradigm that can allow co-creative interactions with the robots and open new ways of creative inquiry.

References

1. AlOthman, S., Im, H.C., Jung, F., Bechthold, M.: Spatial print trajectory. In: Robotic Fabrication in Architecture, Art and Design, pp. 167–180 (2019). https://doi.org/10.1007/978-3-319-92294-2_13
2. Battaglia, C.A., Miller, M.F., Zivkovic, S.: Sub-additive 3d printing of optimized double curved concrete lattice structures. In: Robotic Fabrication in Architecture, Art and Design 2018, pp. 242–255 (2018). https://doi.org/10.1007/978-3-319-92294-2_19
3. Battaglia, C.A., Verian, K., Miller, M.F.: De:stress pavilion. In: Distributed Proximities - Proceedings of the 40th Annual Conference of the Association for Computer-Aided Design in Architecture (2020)
4. Bechthold, M.: The return of the future: a second go at robotic construction. Arch. Des. 80(4), 116–121 (2010). https://doi.org/10.1002/ad.1115
5. Bilotti, J., Norman, B., Liu, J., Rosenwasser, D., Sabin, J.: ROBOSENSE 2.0: Robotic Sensing and Architectural Ceramic Fabrication. CUMINCAD (2018)
6. Brooks, R.A.: Elephants don't play chess. Rob. Auton. Syst. 6(1–2), 3–15 (1990)
7. Brugnaro, G., Baharlou, E., Vasey, L., Menges, A.: Robotic softness: an adaptive robotic fabrication process for woven structures. In: POSTHUMAN FRONTIERS: Data, Designers, and Cognitive Machines (2016)
8. Çapunaman, O.B.: Cam as a tool for creative expression: informing digital fabrication through human interaction. In: RE: Anthropocene, Design in the Age of Humans - Proceedings of the 25th CAADRIA Conference, vol. 1, pp. 243–252 (2020). https://doi.org/10.52842/conf.caadria.2020.1.243
9. Çapunaman, O.B., Dong, W., Gürsoy, B.: A vision-based sensing framework for adaptive robotic tooling of indefinite surfaces. Constr. Rob. (2022). https://doi.org/10.1007/s41693-022-00081-4
10. Chang, H.C., Wang, L.C.: A simple proof of thue's theorem on circle packing (2010)
11. Castro e Costa, E., Duarte, J.P., Bártolo, P.: A review of additive manufacturing for ceramic production. Rapid Prototyp. J. 23(5), 954–963 (2017). https://doi.org/10.1108/RPJ-09-2015-0128
12. Crane, K., Weischedel, C., Wardetzky, M.: The heat method for distance computation. Commun. ACM 60(11), 90–99 (2017)
13. Cuevas, D.D.G., Pugliese, D.G.: Advanced 3D Printing with Grasshopper: Clay and FDM (2020)
14. DeLanda, M.: The new materiality. Arch. Des. 85(5), 16–21 (2015)
15. Dörfler, K., Rist, F., Rust, R.: Interlacing. In: Robotic Fabrication in Architecture, Art and Design 2012, pp. 82–91 (2013). https://doi.org/10.1007/978-3-7091-1465-0_7
16. Emerging Objects: G-code clay. http://emergingobjects.com/project/gcode-clay/
17. Emerging Objects: Seed stitch. http://emergingobjects.com/project/seed-stitch/
18. Ercan Jenny, S., Lloret-Fritschi, E., Gramazio, F., Kohler, M.: Crafting plaster through continuous mobile robotic fabrication on-site. Constr. Rob. 4(3), 261–271 (2020). https://doi.org/10.1007/s41693-020-00043-8
19. Farrokhsiar, P., Gürsoy, B.: Robotic sketching: a study on robotic clay 3d printing. In: SIGraDi 2020 Proceedings of the 24th Conference of the Iberoamerican Society of Digital Graphics, pp. 312–319 (2020)
20. Friedman, J., Kim, H., Mesa, O.: Experiments in additive clay depositions. In: Robotic Fabrication in Architecture, Art and Design 2014, pp. 261–272. Springer, Heidelberg (2014). https://doi.org/10.1007/978-3-319-04663-1_18

21. Gürsoy, B.: From control to uncertainty in 3d printing with clay. In: Computing for a Better Tomorrow - Proceedings of the 36th eCAADe Conference, vol. 2, pp. 21–30 (2018)

22. Im, H.C., Al Othman, S., Garcia del Castillo, J.L.: Responsive spatial print: clay 3d printing of spatial lattices using real-time model recalibration. In: Recalibration: On imprecision and infidelity - Proceedings of the 38th Annual Conference of the Association for Computer-Aided Design in Architecture (2018). https://doi.org/10.52842/conf.acadia.2018.286

23. Kieferle, J.B., Katodrytis, G.: Fabricating semi predictable surfaces - a workshop series on digitally fabricating freeform surfaces with aggregates. In: Complexity & Simplicity - Proceedings of the 34th eCAADe Conference, vol. 1, pp. 329–334 (2016)

24. Kimmel, R., Sethian, J.A.: Computing geodesic paths on manifolds. Proc. Natl. Acad. Sci. **95**(15), 8431–8435 (1998)

25. Ko, M., Shin, D., Ahn, H., Park, H.: Informed ceramics: multi-axis clay 3d printing on freeform molds. In: Robotic Fabrication in Architecture, Art and Design 2018, pp. 297–308 (2019). https://doi.org/10.1007/978-3-319-92294-2_23

26. Lee, Y.C., Llach, D.C.: Hybrid embroidery: exploring interactive fabrication in handcrafts. Leonardo **53**(4), 429–433 (2020). https://doi.org/10.1162/leon_a_01931

27. Menges, A.: The new cyber-physical making in architecture: computational construction. Arch. Des. **85**(5), 28–33 (2015). https://doi.org/10.1002/ad.1950

28. Mohite, A., Kochneva, M., Kotnik, T.: Speed of deposition - vehicle for structural and aesthetic expression in cam. In: Architecture in the Age of the 4th Industrial Revolution - Proceedings of the 37th eCAADe and 23rd SIGraDi Conference, vol. 1, pp. 729–738 (2019)

29. Naboni, R., Breseghello, L., Sanin, S.: Environment-aware 3d concrete printing through robot-vision. In: Co-creating the Future: Inclusion in and through Design - Proceedings of the 40th Conference on Education and Research in Computer Aided Architectural Design in Europe (eCAADe), vol. 2, pp. 409–418 (2022)

30. Nicholas, P., Rossi, G., Williams, E., Bennett, M., Schork, T.: Integrating real-time multi-resolution scanning and machine learning for conformal robotic 3d printing in architecture. Int. J. Arch. Comput. **18**(4), 371–384 (2020). https://doi.org/10.1177/1478077120948203

31. Rael, R., Fratello, V.S.: Printing Architecture: Innovative Recipes for 3D Printing. Princeton Architectural Press, Princeton (2018)

32. Robots. https://github.com/visose/Robots

33. Rosenwasser, D., Mantell, S., Sabin, J.: Clay non-wovens: robotic fabrication and digital ceramics. In: Proceedings of the 37th Annual Conference of the Association for Computer Aided Design in Architecture (ACADIA), pp. 502–511 (2017)

34. Vreedenburgh, C.G.J.: The hyperbolic paraboloid and its mechanical properties. Philips Tech. Rev. **20**(1) (1958)

35. Weichel, C., Hardy, J., Alexander, J., Gellersen, H.: Reform: integrating physical and digital design through bidirectional fabrication. In: Proceedings of the 28th Annual ACM Symposium on User Interface Software & Technology, pp. 93–102 (2015)

36. Zivkovic, S., Battaglia, C.: Rough pass extrusion tooling: cnc post-processing of 3d-printed sub-additive concrete lattice structures. In: Recalibration on Imprecisionand Infidelity (2018)

Biologically Informed Design - Towards Additive Biofabrication with Cyanobacteria

Perla Armaly[1]([✉]) [iD], Lubov Iliassafov[2] [iD], Shay Kirzner[2] [iD], Yechezkel Kashi[2] [iD], and Shany Barath[1] [iD]

[1] Faculty of Architecture and Town Planning, Technion, Technion-Israel Institute of Technology, Haifa, Israel
perla.armaly@campus.technion.ac.il, {shaykirzner,kashi, barathshany}@technion.ac.il
[2] Faculty of Biotechnology and Food Engineering, Technion, Technion-Israel Institute of Technology, Haifa, Israel
lubovili@campus.technion.ac.il

Abstract. As sustainability awareness is increasing within architecture, we witness the emergence of new design approaches seeking to minimize greenhouse gas (GHG) emissions and production of waste. Biodesign addresses such challenges by integrating living organisms within processes of design and manufacturing harnessing their natural performances. This paper aims to outline design principles integrating a living organism, cyanobacteria, in an additive co-fabrication process by utilizing its three main performances: photosynthesis, calcium carbonate precipitation ($CaCO_3$) and carbon dioxide (CO_2) fixation. Cyanobacteria, through photosynthesis, precipitate $CaCO_3$ which enhances the adhesion between cells and surfaces. Through microbially induced $CaCO_3$ (MICP), cyanobacteria can bind sand particles together and enhance soil stability as seen in the formation of biological soil crusts. With the rise of additive manufacturing (AM) techniques, integrating cyanobacteria within additive biofabrication processes can be advantageous in improving sustainable architectural production. Leveraging computational design tools, this paper aims to construct design guidelines that cater to cyanobacterial needs towards optimizing biomass production and, consequently, the architectural performance of the printed components.

Keywords: Biodesign · Additive Manufacturing · Cyanobacteria · MICP · Carbon Dioxide fixation · Material Driven Design

1 Introduction

Buildings, being among the largest consumers of natural resources, account for a significant portion of greenhouse gas (GHG) emissions [1, 15]. To date, the most used unnatural building material is cement concrete which is responsible for 8% of the overall global emissions. The Architecture, Engineering and Construction (AEC) industry is gradually starting to address this concern by developing more sustainable alternatives such as blending cement with recycled materials, designing for minimal material

© The Author(s), under exclusive license to Springer Nature Switzerland AG 2023
M. Turrin et al. (Eds.): CAAD Futures 2023, CCIS 1819, pp. 425–436, 2023.
https://doi.org/10.1007/978-3-031-37189-9_28

use and integrating new biobased materials [1, 6, 9]. Another perspective lies within the field of biodesign which utilizes living organisms within construction materials in order to develop materials that can inherit and potentially perform natural processes observed in nature such as carbon dioxide (CO_2) fixation, degradability, recyclability, self-growth, self-healing, and adaptability to the environment [21, 22]. Such processes can be seen in the formation of ant hills through the natural cementation of sand grains and in the binding of coral reefs by the presence of calcium carbonate ($CaCO_3$) [1]. Contrary to construction procedures, natural processes consume a negligible amount of energy, therefore, adopting such processes within construction has great potential in promoting sustainability. Microbially induced $CaCO_3$ precipitation (MICP) is one of the biological processes that are investigated in relation to construction restoration. It is mainly examined within concrete, being the most used construction material, as it can potentially improve compressive strength, offer self-healing properties, and prevent water penetration damages [20, 27].

Within architecture, the effect of cyanobacterial MICP on sand has also been studied in order to develop a new living building material within casting processes [22]. Expanding on such knowledge, we propose to introduce cyanobacteria within an additive manufacturing (AM) process that utilizes computer-aided design tools (CAD) and adopts a geometrical-driven design method towards enhanced performance of architectural components. CAD tools provide designers with informed performance criteria throughout the design phase and establish a better connection between the digital model to the manufactured outcome [9]. In our case we can design geometrical forms that take into consideration basic cyanobacterial needs such as light exposure within an AM setup. By constructing a new workflow for harnessing the biological data of cyanobacteria ($CaCO_3$ precipitation) as an input in the fabrication process, we can potentially bridge the gap between microbiological and architectural processes to produce printed structures that take advantage of additive properties in both biological (micro) and architectural (macro) scales.

In previous experiments, we demonstrated the behavior of cyanobacteria within a developed biomixture and were able to define suitable environmental conditions for cyanobacterial growth (See Fig. 1) [3]. Our preliminary results determined that a 1:1 quartz sand agar ratio at an initial bacteria concentration of ~2×10^6 CFU/mL is suitable for cyanobacterial growth within regulated environmental conditions for light and temperature (22 ± 1 °C) [3]. Relying on our previous work, here we focus on design guidelines that enhance the potential biological performance of cyanobacteria within 3D printed components. We present a computational strategy to analyze geometrical iterations for increased light exposure to enable fabricating with cyanobacteria towards carbon-efficient architecture.

Fig. 1. Images. (Left) Petri dish containing a sample of the non-living mixture. (Middle) Petri dish containing a sample of the biomixture demonstrating cyanobacterial growth within quartz sand type 1. (Right) Sample of the biomixture demonstrating MICP after 2 weeks of incubation [3].

2 Cyanobacteria

Diverse implementations of living organisms such as algae and mycelium have been demonstrated in various design applications [16, 21]. Similar developments are currently being examined with living bacteria and highlight a wide range of potential applications for construction materials, building processes and products [10, 13, 25]. Cyanobacteria, the organism utilized in this research, demonstrate three main behaviors that can be advantageous for developing sustainable construction processes: 1) photosynthesis, 2) carbon fixation, and 3) calcium carbonate precipitation.

2.1 Microbially Induced Calcium Carbonate Precipitation (MICP)

Biomineralization is among the most notable processes that bacteria can perform. In biologically induced mineralization, microorganisms can secrete metabolic products that react with compounds in the environment resulting in the deposition of mineral particles, mainly carbonate products. These minerals could serve as cementitious materials called biocement [1]. The formation of biocementitious materials relies on the precipitation of microbially induced calcium carbonates (MICP). The MICP process occurs in layers surrounding the cells, mainly in the exopolysaccharide layer (EPS). The secretion and the calcification of the EPS layer serves as a biological glue. The EPS layer increases the cell adhesion to surfaces and the presence of $CaCO_3$ within the EPS layer enhances its mechanical resistance [12, 19] (See Fig. 2).

Cyanobacteria are known for their ability to secrete EPS and for the production of $CaCO_3$ through photosynthesis (See Fig. 2). Cyanobacteria MICP is a natural re-occurring biocementation process in desert soils. Through this biological process, known as biological soil crust formation (BSC), soil stability can be enhanced as the precipitation of $CaCO_3$ interconnects soil particles and binds them together [12]. Recent research indicates that biomediated soils improved the shear strength of soils and prevented wind and water erosion [11, 12, 23]. Within AEC, the biocementation process via MICP, due to its eco-efficiency, has drawn the attention of many researchers for the restoration and reinforcement of construction materials [7, 26]. In limestone and concrete-based materials, MICP has enabled the restoration of structures. MICP for concrete restoration has been favored over expensive chemical treatments and has been reported, in addition to crack repairing, to improve compressive strength and durability [26]. In soils, it strengthened underground foundation conditions for construction. Although the cementation process of porous materials such as soils requires a large amount of $CaCO_3$, the process proved to be less costly than chemical treatments [26]. A recent precedent reported a 25.54% increase in the compressive strength of cement mortar by utilizing cyanobacteria *Synechocystis pevalekii* [24].

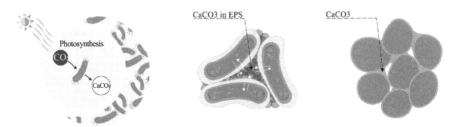

Fig. 2. Diagram. Three target performances of cyanobacteria. (Left) Photosynthesis and carbon fixation. (Middle) Calcification of the EPS layer. (Right) Biocementation of sand particles.

2.2   Cyanobacteria Photosynthesis and Carbon Fixation

Cyanobacteria are a photoautotrophic and oxygen-producing bacteria, and have played a major role in the evolution of different forms of life on earth since over 3500 million years ago [19]. These microorganisms can be found in all potential habitats including freshwater bodies, marine ecosystems, deserts, limestone, soil, and biological soil crusts. Cyanobacteria positively impacts the environment serving as a bioremediation agent that eliminates toxic wastes impact in contaminated sites, and contains multiple strains that can be implemented in human environments [19]. In addition, by harvesting solar energy and performing photosynthesis they fixate CO_2 and generate oxygen in addition to assimilating metabolites and minerals. It is estimated that cyanobacteria accounts for a quarter of the global CO_2 fixation [14, 19].

Effect of Light on Cyanobacteria. Light is the main factor of cyanobacteria metabolism and it directly affects the production of biomass as well as the motility of cells. The motility is regulated by what is known as phototaxis which is defined by the direction, intensity, and wavelength of light [19]. Cyanobacteria are known to move towards or away from the light source depending on its intensity [19]. Cell motility improves with the secretion of the EPS layer which serves as a calcification site for MICP [19]. Thus, increased light exposure could encourage cell motility and consequently encourage MICP through photosynthesis. As we aim to increase the solidification of the printed components through the precipitation of $CaCO_3$, understanding the effect of light on cyanobacterial biomass production is of importance for enhancing the performance of the bacteria. In this context, AM processes could enable increased light exposure due to geometrical articulation and structure porosity of the printed component.

2.3   Cyanobacteria in an Additive Co-fabrication Process

Based on the proven relationship between cyanobacterial MICP, CO_2 fixation and soil stability, we propose to utilize the biosafety level 1 *Synechocystis* being one of the most studied cyanobacterium [19], within an additive manufacturing process of architectural components. Through our previously developed protocols the performance of *Synechocystis* PCC 6803 was examined within different sand-based mixtures and resulted in the binding of quartz sand to a united curved surface [3].

As Belnap [4] suggests that a higher biomass increases soil stability, we aim to leverage CAD and AM to increase the cyanobacterial biomass for the fabrication and solidification of architectural modular components. AM can be advantageous for the photosynthetic nature of cyanobacteria and encourage MICP. For example, through AM techniques we could 3D print porous structures that increase light exposure, therefore encouraging photosynthesis and the production of $CaCO_3$. Contrary to cement concrete production, utilizing cyanobacterial performance within AM processes can promote sustainability since it allows CO_2 fixation throughout the fabrication process. As we recognize that cyanobacteria, throughout their lifespan, are active participants in the fabrication process we will refer to it as a co-fabrication process. Understanding the sustainable advantages of AM, this paper moves forward to the optimization of geometrical forms that cater to both cyanobacterial and fabrication requirements, focusing mainly on light exposure as a critical factor for enhanced *Synechocystis* PCC 6803 performance within the co-fabrication process.

3   Biologically Informed Computational Design

Computational design to manufacturing (CAD to CAM) processes within architecture has accompanied various biodesign processes such as the 3D printing of mycelium-based composites and the fabrication of bioluminescence micro-architectures [18, 25]. With that, to the best of our knowledge, 3D printing of an architectural mixture containing living cells of cyanobacteria has yet to be introduced and still requires further investigation. Utilizing CAD as an investigative tool for designing with and for cyanobacteria, enables a better understanding of the organism's connection to geometry at the architectural scale and within a co-fabrication process [3]. Through CAD techniques we are able to design for increased surface area and light exposure for enhanced cyanobacterial photosynthesis and MICP. As the generated forms of the printed components aim to cater to both biological and architectural requirements, the design process will take into consideration biological constraints such as the motility of the cells, biomass production and $CaCO_3$ precipitation, and architectural performance such as stability through different design parameters including the printing tool path, dimensions of the printed components (width, height), and geometrical properties.

3.1   Incubation and Light Constraints

In our recently founded design biolab, the photosynthetic cyanobacteria are grown within a growth chamber (incubator) with regulated environmental conditions. Therefore, to maintain growth post-fabrication, the printed components will be incubated and kept at the same pre-defined environmental conditions to which the cyanobacteria are accustomed. Consequently, we currently aim to design modular construction components, such as bricks, blocks, and panels within the size constraints (maximum 40 × 40 × 40 cm) of the incubator.

As light is a main factor in cyanobacterial activity, suitable light intensity was also defined to encourage cell growth [3]. Therefore, when studying the designed geometries for increased light exposure we must take into consideration the light direction and distribution within the incubator (See Fig. 3). In incubators, as in the case of our Biochemical Oxygen Demand (BOD) incubator (MRC), lights are positioned above the shelves. While the inner chamber is made of stainless steel which could aid in light distribution, to ensure maximum light exposure, we defined that the main light source origin for the geometrical design process of the printed components is set from above and perpendicular to the incubator shelves.

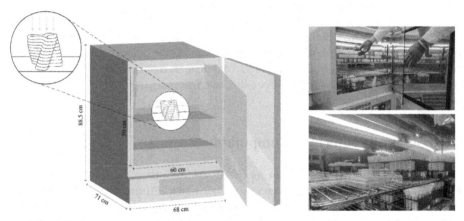

Fig. 3. (Left) Diagram, Post-fabrication incubation of the printed components. The diagram demonstrates the light direction within the BOD incubator. (Right) Images, BOD incubator in the design Biolab containing cyanobacteria cultures (Top) and samples of the biomixture (Bottom), D.D Lab.

3.2 Material-Driven Geometrical Design

Tool Path Design. Tool paths development, many times, is considered a technical process that advances the realization of the modeled design [6]. When fabricating for a specific performance, it becomes a leading factor that alters not only the fabrication process but also the final structural and aesthetic performance of the printed component. For example, in 3D concrete printing, tool paths are designed for minimizing material use, carbon-efficient production and increased strength [6]. While such factors are of relevance also in the case of printing sand-based architectural components, when designing with living cells additional considerations are introduced. Bacterial and fungal growth in angled channels has demonstrated that sharp angles reduce the biomass production of both bacteria and fungi. In the case of bacterial growth, the amount of biomass differed significantly as the sharp angles required movement that was not natural to the bacteria [2].

Knowing that cyanobacteria move towards the light in order to perform viable metabolic activities [19], constraints in motility could risk cell viability in addition to reducing the production of biomass. In order to increase MICP within the architectural components, designing tool paths that encourage cyanobacteria motility needs is of great significance. Therefore, in the tool path design, in addition to increasing light exposure, properties of the habitat geometry are considered to encourage bacterial motility and reduce sharp angles.

Similar to tool paths usually 3D printed with concrete and clay [5, 6, 8, 17], geometrical properties such as component curvature and printing radius are important for the stability of the component throughout the printing process. Through parametric control of such geometrical properties, we developed a tool path design process in order to correlate the thresholds and range of motility for our specific cyanobacterial strain with the geometry of the printing properties. At this point, we examined the link between geometrical form and potential bacterial motility through simple curvature tool paths (See Fig. 4). When establishing the cyanobacterial motility, we can further develop the tool path to examine the motility range by altering the angles and the curvature of the tool path.

Fig. 4. (Left) Diagram, Development of a potential curvature tool path for encouraging cyanobacterial motility using CAD tools such as Rhinoceros and Grasshopper. (Right) Images, Robotic printing of the developed tool path with non-living sand-based mixtures towards printing with cyanobacteria.

Geometrical Analysis. The examination of the components' geometrical design is determined also in relation to microenvironmental conditions that affect cyanobacterial performance. The first step towards designing geometries for photosynthetic activity is to design for light exposure through increased surface area. As a case study, we tested two geometrical principles, rotation and offsetting of the printed layers, in relation to the previously developed extruded tool paths aiming to quantify the increase in surface area and analyze light exposure (See Fig. 5). To maintain a set of base constraints, all geometrical iterations are in the same height and base area. In addition to light exposure these initial geometrical forms, as part of an architectural fabrication process, cater to printing constraints. Therefore, a range was defined to evaluate the structural stability of the overall geometry and resulting overhangs. The rotation of layers was examined in a range of 3–7° and the offsetting of layers was examined in a range of 1–5 mm.

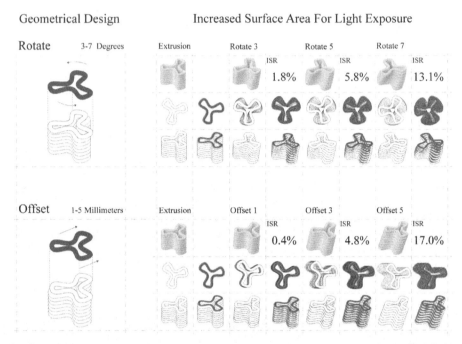

Fig. 5. Geometrical analysis for increased surface area and light exposure. (Left) Applied design principles. (Right) Geometrical variations in the defined ranges for each design princple.

Using Rhinoceros, Grasshopper and its associated plugins Ladybug and Honeybee, the toolpath geometries were calculated for surface area visualizing the areas that are exposed to direct light and simulating the potential lighting conditions in the incubator. The offsetting of layers resulted in a greater increase of surface area; variation "Offset 5" holds the highest surface area of 17% after variation "Rotate 7" with an increase of 13.1% (See Fig. 6). With that, the feasibility of printing such geometries will be tested and optimized before implementing the process with living cyanobacteria cells. This geometrical investigation demonstrates potential guidelines linking biological and geometrical properties to tool path design. This will also guide the optimization of the printed cyanobacteria-based components and increase biological performance.

Rotate 7 (ISA 13.1%)   Offset 5 (ISA 17%)

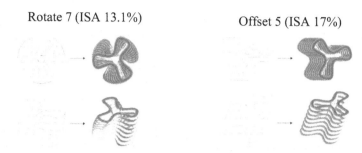

Fig. 6. Top view and isometric view of the light exposure analysis for both variations with the highest increase in surface area, "Rotate 7" and "Offset 5".

4 Prototyping an Additive Co-fabrication Process with Cyanobacteria

Constructing a systematic CAD to CAM setup is essential for the proper implementation of the developed co-fabrication workflow. While this paper demonstrated computational design methods for the optimization of cyanobacteria performance, a corresponding automation workflow has been developed utilizing a UR5e robotic arm for the deposition of a biomixture with cyanobacteria. As printing with living cells presents new obstacles to the architectural world, we aim to expand and upscale biofabrication processes within relatively large-scale architectural additive manufacturing (See Fig. 7).

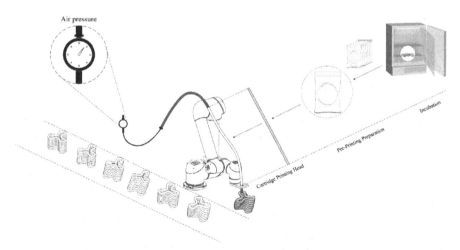

Fig. 7. Diagram. Automation workflow of the upscaled biofabrication process within relatively large-scale architectural production with living cyanobacteria cells.

Currently, based on the geometrical analysis, we have gone on to prototype potential habitat geometries to further understand the fabrication constraints with the selected sand-based mixture using the sand-agar ratios that we developed for our bacterial strain

(See Fig. 8). Expanding on the outlined method for biologically informed computational design, we have drawn inspiration from and for architecture to further design geometrical forms that perform architecturally in terms of stability and function. These architectural components could potentially offer an alternative to common printed concrete blocks and products. Moreover, on the biological level, knowing that cyanobacterial performance will decrease as the rate of cell death naturally increases, we will aim to enhance the cells' activity throughout their lifespan by optimizing biological procedures. In addition, we aim to examine the natural behavior and production of cyanobacteria by identifying the cyanobacterial growth phases, quantifying the fixation of CO_2 and the production of $CaCO_3$.

Fig. 8. Non-living sand-based printing using robotic deposition towards developing architecturally performing components.

5   Conclusion

As the AEC industry is taking a necessary shift towards sustainable design due to rising ecological and environmental concerns, this ongoing research proposes to utilize cyanobacterial performance within a co-fabrication workflow to produce architecturally performing components. Integrating cyanobacteria within a design process can be of great significance as it has the potential to substitute cement concrete with a carbon-efficient biomixture.

As cyanobacteria are active participants throughout the co-fabrication process, the printed geometries should allow the bacteria to maintain its functionality and even encourage its biological performance throughout and post-fabrication. Therefore, in this paper we outlined design guidelines for enhanced cyanobacterial photosynthesis and MICP that can result in the solidification of sand-based biomixtures. Recognizing the importance of light in the mechanisms of cyanobacteria, we leverage CAD tools to design custom tool paths that encourage cyanobacterial motility towards the light in order to quantify and analyze increased light exposure through increased surface area. In addition, future steps will analyze the effect of light penetration on the printed components in relation to geometrical porosity in order to encourage 3D cell spatial distribution and increased biomass production.

Through a CAD to CAM workflow that caters to the three main cyanobacterial performances; photosynthesis, MICP and CO_2 fixation, we aim to enable decision-making in a design process that is informed by cyanobacterial behavior. Such a design process presents new sustainable opportunities to minimize the harmful effects of current production processes in the AEC industry and encourage a positive impact that is harnessed from natural processes.

Acknowledgment. This work is supported by Technion Additive Manufacturing and 3D Printing Center [Grant Agreement ID 86638420]. The authors thank Avraham Cohen and Yoav Dabas from Disrupt.Design Lab for their support.

References

1. Achal, V., Mukherjee, A.: A review of microbial precipitation for sustainable construction. Constr. Build. Mater. **93**, 1224–1235 (2015)
2. Arellano-Caicedo, C., Ohlsson, P., Bengtsson, M., Beech, J.P., Hammer, E.C.: Habitat geometry in artificial microstructure affects bacterial and fungal growth, interactions, and substrate degradation. Commun. Biol. **4**(1) (2021)
3. Armaly, P., Kirzner, S., Kashi, Y., Barath, S.: Biomanufacturing of architectural prototypes with cyanobacteria. In: 28th International Conference on Computer-Aided Architectural Design Research in Asia, CAADRIA 2023. The Association for Computer-Aided Architectural Design Research in Asia (CAADRIA), Asia (2023). https://doi.org/10.52842/conf.caadria.2023.2.149
4. Belnap, J., Phillips, S.L., Witwicki, D.L., Miller, M.E.: Visually assessing the level of development and soil surface stability of cyanobacterially dominated biological soil crusts. J. Arid Environ. **72**(7), 1257–1264 (2008)
5. Breseghello, L., Sanin, S., Naboni, R.: Toolpath simulation, design and manipulation in robotic 3D concrete printing. In: The 26th Annual Conference of the Association for Computer-Aided Architectural Design Research in Asia, CAADRIA 2021, pp. 623–632. The Association for Computer-Aided Architectural Design Research in Asia (CAADRIA), Asia (2021)
6. Breseghello, L., Naboni, R.: Toolpath-based design for 3D concrete printing of carbon-efficient architectural structures. Addit. Manuf. **56**, 102872 (2022)
7. Chuo, S.C., et al.: Insights into the current trends in the utilization of bacteria for microbially induced calcium carbonate precipitation. Materials **13**(21), 4993 (2020)
8. Clarke-Hicks, J.: Grading Light: utilizing plastic deformation to functionally grade ceramic light screens, Master's thesis, University of Waterloo. UWSpace (2021). http://hdl.handle.net/10012/17808
9. Cohen, A., Barath, S.: Integrating large-scale additive manufacturing and bioplastic compounds for architectural acoustic performance. In: 28th International Conference on Computer-Aided Architectural Design Research in Asia, CAADRIA 2023. The Association for Computer-Aided Architectural Design Research in Asia (CAADRIA), Asia (2023). https://doi.org/10.52842/conf.caadria.2023.2.179
10. Dade-Robertson, M.: Living Construction, 1st edn. Taylor and Francis (2020). https://www.perlego.com/book/1828721/living-construction-pdf
11. DeJong, J.T., Mortensen, B.M., Martinez, B.C., Nelson, D.C.: Biomediated soil improvement. Ecol. Eng. **36**(2), 197–210 (2010)
12. Dhawi, F.: How can we stabilize soil using microbial communities and mitigate desertification? Sustainability **15**(1), 863 (2023)

13. Ferral-Pérez, H., Galicia-García, M.: Bioprecipitation of calcium carbonate by Bacillus sub-tilis and its potential to self-healing in cement-based materials. J. Appl. Res. Technol. **18**(5), 245–258 (2020)

14. Gaysina, L.A., Saraf, A., Singh, P.: Cyanobacteria in diverse habitats. In: Cyanobacteria, pp. 1–28. Academic Press (2019)

15. Habert, G., et al.: Environmental impacts and decarbonization strategies in the cement and concrete industries. Nat. Rev. Earth Environ. **1**(11), 559–573 (2020)

16. Heisel, F., Hebel, D.E.: Pioneering construction materials through prototypological research. Biomimetics **4**(3), 56 (2019)

17. Klug, C., Schmitz, T.H.: Examining the interactions of design parameters in the LDM of clay as the basis for new design paradigms. Ceramics **5**(1), 131–147 (2022)

18. Lim, A.C.S., Thomsen, M.R.: Multi-material fabrication for biodegradable structures: enabling the printing of porous mycelium composite structures. In: Stojakovic, V., Tepavce-vic, B (eds.) Towards a new, configurable architecture - Proceedings of the 39th eCAADe Conference (2021), vol. 1, pp. 85–94. University of Novi Sad, Novi Sad, Serbia (2021)

19. Mehdizadeh Allaf, M., Peerhossaini, H.: Cyanobacteria: model microorganisms and beyond. Microorganisms **10**(4), 696 (2022)

20. Mors, R.M., Jonkers, H.M.: Bacteria-based self-healing concrete: evaluation of full scale demonstrator projects. RILEM Tech. Lett. **4**(2019), 138–144 (2020)

21. Myers, D.: BioDesign – nature + science + creativity. Museum of Modern Art, Thames & Hudson (2012). https://www.biology-design.com/

22. Qiu, J., Artier, J., Cook, S., Srubar, W.V., III., Cameron, J.C., Hubler, M.H.: Engineering living building materials for enhanced bacterial viability and mechanical properties. IScience **24**(2), 102083 (2021)

23. Rozenstein, O., Karnieli, A.: Identification and characterization of Biological Soil Crusts in a sand dune desert environment across Israel-Egypt border using LWIR emittance spectroscopy. J. Arid Environ. **112**, 75–86 (2015)

24. Sidhu, N., Goyal, S., Reddy, M.S.: Biomineralization of cyanobacteria Synechocystis pevalekii improves the durability properties of cement mortar. AMB Express **12**(1), 1–12 (2022)

25. Thomsen, M.R., Tamke, M., Mosse, A., Sieder-Semlitsch, J., Bradshaw, H., Buchwald, E.F., Mosshammer, M.: Imprimer la lumiere – 3D printing bioluminescence for architectural mate-riality. In: Yuan, P.F., Chai, H., Yan, C., Leach, N. (eds.) CDRF 2021, pp. 305–315. Springer, Singapore (2022). https://doi.org/10.1007/978-981-16-5983-6_28

26. Zhu, T., Dittrich, M.: Carbonate precipitation through microbial activities in natural envi-ronment, and their potential in biotechnology: a review. Front. Bioeng. Biotechnol. **4**, 4 (2016)

27. Zhu, T., Lin, Y., Lu, X., Dittrich, M.: Assessment of cyanobacterial species for carbonate precipitation on mortar surface under different conditions. Ecol. Eng. **120**, 154–163 (2018)

Minimum Mass Cast Glass Structures Under Performance and Manufacturability Constraints

Anna Maria Koniari[(✉)], Charalampos Andriotis, and Faidra Oikonomopoulou

Faculty of Architecture and the Built Environment,
Delft University of Technology, Delft, The Netherlands
annamariakoniaris@gmail.com

Abstract. This work develops a computational method that produces algorithmically generated design forms, able to overcome inherent challenges related to the use of cast glass for the creation of monolithic structural components with light permeability. Structural Topology Optimization (TO) has a novel applicability potential, as decreased mass is associated with shorter annealing times and, thus, considerably improved manufacturability in terms of time, energy, and cost efficiency. However, realistic TO in such structures is currently hindered by existing mathematical formulations and commercial software capabilities. Incorporating annealing constraints into the optimization problem is an essential feature that needs to be accommodated, whereas the brittle nature of glass invokes asymmetric stress failure criteria that cannot be captured by conventional ductile plasticity surfaces or uniform stress constraints. This paper addresses the approximation problems in the evaluation of principal stresses while concurrently incorporating annealing-related manufacturing constraints into a unified TO formulation. A mass minimization objective is articulated, as this is the most critical factor for cast glass structures. To ensure the structural integrity and manufacturability of the component, the applied constraints refer both to the glass material/structural properties and to criteria that ensue from the annealing and fabrication processes. The developed code is based on the penalized artificial density interpolation scheme and the optimization problem is solved with the interior-point method. The proposed formulation is applied in a planar design domain to explore how different glass compositions and structural design strategies affect the final shape. Upon extraction of the optimized shape, the structural performance of the respective 3D structures is validated with respect to performance constraint violations using the Ansys software. Finally, brief guidelines on the practical aspects of the manufacturing process are provided.

Keywords: topology optimization · structural glass · brittle materials · mass minimization · nonlinear programming · cast glass · reduced annealing

1 Introduction

Cast glass has been recently highlighted as a material with large shaping potential for the design of monolithic load-bearing structures that allow for spatial and light continuity [1, 2]. Particularly, casting allows to create free-form transparent or translucent structural

---

C. Andriotis and F. Oikonomopoulou—These authors contributed equally to this work.

© The Author(s), under exclusive license to Springer Nature Switzerland AG 2023
M. Turrin et al. (Eds.): CAAD Futures 2023, CCIS 1819, pp. 437–451, 2023.
https://doi.org/10.1007/978-3-031-37189-9_29

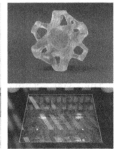

Fig. 1. (from left to right) The Crystal Houses façade made of cast glass bricks/Mirror of Mt. Palomar telescope. Image source: Collection of the Rakow Research Library, The Corning Museum of Glass/Glass node (top) and floor (bottom) designed with TO. Image source: [2].

elements that escape the two-dimensionality of float glass panes and fully exploit the glass properties, such as the great compressive strength, which is higher than that of conventional building materials, including wood, steel, and concrete [1].

Yet, the vast shaping potential of cast glass has, so far, been little explored in structural applications in architecture, hindered mainly by the lengthy annealing[1] process that renders their production unrealistic because of the corresponding high energy and manufacturing costs [1, 2, 3]. The selected structural geometry and glass composition are the most critical factors for the annealing time needed [1]. Essentially, the annealing time increases exponentially when selecting a glass composition with higher thermal expansion coefficient or when enlarging the cross-sectional dimension of a glass component [2, 4]. In the latter case, this results in limiting the existing architectural cast glass applications only to structures made of small glass bricks (Fig. 1, left), so that the cross-sectional dimensions can be cooled down in a reasonable time [5, 6].

However, the optimization of the stiffness-to-weight ratio of cast glass structures and/or the use of glass types with a lower thermal expansion coefficient can greatly reduce the annealing time needed allowing for larger overall dimensions [1]. The benefits of applying these strategies have been well demonstrated by the cast glass honeycomb mirror blanks of the giant telescopes (Fig. 1, middle), since dimensions up to 8.4 m in diameter have been achieved [7, 8] in a considerably reduced annealing time.

In this regard, Topology Optimization (TO) has large potential for the design of massive cast glass structures (Fig. 1, right), since it allows to reach structural forms that maximize stiffness with minimal mass and sparse geometries [2, 3]. This furnishes additional benefits in fabrication time, embodied energy, and cost efficiency making such structures feasible to manufacture. Previous research utilizing commercial TO software successfully demonstrates the ability to engineer glass components of minimum mass, although it highlights the incompatibility of such TO software for the design with glass as structural material [2, 3]. This derives from the fact that such software is developed for conventional, mainly ductile, building materials and, thus, does not fully incorporate

---

[1] The cooling process consists of phases with different cooling rates [4]. In this paper, only annealing is going to be considered since it is the lengthiest of all cooling phases, thus, having the larger effect on the total time needed.

neither manufacturing constraints linked to annealing nor asymmetric principal stress constraints, that reflect the brittle nature of glass, in the formulation [2, 3]. In this light, this study investigates how the optimization problem can be formulated so that it particularly addresses the annealing time constraint, following similar approaches that have recently been developed in the direction of integrating manufacturing limitations in TO formulations for the creation of realistic algorithmic design workflows [9, 10].

Regarding structural performance, the most critical factor in glass structures is tension since, besides its considerably lower strength value, accounting for less than 10% of the respective compressive strength, it can activate different fracture mechanisms in the component leading to failure even before the tensile stress reaches its allowable limit [6]. Therefore, it is essential that an individual evaluation of both principal stresses is incorporated into the TO formulation to converge into feasible results.

The integration of principal stress constraints within TO formulations is considered either with global stress values [11, 12, 13, 14] or with local evaluation of stresses in each finite element. The latter has been proven to be more effective in the elimination of peak values converging, therefore, to more realistic results [13, 15, 16, 17]. However, there are inherent challenges in the integration of stress constraints into the optimization problem, related mainly to the large computational time needed. Thus, the evaluation of stresses is usually linked to the application of material failure criteria and, particularly, the Von mises criterion [11, 12, 13, 14, 15, 17, 18] which refers to ductile materials. Regarding brittle materials, there are approaches which either investigate the asymmetric stress behavior through the application of the Drucker-Prager criterion [19, 20, 21] or apply unified functions that can serve different failure criteria [16].

This paper addresses the aforementioned challenges of (i) annealing-related manufacturing constraints and (ii) asymmetric principal stress criteria within a uniform mathematical formulation for the establishment of a method that will contribute to the efficient design of monolithic glass structures that are feasible to fabricate taking into full consideration the mechanical properties of glass. To do so, we develop a nonlinear programming formulation within the concept of penalized artificial density interpolation [22] to optimize planar structural profiles. Accordingly, the solution algorithm code is coupled with plane stress quadrilateral finite elements and the optimization problem is solved with the interior-point method.

2 Problem Statement

The optimization problem refers to a planar design domain Ω_{des} modeled with quadrilateral finite elements, formulated within the penalized density interpolation scheme as introduced in [22]. In this regard, the stiffness of each finite element is expressed in function of a pseudo-density value ρ_e which reflects the existence or absence of material, such as:

$$E(\rho_e) = E_0 + \rho_e^p(E - E_0) \tag{1}$$

$$0 < \rho_{min} \leq \rho_e \leq 1, \quad e \in \Omega_{des} = \Omega_{mat} \subseteq R^n, \, n = 2$$

where p is the penalization value; E is the Young's modulus of the material; E_0 and ρ_{min} are lower bounds for stiffness and pseudo-density, respectively, in order to avoid singularities of void elements; and Ω_{mat} is the total material domain. Given that non-design domain is not considered, Ω_{des} equals Ω_{mat}.

Additionally, a filtering technique is applied to address inherent numerical issues in TO such as the checkerboard problem [23, 24], which can result into unrealistic shapes. The adopted formulation applies a technique similar to image processing where the pseudo-density value of each element is derived as a weighted average of the element values inside a fixed neighborhood [23]. Instead of the compliance-based approach [14, 21, 23, 24, 25, 26], in this paper the formulation is volume-based since it reflects better the posed problem of minimizing the mass, and has been proven to result in robust solutions [13, 17, 19, 20]. Thus, the objective is formulated as:

$$\min V = \int_{\Omega_{des}} t\rho_e d\Omega, \quad e \in \Omega_{des} = \Omega_{mat} \subseteq R^n, \ n = 2 \tag{2}$$

where t is the thickness of the structure.

In total, six different types of constraints are applied in the TO formulation. The structural constraints refer to global equilibrium, compliance, displacement, and principal stresses (both tension and compression individually) and are formulated as:

$$\mathbf{KU} = \mathbf{F} \tag{3}$$

$$\frac{c(\boldsymbol{\rho})}{c_L} \leq 1, \quad c(\boldsymbol{\rho}) = \sum_{e=1}^{N} \mathbf{U_e^T K_e U_e}, \ c_L = a_c c_0, \tag{4}$$

$$u_{max}^e < \frac{1}{500}l, \quad e \in \Omega_{des} = \Omega_{mat} \subseteq R^n, \ n = 2 \tag{5}$$

$$\rho_e^{(p-q)} \left(\frac{\sigma_{comp,e}}{\sigma_{comp,lm}} \right) \leq 1, \quad e = 1, 2, \ldots, N \tag{6}$$

$$\rho_e^{(p-q)} \left(\frac{\sigma_{ten,e}}{\sigma_{ten,lm}} \right) \leq 1, \quad e = 1, 2, \ldots, N \tag{7}$$

where $\boldsymbol{\rho}$ is the vector of the pseudo-densities[2]; $\mathbf{K}, \mathbf{K_e}$ are the global and element stiffness matrices, which are functions of $E(\rho_e)$; $\mathbf{U}, \mathbf{U_e}$ are the global and element matrices referring to the nodal displacements; $\mathbf{F}$ is the global load matrix; c_0 is the compliance calculated for the full domain; c_L is the allowable compliance limit; α_c is the respective fraction percentage; μ is the Poisson's ratio; u_{max}^e is the maximum nodal displacement; l is the total length of the structure under consideration here; q is the exponent related to the 'qp' approach for stress constraint relaxation as discussed in [18, 19] in order to address the singularity problem [27] and avoid the creation of zero stresses in void elements; $\sigma_{comp,e}$ and $\sigma_{ten,e}$ are the compressive and tensile stresses extracted locally per finite element, respectively; and $\sigma_{comp,lm}$ and $\sigma_{ten,lm}$ are the compressive and tensile strength limits defined according to the glass material properties, respectively.

---

[2] All the variables highlighted in bold refer to vectors and matrices.

It is noted that only the constraints related to the displacement and principal stresses serve to ensure the structural integrity of the design and are directly formulated according to the glass material properties. The compliance constraint is mainly defined by the end user following a strategy as described in [19] and contributes to the overall performance of the algorithm by guiding the simulation faster to an optimal result.

The last constraint refers to the annealing time limit and is formulated according to the maximum length scale approach [26]. The maximum cross-sectional dimension d_{max} is defined through considering primarily the maximum dimension that can be annealed in the set time limit based on the respective annealing rate [4] and the glass composition input. In this light, input based on different glass types is applied to evaluate the extent to which such changes affect the final outcome. Lastly, the need for homogeneous mass distribution in the geometry is also considered as a maximum limit to prevent large cross-sectional differences and uneven cooling that could cause local stress concentrations and breakage right from the cooling process. In total, d_{max} is expressed as:

$$d_{max} = \min(d_{ann}, d_{hom}), \; d_{ann} = \sqrt{\frac{T_{ann,max}\sigma_{res}}{\Delta T \frac{E\alpha_{ex}}{1-\mu} \frac{\rho_{mat}c_p}{\lambda} b}} \tag{8}$$

where d_{ann} is calculated adopting the formula by CelSian Glass & Solar and refers to the maximum cross section to be annealed in the set time limit $T_{ann,max}$; d_{hom} is the maximum cross section to ensure homogeneous mass distribution; σ_{res} is the maximum allowable permanent residual stress in the glass article; ΔT is the annealing temperature range; α_{ex} is the thermal expansion coefficient; ρ_{mat} is the material density; c_p is the specific heat capacity; λ is the thermal conductivity; and b is a factor based on the shape of the cross section and its capability to radiate heat.

3 Numerical Results

3.1 Design Problem

The case study refers to a bridge spanning 4.20 m, whose demand of tensile strength poses an additional challenge to the optimization problem. Moreover, any compromised transparency due to the complexity of the optimized form can serve as an advantage in this case, as it prevents an influence in the depth perception of the visitors, which in turn could decrease their confidence while walking on a completely transparent glass surface. The overall shape, dimensions and boundary conditions are defined based on the needs of an interior bridge placed in the Great Court at the British museum [28]. Additionally, redundancy and safety issues are considered while defining the design strategy. The total slab is divided along the transversal axis into two identical monolithic components, while laminated float glass sheets cover their upper surface [29] (Fig. 2). The latter also prevent from direct impact stresses on the load-bearing glass structure, which can be equally critical with far-field stresses for glass articles due to the risk of activating initial defects and fracture mechanisms [30]. The design domain refers to the characteristic planar longitudinal profile of the individual monolithic component without the top glass sheets.

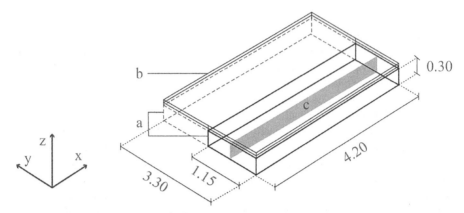

Fig. 2. Design strategies for redundancy and safety: (a) Division into two individual components (b) Float glass sheets (c) Design domain Ω_{des}.

Table 1. Permanent (p) loads, live (l) loads, and safety factors.

Self-weight of slab (p) (kN/m^2)	Float glass sheets (p) (kN/m^2)	People (l) (kN/m^2)	Maintenance (l) (kN/m^2)	Safety factor – permanent loads	Safety factor – live loads
9.8	1.2	5	0.4	1.2	1.5

The respective permanent and live loads are applied uniformly at the upper surface of the design domain according to Eurocode 1, Chapter 6 for museums (Table 1). They are only applied along the vertical direction since any lateral loads, such as due to wind, are eliminated, given that the example refers to an interior slab.

3.2 Optimization Formulation

For the finite element modeling, symmetry in terms of design domain, load application and boundary conditions is exploited, in order to reduce the total computational time and power needed for the algorithm to converge and, thus, improve its performance. The mesh is divided into 3150 quadrilateral finite elements (0.02 m * 0.02 m), but only half of the respective pseudo-density values are inserted as design variables in the optimization solver. The total structure is generated as a reflection of them along the middle transversal axis of the design domain (Fig. 3). Additionally, the constraints are evaluated only on specific critical nodes and elements each time. Particularly, the displacement is evaluated only on the upper middle node, whereas the manufacturing and stress constraints are evaluated locally in each finite element of the half-domain to ensure continuity and efficiently avoid local peak values. Compliance is evaluated as a global constraint for the total structure.

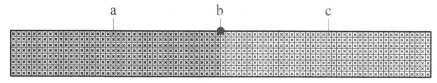

Fig. 3. Design domain Ω_{des} with (a) finite elements related to design variables and evaluation of manufacturing and principal stresses constraints, (b) critical node for displacement constraint, (c) symmetric domain.

3.3 Results

The following examples illustrate the practicality and versatility of the implementation by showcasing how input related to distinct parameters, such as glass composition and design strategies, affects the final outcome (Table 2). Regarding the glass composition, the most prevailing two types are applied: borosilicate and soda-lime glass. They share similar mechanical and structural properties, but they have considerably different thermal properties [6] requiring different annealing durations for the same geometry. Particularly, borosilicate glass has significantly lower thermal expansion coefficient, thus, cooling down approximately three times faster than soda-lime glass (Table 3).

All the examples share the same constraints, though adjusted to reflect the input conditions each time (Table 4). In this light, given that casting is applied, the value of the design tensile strength is compromised comparing to laminated glass because of casting defects and fracture mechanism risks [6]. Regarding the compliance constraint, different fractions are used based on the compliance of the full initial design domain. The rest of the input values related to the optimization are summarized in Table 5.

Table 2. Overview table with input conditions for each example.

Name[a]	Glass composition	Supports	Cross section height (cm)
BR-PN-30	Borosilicate	Point	30
SL-PN-30	Soda-lime	Point	30
BR-FX-30	Borosilicate	Fixed	30
BR-PN-40	Borosilicate	Point	40

[a]The acronyms refer to the glass composition (BR: Borosilicate/SL: Soda-lime); the edge support conditions (PN: Point/FX: Fixed); and the cross section height in cm

Table 3. Input values per glass composition.

	Density (kg/m^3)	Poisson's ratio (−)	Annealing temperature range (K)	Thermal expansion coefficient (1/K)	Thermal conductivity (W/(m*K))	Specific heat capacity (J/(kg*K))
Borosilicate	2500	0.2	70	$3.25*10^{-6}$	1.15	800
Soda-lime	2500	0.2	68	$8.5*10^{-6}$	1.06	870

Table 4. Values used for constraints evaluation.

Design tensile strength[a] (MPa)	Design compressive strength (MPa)	Displacement (m)	Compliance[b] (kNm)	Max annealing time (s)
6.4	500	0.0084 (length/500)	$4\,c_0 = 0.0184$ (h = 0.30 m) $6.5\,c_0 = 0.0182$ (h = 0.40 m)	432000 (5 days)

[a]Calculated based on the German structural design standard in glass constructions (DIN18008), and additional compromises capturing potential casting defects applied according to [6, 30]
[b]The percentages related to the compliance fraction are defined through trial and error for each cross section size. They serve to guide the algorithm faster to feasible solutions [19], but are relaxed to avoid convergence to local minima

Table 5. Input values for optimization setup.

Young's modulus E (GPa)	Young's modulus lower bound E_0 (GPa)	Penalization value p (−)	Stress relaxation value q (−)[a]	Maximum residual stress σ_{res} (MPa)	Shape factor b (−)
70	0.00001	3	2.8	1	0.3

[a]The value used for the q exponent is defined following the approach in [19]

As seen in Fig. 4, the algorithm converges to clear shapes without large grey zones which would be difficult to interpret physically. Therefore, it yields realistic optimal results that have active all the constraints posed in the formulation.

Among the different results, it is evident that the most influential input parameter is the support condition. Hence BR-FX-30 is the only shape variation which differs substantially from the first run BR-PN-30, whereas the rest can be observed as variations of the initial result. Therefore, the main design principles in all optimization runs with point supports stay the same. The resulting shapes consist of a main arc-shape part and a secondary lattice structure at the bottom that increases the total structural stiffness. Additionally, subtle nerves are developed on the top part of the arc to transfer the uniform loads effectively from the top surface to the load bearing structure.

Name	Optimization Result [a]	Volume (m³)	Annealing time (hours : minutes) [b]
BR-PN-30		0,690	35:00
SL-PN-30		0,738	55:20
BR-FX-30		0,419	18:30
BR-PN-40		0,443	23:30
Compact slab (h = 17 cm)		0,821	83:30 (BR) 249:50 (SL)

[a] The resulting shapes are illustrated in a black-white gradient that reflects the existence (black) or absence (white) of material according to the pseudo-density value of each finite element.
[b] The estimated time refers only to the annealing phase proposed by [4] and not to the total cooling time needed.

Fig. 4. Optimization results for different design input variations.

However, there are small adjustments to effectively address the specific input situation in each case. Firstly, in SL-PN-30 the algorithm does not converge to a result with the main arc part as thick as in BR-PN-30, but it details it into a larger number of thinner elements. Therefore, the characteristic cross-sectional dimension of the geometry becomes smaller allowing for annealing time that lies inside the imposed limit, despite the higher thermal expansion coefficient of soda-lime glass.

Additionally, in the case of a larger profile height (BR-PN-40), a clearer formation of the main arc and the respective Y-shaped nerves is achieved. The outcome of this optimization run has the clearest boundary of all counterparts. The only variation that converges to a considerably different result than the initial arc shape is BR-FX-30. In this case, the structure is analyzed into three different parts: two cantilevers[3] on the sides which support a lattice structure placed in the middle part. As earlier, small nerves are created between the different parts to transfer surface loads. Besides the slenderness of the individual elements, the performance of the component lies well inside the limitations related to buckling, which is an important issue for glass articles and can lead to failure.

All the optimization outcomes are considerably more lightweight compared to the reference slab, i.e. the thinnest full material slab that could be applied and still comply with the principal stresses and displacement restrictions (height = 0.17 m). Particularly, the volume reduction achieved through optimization ranges between 10–49% of the reference volume whereas the annealing time needed can be reduced by up to 78%.

---

[3] The geometry of the cantilevers resembles the shape of the classical MBB-Beam problem when similar boundary conditions are imposed [33].

The heaviest outcome comes from the SL-PN-30 variation where, because of the soda lime glass composition, the resulting geometry consists of a larger number of elements compared to BR-PN-30. This is caused by the high thermal expansion coefficient of soda lime glass that renders the thick cross sections not feasible to be annealed in the posed time limit and, thus, they must be analyzed in more elements, which eventually result in increased total structural volume. In contrast, the lighter outcome is related to the BR-FX- 30 variation accounting for almost one half of the volume of the reference slab. Besides the large number of elements in this case, the overall thinner dimensions decrease the overall mass and ensure better performance in terms of annealing time.

In total, this exploration showcases the practicality of the implementation, since the algorithm maintains the design principles that correspond to the optimum result but adjusts the material distribution to respond to the different input conditions. Therefore, it can assist profoundly the design process altering the optimization outcome to meet the specific needs of each space while complying with the set of the posed criteria.

4   Application

4.1   Design Strategy

The optimization outcome from BR-FX-30 variation (Fig. 5) achieves the largest volume and annealing time reduction while at the same time performs efficiently regarding structural performance. Therefore, it is selected to be applied to the slab design.

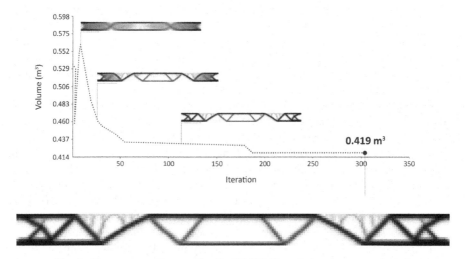

Fig. 5. Convergence diagram for the BR-FX-30 variation outcome selected.

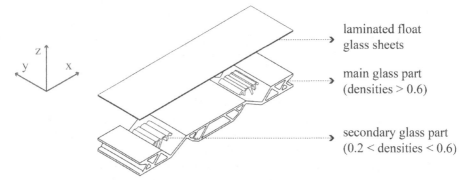

Fig. 6. Design strategy diagram.

To validate the structural performance of the geometry, the planar optimized cross section is translated into a 3-dimensional shape, through its extrusion along the *y* axis. The contribution of each finite element to the total volume is proportional to its contribution to the structural stiffness during the optimization which is reflected to the respective assigned values of the pseudo-densities. Therefore, the elements that correspond to densities above 0.6 are extruded through the whole width, whereas elements with densities between 0.2–0.6 are only extruded through half of it. Lastly, the laminated float glass sheets are applied on the upper surfaces of the components (Fig. 6).

4.2 Structural Evaluation

The performance of the total structure, both the monolithic component and the float glass sheets on top, is evaluated with the use of Ansys software[4]. The factors which are evaluated are the displacement and principal stresses, regarding both tension and compression. The results (Table 6) validate the optimization process since the values of all factors are well inside the allowable limits (Table 4).

Table 6. Results of structural verification with Ansys.

Displacement (m)	Tensile stress (MPa)	Compressive stress (MPa)
0,00012	2,96	5,56

---

[4] Only one of the two monolithic glass components is evaluated structurally in Ansys since the two parts are assumed to perform individually.

4.3 Fabrication

Although permanent steel molds are generally preferred for casting series of high-precision glass elements, the geometrical complexity and customization of the TO geometries renders their use unsuitable in this case[5], pointing towards the direction of using disposable molds as the most promising solution. Particularly, in glass art, large-scale customized castings employ the lost wax technique to produce disposable silica plaster molds which are later used for kiln-casting the components [1]. Yet, this method, is unfavored in our case due to the lengthy, complex, and laborious mold fabrication process [31] as well as compromised dimensional accuracy.

In this light, 3d printed sand molds, which are typically used for metal castings and are recently employed in castings of optimized concrete members [32], are suggested as a solution with large potential for complex glass applications to achieve lower overall cost, higher shape precision and fast fabrication process [2, 3, 8]. Additionally, they are water dissolvable facilitating the unmolding process, while the remaining sand can also be reused. Relevant research by TU Delft [8] already showcases the potential of using 3d printed sand molds made with inorganic binders for kiln glass casting. However, it also highlights the need for further research to refine technical aspects, such as the identification of a coating that allows for a completely transparent surface quality [8], which reduces the need for post-processing and improves the shape precision. Some first promising results in this direction have been recently published by ETH [31], bringing us a step closer to the realization of such complex cast glass structures (Fig. 7).

Fig. 7. Visualization of the final outcome.

---

[5] Although multi-component steel molds can be made for the manufacturing of complex parts, they cannot produce undercuts because the mold must be eventually removed.

5 Conclusions

This paper introduces a new integrated TO formulation combining structural and manufacturing constraints within a unified nonlinear programming statement seeking to enable the design of feasible monolithic large cast glass structures. This can change completely the perception of glass as building material offering unique spatial qualities and enriching profoundly the vocabulary of architectural forms, by introducing aesthetic and structurally sound 3-dimensional glass structures.

The developed formulation is versatile and robust to input alterations. This is showcased through the application of different input conditions regarding the glass composition and the design strategies, which proved that the proposed setting results into robust solutions that comply with all the imposed constraints. Considerably different shapes and volume reduction is achieved based on the glass type and the design strategies applied. Volume reduction ranges between 10–49% compared to the optimal full-material cross section, i.e. the thinnest slab ensuring sufficient structural performance. Similarly, annealing time is reduced up to 78% compared to the reference optimal slab, ultimately rendering the structures more feasible to manufacture.

Overall, this study highlights the potential of using TO as a practical tool in the early design phase leading to better performing and non-intuitive architectural solutions. This diminishes the need for post-processing, shortening the design cycle and allowing for better interconnection between the different specialists involved in the building industry. Future research may focus on incorporating to the formulation practical fabrication-related limitations, such as minimum void dimension to ensure sufficient mold stiffness and integrating additional aspects, such as evaluation of the displacement of the upper laminated float glass sheets or evaluation of second-order structural effects. Moreover, additional design criteria can be considered, such as the establishment of areas in the geometry where the complexity of the form is restricted to allow for increased transparency through minimizing the visual distortions. Finally, to evaluate and improve the accuracy of the developed algorithm, it is important that the numerical results are coupled with mechanical tests on corresponding prototypes. The proposed implementation can be further expanded to other brittle materials, such as unreinforced concrete, and other fabrication methods, such as 3d printing.

Acknowledgements. The authors would like to thank ir. Hans Hoogenboom (Digital Technologies section) at the VR-lab at TU Delft Faculty of Architecture & the Built Environment and Aytac Balci (@Hok Student ICT Support) for offering the facilities and support for the computational needs of this research. Dr. Andriotis would further like to acknowledge the support by the TU Delft AI Labs program.

References

1. Oikonomopoulou, F., Bristogianni, T., Barou, L., Veer, F., Nijsse, R.: The potential of cast glass in structural applications. Lessons learned from large-scale castings and state-of-the art load-bearing cast glass in architecture. J. Build. Eng. **20**, 213–234 (2018)

2. Oikonomopoulou, F., Koniari, A., Damen, W., Koopman, D., Stefanaki, I., Bristogianni, T.: Topologically optimized structural glass megaliths: potential, challenges and guidelines for stretching the mass limits of structural cast glass. In: 8th Eighth International Conference on Structural Engineering, Mechanics and Computation (2022)

3. Damen, W., Oikonomopoulou, F., Bristogianni, T., Turrin, M.: Topologically optimized cast glass: a new design approach for loadbearing monolithic glass components of reduced annealing time. Glass Struct. Eng. (2022)

4. Shand, E., Armistead, W.: Glass Engineering Handbook, New York (1958)

5. Schober, H., Schneider, J., Justiz, S., Gugeler, J., Paech, C., Balz, M.: Innovations with Glass, Steel and Cables. Tampere, Finland (2007)

6. Oikonomopoulou, F.: Unveiling the third dimension of glass. Solid cast glass components and assemblies for structural applications (2019)

7. Zirker, J.: An Acre of Glass: A History and Forecast of the Telescope. JHU Press (2005)

8. Oikonomopoulou, F., Bhatia, I., van der Weijst, F., Damen, W., Bristogianni, T.: Rethinking the cast glass Mould. An exploration on novel techniques for generating complex and customized geometries. In: Challenging Glass 7 Conference on Architectural and Structural Applications of Glass (2020)

9. Langelaar, M.: Topology optimization of 3D self-supporting structures for additive manufacturing. Addit. Manuf. **12**, 60–70 (2016)

10. Luo, Y., Sigmund, O., Li, Q., Liu, S.: Additive manufacturing oriented topology optimization of structures with self-supported enclosed voids. Comput. Methods Appl. Mech. Eng. **372** (2020)

11. Duysinx, P., Sigmund, O.: New developments in handling stress constraints in optimal material distributions. In: 7th AIAA/USAF/NASA/ISSMO Symposium on Multidisciplinary Analysis and Optimization (1998)

12. Le, C., Norato, J., Bruns, T., Ha, C., Tortorelli, D.: Stress-based topology optimization for continua. Struct. Multidiscip. Optim. **41**, 605–620 (2010)

13. Paris, J., Navarrina, F., Colominas, I., Casteleiro, M.: Topology optimization of continuum structures with local and global stress constraints. Struct. Multidiscip. Optim. **39**, 419–437 (2009)

14. Yang, R., Chen, C.: Stress-based topology optimization techniques. Struct. Optimiz. **12**, 98–105 (1996)

15. Duysinx, P., Bendsoe, M.: Topology optimization of continuum structures with local stress constraints. Int. J. Numer. Meth. Eng. **43**, 1453–1458 (1998)

16. Giraldo-Londoño, O., Paulino, G.: A unified approach for topology optimization with local stress constraints considering various failure criteria: von Mises, Drucker–Prager, Tresca, Mohr–Coulomb, Bresler–Pister and Willam–Warnke. Proc. Roy. Soc. A **476** (2020)

17. Senhora, F.V., Giraldo-Londoño, O., Menezes, I.F.M., Paulino, G.H.: Topology optimization with local stress constraints: a stress aggregation-free approach. Struct. Multidiscip. Optim. **62**(4), 1639–1668 (2020). https://doi.org/10.1007/s00158-020-02573-9

18. Bruggi, M.: On an alternative approach to stress constraints relaxation in topology optimization. Struct. Multidiscip. Optim. **36**, 125–141 (2008)

19. Bruggi, M., Duysinx, P.: Topology optimization for minimum weight with compliance and stress constraints. Struct. Multidiscip. Optim. **46**, 369–384 (2012)

20. Luo, Y., Kang, Z.: Topology optimization of continuum structures with Drucker-Prager yield stress constraints. Comput. Struct. **90–91**, 65–75 (2012)

21. Bruggi, M., Duysinx, P.: A stress–based approach to the optimal design of structures with unilateral behavior of material or supports. Struct. Multidiscip. Optim. **48**, 311–326 (2013)

22. Bendsoe, M.: Optimal shape design as a material distribution problem. Struct. Optim. **1**, 193–202 (1989)

23. Sigmund, O., Petersson, J.: Numerical instabilities in topology optimization: a survey on procedures dealing with checkerboards, mesh-dependencies and local minima. Struct. Optim. **16**, 68–75 (1998)

24. Bendsøe, M.P., Sigmund, O.: Topology Optimisation. Theory, Methods and Applications. Springer, Heidelberg (2004). https://doi.org/10.1007/978-3-662-05086-6

25. Sigmund, O.: A 99 line topology optimization code written in Matlab. Struct. Multidiscip. Optim. **21**(2), 120–127 (2001). https://doi.org/10.1007/s001580050176

26. Guest, J.: Imposing maximum length scale in topology optimization. Struct. Multidiscip. Optim. **37**, 463–473 (2009)

27. Cheng, G., Guo, X.: E-relaxed approach in structural topology optimization. Struct. Optim. **13**, 258–266 (1997)

28. Koniari, A.M.: Just Glass. Development of a Topology Optimization Algorithm for a Mass-Optimized Cast Glass Component. Delft University of Technology (2022)

29. Stefanaki, I.M.: Glass Giants. Mass-Optimized Massive Cast Glass Slab. Delft University of Technology (2020)

30. Bristogianni, T., Oikonomopoulou, F., Yu, R., Veer, F., Nijsse, R.: Exploratory study on the fracture resistance of cast glass. Int. J. Struct. Glass Adv. Mater. Res. **5** (2021)

31. Giesecke, R., Dillenburger, B.: Three-dimensionally (3D) printed sand molds for custom glass parts. Glass Struct. Eng. **7**, 231–251 (2022)

32. Jipa, A., Bernhard, M., Meibodi, M., Dillenburger, B.: 3D-printed stay-in-place formwork for topologically optimized concrete slabs. In: TxA Emerging Design + Technology Conference, San Antonio, Texas, USA (2016)

33. Liu, K., Tovar, A.: An efficient 3D topology optimization code written in Matlab. Struct. Multidiscip. Optim. **50**, 1175–1196 (2014)

Spatial Information, Data and Semantics

A Semantic Spatial Policy Model to Automatically Calculate Allowable Gross Floor Areas in Singapore

Ayda Grisiute[1,3](✉) , Heidi Silvennoinen[1], Shiying Li[1], Arkadiusz Chadzynski[2] ,
Martin Raubal[3] , Markus Kraft[2,4,5,6] , Aurel von Richthofen[7] ,
and Pieter Herthogs[1]

[1] Singapore-ETH Centre, Future Cities Lab Global Programme, CREATE Campus, 1 CREATE
Way, #06-01 CREATE Tower, Singapore 138602, Singapore
aydagris@gmail.com

[2] CARES, Cambridge Centre for Advanced Research and Education in Singapore, Singapore,
Singapore

[3] Institute of Cartography and Geoinformation, ETH Zurich, Zurich, Switzerland

[4] Department of Chemical Engineering and Biotechnology, University of Cambridge, Philippa
Fawcett Drive, Cambridge CB3 0AS, UK

[5] School of Chemical and Biomedical Engineering, Nanyang Technological University, 62
Nanyang Drive, Singapore 637459, Singapore

[6] The Alan Turing Institute, London, UK

[7] Arup, Berlin, Germany

Abstract. Urban data analytics is helping to shape current and future cities, but
the process of generating urban analytical indicators is often difficult to scale and
automate. For instance, planners determine allowable Gross Floor Area (GFA) on
a plot by manually cross-referencing multi-domain policies. As allowable GFA
governs potential future developments, it is imperative to quantify and understand
its values city-wide.

This paper presents the first steps of a research effort to develop an automated
semantic spatial policy model to estimate allowable GFA for plots in Singapore.
We use ontologies and Knowledge Graph (KG) platforms to address regulatory
data interoperability and automation challenges. We filtered regulation concepts
that determine buildable area and volume at Level of Detail 1 (LoD1) and standardised these concepts across different regulatory sources. Then, we modelled
concept-related policies and automated the generation of possible GFA values
per plot. Finally, we developed an ontology to store these values in a dynamic
geospatial KG. Our approach presents two key benefits: 1) a generated dataset of
allowable GFA eliminates the need for manual calculation by field experts, and
2) a graph data structure is ideally suited for unstructured regulatory data, like
planning regulations.

We conclude that semantic spatial policy models improve the interoperability
between multi-domain regulatory data and plan to generate a dataset for the entire
Singapore as well as integrate regulatory data for mixed-use plots.

Keywords: Regulatory Data · Knowledge Graph · Urban Indicators · Ontology ·
Semantic Web · Land Use Planning

© The Author(s), under exclusive license to Springer Nature Switzerland AG 2023
M. Turrin et al. (Eds.): CAAD Futures 2023, CCIS 1819, pp. 455–469, 2023.
https://doi.org/10.1007/978-3-031-37189-9_30

1 Introduction

Urban data analytics rely on varying size, quality and granularity datasets for helping planning professionals to shape current and future cities [1]. Thus, diverse urban datasets are at the core of digitally-enabled city planning. For instance, transparent AI methods for advancing city planning require good-quality domain datasets [2]. Similarly, having a ready-to-use urban indicator dataset may save substantial effort for researchers, aid less computationally-driven urban planners and enable quick comparative urban studies [3]. However, the process of generating such datasets is often difficult to scale and automate due to heterogeneous data formats and inconsistent sources [4, 5]. Furthermore, future-oriented urban analytics require operating in modality as well as reality, i.e. to evaluate both, the existing built environment and its desired vision, typically embodied in spatial planning regulations [6].

However, generating urban indicators from regulatory data can be a complex and repetitive task, as correlating the information across domains requires significant effort [7]. For instance, planners determine allowable Gross Floor Area (GFA) on a plot by manually cross-referencing diverse planning regulations. These often non-standardised and siloed urban planning documents hinder the generation of an allowable GFA dataset and the integration of related regulatory data into urban data analytics [8, 9].

In Singapore, where space for new developments is scarce, land reclamation processes remain slow [10] and climate change effects, like sea level rise, urban heat island effects or the risk for mosquito-borne deseases [11], are evident, analysing the effects of planning regulations on the present and future built environment in an automated way could be particularly relevant for efficient and resilient urban development. Therefore, it is imperative to quantify and understand allowable GFA values city-wide and to automate allowable GFA indicator calculation at scale.

This entails the integration of diverse modelling frameworks and methods to effectively process, exchange and structure spatial planning regulations and simulate the allowed built form [12]. Semantic Web Technologies (SWT) offer promising approaches to improve interoperability in the urban planning domain by formalising the terms of interest with ontologies [13], while Knowledge Graphs (KGs) address inefficient data exchange and automation challenges [14]. The Cities Knowledge Graph (CKG) project, to which the presented research belongs, applies SWTs and KGs to address the aforementioned needs [13]. The CKG is also part of a broader, dynamic knowledge graph called The World Avatar that consists of multi-domain knowledge representation and an ecosystem of autonomous software agents operating on it [15].

This paper presents the first steps of a research effort to develop an automated semantic spatial policy model that utilises standardised regulatory data to estimate the maximum buildable space and allowable GFA for plots in Singapore using SWTs. Additionally, we developed the OntoBuildableSpace ontology to formalise allowable GFA and store its values in our developed KG. With our approach, professionals could superimpose regulatory outcomes for building plots with the existing building stock and perform comparative analyses over large built environments [4] or analyse which land uses are more affected by the climate change effects, amongst other applications. Additionally, using SWTs for estimating such urban indicators could potentially supplement natural language-based planning and building regulations, reducing uncertainty in urban development outcomes [8].

The remainder of this paper consists of four sections. The Background Sect. 2 presents the broader research topics. It discusses a variety of known urban indicators, existing policy models and potential improvements when used with SWTs, and introduces the allowable GFA urban indicator with the related regulatory data that govern its value in Singapore. The Methodology Sect. 3 describes the methods for standardising and filtering planning regulations. The section provides details on the development of a semantic spatial policy model to generate a dataset with allowable GFA values per plot and describes an ontology created to formalise GFA values and store them in a KG. The Results and Discussion Sect. 4 presents the generated GFA dataset, the benefits of our automated urban indicator generation approach, as well as encountered limitations. Finally, the Conclusions and Future Work Sect. 5 summarises the outcomes and lists directions for future work.

2 Background

2.1 Large Scale Datasets of Urban Indicators

City development is guided by existing planning regulations [10], various city performance targets [1], remote sensing [16] or forecast predictions [17]. Therefore, it is evident that urban analytics need to be able to operate across modalities to build a rich understanding of cities across temporal scales [6]. For instance, there exists an abundance of available datasets with urban indicators for the built-form properties that mainly present the current state of the city [3]. Different stakeholders, like planning authorities [18], research institutions [19] or private companies [20], produce such datasets. However, planning regulations embodying a city vision is an example of a city's modality—a state of affairs that does not exist or may never exist but is prescribed by a plan or specification [21]. Thus, data analytics based on modality require different kinds of datasets, like urban simulation outputs or datasets based on planning regulations [6]. While machine-readable urban simulation models create an abundance of data that can support urban indicator development, they typically answer only domain-specific questions and remain siloed [22]. On the contrary, it is particularly challenging to retrieve modal urban indicators based on planning regulations, e.g. site coverage or GFA, because regulations rarely exist in machine-readable formats and require high-level interpreting capabilities [5, 12]. Thus, quantifying and evaluating the envisioned future city remains challenging.

Furthermore, inconsistent terminology, granularity and relations across urban indicators hinder the development of comparative studies [5]. For instance, it could be questioned whether the building volume, commonly computed by multiplying footprint area by height, should be considered as a distinct urban indicator or simply a derivative [3]. Thus, the current lack of comprehensive systematisation and comparability of urban indicators and urban form characteristics could generally benefit from formal frameworks [5].

2.2 Semantic Spatial Policy Models

A spatial policy model is one way to generate regulatory (modal) urban indicators and to support city planning across modalities by enabling predictable and controllable regulation outcomes [4]. The following spatial policy models serve as appropriate example cases for our work, either in terms of the methods and tools they use or in terms of the goals they set [23]. These examples focus on rule-based urban form generation, improving regulatory data interoperability or analysing urban indicators.

Spatial policy models have been applied in code compliance checking systems [24, 25]. At the urban scale, [4] designed an automated spatial policy model focusing on allowable building density in Switzerland for comparing the current and the envisioned built environments. A large-scale spatial policy model *'LandBook'* for Seoul contains synthesised building regulations, allowing the basic generation of the built form [26]. Zhang and Schnabel [8] integrated urban regulatory data and parametric modelling primarily focusing on form generation. However, the planning rules in the mentioned examples are often hardcoded in the model, limiting the transparency of such models and hindering the integration of regulation changes over time. Meanwhile, semantic spatial policy models supported by formal vocabularies, i.e. ontologies, can address the aforementioned issue by improving human and machine readability.

An interface of urban regulatory data and SWT is an example of a Semantic City Planning System (SCPS)—an umbrella term to refer to the integrations of city planning and SWT, where different SWT technologies support different actions performed in planning practices [13]. For instance, shape grammars, which essentially encode the desired design language in a rule-based way, have been extensively coupled with ontologies in order to provide contextual information for rule selection processes [27, 28]. Iwaniak et al. suggest HTML semantic annotations for heterogeneous spatial planning documents in Poland, enabling efficient web scraping of regulatory data [9]. In the Singapore context, Silvennoinen [29] engineered an ontology that represents land use regulatory data, enabling site search for allowed land uses (including mixed-use).

However, it is necessary to align and fuse separate aspects from existing examples to meaningfully integrate long-term urban planning practices and urban analytics [6]. Therefore, we propose a semantic spatial policy model that synthesises formalised regulatory data and simulates allowable buildable space, enabling the generation and analysis of modal urban indicators.

2.3 Regulatory Urban Indicators: Allowable Gross Floor Area for Plots in Singapore

Gross Floor Area (GFA) is a universal planning concept with slightly varying definitions across countries and is used to control efficient infill practices and sustainable land management [30]. In essence, GFA is the total floor area contained within a building's outer edges that excludes empty spaces, like atriums or stairwells [31]. In Singapore, authorities define GFA as 'the total area of covered floor space measured between the centre line of party walls, including the thickness of external walls but excluding voids' [32].

Urban data analytics can be performed on existing GFA values using available datasets for building footprints [3]. However, performing urban analyses in modality requires datasets with allowed GFAs based on planning regulations. Such a dataset currently does not exist for Singapore. Planning stakeholders typically process regulatory data for individual plot developments instead of evaluating allowed GFAs at scale.

The allowed GFA is linked to another urban indicator–Gross Plot Ratio (GPR). The GPR controls city density by governing the amount of GFA that could be built on a building plot, relative to the size [32]. We focus on allowable GFA as a more detailed regulatory unit—plots may have multiple allowable GFAs for different land uses or programmes but only one GPR (plots in Singapore sometimes miss the latter).

At a low level of detail, GFA calculation requires knowing the allowed floor area for every storey. Floor area is defined by applying required setbacks and allowed site coverage to the plot, while the number of storeys is retrieved directly from the regulatory data, estimated from the total allowed building height or from the existing context. Two types of planning regulations, affecting allowed GFA, floor area and development height, exist in Singapore:

- Planning regulations that depend on the area. These regulations are associated with a particular spatial boundary. Street Block Plans, Urban Design Guidelines, Height Control Plans and Landed Housing Areas fall in this category.
- Planning regulations based on the development type. These regulations are associated with particular land use types or other development properties. Development Control Plans fall in this category.

The described planning regulations are heterogeneous in nature. They are distributed across different sources and formats and can be accessed via Singapore's planning authority's websites [18, 32]. For instance, Development Control Plans are primarily accessible as online text. Street Block Plan information is stored in separate PDF documents for every street block plan and its boundaries can be downloaded as a GIS layer. Height Control Plans and Landed Housing Areas are available online as GIS layers, while Urban Design Guidelines are stored as PDF maps with supporting information accessible as online text. Furthermore, observed changes towards a more visual planning regulation representation pose additional challenges for machine readability.

3 Methodology

3.1 Aims and Scope of Our Methodology

Planning regulations are typically formulated in ways that are easier to interpret for humans than for machines. Therefore, as the first step in our methodology, we explicitly defined the scope of our semantic spatial policy model for geographic location, ontology development and the level of detail of our algorithmic implementation.

Location. As a proof-of-concept, we applied our automation approach to the Singapore River Valley Area (a sub-part of Singapore), as the overlapping planning regulations provide a sufficient overview of potential challenges while limiting the manual work for regulatory data digitalisation.

Ontology Development. We developed an ontology that specifically focuses on supporting allowed GFA calculation and instantiation in our KG. This entailed systematic digitalisation of regulatory data to address datatype complexities and the introduction of a uniform vocabulary with key planning concepts. While the ontology could formalise more known indicators, we assume that estimating urban indicators from other domains may require the integration of additional concepts and taking into account new regulatory data types and governance principles, which is out of scope for the present paper.

Level of Detail. Additionally, we restricted our GFA calculations to regulation concepts that determine buildable area and volume (both characteristics contributing to GFA calculation) at Level of Detail (LoD) 1. We used LoD as means to introduce a meta-structure for regulation modelling, also enabling compatibility with existing international standards for city data exchange, such as CityGML [33]. For example, regulations for setbacks or absolute height affect the buildable space massing at LoD 1, while regulations for e.g. site layout, daylighting, or placement of ancillary structures govern aspects of building design at a higher LoD. Hence, it is important to emphasise that our calculation method results in estimations of allowed GFA values, which would almost certainly differ when taking higher LoD regulations into account as well.

3.2 Standardising and Filtering Planning Regulations

We examined publicly available planning regulations and harmonised them by introducing consistent concept names, semantic structure and uniform data formats across regulations (Fig. 1). Such encoding of regulatory data was imperative to address common challenges in urban data integration, like syntactic and semantic discrepancies due to spatial, temporal, and thematic diversities [7]. The standardisation process helped achieve a higher degree of consistency [5] and ensured data interoperability.

First, we established consistent names for occurring planning concepts across relevant regulations and grouped them into distinctive typological groups. For example, absolute height, number of storeys or floor-to-floor height define different types of height. Second, we filtered out concepts that do not determine buildable space and volume at LoD 1. Third, we digitised and synthesised regulatory data in order to align planning regulation formats. This resulted in two regulatory data representations: GIS layers with attributes for area-based regulations and spreadsheets for development type-based regulations. It is noteworthy that this data transformation process was necessary to understand regulatory data from a computational standpoint and the resulting formats are only an intermediate step in formalising planning regulations (see Sect. 3.1). This grouping process enabled us to better understand regulatory data impact across different LoDs. For instance, Development Control Plans govern concepts across several LODs while Height Control Plans mainly govern building height at LoD 1.

Planning regulations that depend on development types were transformed into spreadsheets. More precisely, we synthesised Development Control Plan online texts for every zoning type into rows containing all selected planning concepts as columns

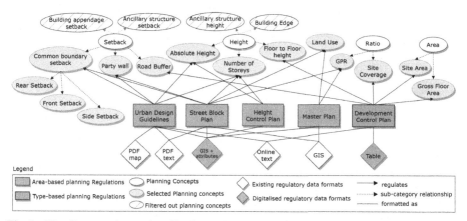

Fig. 1. Visualises regulatory data filtering and standardisation: it maps the grouping of planning concepts in higher degree typological groups, the indication of sub-concepts that relate to LoD 1, the mapping of links between applicable planning regulations and LoD 1 concepts and the resulting data formats after digitising planning regulations.

and their applicable rules as values (Fig. 2). Regulatory exceptions, related to particular land uses or contexts were formalised as additional rows in the table. More detailed requirements are particularly common among residential and mixed residential zoning types, which regulate the specific kinds of residential developments (e.g. bungalows, flats, condominiums) and their allowed GFAs based on a wide range of criteria, such as the width and depth of the plot or its location.

For planning regulations that depend on the area, we linked concept-related regulations with corresponding GIS layers. For instance, we manually linked the publicly available GIS layer with Street Block Plan boundaries and regulatory PDF documents by adding related regulations for selected concepts as attributes for every boundary. Height Control Plans, already represented as a GIS layer with an attribute for allowed heights, only required attribute re-naming. We transformed Urban Design Guidelines, represented as map drawings stored in PDF format, into GIS layers using the Feature Manipulation Engine (FME), and removed the remaining graphical artefacts manually. Colour or line-coded rules in the map were expressed as attributes and linked to resulting geometries.

3.3 Generating the Allowable GFA Dataset

Our implementation to automatically generate an allowable GFA dataset in a form of Python script consisted of three tasks: 1) processing plot and regulatory data, 2) determining the allowed height and required setbacks for every plot, and 3) calculating the allowable GFA. As a base for allowable GFA generation, we used Singapore's Masterplan 2019 dataset, containing plots with two key attributes: assigned zoning type and allowed GPR [34]. The dataset was formally represented and stored in our KG using the OntoCityGML ontology [35].

Zone	Context Exception (Neighbor)	Development Type Exception	Setback (m)	Storeys	FloorToFloor Height (m)	Site Coverage
RESIDENTIAL WITH COMMERCIAL AT 1ST FLOOR			3		3.6;5.0	
COMMERCIAL & RESIDENTIAL			3		5.0; 3.6	
COMMERCIAL			3		5	
HOTEL			3		5	
BUSINESS PARK			4.5			
BUSINESS PARK - WHITE			4.5		4	
BUSINESS 1		WorkersDormitory	3		4	
BUSINESS 1			4.5		4	
BUSINESS 1	BUSINESS 1; BUSINESS 2; PLACE OF WORSHIP		0		4	
BUSINESS 2		WorkersDormitory	3		4	
BUSINESS 2			4.5		4	
BUSINESS 2	BUSINESS 1; BUSINESS 2; PLACE OF WORSHIP		0		4	
BUSINESS 1 - WHITE		WorkersDormitory	4.5		4	
BUSINESS 1 - WHITE			3		4	
BUSINESS 1 - WHITE	BUSINESS 1; BUSINESS 2; PLACE OF WORSHIP		0		4	
BUSINESS 2 - WHITE		WorkersDormitory	3		3.6	
BUSINESS 2 - WHITE			4.5		4	
BUSINESS 2 - WHITE	BUSINESS 1; BUSINESS 2; PLACE OF WORSHIP		0		4	
RESIDENTIAL / INSTITUTION COMMERCIAL / INSTITUTION						
HEALTH & MEDICAL CARE		PrivateMedicalClinic	4.5		5	
HEALTH & MEDICAL CARE			4.5		5	
EDUCATIONAL INSTITUTION		SpecialEducationAmbulatorySchool	4.5	4	5	0.5
EDUCATIONAL INSTITUTION		SpecialEducationNonAmbulatorySchool	4.5	2	5	0.5
EDUCATIONAL INSTITUTION		StudentHostel	4.5		5	
EDUCATIONal INSTITUTION			4.5		5	
PLACE OF WORSHIP	BUSINESS 1; BUSINESS 2		0		5	
PLACE OF WORSHIP			4.5		5	

Fig. 2. Illustrates an excerpt from a digitised Develoment Control Plan, where colour-coded repeating rows refer to exceptions linked to these zoning types. For instance, the *'BUSINESS 1'* zoning type has three rows formalising related planning regulations because it has a type-based exception for *'WorkersDormitory'* and context-based exceptions for plots that are adjacent to *'Business 1'*, *'Business 2'* and *'Place of Worship'*. Columns on the right refer to applicable regulatory rules for selected concepts.

Determining the Allowed Number of Storeys. For every plot, we iterated applicable regulations, checked the standardised attribute called *'Storeys'* and selected the lowest (most conservative) value. In cases where planning regulations allow a different number of storeys for specific plot parts, which can be the case in e.g. Urban Design Guidelines, the script divided the plot into individual parts, and iterated individual parts selecting the lowest allowed height value for each part.

Determining Required Setbacks. The script checked *'Setback'* attribute values across applicable urban regulations, as well as linked road buffers based on plot adjacency to roads, and assigned the most conservative (highest) setback value for every plot edge. Additionally, our script checked the *'PartyWall'* attribute value to determine whether a plot requires a party wall development. For those plots, edges that are adjacent to another party wall plot were assigned a setback of zero. Furthermore, as some setback regulations vary for different storeys (floor levels), our script assigned setbacks per edge for every allowed floor level.

Allowed GFA Calculation. From the assigned number of storeys for every plot or plot part and the setbacks for every plot edge, our script generated a set of footprints at every storey. Site coverage values were calculated for each plot from generated footprints and plot areas and checked against allowed *'SiteCoverage'* attribute values. For plots exceeding allowable site coverage, the footprint areas were reduced to meet site coverage requirements. Resulting footprint areas were summed into an aggregate allowed GFA value (estimate). Similarly, the script derived GPR values from our GFA estimates and checked it against allowed 'GPR' attribute values. For plots exceeding the allowed GPR, the generated GFA value was reset by multiplying the plot area by the applicable GPR. The resulting allowable GFA values for every plot were written into a JSON file, an established format for data exchange. Thus, the generated dataset can be used with existing databases and planning tools; however, these files do not store (our) data semantics.

Exclusions. As this paper describes the first steps of our research effort to develop a semantic spatial policy model, we did not estimate allowed GFAs exhaustively for all plots in the Singapore River Valley Area. Particularly, we excluded plots in conservation areas, as we only model regulations for new developments, which are unlikely in conservation areas. Additionally, plots with non-specific regulations, like *'Subject to Detailed Control'* or plots with zoning types that did not have publicly available planning regulations, like *'Mass Rapid Transit'*, *'Port/Airport'*, and *'Transport Facilities'* were also excluded from our spatial policy model, as they require manual assessment or follow different governing principles. Similarly, plots with zoning types like *'White'* or *'Community OR Institution'* were also omitted for now, because they allow a combination of different land uses and programs, potentially resulting in a myriad of combinatorial regulation options and allowable GFA instances.

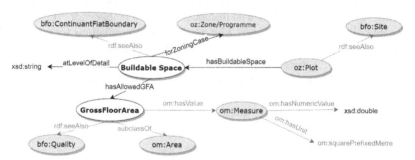

Fig. 3. Shows an excerpt of our OntoBuildableSpace ontology used to represent allowed GFA (represented in black). Blue nodes and arrows refer to imported ontology classes and relationships (from OntoZoning, the Units of Measures ontology and Basic Formal Ontology). With Gross-FloorArea defined as a sub-class of Area (OM ontology), we could formalise GFA values in OM terms, i.e. assigning a unit and a numeric value.

3.4 OntoBuildableSpace Ontology for Allowable GFA Instantiation

We developed the OntoBuildableSpace ontology to formalise measurable characteristics of buildable space on a plot, with a particular focus on allowed GFA, in order to store modal data in our dynamic geospatial KG (see Sect. 3.1). In this ontology, the BuildableSpace concept is defined as a 3D volume allowed by planning regulations – a type of modal space that has not been previously defined in existing space classification literature within spatial sciences [36].

The OntoBuidableSpace ontology makes use of the existing OntoZoning ontology [29], which explicitly formalises land-use planning concepts and regulations in Singapore, and the Units of Measure ontology (OM) [37], which represents the different concepts used to define measures and units. Moreover, OntoBuildableSpace ontology concepts were linked to Basic Formal Ontology (BFO)—an upper-level ontology—in order to enable interoperability with other scientific research ontologies [38]. Although OntoBuildableSpace could be linked to other built-form standards, we decided to adopt BFO as we anticipate its domain-agnostic nature will help extend the ontology with other concepts and processes related to city planning, from other (regulatory) domains, which aligns with the scope of our methodology (see Sect. 3.1).

In this paper, we instantiated allowable GFA values using a sub-set of concepts from the OntoBuildableSpace ontology (Fig. 3). The ontology represents allowed GFA by linking *Plot* and *BuildableSpace* classes via the *allowsBuildableSpace* predicate, *BuildableSpace* with *GrossFloorArea* via the *hasAllowedGFA* predicate, and defining *GrossFloorArea* in terms of OM. Additionally, *BuildableSpace* was linked to Onto-Zoning *ZoningType* or *Programme* classes via the *forZoningCase* predicate to express allowed GFAs for land use-specific regulation exceptions. For instance, plots with the *Educational Institution* zoning type have regulation exceptions for student hostels. Thus, a plot with this zoning type would have an additional *BuildableSpace* instance with an allowed GFA value for the OntoZoning programme *StudentHostel*. Similarly, plots with a residential zoning type would have an instance of *BuildableSpace* for every residential type possible on that plot, e.g. *Condo, Flat,* or *Terrace* as regulations for each residential development type vary. Finally, OntoBuildableSpace concepts were linked to BFO concepts via the RDF schema *seeAlso* relationship. We used the OntoBuildableSpace as a schema for a Python script that transformed generated allowable GFA values from the JSON file into semantic triples (two nodes connected with a relation) and uploaded them to our KG via HTTP POST request.

4   Results and Discussion

In this paper, we presented an SWT-driven approach to comprehensively standardise urban regulation concepts and to develop an automated semantic spatial policy model to estimate allowable GFAs for plots in Singapore. Allowable GFA quantities previously were not measurable at scale due to interoperability issues and fragmentation. For the first time, a subset of Singapore's urban planning policy interconnections can be measured at scale (Fig. 4). For instance, we synthesised and digitised planning regulations in order to calculate allowable buildable space and volume, to then retrieve allowable GFA value estimates for 6590 plots in the Singapore River Valley Area, resulting in 3869 allowed GFA values linked to 1481 plots (22.5%). The remaining 5109 plots have been excluded from the allowable GFA calculation as described in Sect. 3.3. Thus, the allowable GFA dataset works as a concrete example of how Singapore's planning policies can be quantified and studied using urban analytics.

Our methodology to automatically generate modal urban indicators at scale introduces several potential benefits for future city governance and planning. Aside from manually digitising planning regulation inputs, our allowed GFA calculation is fully automated, eliminating the need for manual calculation by domain experts. This enables efficient city-wide urban analytics and can support planners to better understand the maximum allowable buildable areas per land use type (combination) spatially (Fig. 4a), which is arguably imperative for a city's present and future development. For instance, planners could superimpose the allowed GFAs on existing building stock data to detect unused development potential or could assess and explore plots with novel combinations of particular amounts of land uses and programmes.

A general correlation between GFA values and plot size, except for zoning types *'Commercial & Residential'*, *'Hotel'* and *'Commercial'*, is apparent in Fig. 4b where GFAs increase with plot size. The mentioned exceptions may stem from the unique Singapore-specific governing rules allowing higher GFAs in these zones. Hence, replacing solely *'Residential'* plot zoning type with a mixed-use *'Commercial & Residential'* zoning type may yield significantly larger available GFAs in the future for mitigating Singapore's land scarcity. It is worth noting that the implemented model overestimates the allowable GFA values for larger plots (more), because we do not model regulations that govern building depth to ensure sufficient daylight.

Finally, more than half of the plots (58.2%) have multiple allowable GFAs linked to them. This offers a more differentiated view of allowed buildable spaces compared to the estimates using a plot's GPR only (Fig. 4c). The increased differentiation could inform planners about novel and viable mixed-use combinations, as performance and efficiency assessments are difficult to carry out due to their combinatorial nature and are therefore often ignored. For example, having access to GFA values for every allowed development type on a plot could offer more detailed insights into the potential impact of land use changes on city infrastructure, like its energy grid.

Because of the regulatory data digitisation and semantic standardisation required for the presented spatial policy model, it is now feasible to perform systematic analyses on a subset of Singapore's planning policies and to check for any contradictions or discrepancies. For instance, as future work, semantic reasoning engines could be leveraged to ensure that proposed policy changes can be assessed for potential contradictions. This could ultimately enhance the human-machine readability and improve the management of urban regulatory data, thus increasing the transparency of planning decision-making. In addition, this work illustrates that Singapore's regulatory data contains a broad range of land use and context-based exceptions (which we assume is common in land use regulatory systems world-wide). The malleable data structure of a KG is ideally suited for the integration of such exceptions, and can thus support incremental changes to the policy, enabling iterative and hypothetical manipulation and analysis of potential planning policies at scale. For example, changes in terms of floor area gains or urban density increase could be explored easily by changing particular height or setback rules.

Despite the presented benefits, the current state of our spatial policy model and generated allowed GFA has limitations. Without standardisation efforts for diverse regulatory data formats in use, digitising regulatory data for our proposed model will require manual work to input and clean this data. In addition, the calculation, interpretation, and representation of planning regulations in our semantic spatial policy model should be

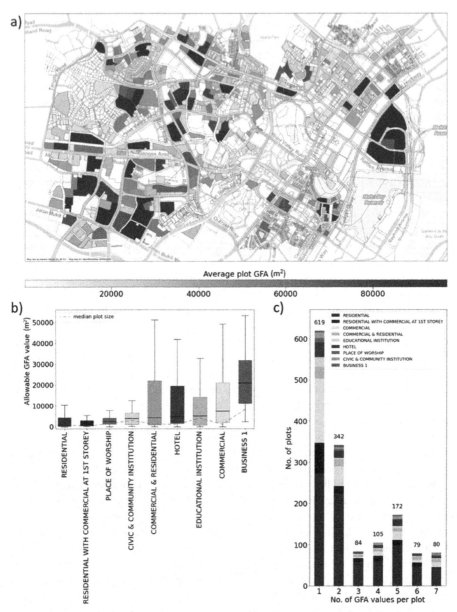

Fig. 4. Describes the main generated dataset characteristics. a) A gradient map showing the spatial distribution of average allowable GFA values for plots in the Singapore River Valley Area. b) The distribution of allowable GFA values and median plot size for individual zoning types c) Stacked bar chart illustrates how many plots have a certain number of allowable GFA values.

validated by government planning actors. While we manually verified some of the generated GFAs as a form of spot check, validating all GFA values across the chosen area is outside of this paper's scope. Finally, our semantic spatial policy model is a starting point for a more comprehensive implementation, and we are yet to include planning regulations for certain mixed-use zoning types.

5 Conclusions and Future Work

The paper introduced an SWT-driven approach to generate modal urban indicators at scale with an automated semantic spatial policy model and the allowable GFA dataset for plots in Singapore as a concrete example. We conclude that the presented approach can support city planning by improving the interoperability of siloed regulatory data, enabling urban indicator calculation at scale and urban analytics across modalities. However, to apply such automation processes to cities worldwide for a variety of urban indicators, one would need to balance planning regulation representation in such a way that they are machine-readable yet comprehensible to urban planning professionals.

We see several directions for future work: 1) we plan to upgrade our allowable GFA calculation to integrate regulations for the multi-use zoning types; 2) we aim to generate a complete GFA dataset for the whole of Singapore; 3) we are formalising a broader range of related planning regulation concepts in ontologies (e.g. allowed number of storeys, floor-to-floor height or each floor footprint area) and instantiating them in our dynamic geospatial KG to increase the range of possible planning applications of our approach; 4) we envision formalising other domain regulatory data using ontologies, in order to enable access to a broader range of modal urban indicators; 5) ultimately, instead of a collection of manually triggered scripts, the GFA calculation functionality could be implemented as an autonomous software agent [15].

References

1. Massaro, E., Athanassiadis, A., Psyllidis, A., Binder, C.R.: Ontology-based integration of urban sustainability indicators. In: Binder, C.R., Massaro, E., Wyss, R. (eds.) Sustainability Assessment of Urban Systems, pp. 332–350. Cambridge University Press, Cambridge (2020). https://doi.org/10.1017/9781108574334
2. Wang, J., Biljecki, F.: Unsupervised machine learning in urban studies: a systematic review of applications. Cities **129** (2022). https://doi.org/10.1016/j.cities.2022.103925
3. Biljecki, F., Chow, Y.S.: Global building morphology indicators. Comput. Environ. Urban Syst. **95** (2022). https://doi.org/10.1016/j.compenvurbsys.2022.101809
4. Walczak, M.: A multi-dimensional spatial policy model for large-scale multi-municipal Swiss contexts. Environ. Plan. B: Urban Anal. City Sci. **48**(9), 2675–2690 (2021). https://doi.org/10.1177/2399808320985854
5. Fleischmann, M., Romice, O., Porta, S.: Measuring urban form: overcoming terminological inconsistencies for a quantitative and comprehensive morphologic analysis of cities. Environ. Plan. B: Urban Anal. City Sci. (2020). https://doi.org/10.1177/2399808320910444
6. Kandt, J., Batty, M.: Smart cities, big data and urban policy: towards urban analytics for the long run. Cities **109** (2021). https://doi.org/10.1016/j.cities.2020.102992
7. Psyllidis, A., Bozzon, A., Bocconi, S., Titos Bolivar, C.: A platform for urban analytics and semantic data integration in city planning. In: Celani, G., Sperling, D.M., Franco, J.M.S. (eds.) CAAD Futures 2015. CCIS, vol. 527, pp. 21–36. Springer, Heidelberg (2015). https://doi.org/10.1007/978-3-662-47386-3_2

8. Zhang, Y., Schnabel, M.A.: A workflow of data integrating and parametric modelling in urban design regulation. In: The 51st International Conference of the ASA (2017)
9. Iwaniak, A., et al.: Semantic metadata for heterogeneous spatial planning documents. ISPRS **IV-4-W1**, 27–36 (2016). https://doi.org/10.5194/isprs-annals-IV-4-W1-27-2016
10. Urban Redevelopment Authority. https://www.ura.gov.sg/Corporate/Planning/Long-Term-Plan-Review/Space-for-Our-Dreams-Exhibition. Accessed 29 Aug 2022
11. National Climate Change Secretariat Singapore. https://www.nccs.gov.sg/singapores-climate-action/impact-of-climate-change-in-singapore/. Accessed 19 Feb 2023
12. Guler, D., Yomralioglu, T.: Reviewing the literature on the tripartite cycle containing digital building permit, 3D city modeling, and 3D property ownership. Land Use Policy **121** (2022). https://doi.org/10.1016/j.landusepol.2022.106337
13. von Richthofen, A., Herthogs, P., Kraft, M., Cairns, S.: Semantic city planning systems (SCPS): a literature review. J. Plan. Lit. (2022). https://doi.org/10.1177/08854122211068526
14. Farazi, F., et al.: Knowledge graph approach to combustion chemistry and interoperability. ACS Omega **5**, 18342–18348 (2020). https://doi.org/10.1021/acsomega.0c02055
15. Chadzynski, A., et al.: Semantic 3D city agents - an intelligent automation for dynamic geospatial knowledge graphs. Energy AI **8** (2022). https://doi.org/10.1016/j.egyai.2022.100137
16. Smart Nation Singapore. https://www.smartnation.gov.sg//initiatives/transport/open-data-analytics. Accessed 05 Sept 2022
17. CHAOS. https://chaosarchitects.com/. Accessed 03 Sept 2022
18. GovTech. https://data.gov.sg/about. Accessed 31 Aug 2022
19. Wu, A.N., Biljecki, F.: Roofpedia: automatic mapping of green and solar roofs for an open roofscape registry and evaluation of urban sustainability. Landsc. Urban Plan. **214** (2021). https://doi.org/10.1016/j.landurbplan.2021.104167
20. CITYDATA.AI. https://univercity.ai/mobility-trip-patterns-for-greater-sydney-metropolitan-area/. Accessed 29 Aug 2022
21. Rudnicki, R.: An Overview of the Common Core Ontologies. CUBRC. Inc. (2016)
22. Grisiute, A., et al.: Unlocking urban simulation data with a semantic city planning system: ontologically representing and integrating MATSim output data in a knowledge graph. In: the 40th eCAADe Conference Proceedings, Belgium, vol. 2, pp. 257–267 (2022). https://doi.org/10.52842/conf.ecaade.2022.2.257
23. Schevers, H., Trinidad, G., Drogemuller, R.: Towards integrated assessments for urban development. ITcon **11**, 225–236 (2006)
24. Borrmann, A., König, M., Koch, C., Beetz, J. (eds.): Building Information Modeling. Springer, Cham (2018). https://doi.org/10.1007/978-3-319-92862-3
25. Lee, H., Lee, S., Park, S., Lee, J.-K.: An approach to translate Korea building act into computer-readable form for automated design assessment. In: ISARC, Finland (2015)
26. Spacewalk. https://spacewalk.tech/. Accessed 15 Sept 2022
27. Beirão, J., Duarte, J.P., Stouffs, R.: Structuring a generative model for urban design: linking GIS to shape grammars. In: The 26th eCAADe Conference Proceedings, Belgium, pp. 929–938 (2008). https://doi.org/10.52842/conf.ecaade.2008.929
28. Grobler, F., et al.: Ontologies and shape grammars: communication between knowledge-based and generative systems. In: Gero, J.S., Goel, A.K. (eds.) Design Computing and Cognition 2008, pp. 23–40. Springer, Dordrecht (2008). https://doi.org/10.1007/978-1-4020-8728-8_2
29. Silvennoinen, H., et al.: A semantic web approach to land use regulations in urban planning: the OntoZoning ontology of zones, land uses and programmes for Singapore. J. Urban Manag. (2023). https://doi.org/10.1016/j.jum.2023.02.002. ISSN 2226-5856
30. Morosini, R., Zucaro, F.: Land use and urban sustainability assessment: a 3D-GIS application to a case study in Gozo city. Territ. Archit. **6**(1), 1–20 (2019). https://doi.org/10.1186/s40410-019-0106-z

31. European Commission. https://ec.europa.eu/info/sites/default/files/measuring-code_en.pdf. Accessed 05 Oct 2022

32. Urban Redevelopment Authority. https://www.ura.gov.sg/Corporate/Guidelines/Development-Control/gross-floor-area/GFA/Advisory-Notes. Accessed 29 Aug 2022

33. Noardo, F., et al.: Tools for BIM-GIS integration (IFC georeferencing and conversions): results from the GeoBIM benchmark 2019. ISPRS Int. J. Geo-Inf. **9**(9) (2020). https://doi.org/10.3390/ijgi9090502. Art. no. 9

34. Urban Redevelopment Authority. https://www.ura.gov.sg/Corporate/Planning/Master-Plan. Accessed 05 Oct 2022

35. Chadzynski, A., et al.: Semantic 3D City Database - an enabler for a dynamic geospatial knowledge graph. Energy AI **6** (2021). https://doi.org/10.1016/j.egyai.2021.100106

36. Zlatanova, S., et al.: Spaces in spatial science and urban applications—State of the art review. ISPRS Int. J. Geo-Inf. **9**, 58 (2020). https://doi.org/10.3390/ijgi9010058

37. Rijgersberg, H., van Assem, M., Top, J.: Ontology of units of measure and related concepts. Semant. Web **4**, 3–13 (2013). https://doi.org/10.3233/SW-2012-0069

38. Arp, R., Smith, B., Spear, A.D.: Building Ontologies with Basic Formal Ontology, Cambridge, MA (2015). https://doi.org/10.7551/mitpress/9780262527811.003.0005

Building Information Validation and Reasoning Using Semantic Web Technologies

Diellza Elshani[1]([✉]) [ID], Daniel Hernandez[2] [ID], Alessio Lombardi[3],
Lasath Siriwardena[4] [ID], Tobias Schwinn[4] [ID], Al Fisher[3], Steffen Staab[2,5] [ID],
Achim Menges[4] [ID], and Thomas Wortmann[1] [ID]

[1] Chair for Computing in Architecture, Institute for Computational Design and Construction
(ICD/CA), University of Stuttgart, Keplerstraße 11, 70174 Stuttgart, Germany
diellza.elshani@icd.uni-stuttgart.de
[2] Department for Analytic Computing (AC), Institute for Parallel and Distributed System
(IPVS), University of Stuttgart, Keplerstraße 11, 70174 Stuttgart, Germany
[3] Buro Happold, 17 Newman St, London W1T 1PD, UK
[4] Institute for Computational Design and Construction (ICD), University of Stuttgart,
Keplerstraße 11, 70174 Stuttgart, Germany
[5] Electronics and Computer Science, University of Southampton, University Road,
Southampton SO17 1BJ, UK

Abstract. The integration of data from various disciplines, including requirements and regulations, is essential in the co-design of buildings. Constraints that
arise during design, fabrication, or construction are mainly considered in the later
stages of the design process. This often leads to costly revisions during fabrication
or construction. While there are some works proposing Semantic Web approaches
or ad-hoc methods to identify constraint violations in the early stages of design,
they mainly rely on data derived from monolithic data schemas. This paper presents
a Semantic Web approach that validates and checks building data derived from
federated data schemas. The proposed approach is exemplified through the design
process of a segmented timber shell, a complex structure requiring negotiations
across architectural, engineering, and fabrication parameters. The Building Habitat object Models (BHoM) framework is used to represent federated building data,
and the approach enables a seamless movement between the graph environment
and object models to support the design process. Additionally, the study discusses
various technologies that aid in knowledge inference or data validation. The results
of this study suggest that the use of Semantic Web technologies in conjunction with
federated data schemas has significant potential to enhance co-design processes in
the building industry. While further research is needed to assess its effectiveness in
other scenarios, our application of this approach in checking and validating building data for a complex segmented timber shell structure demonstrates its promise
as a means to streamline the design process.

Keywords: Data Validation · Reasoning · Data Modelling · Semantic Web
Technologies

© The Author(s), under exclusive license to Springer Nature Switzerland AG 2023
M. Turrin et al. (Eds.): CAAD Futures 2023, CCIS 1819, pp. 470–484, 2023.
https://doi.org/10.1007/978-3-031-37189-9_31

1   Introduction

Co-designing buildings requires the consideration of design and analysis methods, manufacturing and construction processes, and material and building systems simultaneously while integrating data from multiple disciplines [1]. This includes not only design data but also requirements and regulations because limitations arising from one discipline might impact design decisions in another. However, limitations arising from design, fabrication, or construction are often only taken into account in the later stages of a linear design process, resulting in costly revisions during construction [2]. A system that integrates data, provides reasoning, and validates models across all disciplines during the design phase would reduce the back-and-forth iterations of data among disciplines. Many model-checking approaches exist in the Architecture Engineering and Construction (AEC) industry, but they mainly rely on data derived from monolithic data schemas. Other methods that focus on design for manufacturing and assembly (DfMA [3]) do not take into account some specific product features and production capabilities [4].

Work presented in [4] on connecting design and manufacturing knowledge using an ontology-based approach demonstrates that Semantic Web Technologies (SWT) provide a solid ground for connecting interdisciplinary knowledge and allow for constraint checking in the AEC industry. SWTs are built on top of a data model, called the Resource Data Framework (RDF), that allows for building knowledge bases in the form of graphs, called knowledge graphs [5]. A knowledge graph is essentially a knowledge base that describes entities and their relationships. A fundamental characteristic of the RDF data model is that entities are identified with global Web identifiers, called IRIs. The use of global identifiers allows connecting and integrating multiple knowledge graphs because global identifiers can be referred from other knowledge graphs, that is, stating links across knowledge graphs. Aside from linking knowledge graphs, SWTs include languages as OWL that are used to enrich the data with semantics in the form of ontologies. This ontological knowledge about the data can be used to reason, infer new facts, and detect inconsistencies. In AEC, this semantics would allow early recognition of design constraints from involved disciplines such as structural or fabrication.

Reasoning and data validation using SWTs have been demonstrated on AEC data derived from monolithic data schemas such as the Industry Foundation Classes (IFC) [6–9], but not on federated data schemas such as The Buildings and Habitats object Model (BHoM) [10]. Compared to federated data schemas, the limitations of monolithic data schemas while designing restrict designers from defining multiple discrete representations of designs [2]. Furthermore, to our knowledge, current converters of building data to Semantic Web standards do not provide a read-back from the graph to design tools or objects. Such a limitation prevents designers from considering inferred knowledge from combined data while designing. BHoM applies a federated data schema interoperability that allows designers to model multiple discrete representations of buildings. BHoM's converter to Semantic Web languages allows not only exporting design data to knowledge graphs but also reading back these files. All this functionality is already integrated into design tools, including Grasshopper 3D.

Therefore, this paper presents how SWTs assist in validating building information and reason on building data during the design phase using BHoM as a federated data schema framework. It uses BHoM RDF Toolkit [11, 16] to convert back and forth between BHoM object models and knowledge graphs. It validates the graph using Semantic Web rule checking and validation technologies. To exemplify the research, the paper uses data of a large timber shell with cassettes as a roof for a large research building which will be fabricated and assembled in a cyber-physical setup. Such a structure and design system is a good example to test the workflow because of its complex data communication arising from the building system and the limitations of the fabrication or construction.

Structure of This Paper. In Sect. 2, we provide an overview of the difficulties encountered in designing complex structures, along with an exposition of SWTs that may potentially address these challenges. Section 3 presents the process of generating a data graph from a timber shell using the BHoM Framework, detailing several feasible approaches for data validation and reasoning, and supporting them with examples. In Sect. 4, we present the results of our research, which includes demonstrations of SWTs for ensuring data completeness and inferring knowledge about the timber shell. Additionally, we show how this methodology can be incorporated into the design process of the timber shell. Finally, Sect. 5 offers our concluding remarks and outlines potential areas for future research.

2   Background

This section outlines the challenges involved in co-designing timber shells with respect to data communication and the application of SWTs for reasoning and data validation in the building industry.

2.1   Challenges of Co-designing Timber Shells in Terms of Data Communication

Co-designing buildings requires negotiating and interacting with multiple disciplines, as well as complying with numerous norms and regulations. The work presented in planar segmented timber shells [12, 13] emphasizes the need to integrate data digitally and understand the relations between different actors and disciplines. Each segment in a segmented timber shell includes interlinked information on design norms and requirements from various disciplines. This information includes geometric and topological information, properties related to structural design, material properties, and fabrication data related to the robotic fabrication of the components, among others [5].

Reflecting on the BUGA Wood Pavilion [13], the authors indicate that to automate the process and reduce manual labor, designers need to understand the standing of each data and parameter in terms of their relations across different actors and different schedules of the process. Furthermore, they argue that mapping the relations and links between data from different domains requires knowledge developed through experience.

Arguably, two key points can be discussed to improve design processes for such structures. First, a design methodology is needed that allows representing building data and the relations between data dynamically. Second, the design methodology must include rules that allow for constraint-checking and data validation during the design process. Additionally, a collaborative framework that allows designing custom objects, including semantics, is needed to define multiple discrete disciplinary data and information for designing such building parts.

2.2 Semantic Web Technologies for Reasoning and Data Validation for AEC

SWTs are a promising approach to dynamically linking and integrating data from heterogeneous sources, including rule and constraint languages [14]. These technologies use knowledge graphs, also known as Knowledge Bases (KBs), to represent data and make it machine-readable. KBs contain both facts and rules. In the Architecture, Engineering, and Construction (AEC) field, facts represent building data, while rules cover regulations and design limitations. Leveraging rules and data constraints in knowledge graphs helps to dynamically infer knowledge while working with the data. There are three main categories of rule systems: first-order, logic-programming, and action rules, and each can be used to describe different AEC design regulations or construction constraints. The SWT stack includes several technologies, such as the Resource Description Framework (RDF), the Web Ontology Language (OWL), the Semantic Web Rule Language (SWRL), Rule Interchange Format (RIF), Shapes Constraint Language (SHACL), and the query language SPARQL. Although some data can be expressed using more than one technology, they are designed for different purposes and complement each other. For instance, SWRL is an extension to OWL.

An approach to reasoning (left) and data validation (right) of RDF graphs is presented in Fig. 1. While reasoning with OWL and Semantic Web rule languages such as SWRL follows an Open World Assumption (OWA), the concept of data completeness and validation is related to the Closed World Assumption (CWA) [14]. In OWA, new information can be added to the graph at any time. Furthermore, by defining rules in a data graph, one can infer new facts and include that information on the graph. Contrastingly, SHACL follows the CWA. According to CWA, a fact is false if it is not present in the database (and cannot be inferred from it).

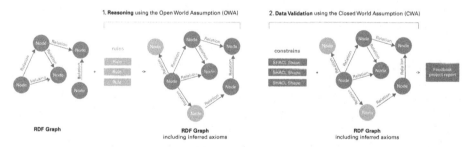

Fig. 1. AEC data validation supported by reasoning methods

SWTs are already being used in the AEC industry, and many AEC ontologies currently support data integration, such as ifcOWL and BOT ontology. Currently, efforts are being made to suggest the use of Linked Data for semantically connected Common Data Environments (CDE) using the Solid ecosystem, which results in decentralized CDEs [15]. These CDEs are significant for modular applications of data integration, not only for construction activities but also for historical data, geographical data, circular economy, etc.

SWTs, specifically ontologies, have been increasingly used to validate various types of data in the building industry. Ontology-based approaches allow for the development of formal and explicit descriptions of domain knowledge, which can help to ensure data consistency, completeness, and accuracy. For example, [4] developed an ontology-based cloud manufacturing framework for industrialized construction. Their methods include semantic reasoning, which combines feature-based modeling, production capability modeling, and manufacturing rules to analyze and validate manufacturability.

Liu et al. [23] propose an ontology-based semantic approach to extracting construction-oriented quantity take-off information from a BIM design model, using a domain vocabulary to capitalize on building product ontology formalized from construction perspectives. Kovacs et al. [17] propose a model for requirements as Linked Data and provide an example for quality checking using SHACL. The proposed system aims to improve the quality and reduce the cost of BIM design methods in the AEC industry.

The aforementioned methods rely on creating data using the IFC schema for validation, which represents data in a monolithic approach. Once the design has been finalized, and no further changes are expected, a centralized data schema that holds all the information about the building is indeed helpful for analytics. However, in this paper, we argue that during a co-design process, such monolithic data representation contrasts with the multi-disciplinary nature of building design. The focus of the research is to enable co-design, exchange data continuously, and generate disciplinary data graphs while designing, with federated schemas similarly with the federated nature of the AEC sector. Therefore, we adopt a non-centralized data schema framework instead of a centralized one. To demonstrate the practicality of these technologies, we utilize BHoM as a federated data representation tool in our case study. BHoM is an open-source collaborative framework that allows customizing object models, permits disciplinary representation of building elements, and has an open and extendable data dictionary [2]. Moreover, BHoM objects are convertible to Semantic Web Standard data models, including their ontologies, as presented in [11].

3 Method

Addressing data interoperability challenges without requiring file conversions and enhancing decision-making via inferential reasoning are two important advantages of SWTs, as previously argued. In addition, the BHoM framework enables the conversion of object-oriented data models to OWL/RDF within design platforms like Grasshopper 3D. This section aims to demonstrate how SWTs can support design processes by utilizing design data modeled in BHoM and combining it with fabrication constraints in Semantic Web languages. The section is divided into two parts. The first part describes the method of design data generation using the BHoM framework, which allows the conversion of object-oriented data models to OWL/RDF within design platforms such as Grasshopper 3D. The second part focuses on the methods for reasoning and data validation using SWTs.

3.1 Design Data Generation

As a case study, this paper presents a large timber shell with cassettes used as a roof for a large research building, as shown in Fig. 2. This interdisciplinary project is still under development and will demonstrate several streams of research by the Cluster of Excellence Integrative Computational Design and Construction for Architecture (IntCDC). Previous research [13, 18] has demonstrated that high-performing timber structures can be designed and produced through computational design, simulation, and digital robotic fabrication. The roof selected for this study case consists of several segmented timber shells that are joined together to create a lightweight, long-spanning timber structure. Each shell has doubled curvature that is translated into fabricable planar elements through a segmentation process. This process is performed using a custom add-on for the interactive agent-based modeling framework, ABxM [19].

For this case study, the developed graph will focus on a single shell of the roof, which comprises of 213 segments. Each segment has a base polygonal shape from the segmentation but is generally unique in its geometry, material, and compositional buildup. There are two fundamental types of segments, a solid CLT cassette, and a hollow cassette. A hollow cassette typically consists of a top plate, bottom plate, and a series of edge beams. Additional elements such as acoustic panels can also be integrated into the cassettes.

3.2 Methods for Reasoning and Data Validation

In BHoM, rule-based verifications are resolved via Engine methods, which are C# functions. The Engine methods that perform verifications target C# Interfaces rather than C# Classes. While such rules can be applied to particular representations of buildings within a design software, we argue that such checking does not include logical inferences and implicit information that can be explicit through SWTs.

Semantic Web utilizes several technologies to infer knowledge. This paper combines some of these technologies with the data graph from the case study. Table 1 exemplifies how OWL, SWRL, roles, and SHACL can be used to infer knowledge or validate data. OWL helps to make inferences about classes or individuals in an ontology, while SWRL extends OWL's vocabulary by including mathematical expressions, certain string formatting properties, datatypes related to time and duration, datatypes related to URIs, and RDF-style lists [20]. Roles define relationships between nodes in a graph, and SHACL assists in validating the graph against a set of conditions.

On the other hand, although SPARQL is a query language for RDF graphs, its ability to add, remove, and retrieve data from graphs makes it applicable for inferring knowledge. SPARQL queries can match subject-predicate-object triple patterns and create filters and new variable bindings using mathematical operations and a variety of utility functions [21]. For example, the CONSTRUCT query matches graph patterns and returns a new graph. Although SPARQL is not a rule language, one could implement a simple rule-based system using CONSTRUCT queries to represent rules [22]. The last row of Table 1 presents an example of such a query.

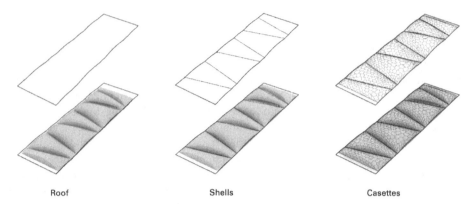

Roof   Shells   Casettes

Fig. 2. Timber roof, which consists of 7 shells, where each shell has around 200 cassettes

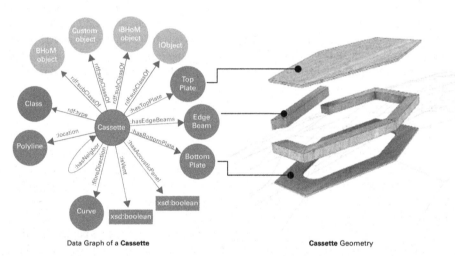

Data Graph of a **Cassette**   **Cassette** Geometry

Fig. 3. The ontology and the corresponding geometrical representation of a Cassette

4 Results

In this section, we describe the resulting graph from the design data of the timber shell, which includes the generated timber shell ontology based on the objects and the actual data. This graph, along with the generated ontology, can be integrated into different graph-based databases, and for this study, we use GraphDB[1] to navigate and work with the data. We apply reasoning and data validation methods to the data graph discussed in Sect. 3 and provide feedback either in the form of a report or the graph with inferred facts to BHoM objects, using this knowledge to make design decisions. Subsection 4.1 explains the generation of the data graph, while Subsect. 4.2 presents cases of Timber shell data completeness, validation, and reasoning.

---

[1] A graph database compliant with RDF and SPARQL.

Table 1. Methods for reasoning and data validation.

Method	Purpose	Example
OWL	To make inferences about classes and individuals in an ontology	SubClass(Casette, Structure), hasMaterial(c, m), isA(c, Casette), hasRange(hasMaterial, Material) $\Rightarrow$ isA(m, Material)
SWRL	To infer new data; to define rules for automated reasoning and inferencing in knowledge graphs	CassettePolyline (c1) $\wedge$ StartNode (c?, ?p1) $\wedge$ EndNode (?c, ?p2) $\wedge$ subtract (?Length, ?p2, ?p1) $\Rightarrow$ PerimeterLength (?c, ?length)
Role in DL (Description Logics)	To define the relationships between entities in a knowledge graph	locatedIn(x,z), partOf(z,y)
SHACL	To validate the graph against a shape	:CassetteShape a sh:NodeShape; sh:targetClass:Cassette; sh:property [sh:path:GUID; sh:maxCount 1; sh:datatype xsd:string;]
SPARQL CONSTRUCT	To create a graph derived from the knowledge in the initial graph (e.g. a graph that includes each individual cassette and its neighboring cassettes.)	CONSTRUCT{ ?casette1 :neighbour ?casette2 } WHERE { ?casette1 a :Casette, :has ?edge. ?casette2 a :Casette, :has ?edge. FILTER (?casette1 != ?casette2) }

4.1 Conversion of the Design Data to the Knowledge Graph

The design data for the timber shell is composed of interconnected classes with custom properties. To represent the different parts of the shell, we used BHoM. This allowed us to describe the objects and provided a solid foundation for storing the design data into an OWL/RDF representation using an existing converter [11]. Following the custom classes and BHoM conventions, the resulting ontology was divided into three types of classes, as shown in Fig. 4: (1) classes related to the topology of the timber shell (Roof, Shell, Cassette, TopPlate, etc.); (2) classes related to the geometry of the shell (point, line, polyline, NURBScurve, etc.); and (3) classes related to BHoM conventions (BHoMObject, IGeometry, IElement2D, etc.), as shown in Fig. 3. This categorization is helpful when querying specific objects, such as IElement2D, which returns all objects with a 2D representation. Additionally, in Subsect. 4.2, we present cases of timber shell data completeness, validation, and reasoning.

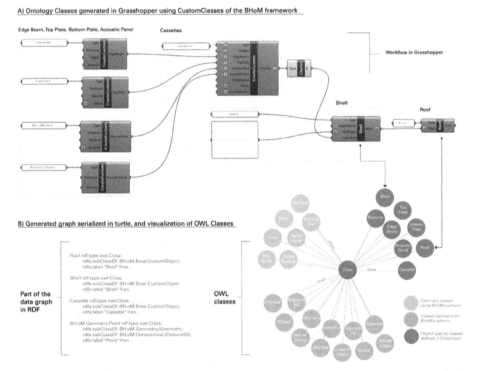

Fig. 4. The roof elements described in Custom BHoM objects (A) and the owl:Classes of the generated ontology (B).

The conversion to OWL/RDF is performed multiple times to address bottleneck and chokepoint issues with different datatypes. As expected, heavy meshes and complex geometries increase the conversion time. Therefore, only the minimum required information for each shell is converted to OWL/RDF. For example, for the geometric representation of shells, only a polygon and shell thickness are stored in the graph. The resulting ontology with the actual data is imported into GraphDB to store and navigate the data and their relationships. Figure 5 shows a simplified graph of the generated knowledge graph, where the main ontology classes are visualized in green, classes derived from the BHoM framework in yellow, and the instances (named individuals) in blue.

4.2 Timber Shell Data Completeness, Validation and Reasoning

Live graph data storage of building information can improve design and decision processes by including raw data and implicit inferred knowledge. In the following examples, we use SHACL shapes or rules expressed in SWRL to express regulations of restrictions from different disciplines against the data graph. The updated graph is fed back

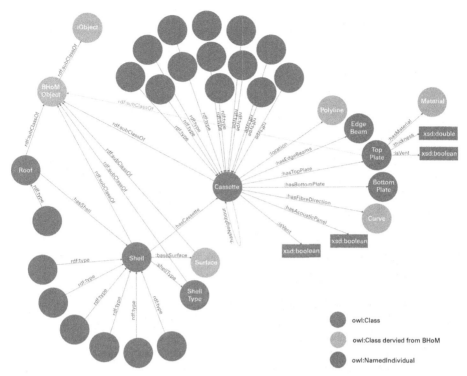

Fig. 5. A simplified visualization of the knowledge graph of the timber shell. (Color figure online)

to design platforms closing so the loop of designing. The process of such a co-design loop is illustrated in Fig. 6. The diagram depicts a sequence of five steps involved in the loop. Initially, the process commences with "designing with BHoM objects in a design platform", which enables the creation of a design with flexibility in selecting the design tool and disciplinary design. This approach is restricted to designing platforms compatible with the BHoM framework. The second step involves converting BHoM objects to OWL/RDF through the utilization of BHoM's converter. This component is accessible within the design tool employed for the design. In the subsequent third step, constraints are imposed on the data to meet specific criteria and regulations, enabling the derivation of new knowledge. At this juncture, the task is conducted externally to the design platform and lacks automation. It necessitates a comprehensive understanding of SWRL or SHACL to construct such limitations. The fourth step involves assessing the consistency of the graph and identifying any discrepancies. From the rules we added, we expect to infer new facts on the chart and enrich our data graph with more nodes. In the fifth step, we can get two outputs: a graph that includes inferred knowledge and a report. The new graph can be imported to the design platform, and BHoM's back converter can convert the graph to BHoM objects, allowing further manipulation of data. The report would give information if the graph satisfied our set of conditions. It has a text format we can read and apply manually to the design platform.

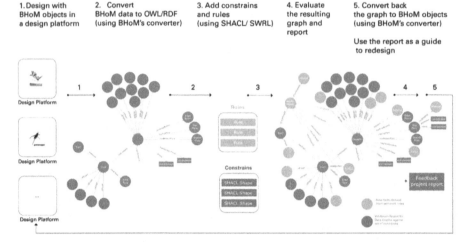

Fig. 6. Building information validation and reasoning using Semantic Web Technologies while co-designing using the BHoM framework

SHACL Shapes and Validation Report. This subsection outlines a method for evaluating data completeness in the design process using SHACL shapes. Checking data completeness is a critical aspect of the design process, as it provides designers with insight into the current stage of the design and the areas that require further attention, such as unfinished objects or incomplete data. To demonstrate this method, we apply it to timber shell data and verify that each Cassette object has a corresponding material assigned to it. Specifically, we use a SHACL shape to check if every Cassette class has at least one "material" property. This approach enables designers to effectively monitor data completeness:

```
:CassetteShape a sh:NodeShape;
    sh:targetClass :Cassette;
    sh:property [
    sh:path :Material;
    sh:minCount 1
    ].
```

The validation report from the data we used is:
Validation Report
Success
No
Errors found
<http://www.uni-stuttgart.de/B75CE802FB27A2D0214BC60182590E7FF31610F
E1F5E62B1F80CB0892697D83A>:
    sh:path:Material:

No Value

The report that is returned from such validation contains information on the Cassette IDs that do not have a property named:Material. This helps us to detect which Cassette is missing information on the Material. The URI of the Cassette is also its ID; therefore, while building back BHoM objects from the graph, the designers can query the requested Cassette and augment it with new information.

SPARQL Queries to Infer Knowledge. As discussed in Subsect. 3.2, SPARQL not only facilitates querying, but also enables knowledge inference. The current data graph only contains the minimum necessary information, lacking data on the weight and volume of the Cassettes. However, the volume of building elements is essential to determining several properties of a cyber-physical fabrication setup, as well as the representation of the building in the fabrication discipline. Fortunately, we can calculate and infer the volume and weight of the Cassettes by leveraging the information available in the current BHoM OWL/RDF graph, which includes design properties of the roof, such as the polygon coordinates, thickness, and material properties of the Cassettes. To derive this information, we utilize SPARQL queries. While the area of the surface can be calculated from the polygon's coordinates, the volume is calculated by multiplying the area and thickness of the object. We use SPARQL to calculate the weight of each Cassette by multiplying its volume with its material density and bind this weight information to the graph. Furthermore, we verify whether any Cassettes exceed the lifting capacity of robots, and reference the limitations of an industrial robot, KUKA, for this purpose. This finding may be amalgamated within the report, accompanied by communication that specifies the Cassette identification number that cannot be lifted, in addition to the particular parameter that requires modification, such as the surface area or material of the Cassette:

```
SELECT ?Cassette ?GUID ?volume ?weight
WHERE
{
    ?Cassette bhom:BH.oM.Base.BHoMObject.BHoM_Guid ?GUID.
    ?Cassette bhom:Volume ?volume.
    bhom:Timber bhom:MaterialDensity ?x.
    BIND ( (xsd:integer(?x) * xsd:double(?volume)) AS ?weight).
    FILTER (?weight > 360)
}
```

SWRL for Data Models Checking and Knowledge. Upon running the timber shell ontology with SWRL rules, new facts are added to the graph. In this instance, we utilized SWRL to examine the number of neighbours associated with the Cassettes. We identify Cassettes having two or fewer neighbours as EdgeCassettes. The data regarding the number of neighbours can aid in deciding suitable approaches for isolating or selecting materials for the cassettes.

```
Cassette (?c) ^ hasNeighbor (?c, n?) ^ sqwrl:makeSet(?s, ?n) ^
swrlb:greaterThanOrEqual(?s, 2) -> isEdgeCassette(?c, true)
```

Other rules can be defined, which may have an impact on cross-disciplinary data models and design decisions, similar to the example presented earlier.

5   Conclusion

This paper demonstrates how SWTs can support the design of a timber shell by validating data and reasoning on multidisciplinary information. It uses design data of a timber shell modeled in BHoM objects and converts them to OWL/RDF. We explore OWA and CWA approaches to infer knowledge as well as validate data. In terms of the rule complexity found in knowledge graphs, parametric design software such as Grasshopper3D seems better equipped to handle reasoning techniques involving calculations and conditionals for decisions, particularly those derived from geometry. However, while ad-hoc solutions implemented in design software can be useful, complex queries and reusable rules in graphs remain more powerful as seen in the SPARQL query in Sect. 4.2.

A limitation of this approach is the lack of alignment between the generated ontology and existing ontologies, which can hinder the reuse of concepts or rules. Future work will investigate how to develop modular BHoM ontologies that can align with other AEC ontologies. Integrating BHoM ontology with existing ontologies can improve alignment, facilitate querying, enable concept reuse, and promote interoperability, enhancing its usefulness, accuracy, and overall representation and understanding of the domain. So far, we have discussed semantics in multidisciplinary design, including restrictions from several disciplines, but not geometry computation and geometry-related calculation. Due to the complexity of geometrical information in building design, the Semantic Web's standard data model seems unsuitable for the representation of heavy complex geometries. From the graph generation discussed in Subsect. 4.1, and weight inference discussed in Subsect. 4.2, we state that the complexity of such operation increases when working with geometry operations on the graph database environment. Further work will consider splitting the generation of the geometric information from the semantic one when converting to the RDF graph. We plan to investigate heterogeneous database systems with a single, unified query interface. Such a study requires addressing Geometry Kernels and approaches to store geometric information that relate to industry standards (e.g. IFC uses the STEP format).

Besides the complexity of geometries, in conclusion, SWTs offer a solid foundation for connecting interdisciplinary knowledge and enable constraint checking in the AEC industry. Linking disciplinary representations of buildings and disciplinary regulations and constraints from federated data schemas using Semantic Web standards while designing allows designers to have a holistic and integrated overview of the design and to validate or check building data simultaneously.

Acknowledgements. Partially supported by the Deutsche Forschungsgemeinschaft (DFG, German Research Foundation) under Germany's Excellence Strategy - EXC 2120/1 – 390831618 and COFFEE - STA 572_15-2. Part of this paper was inspired by a collaborative effort with Dimitris Mavrokapnidis, Adam Tamas Kovacs, and Detlev van Loenhout during SSoLDAC2022.

References

1. Knippers, J., Kropp, C., Menges, A., Sawodny, O., Weiskopf, D.: Integrative computational design and construction: rethinking architecture digitally. Civ. Eng. Des. **3**(4), 123–135 (2021). https://doi.org/10.1002/cend.202100027
2. Elshani, D., Wortmann, T., Staab, S.: Towards better co-design with disciplinary ontologies: review and evaluation of data interoperability in the AEC industry. Presented at the Linked Data in Architecture and Construction 2022 (2022)
3. Tan, T., et al.: Construction-oriented design for manufacture and assembly guidelines. J. Constr. Eng. Manag. **146**(8), 04020085 (2020). https://doi.org/10.1061/(ASCE)CO.1943-7862.0001877
4. Cao, J., Vakaj, E., Soman, R.K., Hall, D.M.: Ontology-based manufacturability analysis automation for industrialized construction. Autom. Constr. **139**, 104277 (2022). https://doi.org/10.1016/j.autcon.2022.104277
5. Hogan, A., et al.: Knowledge graphs. Synth. Lect. Data Semant. Knowl. **12**(2), 1–257 (2021). https://doi.org/10.2200/S01125ED1V01Y202109DSK022
6. Pauwels, P., et al.: A semantic rule checking environment for building performance checking. Autom. Constr. **20**(5), 506–518 (2011). https://doi.org/10.1016/j.autcon.2010.11.017
7. Werbrouck, J., Senthilvel, M., Beetz, J., Pauwels, P.: A Checking Approach for Distributed Building Data (2019)
8. Oraskari, J., Beetz, J., Senthilvel, M.: SHACL is for LBD What mvdXML is for IFC (2021)
9. Dimyadi, J., Solihin, W., Eastman, C., Amor, R.: Integrating the BIM Rule Language into Compliant Design Audit Processes (2016)
10. BHoM. The Buildings and Habitats object Model (n.d.). https://bhom.xyz/
11. Elshani, D., Lombardi, A., Fisher, A., Hernandez, D., Staab, S., Wortmann, T.: Knowledge graphs for multidisciplinary co-design: introducing RDF to BHoM. Presented at the Linked Data in Architecture and Construction 2022 (2022)
12. Krieg, O., Bechert, S., Groenewolt, A., Horn, R., Knippers, J., Menges, A.: Affordances of Complexity: Evaluation of a Robotic Production Process for Segmented Timber Shell Structures (2018)
13. Wagner, H.J., Alvarez, M., Groenewolt, A., Menges, A.: Towards digital automation flexibility in large-scale timber construction: integrative robotic prefabrication and co-design of the BUGA Wood Pavilion. Constr. Robot. **4**(3–4), 187–204 (2020). https://doi.org/10.1007/s41693-020-00038-5
14. Reiter, R.: On closed world data bases. In: Gallaire, H., Minker, J. (eds.) Logic and Data Bases, pp. 55–76. Springer, Boston (1978). https://doi.org/10.1007/978-1-4684-3384-5_3
15. Werbrouck, J., Pauwels, P., Beetz, J., Mannens, E.: Data patterns for the organisation of federated linked building data. In: Conference: CIB W78 - LDAC 2021 (2021)
16. Elshani, D., Lombardi, A., Fisher, A., Hernandez, D., Staab, S., Wortmann, T.: Inferential reasoning in co-design using semantic web standards alongside BHoM. Presented at the 33. Forum Bauinformatik 2022, Munich (2022)
17. Kovacs, A.T., Micsik, A.: BIM quality control based on requirement linked data. Int. J. Archit. Comput. **19** (2021). https://doi.org/10.1177/14780771211012175

18. Groenewolt, A., Schwinn, T., Nguyen, L., Menges, A.: An interactive agent-based framework for materialization-informed architectural design. Swarm Intell. **12**(2), 155–186 (2017). https://doi.org/10.1007/s11721-017-0151-8
19. Nguyen, L., et al.: ABxM.Core: The Core Libraries of the ABxM Framework. DaRUS (2022)
20. SWRL: A Semantic Web Rule Language Combining OWL and RuleML. World Wide Web Consortium (W3C), 21 May 2004. https://www.w3.org/Submission/SWRL/
21. SPARQL Query Language for RDF: W3C Recommendation. W3C Recommendation (2008). https://www.w3.org/TR/rdf-sparql-query/
22. Staab, S.: Knowledge Graphs: Query Languages. Presented at the Knowledge Graphs Lecture, Stuttgart (2022)
23. Liu, H., Lu, M., Al-Hussein, M.: Ontology-based semantic approach for construction-oriented quantity take-off from BIM models in the light-frame building industry. Adv. Eng. Inform. **30**, 190–207 (2016). https://doi.org/10.1016/j.aei.2016.03.001

A Visual Support Tool for Decision-Making over Federated Building Information

Alex Donkers(✉) ⓘ, Dujuan Yang ⓘ, Bauke de Vries ⓘ, and Nico Baken

Eindhoven University of Technology, Groene Loper 6, 5612AZ Eindhoven, The Netherlands
a.j.a.donkers@tue.nl

Abstract. Interconnecting building information on multiple scales and lifecycle stages enables designers to improve their decision-making and tackle ecological, societal, and economical challenges. However, this information is often generated in different software tools, saved in different file formats, using a stakeholder's own domain language, and is stored in decentral storage locations. This paper presents a tool that integrates data from different stakeholders into a single viewpoint. The tool – LBDviz – combines a browser-based IFC viewer and semantic web technologies to combine geometry with other data that falls outside the scope of the IFC schema. The tool is tested in multiple case studies. These case studies range from requirement checking during the design phase, generating design feedback during the operational phase, and maintaining building performance in a sustainable and economically viable manner. The results of these studies can be reused in future designs, stimulating a more circular design process. The method used in LBDviz enables the interconnection of data within and without the architectural design domain and reduces data interoperability challenges in many lifecycle phases of a construction project.

Keywords: Linked Data · Semantic Web Technologies · Federated Building Information · LBDviz · IFC.js

1 Introduction

Visualizing heterogeneous building information enhances the decision-making processes of stakeholders in a building's lifecycle and is essential for (non-)professionals to understand a building's performance [1]. Designers use visualization tools to improve their designs. Technicians use them to run simulations and perform calculations. Once a building is built, facility managers and maintenance workers could use visual information to monitor their buildings and improve the maintenance process. As state-of-the-art architectural design challenges are interdisciplinary by nature, interconnecting heterogeneous building information into a single viewpoint would stimulate cross-domain decision-making and change the rather linear way of design and planning to a more circular approach that enables improving the building throughout multiple lifecycle stages [2].

© The Author(s) 2023
M. Turrin et al. (Eds.): CAAD Futures 2023, CCIS 1819, pp. 485–500, 2023.
https://doi.org/10.1007/978-3-031-37189-9_32

The heterogeneous nature of data in the AEC (architecture, engineering, and construction) industry introduces challenges to integrating building information in a single viewer. Data comes in different formats, is often stored at different locations, and is owned and maintained by different stakeholders. Next to this, data is generated in different phases of the building's lifecycle [3].

The IFC (Industry Foundations Classes) schema is the de facto standard in the AEC industry and allows stakeholders to exchange building information across multiple BIM authoring tools. However, IFC has some challenges related to data integration in various building lifecycle stages. IFC was never created to support the wide range of input data necessary for complex building performance simulations during the design phase [4]. Simultaneously, IFC does not easily integrate data that is being generated in the operational phase of a building [2], such as survey results, interviews, sensor measurements, IoT data, and time-series data [1]. Finally, reasoning over IFC models is difficult as the IFC schema lacks formal semantics [5].

Semantic web technologies have successfully opted as a solution to those issues in the past. Pauwels et al. [3] conducted an extensive literature review on semantic web technologies in the AEC industry. Boje et al. [2] researched how these semantic web technologies could lead to digital twins. These semantic digital twins are proposed as a solution to integrate data related to building energy performance [6], indoor environmental quality [7], and building management systems [8].

Some projects have already developed user interfaces so that end-users can interact with the integrated data. Rasmussen et al. [9] integrated sensor data and BIM data using semantic web technologies and visualized their results in a 2D viewer. Valra et al. [10] created a 3D viewer for managing BIM that applies semantic web technologies to link IFC files to other relevant files. Work of O'Donnell et al. [11] shows how semantic web technologies could be used to create interactive dashboards for building energy management. The examples mentioned here store sensor data in the RDF format using SSN [12], SAREF [13], or similar ontologies.

However, these interfaces are rather one-dimensional, while interconnecting cross-domain knowledge promises to solve both ecological, societal, and economical challenges. To enable interdisciplinary decision-making at all scales and lifecycle stages of architectural design, a flexible framework is necessary that can integrate heterogeneous and federated building information, spread across multiple sources, into a single viewpoint. Boje et al. [2] state that such a dynamic, web-based digital twin of the built environment would improve lifecycle costs, reduce carbon emissions, and deliver better construction services, but that the current interconnection between the various stakeholders, their use cases, and their data is weak. This work aims to fill this gap by presenting a web application – LBDviz – that dynamically integrates data from different knowledge domains scattered across the internet.

2   Method

LBDviz combines a graphical user interface with a method to query federated knowledge graphs (Fig. 1). This paper first proposes a method to integrate heterogeneous building information into a web of data. A generic framework to represent building information, sensor data, and occupant feedback using the RDF (Resource Description Framework) model is presented in Sect. 3. The result of this framework is a set of RDF graphs that contain information from different stakeholders in the AEC industry. Comunica is then used to query the decentralized RDF data.

The graphical user interface is built upon the knowledge graphs and uses the IFC.js library as a basis. The library enables interaction with IFC models on the web. As shown in Fig. 1, a link between the web geometry and the linked data is created by consistently converting the GUIDs in the IFC file to literals in the RDF data so that a direct link between the web geometry and the linked data is established.

The graphical user interface serves as a gateway to interact with the data in the decentralized knowledge graphs. Multiple methods of querying data are provided, such as buttons, custom SPARQL queries, and interaction with the web geometry. Next to this, LBDviz enables querying time-series data from time-series databases.

To test this new tool, multiple case studies are presented in Sect. 5. These case studies aim to show how interconnected building information can enhance decision-making in the design and operational phase of a building.

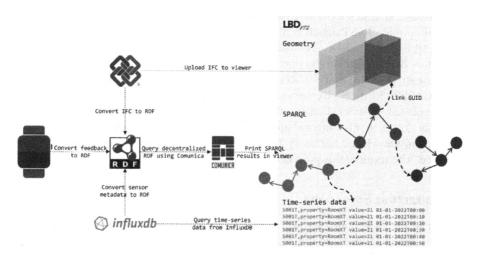

Fig. 1. System architecture of LBDviz

3   Integrating Heterogeneous Building Information

To solve the issues related to IFC, as mentioned in Sect. 1, semantic web technologies are proposed to represent building information. Earlier research successfully integrated heterogeneous building information using these technologies [7, 14]. Data is stored using the Resource Description Framework (RDF) format, resulting in labeled and directed graphs composed of triples. These triples are constructed using a subject, predicate, and object (i.e.:Building:hasRoom:LivingRoom). All resources have unique identifiers (URIs), enabling the integration of heterogeneous data on the web. RDF data can be enriched by linking the resources to ontologies: formal and explicit specifications of shared conceptualizations. In the AEC industry, ontologies typically contain snippets of domain knowledge that can be reused by other stakeholders.

3.1 Converting Building Information Models to RDF

While early-age ontologies in the AEC industry were rather monolithic ontologies based on the IFC schema (such as ifcOWL [15, 16]), recent ontologies focus on creating smaller, modular domain ontologies. The Building Topology Ontology (BOT) is developed as a core ontology in the AEC industry [17] and describes core topological concepts of a building, such as spaces, levels, and elements. Various researchers designed extensions to BOT that can be used to describe domain knowledge. Building properties could, for example, be described using BOP [18], PROPS[1], and OPM [19], while taxonomies of building elements are described using the PRODUCT[2], BEO[3], and MEP[4] ontologies. The result of the existing work is a modular set of ontologies that can be used in a plug-and-play-like approach to describe building information in the RDF format based on the needs and knowledge of the user.

Various conversion tools have been created to enhance the conversion of BIM files to the RDF format. Both Pauwels & Terkaj [20] and Hoang and Törmä [21] created a tool to convert IFC files to RDF following the ifcOWL ontology. Later initiatives introduced converters that implement the domain ontologies, such as the IFC-to-LBD converter by Oraskari et al. [5] and the Revit-bot-exporter[5] that creates RDF exports directly from Autodesk Revit. Both converters implement the BOT ontology as a core ontology, complemented by other domain ontologies. Converting BIM files to RDF following other ontologies is still a manual process. As the existing ontologies do not fully cover the domain knowledge necessary for this project, a manual conversion procedure is executed. The result of this procedure is visible in Fig. 2.

3.2 Integrating Sensor Data

Similar to the BIM domain, researchers investigated how semantic web technologies could be applied to describe sensors and the data they produce. This resulted in various ontologies related to sensors, such as SSN/SOSA [12, 22] and SAREF [13]. Although some researchers argue that sensor data could be converted to RDF, best practices suggest that sensor data could be best stored in time-series databases [23, 24].

To link the different databases to the linked building data, we propose storing some metadata of the sensor in the RDF file. The Building Performance Ontology [18] was developed to represent this metadata in an RDF format. Figure 2 shows how this metadata is linked to other building information. The query language SPARQL could now be used to query relevant information about the sensor, and the results of this query can be used

---

[1] https://github.com/w3c-lbd-cg/lbd/blob/gh-pages/presentations/props/.

[2] https://github.com/w3c-lbd-cg/product.

[3] https://pi.pauwel.be/voc/buildingelement.

[4] https://pi.pauwel.be/voc/distributionelement.

[5] https://github.com/pipauwel/revit-bot-exporter.

to access the right data point in the right database. This method allows sensor data stored at different locations to be integrated with the linked building data [14].

3.3 Integrating Occupant Feedback

In earlier work, we developed a smartwatch application – Mintal – that can be used to obtain occupant feedback on building performance during the operational phase. The application, running on Fitbit Sense and Fitbit Versa 3, automatically converts this feedback to small RDF snippets. As the URIs of the given feedback overlap with the URIs of rooms in the RDF file of the building, this feedback will automatically be integrated with the knowledge graph. This enables querying both the measured building performance (such as sensor data) and the perceived performance (e.g., the feedback) using a single SPARQL query. The Occupant Feedback Ontology (OFO) [25] was developed to structure this feedback. The ontology is firmly attached to the Building Performance Ontology, enhancing query functionalities. Figure 2 shows how the occupant feedback is integrated with the other data.

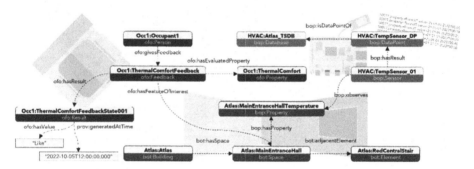

Fig. 2. Integrating BIM, sensor data, and occupant feedback using the RDF model

4 LBDviz: Visualising Linked Building Data on the Web

4.1 Overview

LBDviz is a single-page web app that consists of a full-page IFC viewer with various menu windows on top of the viewer (Fig. 3). This application aims to enable users to interact with interconnected building information from multiple sources in a single viewpoint, with a low software skill threshold for new users.

4.2 The 3D Model

LBDviz uses the IFC file extension to import geometries. IFC is the de facto standard in the AEC industry and can be exported by most proprietary software tools. It, therefore, enables importing geometries created by different stakeholders and different tools.

To build web applications around the IFC format, LBDviz uses IFC.js[6]. IFC.js is a JavaScript library that enables reading and writing IFC files in the browser. IFC.js uses Three.js[7] to represent the geometry in the web viewer. Next to that, the library parses the IFC file so that one can directly apply JavaScript functions to the file.

These JavaScript functions were used to add typical BIM software functionalities to the viewer. After uploading an IFC file, a user can zoom, pan, or rotate the model using the mouse wheel and the right and left mouse buttons. One can crop the 3D model in all directions using cropping panes, allowing the user to navigate through the IFC model. Interaction with elements of the 3D geometry is enabled by simply clicking on them. After selecting the element, various options for advanced interaction are available in the menu bar.

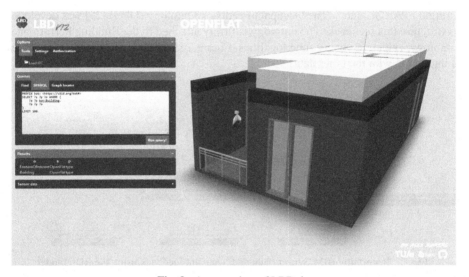

Fig. 3. An overview of LBDviz

4.3   Querying Federated Building Information

End-users would typically need information from multiple files. A maintenance engineer that needs to check the heating installation would, for example, need information from both the architect's and the HVAC engineer's models. In the semantic web context, interconnected building information would typically be maintained decentrally on the web or in various databases. To query this federated information, Comunica[8] is applied. Comunica is a meta-query engine that can be used to execute SPARQL queries over multiple SPARQL endpoints or files.

---

[6] https://ifcjs.io/.
[7] https://threejs.org/.
[8] https://comunica.dev/.

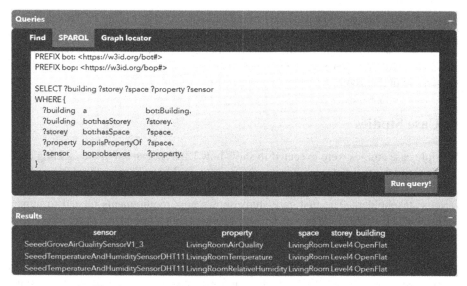

Fig. 4. Querying federated building information in LBDviz

The LBD viewer has multiple ways to execute SPARQL queries. First, the user can travel to the *SPARQL* tab, which has a text box in which the user can type custom SPARQL queries. A second tab – the *Graph Locator* tab – can be used to select one or more data sources. After executing the query, the results will be printed in a table in the *Results* tab. The table will dynamically create headers based on the variables that are passed on in the query (Fig. 4).

As many stakeholders in the AEC industry are unlikely to be able to create custom SPARQL queries, LBDviz also has a *Find* tab. This tab contains buttons with predefined SPARQL queries behind them so that every end user can run basic queries without needing expertise in semantic web technologies. These buttons can be tailored for specific roles or stakeholders. Current buttons contain basic queries, such as finding all walls, all doors, all windows, and all sensors.

Finally, end-users can enrich queries by interacting with the 3D model. After clicking on an element in the 3D model, the GUID of the element is extracted. This GUID is then inserted in predefined SPARQL queries to filter results for this specific element. This method can be used to query all the properties of a selected element or all the sensors in a selected room. Using this same GUID, LBDviz has the option to query occupant feedback on elements or rooms. These queries are also executed using a button in the *Find* tab. After clicking the button, the viewer will print the feedback in the *Results* tab.

4.4 Querying Time-Series Data

Where most data in the design phase of a building is typically static information (such as drawings), data produced in the operational phase of a building has a more dynamic nature. InfluxDB – a time-series database – can be used to store the data produced by sensors. Querying time-series data in LBDviz is a two-step procedure. First, metadata related to the sensor device is stored in the knowledge graph (Fig. 2). This metadata is queried and imputed in a Flux query, that returns the sensor readings from the time-series database. A module in LBDviz automatically creates these queries based on the interaction with the viewer. If the user selects a specific room, LBDviz will only select the sensors in that room.

5   Case Studies

Multiple case studies were executed to show how LBDviz can stimulate tackling challenges throughout different lifecycle phases of a building. During the design phase, complex building requirements can only be validated after combining information from various stakeholders. The first case study shows how requirements related to human health and well-being could be checked during the design phase so that designers receive real-time feedback on the current state of their design. These requirements are based on the WELL v2 standard [26].

Even if a building scores high on those standards, research suggests that buildings do not necessarily improve occupants' comfort levels [27, 28]. Therefore, a second case study compares the as-designed information with the as-built information. First, designed thermal performance is compared with measured thermal performance using the PMV/PPD method. This method calculates how many people are likely to complain about the thermal comfort in a space. Afterward, measured performance is compared with perceived performance, by integrating the occupant feedback from the smartwatch app Mintal.

LBDviz promises to help tackle ecological and economical challenges. The real-time energy consumption of buildings and devices is added to the platform so that the building performance can be compared with energy consumption, and thus CO_2 emissions and costs.

Finally, a feedback loop is created to show how knowledge from the past can be imputed into new design processes.

5.1   Requirement Checking

WELL v2 is a standard for buildings that assesses the human health and well-being in a building. Various parameters of this assessment relate to the building's design and should therefore be part of the design requirements. To assess those requirements, certain rules in WELL v2 were manually converted to SPARQL queries. The Atlas building of the TU/e campus was used for this case study. The building was renovated between 2014 and 2019, which is why the design phase contained multiple BIM files, some representing the existing building and some representing the newly built parts. As many subcontractors were responsible for specific parts of this building, requirement checking is a complex task that involves information from many parties.

Figure 5 shows how LBDviz integrates various IFC models into a single viewer: one model contains existing parts of the building, one model contains newly designed walls and floors, one model contains the façade, and one model contains newly designed stairs. The IFC-to-LBD converter [5] was used to convert the IFC models to an RDF format.

They were uploaded as separate files to a GitHub repository, after which we inserted their web locations into the *Graph locator* tab.

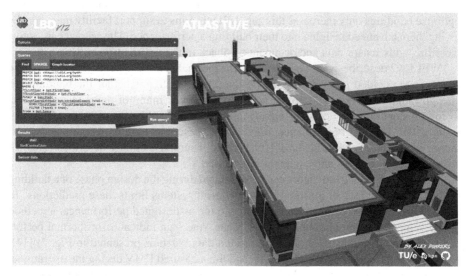

Fig. 5. Validating a WELL v2 requirement using SPARQL

One could then use the SPARQL queries to test individual WELL v2 rules and assess certain features of human health and well-being. The query in Fig. 5 returns all stairs that comply with feature V03 Circulation Network in WELL v2, requiring that at least one staircase open to regular occupants serves all floors of the project and:

- is designed through at least two of the following strategies: music, artwork, light levels of at least 215 lx when in use, windows that provide access to daylight, natural design elements or gamification,
- is located physically or visibly before elevators and escalators,
- and has a point of decision signage near the main entrance or reception desk.

Validating these requirements during the design phase requires a lot of coordination and communication between stakeholders. In the case of Atlas, four files needed to be combined to check this feature. Considering that WELL v2 has 122 feature sets, checking them during the design phase would be too time- and knowledge-intensive.

The *Results* tab in Fig. 5 successfully returns the stairs that comply with feature V03. It integrates the information from multiple RDF files through the Comunica engine. As LBDviz directly queries from web sources, an update of these web sources will also result in updated query results in LBDviz. This has been validated by deleting the stair from the RDF file of one of the subcontractors and deleting the reception desk from the RDF file of the architect. In both cases, the query in LBDviz did not return the stair anymore, as it did not comply with V03.

5.2 Designing High-Performing Buildings

Researchers found a gap between the as-designed and as-built performance of LEED [28] and BREEAM [27] certified buildings. The lack of tools to monitor, control and optimize buildings only increases this gap. LBDviz aims to support facility managers by giving them advanced insights into their building's performance. The second case study, therefore, adds sensor data and occupant feedback to the information in Sect. 5.1.

WELL v2 describes various features in the thermal comfort category. Some of those features are decided upon during the design phase, such as whether the building has enough radiant heating and cooling capacity. However, many of those features relate to complex operational decision-making and are influenced by the users of a building. Feature T01 Thermal Performance, for example, describes requirements related to the indoor temperature, while feature T02 Verified Thermal Comfort describes rules for post-occupancy surveys and occupant satisfaction.

These performance indicators can be simulated during the design phase of a building [29], although the lack of information on technical systems limits these predictions. To test how the as-built performance compares to the as-designed performance, a method was developed to calculate the Predicted Mean Vote – an indicator for thermal performance – in real-time using the interconnected data (such as presented in Fig. 2) [14]. As we can store both the predicted PMV and the measured PMV during the operational phase in the knowledge graph, both can be queried, next to other parameters. Figure 6 shows a comparison of the simulated and measured thermal performance of the building. It shows both the final assessment score (based on [30]) as well as individual components of the assessment. Designers can use this as a direct feedback loop to their designs.

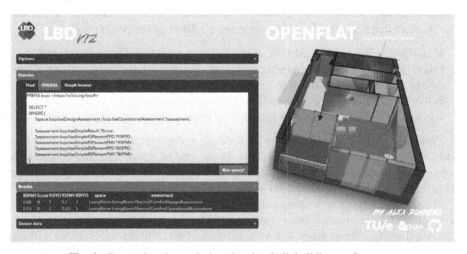

Fig. 6. Comparing the as-designed and as-built building performance

Next to this, LBDviz can help in finding the gap between measured and perceived building performance. Figure 7 shows how LBDviz integrates the web geometry with temperature readings and occupant feedback. The sensor data is acquired through the method described in Sect. 4.4. The occupant feedback is provided using a smartwatch and translated to RDF as described in Sect. 3.3. Simultaneously querying both the sensor data and the feedback results in a mixed chart that designers and facility managers can use to better understand positive and negative occupant feedback in relation to measured data. Next to checking features in the WELL v2 certificate, the interconnected data can be used to 1. Find anomalies in the sensor data streams, 2. Group rooms with similar complaints, 3. Find the rooms with the most complaints, and 4. Find HVAC systems that relate to properties that people complain about.

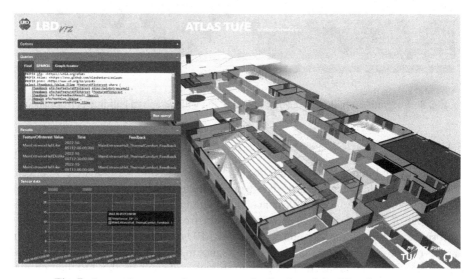

Fig. 7. Integrating sensor data, occupant feedback, and the 3D web geometry

5.3 Ecological and Economical Challenges

As the current energy crisis in The Netherlands is causing an extra financial burden on real estate finances, gaining insights into the costs of maintaining high building performance is crucial. Simultaneously, keeping energy consumption low will have positive effects on greenhouse gas emissions.

LBDviz integrates real-time energy consumption of buildings and individual devices with the building information. It allows facility managers to monitor energy consumption and includes ecological and economical parameters in their decision-making processes. Figure 8 shows how feedback on illuminance and energy consumption of the building and a lighting source (using a smart energy socket) can be integrated in LBDviz. Facility managers can now see what the energy consumption (and thus costs) of certain interventions are. As the data is integrated into the knowledge graph, advanced algorithms would also be able to find an optimal pay-off between (ecological and economical) costs and indoor comfort.

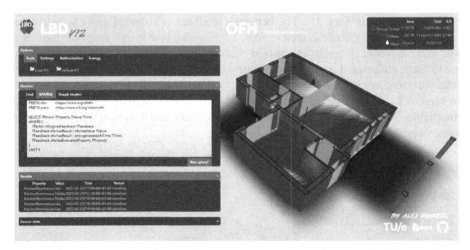

Fig. 8. Integrating a building's energy consumption with perceived building performance

5.4 Design Feedback

Interconnecting building information doesn't stop at one building. Semantic web technologies enable the integration of data from multiple buildings within (or without) a designer's portfolio. As the performance of a building is crucial to determine the optimal layout of buildings, designers would benefit from historical feedback about this performance [29]. As the Comunica engine in LBDviz is built to integrate data from various buildings, a designer can query performance indicators and other building parameters from various buildings in the portfolio (Fig. 9) and compare these to their current design. Figure 9 shows a comparison of thermal comfort scores from multiple buildings and some possible influencers. This can be applied for many use cases, including the ones mentioned in Sect. 5, and leads to a more circular design process.

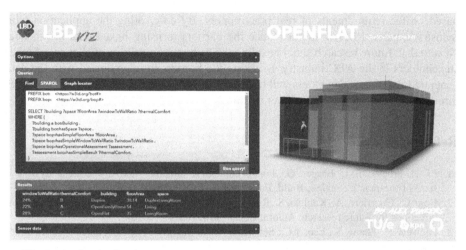

Fig. 9. Design feedback: comparing thermal comfort with previously designed buildings

6 Conclusion

Interconnecting building information on multiple scales and lifecycle stages enables designers to improve their decision-making and tackle ecological, societal, and economical challenges. However, this information is often generated in different software tools, saved in different file formats, using a stakeholder's own domain language, and is stored in decentral storage locations. Since communication between all the stakeholders in a construction project is a complex and time-intensive task, decision-making is often based on incomplete information, or worse: gut feeling.

This paper presents LBDviz, a tool that integrates heterogeneous and federated building information spread across multiple sources, into a single viewpoint. LBDviz combines a web viewer (based on IFC.js), a meta-query engine to query linked data (using Comunica), and a time-series query engine (using Flux).

The software tool was tested in multiple case studies. The first study shows how interconnecting data can help requirement checking during the design phase. First, relatively static requirements such as staircases were checked, after which more dynamic building performance indicator checks were performed. Interconnecting data across multiple lifecycle stages enabled comparing the as-designed with the as-built building performance. The perceived performance was also integrated, showing designers and facility managers whether occupants actually like their buildings. Finally, the energy consumption of buildings and devices was integrated into LBDviz, enabling decision-makers to take ecological and economic considerations into account when maintaining the performance of their buildings. Interconnecting data goes further than just one building, and we showed that our methodology can lead to a more circular design process, where designers learn from their previous buildings.

LBDviz is an ongoing project. Combining web geometry with linked data is promising to solve data interoperability challenges in all phases of a construction project. Future research should therefore focus on further developing the application's user interface based on the requirements of real practitioners. By co-creating the application with industry experts, LBDviz might reduce the barriers to using linked data in practice. Next to that, future research should explore the data interoperability challenges of other stakeholders in the AEC industry. Breaking the barriers to interconnect with the data generated by other stakeholders will hopefully improve the decision-making processes throughout the entire lifecycle of our buildings.

References

1. Li, P., Froese, T.M., Brager, G.: Post-occupancy evaluation: state-of-the-art analysis and state-of-the-practice review. Build. Environ. **133**, 187–202 (2018)
2. Boje, C., Guerriero, A., Kubicki, S., Rezgui, Y.: Towards a semantic construction digital twin: directions for future research. Autom. Constr. **114**, 103179 (2020)
3. Pauwels, P., Zhang, S., Lee, Y.C.: Semantic web technologies in AEC industry: a literature overview. Autom. Constr. **73**, 145–165 (2017)
4. Elagiry, M., et al.: IFC to building energy performance simulation: a systematic review of the main adopted tools and approaches. In: Monsberger, M., et al. (eds.) Proceedings of BauSim Conference 2020. Verlag der Technischen Universität Graz, Graz (2020)
5. Bonduel, M., Oraskari, J., Pauwels, P., Vergauwen, M., Klein, R.: The IFC to linked building data converter - Current status. In: Poveda-Villalón, M., Pauwels, P., Roxin, A. (eds.) Proceedings of the 6th Linked Data in Architecture and Construction Workshop. CEUR-WS.org, Aachen (2018)
6. Hu, S., Wang, J., Hoare, C., Li, Y., Pauwels, P., O'Donnell, J.: Building energy performance assessment using linked data and cross-domain semantic reasoning. Autom. Constr. **124**, 103580 (2021)
7. Donkers, A., de Vries, B., Yang, D.: Knowledge discovery approach to understand occupant experience in cross-domain semantic digital twins. In: Pauwels, P., Poveda-Villalón, M., Terkaj, W. (eds.) Proceedings of the 10th Linked Data in Architecture and Construction Workshop Co-located with 19th European Semantic Web Conference (ESWC 2022). CEUR-WS.org, Aachen (2022)
8. Chamari, L., Petrova, E., Pauwels, P.: A web-based approach to BMS, BIM and IoT integration: a case study. In: CLIMA 2022 (2022)
9. Rasmussen, M.H., Frausing, C.A., Hviid, C.A., Karlshoj, J.: Demo: integrating building information modeling and sensor observations using semantic web. In: Lefrançois, M., et al. (eds.) Proceedings of the 9th International Semantic Sensor Networks Workshop Co-located with 17th International Semantic Web Conference (ISWC 2018). CEUR-WS.org, Aachen (2018)
10. Valra, A., Madeddu, D., Chiappetti, J., Farina, D.: The BIM management system: a common data environment using linked data to support the efficient renovation in buildings. In: Lennard, Z. (eds.) Proceedings of the 8th Annual International Sustainable Places Conference (SP 2020). MDPI, Basel (2021)
11. O'Donnell, J., Corry, E., Hasan, S., Keane, M., Curry, E.: Building performance optimization using cross-domain scenario modeling, Linked data, and complex event processing. Build. Environ. **62**, 102–111 (2013)
12. Haller, A., et al.: The modular SSN ontology: a joint W3C and OGC standard specifying the semantics of sensors, observations, sampling, and actuation. Semant. Web **10**(1), 9–32 (2018)
13. Daniele, L., Solanki, M., den Hartog, F., Roes, J.: Interoperability for smart appliances in the IoT world. In: Groth, P., et al. (eds.) ISWC 2016. LNCS, vol. 9982, pp. 21–29. Springer, Cham (2016). https://doi.org/10.1007/978-3-319-46547-0_3

14. Donkers, A., Yang, D., De Vries, B., Baken, N.: Real-time building performance monitoring using semantic digital twins. In: Poveda-Villalón, M., Pauwels, P. (eds.) Proceedings of the 9th Linked Data in Architecture and Construction Workshop. CEUR-WS.org, Aachen (2021)
15. Beetz, J., Van Leeuwen, J., De Vries, B.: IfcOWL: A case of transforming EXPRESS schemas into ontologies. Artif. Intell. Eng. Des. Anal. Manuf. AIEDAM **23**, 89–101 (2009)
16. Pauwels, P., Krijnen, T., Terkaj, W., Beetz, J.: Enhancing the ifcOWL ontology with an alternative representation for geometric data. Autom. Constr. **80**, 77–94 (2017)
17. Rasmussen, M.H., Lefrançois, M., Schneider, G.F., Pauwels, P.: BOT: the building topology ontology of the W3C linked building data group. Semant. Web **12**, 143–161 (2020)
18. Donkers, A., Yang, D., de Vries, B., Baken, N.: Semantic web technologies for indoor environmental quality: a review and ontology design. Buildings **12**(10), 1522 (2022)
19. Rasmussen, M.H., Lefrançois, M., Bonduel, M., Hviid, C.A., Karlshø, J.: OPM: an ontology for describing properties that evolve over time. In: Poveda-Villalón, M., Pauwels, P., Roxin, A. (eds.) Proceedings of the 6th Linked Data in Architecture and Construction Workshop. CEUR-WS.org, Aachen (2018)
20. Pauwels, P., Terkaj, W.: EXPRESS to OWL for construction industry: towards a recommendable and usable ifcOWL ontology. Autom. Constr. **63**, 100–133 (2016)
21. Hoang, N.V., Torma, S.: Implementation and experiments with an IFC-to-linked data converter. In: Proceedings of the 32nd CIB W78 Conference 2015 (2015)
22. Janowicz, K., Haller, A., Cox, S.J.D., Le Phuoc, D., Lefrançois, M.: SOSA: a lightweight ontology for sensors, observations, samples, and actuators. J. Web Semant. **56**, 1–10 (2019)
23. Esnaola-Gonzalez, I., Javier Diez, F.: Integrating building and IoT data in demand response solutions. In: Poveda-Villalon, M., Pauwels, P., De Klerk, R., Roxin, A. (eds.) Proceedings of the 7th Linked Data in Architecture and Construction Workshop. CEUR-WS.org, Aachen (2019)
24. Petrova, E., Pauwels, P., Svidt, K., Jensen, R.L.: In search of sustainable design patterns: combining data mining and semantic data modelling on disparate building data. In: Mutis, I., Hartmann, T. (eds.) Advances in Informatics and Computing in Civil and Construction Engineering, pp. 19–26. Springer, Cham (2019). https://doi.org/10.1007/978-3-030-002 20-6_3
25. Donkers, A., De Vries, B., Yang, D.: Creating occupant-centered digital twins using the occupant feedback ontology implemented in a smartwatch app. Semant. Web (2022)
26. International WELL Building Institute: The WELL Building Standard V2.0
27. Altomonte, S., Saadouni, S., Kent, M.G., Schiavon, S.: Satisfaction with indoor environmental quality in BREEAM and non-BREEAM certified office buildings. Archit. Sci. Rev. **60**, 343–355 (2017)
28. Altomonte, S., Schiavon, S.: Occupant satisfaction in LEED and non-LEED certified buildings. Build. Environ. **68**, 66–76 (2013)
29. Anand, P., Deb, C., Alur, R.: A simplified tool for building layout design based on thermal comfort simulations. Front. Architectural Res. **6**(2), 218–230 (2017)
30. Loomans, M.G.L.C., et al.: D1.6: Optimal indoor performance indicators (KIPI Framework). Eindhoven University of Technology, Eindhoven, The Netherlands (2011)

Building Data Analysis, Visualisation, Interaction

BIMThermoAR: Visualizing Building Thermal Simulation Using BIM-Based Augmented Reality

Kifah Alhazzaa[1,3]([✉]) [iD], Manish Dixit[2] [iD], and Wei Yan[1] [iD]

[1] Department of Architecture, College of Architecture, Texas A&M University, College Station, TX 77840, USA
{Alhazzaa,Wyan}@TAMU.edu

[2] Department of Construction Science, College of Architecture, Texas A&M University, College Station, TX 77840, USA
Mdixit@TAMU.edu

[3] Department of Architecture, College of Architecture and Planning, Qassim University, Qassim 52571, Saudi Arabia
K.alhazzaa@QU.edu.sa

Abstract. The complexity of energy modeling concepts, the number of factors that influence the outcome, and their interconnections and correlations all combine to make building energy simulation a difficult subject to grasp. As a result, the preponderance of the time, the decision makers base their decisions on construction cost. Even if we show decision makers a 2D representation of the simulation outcomes, they will struggle to understand and, more crucially, relate simulation outcomes to the actual physical structure. The goal of this research is to see if augmented reality (AR) technology may be used as a new medium for visualizing the results of energy simulations. The goal of creating an AR mobile app prototype (BIMThermoAR) is to improve audience understanding of building energy simulations and the influence of material properties on interior temperature and annual energy consumption by superimposing the simulation outcomes on the actual physical structure. BIMThermoAR was created using a dynamic interaction between users, Building Information Modeling (BIM), and building materials selection which is based on materials' construction types and insulation capabilities (e.g., R-value), resulting in a variety of building performance scenarios. The BIM is used by the AR app to present the Energy Plus engine's simulated energy condition results, and the AR app has access to the BIM's metadata. AR engagement is performed by changing the properties of building components and viewing how each change impacts interior temperatures and total building energy consumption. The visualization methods of BIMThermoAR are expected to enhance the understanding of building performance simulation by designers, engineers, and students.

Keywords: Augmented reality · Energy Simulation Augmented Reality · Augmented Reality BIM

M. Turrin et al. (Eds.): CAAD Futures 2023, CCIS 1819, pp. 503–517, 2023.
https://doi.org/10.1007/978-3-031-37189-9_33

1 Introduction

One of the most important aspects of reducing the effects of climate change on the world is high-performance buildings. According to the International Energy Agency (IEA) Buildings account for almost 38% of worldwide CO_2 emissions each year. Building operations account for 28% of total emissions each year, while building materials and construction (also known as embodied carbon) account for another 10%.

To understand and visualize the building energy simulation findings, architects and engineers usually utilize a 2D graphic. The traditional medium (e.g., papers and screens) is missing the link between the produced 2D or 3D drawing and the actual structure. As a result, it is difficult to understand the drawing and relate it to the project while using the traditional medium. It is more difficult for non-experts to perceive the complicated parameters in drawings and models created with Building Information Modeling (BIM); therefore, it is expected to be comparatively easy to gain intuitive impressions by a walkthrough in an augmented building with simulation results. An icon says you are here, always appearing in public outdoor areas and buildings navigation floor plan to help users navigate and familiarize themselves with space [1]. On the other hand, Augmented Reality (AR) technology offers a more immersive relationship by placing virtual objects in the real environment [2]. The AR immersion and interactivity increase learning outcomes and knowledge retention [3–5].

The complexity of energy simulation principles, the number of elements that impact the finding, and their linkages and relationships to one another all combine to make building energy simulation a challenging subject to absorb. Thus, the majority of the time, project clients make decisions primarily based on cost. Even if we show them a 2D representation of the simulation findings, they will struggle to comprehend them and, more importantly, relate them to the actual physical structure, which may help them link the simulation findings to the existing buildings [6].

This study explores AR technology's ability to be a new medium for visualizing energy simulation findings. Is AR platform capable of conveying the building energy simulation results? This is the main question of this study. The purpose of developing an AR mobile prototype (BIMThermoAR) is to enhance users' understanding of building energy simulations and the impact of materials properties on the indoor temperature and the yearly energy consumption by superimposing the simulation outcomes on the existing physical structure. The BIMThermoAR has been developed with a dynamic interaction between the users, Building Information Modelling (BIM), and the building materials selection based on their construction type and insulation capability (R-value), which create various building performance scenarios. The prototype contains multiple pre-simulated conditions that are visualized by summer and winter indoor temperature mapping, along with a whole year simulation for overall energy consumption.

1.1 Augmented Reality

AR uses a variety of digital components in the domains of hearing, sight, touch, and scent to improve people's perception. Enlarging, extending, or increasing the characteristics of physical components can all be understood as augmented reality [7]. AR is a sort of computer-generated information that is superimposed on the actual environment. Our environment has been augmented to make it easier for the user to complete the work at hand [8]. The world in AR is genuine, but it has been enhanced with data and pictures from the system [9]. Rather than consisting only of virtual scenes, AR improves the physical world and is built on top of reality [10].

The requirement of obtaining more comfortable and inexpensive living standards, according to AR practitioners, makes the incorporation of AR into our life unavoidable.[2, 8]. The three fundamental components of AR, according to Azuma [10, 11] are:

- The merging of real and virtual items
- Real-time interactivity
- The alignment of both real and virtual objects with one another.

By comparing 3D AR displays to standard 2D graphics presented on a computer screen, researchers discovered that, despite the increased mental load experienced by participants during the 3D AR feedback, participants considered the 3D AR feedback to be more interesting and inspiring [12].

2   Related Work

Using AR technology, researchers (H. Rashed-Ali et al.) are developing an innovative teaching technique for passive design and energy efficiency. The tablet-based AR application simulated the influence of various residential building design characteristics. Because the learner may alter the architectural design parameters, the prototype is interactive (e.g., form, orientation, material choices, window size, placement, etc.). The real-time interaction aids students in comprehending the relationship between architectural design parameters and the science that underpins them [13].

The simulation uses complicated forms and parameters. Immersive technologies are not employed in building analysis and simulation to enhance the building design process. Visualization improves with understanding. Zhao et al. presented their method for two applications in their study [14]. The first scenario comprised an AR high-rise simulation. The second demonstrated a two-story home daylight simulation in Mix Reality (MR), which is for retrofitting. The research aimed to help building designers grasp this new technology [14].

A team of researchers has developed a new approach for intuitively viewing indoor thermal settings using AR and computational fluid dynamic (CFD) analysis [15]. The combination of CFD and AR improves the user experience by giving them a more realistic sense of the future thermal environment, which helps them understand the simulations better. The research focused on remodeling projects, which have become more popular since they aim to improve the building's thermal efficiency.

Existing research rarely offers a provide full-scale building experience that help users visualize an actual building's energy simulation outcome. The interactivity is also limited and not reflected in simulation. This study will close these gaps by enabling interactive visualization of the building energy simulation results using multiple visualization methods and creating a full simulation for each design alternative.

3   Methods and Prototyping

This study focuses on creating an AR mobile app that overlays virtual Building Information Modeling (BIM) and simulation (SIM) results on the existing building using the Unity game engine. The AR app will make use of the BIM to present the EnergyPlus simulation engine's simulated energy condition outcomes, and the embedded information (metadata) on BIM will be accessible through the AR app. The interaction with AR will be achieved by altering the attributes of the building components and seeing how each change affects the indoor temperatures and total building energy consumption.

3.1   BIMxThermoAR Prototype Workflow

The BIMThermoAR app involves multiple software tools and platforms. Figure 1 illustrates the general development workflow. Starting from preparing and building the BIM model, the researchers utilized the power of Autodesk Revit software based on field measurements of the space of study. This study harnessed the power of 3D scanned points cloud by the Matterport Pro2 camera for two reasons: first, correcting the BIM model of the space of study to make sure the BIM model is accurate to present the existing room, and second, utilizing the Area Target registration method of Vuforia Engine for Unity [16] which requires Matterport 3D scanned points cloud or generated meshes for best results of registering the virtual model onto the physical environment in accurate position. The study took advantage of a link between Grasshopper – Ladybug tools and EnergyPlus simulation engine to build the energy model using Rhinoceros 3d software interface along with its embedded Grasshopper visual programming extension. The Unity game engine is the main AR development platform that carries the generated assets and user interface. The researchers chose Unity since it supports the Vuforia Engine, which provides many advanced AR technologies to ease and improve the development of AR applications. The Area Target feature from Vuforia Engine is a markerless AR registration method that uses the 3D scans of space for recognizing the space, and it is developed for large spaces [16]. This proposed prototype utilized the Area Target feature as the main registration method which enhances the user experience and improves the accuracy of superimposing the virtual model (BIM and the simulation results) onto the real environment. BIMThermoAR has been developed and optimized to work with iOS and Android mobile devices since smartphones are the most spread and affordable AR devices compared to other AR devices in the market.

3.2   The Space of Study

The space of study is the Ph.D. student room on the fourth floor of Langford building A in the school of Architecture at Texas A&M University, which is in College Station, Texas. The Ph.D. room has a length of ≈39 ft 3 in (11.96 m) and a width of ≈36 ft 9 in (11.20 m) (Fig. 2). The room is facing southeast with only one façade which is constructed with single non-reflected glazing widows and precast uninsulated concrete wall (Fig. 3). The room is adjacent to studio design spaces from two sides and classroom from the third side. The space is equipped with furniture equipment such as desks, chairs, computers, monitors, etc. (Fig. 4). Figs. 5 and 6 show the developed BIM model, which will be used as a base for the simulation outcomes, for the studied space One of the reasons of choosing the Ph.D. student room to examine the Vuforia Engine Area Target feature capability of recognizing and working with a space is that the space is filled with different sizes of static components (e.g., lockers), which are hard to move by students, and dynamic components (e.g., chairs, computers), which can be moved by students. The other reason is to create a non-restricted AR lab for graduate students in the architecture school who are curious to explore the world of AR.

Revit Software	Preparing the BIM model based on field measurements and 3D scanning.
Matterport	3D scanning to facilitate the construction of the BIM model and utilize the area target feature in Vuforia.
Grasshopper - Ladybug tools	Creating the energy model, thermal spatial mapping algorithm, and running the energy simulations.
Unity Gaming Engine	Developing AR app for iOS and Android platforms. Implementing UI and combining the energy simulation and thermal mapping with the BIM model.
Vuforia Engine for Unity	Using the Area Target registration method utilizing the Matterport 3D scanned point cloud data.
Mobile Device	Visualizing the final combination of the BIM model and simulation outcomes superimposed on the existing physical room.

Fig. 1. The development workflow

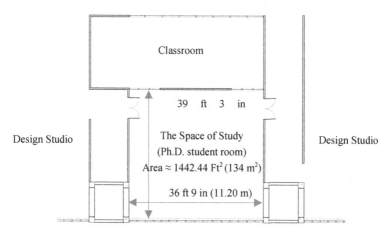

Fig. 2. Space of study floor plan

Fig. 3. The exterior wall of Ph.D. student room.

Fig. 4. Overview of the space showing the students' equipment.

Fig. 5. The BIM model of the Ph.D. room

Fig. 6. The interior of Ph.D. room's BIM model

3.3 Energy Simulation Setup and Scenarios

The building was built in 1977, and the actual building material design was not available. However, based on the researchers' observation and that an assumption could be representative of demonstrating the developed AR app, the researchers assumed that the existing space has single non-reflected glazing exterior windows with $0.166 \, m^2K/W$ $(0.0942 \, Ft^2 \cdot {}^\circ F \cdot h/BTU)$ Thermal Resistance (R-value) and 55% solar heat gain coefficient (SHGC), and precast uninsulated concrete wall for exterior walls with $0.132 \, m^2K/W$ $(0.749 \, Ft^2 \cdot {}^\circ F \cdot h/BTU)$ R-value. The thermal properties of the assumed materials for this building are obtained from EnergyPlus.

The BIMThermoAR allows users to interact with two alternatives of high-performance building materials which have higher R-values than the existing materials. One glazing material for the exterior windows has $1.419 \, m^2K/W$ $(8.059 \, Ft^2 \cdot {}^\circ F \cdot h/BTU)$ R-value and 55% SHGC. One opaque material for the exterior consists of three layers starting from the interior layer, 12 in. Normal weight concrete floor, which is the existing wall material, insulation- R19, and metal siding. The total R-value is $3.478 \, m^2K/W$ $(19.749 \, Ft^2 \cdot {}^\circ F \cdot h/BTU)$. The combination of the two existing and two high-performance materials results in four energy simulations conditions:

- Existing condition

  o Walls and windows are all constructed with the existing materials.

- High performance walls

  o Walls are made of the proposed high performance opaque material, and the windows are from the existing condition.

- High performance windows

  o Windows are constructed of the proposed high performance glazing materials, and the walls are from the existing condition.

- High performance walls & windows

  o Walls and windows are all made of the proposed high-performance materials.

  BIMThermoAR provides each simulation condition with seven simulations results:

- Overall energy consumption
- Summer – interior wall temperature map
- Summer – Vertical map
- Summer – Volumetric map
- Winter – interior wall temperature map
- Winter – Vertical map
- Winter – Volumetric map

The overall energy consumption is based on the yearly total energy usage of only the study space utilizing the College Station weather data [17]. The summer and winter simulation periods are, according to the College Station weather data, the warmest and coldest times of the year [17]. The summertime is August 3rd at 3:00 PM, and the wintertime is January 1st at 7:00 AM. The interior wall temperatures are the surface temperatures of the interior layer of walls, which have been visualized by simple single surfaces that are overlaid on the BIM model. The vertical interior temperature maps are a series of 250 sensors that are stacked vertically to form a surface. The volumetric interior temperature maps are a collection of 360 sensors that are evenly distributed in the room, and each sensor is represented by one box floating in the space. The indoor temperature is calculated by the Ladybug plugin for the Grasshopper spatial thermal map tool, which is an integration of three methods of spatial thermal map simulation [18–20]. The Ladybug simulation method has been validated and showing an acceptable range of consistency compared to ENVI-met software and field measurements [21, 22]. The energy consumption and walls temperature have been simulated using EnergyPlus engine. The simulation time could vary depending on the computing power. The computer used for this study is a MSI laptop with Intel i7-7700HQ CPU, 32 GB Ram, and Nvidia 1070 RTX GPU. The simulation time for each run is approximately 25 min for the entire year simulation and around 30 to 45 min for the thermal mapping.

3.4   BIMThermoAR Prototype Implementation

Registration

Vuforia is one of the most widely used and user-friendly augmented reality software development kits (SDKs). Vuforia is able to identify 2D/3D objects, words, barcodes, or QR codes using computer vision technology, and it has the capability to track from a distance of between 1 and 4 m. BIMThermoAR takes advantage of the power of Vuforia Area Target in order to register and overlay the BIM on the actual world. Then the results of the simulation are implemented into the BIM. The Area Target needs a scanned version of the real items, which is then registered in an Area Target generating program. Through Unity, our app can access the visual components and then apply digital material to a virtual layer. Following the completion of this procedure, digital material is generated in real-time while 3D objects are scanned using a camera. Users are given the opportunity to see 3D information from a variety of perspectives thanks to motion tracking and enhanced orientation capabilities. Touchscreens may also have buttons added to the virtual layer, allowing users to interact with and alter the objects parametrically shown on the screen.

Registration Evaluation

The quantifiable registration assessment is challenging due to the depth of perspective; thus, taking a direct measurement will not result in an accurate reading. In this study, registration evaluation consisted of two stages: initially, visually examining a number of images. Secondly, manually measuring the distances of the sampled points on the edges between the virtual model and the physical building. The manual measurement was performed in Adobe InDesign by initially drawing a vertical or a horizontal line on the image of known physical height or width, which is rotated to match the perspective lines in the screenshot images of the structure that was being measured. The next step was to scale the image of the screenshot so that it corresponded with the vertical/horizontal line. The final step was to measure the difference between the virtual model and the actual space, as illustrated in Figs. 7, 8, and 9. The inaccuracy across the entirety of the virtual model is approximately 1.7 cm on average. This method creates an approximate reading of the misalignment between the virtual model and the physical building, which gives an indication of the registration accuracy along with visual observations.

User Interface (UI)

BIMThermoAR used the graphical user interface (GUI) idea by integrating the GUI on the two-dimensional canvas that is the screen of the mobile device. The user interface of BIMThermoAR is built using buttons, toggles, and sliders to regulate how the user interacts with the software. The BIMThermoAR has a total of three different user interface menus. The application begins with the main menu (shown in Fig. 10), which is composed of five elements in ascending order. The first button launches of the registration procedure and activates the BIM Metadata panel, which is located in the upper left-hand corner of the screen. The second button brings up a selection of several simulation scenarios (Fig. 11). The third button shows up a window that provides definitions of the R-Value and the solar heat gain coefficient (SHGC), in addition to providing basic information on the simulation time (Fig. 12). The fourth component is a slider that regulates the degree to which the overlaid BIM is see-through. When there is a lot of light coming in from different directions, the fifth slider, which regulates the transparency of the map's colors, comes in help. The user may lower the transparency to see the colors of the map more clearly.

The simulation conditions are broken down into the following four categories on the menu for the simulation scenarios: 1- Existing Condition 2- High performance exterior walls 3- High performance exterior windows 4 – high performance exterior windows and walls (Fig. 11). Each condition toggle makes active three sets of linked data, the first of which is the thermal maps menu (Fig. 13). Second, the R-value panel illustrates

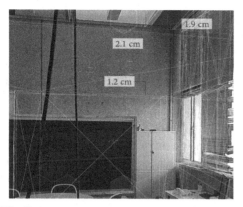

Fig. 7. The corner of the intersection between the exterior and interior walls

Fig. 8. The edges of the exterior wall

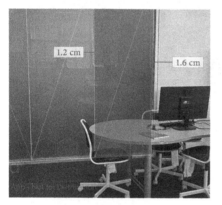

Fig. 9. The edges of the interior curtain wall

Fig. 10. BIMThermoAR main menu

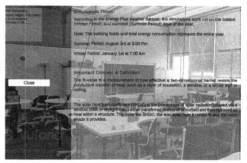

Fig. 11. Simulation scenarios menu

Fig. 12. General information panel

how the properties of the materials change in response to each condition. Third, the energy consumption table presents the total energy consumption along with the cooling load, heating load, and a comparison of those loads to the condition that is currently in place (Fig. 13). Proceed to the thermal map menu (Fig. 13), which primarily consists of two submenus: summer and winter simulation period. Vertical maps and map values (Fig. 14), Volumetric maps and map values (Fig. 15), and wall temperature are the fifth distinct alternatives that are available throughout each simulation time (Fig. 16). Every choice makes the corresponding color legend come into effect. Because of this, the entire application's summer vertical and volumetric maps use the same color legend. This concept is then applied to the application's winter vertical and Volumetric maps, summer wall temperatures, and winter wall temperatures in order to make reading the simulation outcomes easier and to compare them.

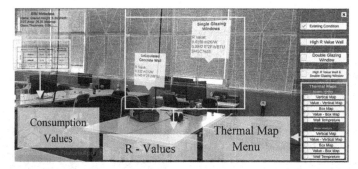

Fig. 13. The linked data to each simulation condition (Consumption table, R-values, and thermal menu)

Fig. 14. The existing condition vertical map with its values. Note: The same visualization is applied for all simulation conditions

Design for User Experience (UX)

BIMThermoAR was developed with performance, a seamless user experience, and the overall user experience and satisfaction in mind throughout development. The application has a number of useful features that improve the experience of using it. The users

Fig. 15. The existing condition volumetric map with its values.

Fig. 16. The existing condition wall temperatures.

will be clearly informed about their choices by the BIM selection display, which converts the chosen component into a transparent red color (Fig. 17). This helps to minimize confusion. The simulation scenarios menu (Fig. 11) has a feature that, when a user selects a simulation condition, displays a few map possibilities. The user's selection is then saved, which enables the user to compare maps more effectively in various simulation setting. Both the user's experience and the app's overall speed may be improved because to two aspects that BIMThermoAR integrates into the map's value. The map values are presented in relation to the position of the mobile camera in the room; as a result, this feature generates a sphere with a center point at the mobile camera and a circumference of three meters. Every value that falls within the spherical range will be displayed, however any value that falls outside of that range will not be shown (Figs. 14 and 15). This feature is one of the crucial aspects of the user experience and app performance because of 250 values in each vertical map and 360 values in each Volumetric map; consequently, it will be challenging for mobile hardware to render all of these meshes. When using the initial number of values, the value geometries obscure the view of the surrounding physical world, making it difficult to read since they are either physically far from the user or clearly overlapping one another in the volumetric maps. By lowering the total number of values, users will have a more streamlined experience, as well as increased engagement and immersion in the surrounding physical world. The map values in BIMThermoAR are actively tracked by the mobile camera to guarantee that users can read the values from any angle and location, particularly the values near the ceiling that are near the top of the map. This feature makes use of the location of the camera to update the x or y axis rotation of the meshes. In any given position, the normal of these meshes will be directed toward the center of the camera (Fig. 14, Fig. 15).

Fig. 17. BIM selection technique.

4   Conclusion and Future Work

The novel BIMThermoAR has shown the potential for augmented reality (AR) to be used in the area of energy simulation and visualization. It can visualize the otherwise invisible information in a built environment. This can relate simulation results to the real world, which in turn could improve the users' comprehension of the simulation results. The introduction of the thermal properties of the materials into the AR environment, together with the possibility of altering the material, exemplifies the close connection that exists between the thermal properties of the materials and the performance of the building. It is made easier to grasp the connection between the input and output of the building energy simulation by providing a visual representation of the thermal impact caused by each material in the form of spatial thermal maps and colored wall temperature. Additionally, the users are better prepared to read the numerical result of the simulation's energy consumption, which is typically the most difficult component to comprehend for the majority of students, engineers, and architects. This is because the visual representation of the simulation outcomes displays the outcomes in a way that is easily understood. By utilizing the BIMThermoAR, they will have a better understanding of the overall pattern as well as the kind of material they should use for improved performance. A portion of the numerical data has been laid out in a table inside BIMThermoAR. This has been done to enable the user to quickly correlate the thermal visual display of the simulation with the numerical results about the amount of energy used. The prototype displays pre-simulated results. This project intends to improve the understanding of building energy simulations and performance. It would be a great addition to the future development of this prototype to include real-time sensor data display for understanding the building's existing condition [23, 24] and comparing it with design alternatives that can be visualized in AR.

The BIM metadata display in BIMThermoAR connects the energy modeling with the building materials and assemblies, allowing the user to get familiarized with the current structure. Users are able to check their selection and understand the assistance of a choosing method that can be found in the BIM metadata panel. This method converts the chosen component into one with a transparent red color. The test case in this research is an example building room and is limited in representing a large building scale. The AR registration platform used in this project (Vuforia Area Target) could support and cover large areas with multiple rooms. As AR-enabled mobile device or future AR glasses are becoming more powerful in hardware and software capabilities, we expect large scale and complex buildings can utilize BIMThermoAR, either through a single, complete building model, or connected separate building area models (sub-models).

AR applications have the potential to be helpful and transformative after overcoming the challenges, e.g., the time consuming AR registration process using 3D scanning to correct BIM to match with the physical reality. It is necessary to develop a robust registration method in order to make applications resistant to the slight modifications that may be made to the physical environment. BIMThermoAR has various potential applications for students, architects, engineers, and other users. The tool may be used for educating building energy simulation and performance, Building Information Modeling, building rehabilitation design, and more. The prototyping workflow can be adapted to each user group with different emphases, for example, for students, visualizing the fundamental building physics principles for them to improve learning, for architects, visualizing the design performances of different design options for enhancing architectural design, and for engineers, visualizing detailed simulation outcomes for validating and optimizing engineering design.

It is anticipated that the visualization approaches provided by BIMThermoAR will contribute to an improved comprehension of building performance simulation among architects, engineers, and students. Students studying architecture and engineering will be included in our planned assessment of the BIMThermoAR visualization tools. In the study, a conventional method of studying the building energy simulation and the BIMThermoAR will both be used, and the researchers will compare the learning gains that each method produces from the students. In order to provide a complete assessment of the impact of the prototype on the learning curve of students, the research will involve students with a variety of education levels.

Acknowledgements. This material is based upon work partially supported by the National Science Foundation under Grant No. 2119549 and Mattia Flabiano III AIA/Page Southerland Design Professorship at Texas A&M University.

References

1. Dünser, A., Billinghurst, M., Wen, J., Lehtinen, V., Nurminen, A.: Exploring the use of handheld AR for outdoor navigation. Comput. Graph. **36**(8), 1084–1095 (2012). https://doi.org/10.1016/j.cag.2012.10.001
2. Milgram, P., Kishino, F.: A taxonomy of mixed reality visual displays. IEICE Trans. Inf. Syst. **77**(12), 1321–1329 (1994)

3. Sommerauer, P., Müller, O.: Augmented reality in informal learning environments: a field experiment in a mathematics exhibition. Comput. Educ. **79**, 59–68 (2014). https://doi.org/10.1016/j.compedu.2014.07.013

4. Perez-Lopez, D., Contero, M.: Delivering educational multimedia contents through an augmented reality application: a case study on its impact on knowledge acquisition and retention. Turk. Online J. Educ. Technol. TOJET **12**(4), 19–28 (2013)

5. Adedokun-Shittu, N.A., Ajani, A.H., Nuhu, K.M., Shittu, A.K.: Augmented reality instructional tool in enhancing geography learners academic performance and retention in Osun state Nigeria. Educ. Inf. Technol. **25**(4), 3021–3033 (2020). https://doi.org/10.1007/s10639-020-10099-2

6. Diao, P.-H., Shih, N.-J.: Trends and research issues of augmented reality studies in architectural and civil engineering education—a review of academic journal publications. Appl. Sci. **9**(9), 1840 (2019)

7. Furht, B.: Handbook of augmented reality. Springer Science and Business Media, New York (2011)

8. Carmigniani, J., Furht, B., Anisetti, M., Ceravolo, P., Damiani, E., Ivkovic, M.: Augmented reality technologies, systems and applications. Multimedia Tools Appl. **51**(1), 341–377 (2011)

9. Lee, K.: The future of learning and training in augmented reality. InSight J. Sch. Teach. **7**, 31–42 (2012)

10. Azuma, R., Baillot, Y., Behringer, R., Feiner, S., Julier, S., MacIntyre, B.: Recent advances in augmented reality. IEEE Comput. Graphics Appl. **21**(6), 34–47 (2001)

11. Azuma, R.T.: A survey of augmented reality. Presence Teleoperators Virtual Environ. **6**(4), 355–385 (1997)

12. Chin, Z.Y., Ang, K.K., Wang, C., Guan, C.: Online performance evaluation of motor imagery BCI with augmented-reality virtual hand feedback. In: Proceedings of the 2010 Annual International Conference of the IEEE Engineering in Medicine and Biology, pp. 3341–3344 (2010)

13. Rashed-Ali, H., Quarles, J., Fies, C., Sanciuc, L.: Use of augmented-reality in teaching energy efficiency: prototype development and testing. In: ARCC Conference Repository (2014)

14. Zhao, S., Zhang, L., de Angelis, E.: Using augmented reality and mixed reality to interpret design choices of high-performance buildings (2019)

15. Fukuda, T., Yokoi, K., Yabuki, N., Motamedi, A.: An indoor thermal environment design system for renovation using augmented reality. J. Comput. Des. Eng. **6**(2), 179–188 (2019)

16. AR App Development With Vuforia Engine — PTC. https://www.ptc.com/en/products/vuforia/vuforia-engine/ar-app-development. Accessed 03 Jun 2022

17. EnergyPlus. https://energyplus.net/weather-search/College%20Station. Accessed 03 Jun 2022

18. Webb, A.L.: Mapping comfort : an analysis method for understanding diversity in the thermal environment. Thesis, Massachusetts Institute of Technology (2012). https://dspace.mit.edu/handle/1721.1/72870. Accessed 01 Jun 2022

19. Brandan, M., Alejandra, M.: Study of airflow and thermal stratification in naturally ventilated rooms. Thesis, Massachusetts Institute of Technology (2012).https://dspace.mit.edu/handle/1721.1/74907. Accessed 01 Jun 2022

20. Arens, E., Hoyt, T., Zhou, X., Huang, L., Zhang, H., Schiavon, S.: Modeling the comfort effects of short-wave solar radiation indoors, June 2015. https://doi.org/10.1016/j.buildenv.2014.09.004

21. Ibrahim, Y.I., Kershaw, T., Shepherd, P.: A methodology for modelling microclimate: a ladybug-tools and ENVI-met verification study. In: Proceedings of the 35th PLEA Conference Sustainable Architecture and Urban Design: Planning Post Carbon Cities (2020)

22. Elwy, I., Ibrahim, Y., Fahmy, M., Mahdy, M.: Outdoor microclimatic validation for hybrid simulation workflow in hot arid climates against ENVI-met and field measurements. Energy Procedia **153**, 29–34 (2018). https://doi.org/10.1016/j.egypro.2018.10.009
23. Shahinmoghadam, M., Natephra, W., Motamedi, A.: BIM-and IoT-based virtual reality tool for real-time thermal comfort assessment in building enclosures. Build. Environ. **199**, 107905 (2021)
24. Nytsch-Geusen, C., Ayubi, T., Möckel, J., Rädler, J., Thorade, M.: BuildingSystems_VR–A new approach for immersive and interactive building energy simulation. In:Proceedings of the Fifteenth IBPSA Conference on Building Simulation 2017, pp. 628–634 (2017)

A Framework for Monitoring and Identifying Indoor Air Pollutants Based on BIM with IoT Sensors

Jihoon Chung(✉) ⓘ, Alexandros Tsamis ⓘ, and Dennis Shelden ⓘ

Rensselaer Polytechnic Institute, Troy, NY 12180, USA
chungj11@rpi.edu

Abstract. Americans spend 86.9% of their life in buildings; however, about 1.64 million people died in 2019 due to diseases related to indoor air pollution. In the indoor air, thousands of chemical substances exist, and we have limited data to identify the source of the pollutants. Internet of Things (IoT) technology recently has addressed low resolution of spatio-temporal air quality data; but there remain limitations in connecting the sensor data to spatial information. This study integrates IoT sensors integrated with a Building Information Modeling (BIM) database, tracking indoor air quality in real-time, providing higher fidelity assessments of the pollutant sources using the locations and properties of building components. This paper proposes a framework for an indoor air quality monitoring system to achieve the following objectives: 1) Integrate IoT sensor data with BIM data sources into an integrated database, 2) Analyze the sensor data and estimate the probable area where the air pollutant sources can be located, 3) Suggest viable solutions for mitigating the air pollution. To demonstrate the framework, a system prototype has been developed, and two pilot tests and a case study have been implemented as proof-of-concept in a university laboratory. This result can be a basis to develop air quality monitoring infrastructure for understanding where the indoor air pollutants come from and how to deal with the problems in real-time.

Keywords: Building Information Modeling · Internet of Things · Indoor Air Quality · Air Quality Monitoring · Building Infrastructure

1 Introduction

According to a result of a two-year national telephone survey in the 48 contiguous United States, residents spend 86.9% of their life in buildings (Klepeis 2001); however, the Global Burden of Disease Study in 2019, conducted by Institute for Health Metrics and Evaluation (IHME), reported that about 1.64 million people annually died in 2019 due to indoor air pollution (OurWorldData 2022). Saini et al. (2020) review major indoor air pollutants, such as sulphur dioxide (SO_2), carbon oxides (CO), nitric oxides (NOx), particulate matters (PM), Ozone, Volatile Organic Compounds (VOCs) (Saini et al. 2020). These compounds cause various health issues, from minor symptoms like headache, dizziness, tiredness, or poor concentration to health diseases like exacerbation of asthma, allergic skin reaction, or lung function; however, the health issues are difficult to identify its cause or source after having the symptoms or getting the diseases, because thousands of chemical substances exist in the air and most of them are colorless and odorless (Guyot et al. 2018). Since the 1990s, sick building syndrome, which means the confusion that comes from the diversity of ill health effects and the difficulty of finding their specific causes in a building, has increased attention (Murphy 2006). Many empirical studies were conducted to untangle this complexity; however, it is necessary to analyze quantitative data using sensors to identify the specific causes.

As an initial step to reach the ultimate goal, this proposes a framework for monitoring IAQ in real-time and estimating the probable area, where the air pollutant sources can be located, based IoT sensors and a BIM database. The framework consists of the following major parts: air quality monitoring module, data analysis module, and ventilation module. To demonstrate the proposed framework, two pilot tests and a cast study are conducted by developing a prototype of a BIM and IoT-based system. Furthermore, the result and its limitations were discussed for future works.

2 Backgrounds

Conventional air quality monitoring systems automatically measure air pollutants; nevertheless, there are still several limitations: cost & maintenance, accuracy, 3D data attainment, absence of active monitoring, flexibility/scalability, and power consumption (Idrees et al. 2018). Particularly, many sensors are needed for dense networks to acquire high resolution of spatio-temporal information in real-time; however, high cost, limited scalability, and high-power consumption are major obstacles to expanding and maintaining the sensor network (Yi et al. 2015; Morawska et al. 2018; Zauli-Sajani et al. 2021; Castell et al. 2017).

Last two decades, Internet of Things (IoT) technology has attracted attention from academia and industry due to its low-cost, compact size, wireless network, and easiness of interoperability (Marques et al. 2019; Senthilkumar et al. 2020; Idrees et al. 2018; Floris et al. 2021; Mahbub et al. 2020). This technology has been also applied to the air quality monitoring field and enabled lower cost, less power consumption, as well as wide-scale and dense network system (Spandonidis et al. 2020; Wesseling et al. 2019; Zimmerman et al. 2018; Kang and Hwang 2016; Saini et al. 2020; Luo et al. 2018). Nevertheless, lots of effort is still required to identify the source of air pollution because the pollutants are caused by many different factors. According to Tham (2016), the Indoor Air Quality (IAQ) is affected by materials (wall finishes, furniture, etc.), equipment, Heating, Ventilation, and Air Conditioning (HVAC) system, occupant activities, respiratory contamination from humans, and so on (Tham 2016). Too many factors make it difficult to understand where the air pollutants come from, even if advanced sensing technologies are used.

How can we identify and resolve indoor air pollutants? The Environmental Protection Agency (EPA) in the United States provides practical guidelines for building owners and facility managers to deal with this issue. Nonetheless, it still requires much effort and time for a series of manual investigations, such as reviewing existing records, walkthrough investigations, developing hypotheses, and conducting tests (EPA 2014, 1998).

This is because all the related information is not stored and analyzed in one database. In the architecture, engineering, and construction (AEC) industry, since the 2000s, BIM has developed as a multidisciplinary data repository including 3D geometries, building properties, and time variables. This database is mainly built to contain detailed information of the designed building for construction, but it can be also used for thermal-, daylighting-, occupant-behavior simulation, and fault detection (Micolier et al. 2019; Kota et al. 2014; Bahar et al. 2013; Golabchi et al. 2016).

Based on the literature review, we hypothesize that if a BIM model from the design and engineering phases is reused and combined with IAQ data from IoT sensors in the operating & maintenance phase, it could provide rich information to investigate the location of air pollution sources, such as XYZ coordinates of elements, building materials, maintenance history, and so on.

3   Research Method

IoT sensors and BIM database have emerged as an innovative technology in each field; nevertheless, they still have several limitations as outlined in the previous section. To overcome them, this study aims to integrate an IoT-based IAQ monitoring system with a BIM database. The integration of the IoT technology and the BIM database can enable analyzing relationships among sensors based on their position and to update building information in real-time. Furthermore, these analyses can be used to evaluate the influence of the building's components. This paper proposes an IAQ monitoring system framework to achieve the following objectives: 1) to integrate IoT sensor data with the BIM database, 2) to analyze and estimate the probable area of an air pollutant source, 3) to suggest viable solutions for mitigating air pollution. This paper presents a methodology that can be divided into three parts, as shown in Fig. 1:

3.1   Air Quality Monitoring Module

Firstly, in the air quality monitoring module, air quality data from sensors is integrated with a BIM database and is visualized on a 3D BIM model. If the system begins working, statuses of monitored window, door, HVAC are updated through IoT sensors. This data is needed to calculate wind vectors from Computational Fluid Dynamics (CFD) simulation, and it should be calculated and stored as a file in advance because it requires much computational cost. Then, air quality data from sensors is collected and forwarded to a BIM database. Each sensor data has a unique identification of a physical sensor module, and it is combined with an identical identification of a digital sensor element in the BIM database. Each sensor element has XYZ coordinates, and the measured data can be visualized on the 3D model based on the location information in real-time. If the air pollution data exceeds a threshold, such as values specified in United States Environmental Protection Agency (EPA) or Leadership in Energy and Environmental Design (LEED) guidelines, the data is forwarded to the next module.

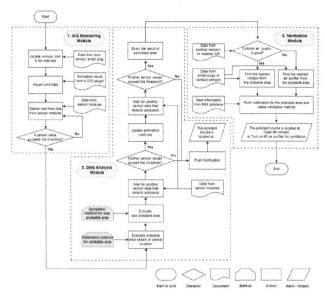

Fig. 1. Proposed Framework for Monitoring Indoor Air Quality

3.2 Data Analysis Module

In the analysis module, the system analyzes the collected data from sensors and building elements in a BIM database to identify the location of air pollutant sources. It estimates the probable area, where the air pollutant sources can be located, based on information from the BIM database and IoT devices, such as detection time, sensor location, equipment status, HVAC system, gaseous concentration, or wind data. This paper suggests a method to estimate the probable area rather than the exact location of the pollutant sources because, in practice, it is unfeasible to precisely get many parameters for applying gas diffusion laws, such as diffusion coefficients or molar masses of the pollutants. Figure 2 represents the method for estimating the probable area of air pollution sources in detail. It assumes that without external force like wind flow, air pollutants are statistically evenly diffused in all directions in the air, following Brownian motion theory. In this assumption, the closer the source of the air pollutants, the earlier the sensor detects the pollutants. Based on the location of the triggered sensors, the probable area can be estimated. To divide the whole area based on sensor locations, this method utilized Voronoi tessellation which partitions a plane with given points into polygons. Each polyline is generated as a centerline between two adjacent sensor points, and each polygon consists of the centerlines. However, in the real world, wind flow is a major factor that prevents the air pollutant from being evenly diffused in all directions, and it causes a large gap between the estimated and actual locations. To reduce the gap, in this paper, the generated Voronoi diagram is adjusted using wind vectors from CFD simulation or wind sensors. The system relocates the sensor points of the polygons along the reverse direction of the wind vectors. And then, it re-generates the polygons based on the relocated points because air pollutants move around in a space along the wind directions. As a result, the estimated probable area is changed depending on wind velocity, the location, and the number of sensors because these factors affect the uncertainty of the estimation. As shown in Fig. 3, based on the assumption, the area with less probability can be identified and excluded even if it is close to the first triggered sensor. Because the area is relatively far from the second and third triggered sensors. Using the coordinates and sequence of the triggered sensors, the system can narrow down the probable area.

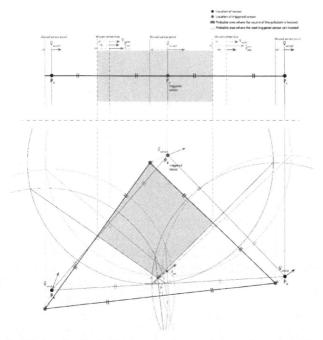

Fig. 2. Method of Estimating Probable Area in One- and Two-Dimensional Space with Wind Flow

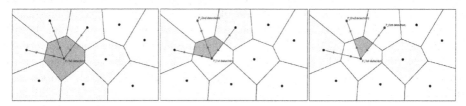

Fig. 3. Method of Excluding the Area with Less Probability using the Sequence of Triggered Sensors

3.3 Ventilation Module

Finally, the ventilation module suggests viable solutions for remediating air pollution by analyzing building elements' status and outdoor air quality. Based on the estimated probable area derived from the data analysis module, the system determines viable solutions considering the outdoor air quality and building information such as statuses of windows and air purifiers. If the outdoor air quality is good, it suggests opening the nearest windows. Otherwise, turning on the nearest air purifier is recommended. Using smart plugs that collect power consumption data, the system can identify which equipment is turned on or off. Furthermore, if indoor air pollution occurs, the system pushes notifications on the nearest occupant's phone via a mobile application or SMS to urge action for mitigating the pollution.

3.4 Prototype of Air Quality Monitoring System

To demonstrate the proposed framework, a system prototype was developed for integrating BIM and IoT sensors, as shown in Fig. 4. The developed system consisted of three main components: BIM software, a central server, and IoT sensor modules.

Autodesk's BIM authoring tool, Revit, was used to combine the sensor data from the server with building elements on a BIM model. Dynamo was a visual programming tool in Revit and was used to automate extracting, processing, and updating building component data from the BIM model. In this environment, real-time data and a chunk of historical data with timestamps were forwarded from the server to the BIM software. All the data was transmitted in JavaScript Object Notation (JSON) format which was an open-standard data interchange and language-independent format. The real-time data from the server was monitored and visualized on the BIM model, and if air pollution occurred, the historical data was analyzed for identifying the location of air pollutant sources. In the BIM database, seat information was used to contact occupants, including names, email addresses, or phone numbers. Based on XYZ coordinates, equipment status, outdoor air quality, and seat information, the system suggested a viable solution to the nearest occupants in the space by pushing notifications on their phones through API, as described above.

The central server was established based on Node-RED that was built on Node.js and was a flow-based programming tool like Grasshopper in Rhinoceros or Dynamo in Revit. This visual programming tool provided diverse packages for wiring hardware devices, APIs, and online services together, such as a dashboard, machine learning package, python shell, or MongoDB. Communication between the server and sensors modules was based on Message Queuing Telemetry Transport (MQTT) protocol which was a lightweight messaging protocol using publish or subscribe method, and it was suitable to run low-cost hardware. On the Node-RED server, the data collected from the sensor modules were merged with sensor IDs and timestamps and was stored on a NoSQL database, MongoDB. This type of database stored data without defining schema upfront, required less transformation for moving data, and allowed a faster query execution process than relational databases (Li and Manoharan 2013). In addition, it could support increased traffic without downtime and is ideal for IoT-based applications (Rautmare and Bhalerao 2016).

To get measured values from sensors, ESP-WROOM-32 development boards were deployed, and they forwarded it to a central server via a WiFi network. Each board was connected to a sensor, such as Sensirion SEN54 sensors for Total Volatile Organic Compounds (TVOC), temperature, humidity, and FS3000 sensors for wind flow. To measure power consumption and opening statuses of the windows or door, commercial smart plugs and door sensors were installed, such as Sengled E1C-NB7 and E1D-G73. These sensors forwarded the data to the server via Zigbee and MQTT protocol.

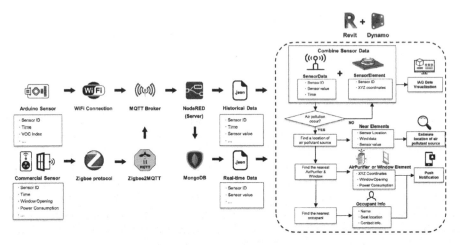

Fig. 4. System Architecture of indoor air monitoring system

4   Results and Discussion

The analysis module was the main part of the system in this study, and two pilot tests were conducted to ensure that the prototype could appropriately estimate the probable area of air pollutant sources based on sensor information and a BIM database. The tests were conducted in a meeting room of an academic research center. A BIM model was set up using Revit software, including sensor, furniture, and HVAC system elements, as shown in Fig. 5. Using the 'Butterfly' CFD plugin in the BIM software, airflow vectors were generated in advance and stored in a comma-separated values (CSV) file to avoid computational overload, as shown in Fig. 6. For these pilot tests, SEN54 sensors for TVOC and E1D-G73 sensors for door and window opening were deployed, as shown in Fig. 7. Ten sensor modules were installed on the table, and a source of air pollutants, nail polish remover, was placed in-between the devices, as shown in Fig. 8.

The pilot tests were conducted for two cases: 1) a case that the windows and door were closed and 2) the other case that they were opened. The tests begin when a box containing the source of air pollutants was opened; the tests ended when all the sensors detected the pollutants. During the experiment, all the sensor values and the sequence of the triggered sensors were recorded. SEN54 sensor provided a VOC index that describes the relative amount of TVOC compared to the recent history of sensors. The average value is 100, and a value above 100 means more amount of TVOC compared to the average. In the tests, the VOC index above 150 was regarded as the detection of air pollution.

Figure 9 represents the results of the first case without opening the windows and the door. The sequence of the triggered sensors was Sensor 10, Sensor 6, Sensor 9, Sensor 7, Sensor 3, Sensor 1, Sensor 2, Sensor 5, Sensor 8, and Sensor 4. Figure 10 represents the result of the second case with opening the windows and the door. In this case, wind vectors were imported from CFD simulation due to the opened windows and door. The measured air speeds at the windows were 0.02 m/s to 0.04 m/s, and the speed at the door was 0.03 m/s to 0.04 m/s. The sequence of the triggered sensors was Sensor 10, Sensor 6, Sensor 9, Sensor 7, Sensor 3, Sensor 5, Sensor 2, Sensor 4, Sensor 1, and Sensor 8. As a result, for both cases, the first three triggered sensors were Sensor 10, Sensor 6, and Sensor 9, and the system identified the location of the pollutant source well based on the proposed method, as shown in Fig. 11.

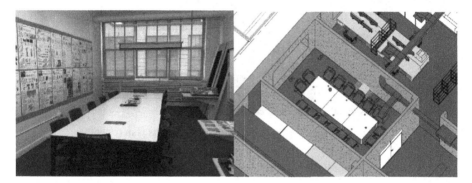

Fig. 5. Image of the Meeting Room (left) and its BIM Mode (right)

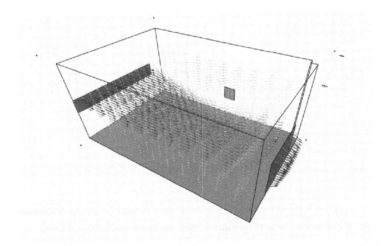

Fig. 6. Indoor Air Flow Simulation using CFD plugin 'Butterfly'

Next, a case study was conducted in the laboratory space to test the entire framework. The prototype system was deployed with 13 sensor modules, as shown in Fig. 12. Three

Fig. 7. Image of prepared sensor modules (left), door sensor (middle), and window sensors (right)

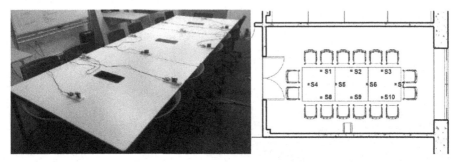

Fig. 8. Image of experiment setting (left) and its plan (right)

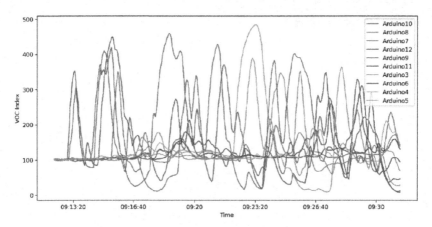

Fig. 9. Result of Experiment without opening the windows and the door

air purifiers were connected to smart plugs and were placed for mitigating indoor air pollution. In the BIM model, seat information was set up on each desk including occupants' names, phone numbers, and email addresses. This information was used for pushing notifications on the occupants' phones via message, mobile application, or email, as shown in Fig. 13. In the same way as the pilot tests, the source of air pollution was placed on a desk, and after detecting air pollution, and the excess sensor values were visualized on the BIM model as red color, as shown in Fig. 14. At the same time, the probable area was

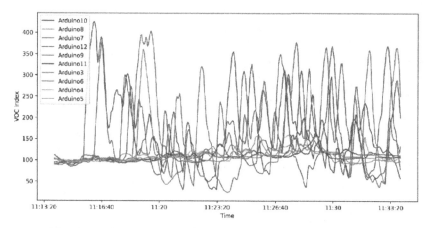

Fig. 10. Result of Experiment with opening the windows and the door

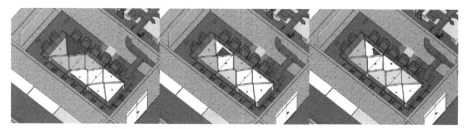

Fig. 11. Pilot Test of Data Analysis Module: First Detection (left), Second Detection (middle), Third Detection (right)

presented based on sensor location and wind data from CFD simulation, and the system pushed a notification to the occupant whose seat was the closest to the area. The message was sent via the IFTTT application and included a type of pollution, VOC index, and a viable solution depending on outdoor air quality data from OpenWeatherMap API. In this case, the system suggested opening the nearest window because the outdoor air quality index was 45, 'Good'. For getting more accurate data, outdoor sensor modules could be installed near windows in future works. The alert information is stored in an excel file with the time, sensor data, location of the estimated probable area, outdoor air quality data, suggested action, notification receiver's name, and inspector's note. The default value of the inspector's note is empty; but, after mitigating the air pollution, the inspector would provide detailed pollution information, such as the pollutant source, the actual location of the source, the cause of the air pollution, and the actual method to mitigate the pollution. This database, including sensor data and the BIM model, can be reused for re-designing, renovating, or retrofitting the building by architects or engineers in the future.

Fig. 12. BIM Model of Academic Research Center

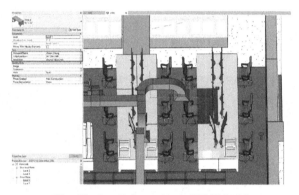

Fig. 13. Seat Information in BIM Model

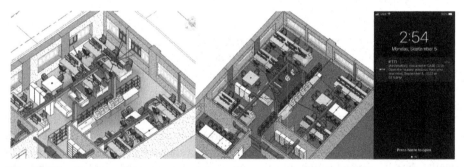

Fig. 14. Visualization of Sensor Values (left), Estimated Probable Area (middle), and Notification on Occupant's Mobile Phone (right)

5 Conclusion

In indoor air, too many kinds of air pollutants are there, and they cause diverse symptoms. Investigating the sources of indoor air pollution requires much time and effort. IoT technology has recently emerged to improve the resolution of temporal and spatial air quality information; however, it is still difficult to interpret the measured data from many sensors. As a multidisciplinary repository, BIM mainly contains detailed information about the designed building, but it can be reused for various purposes. Through the literature review on IAQ, IoT, and BIM. We hypothesize that a BIM model from the design and engineering phases could be reused and combined with IoT sensor data and in the operating & maintenance phase for investigating the location of air pollution sources. This study proposes a framework for integrating IoT with BIM that can provide useful information to interpret the sensor data. The purpose of the framework is to monitor IAQ in real-time and identify the probable area where the air pollutant sources can be located. It is comprised of three parts: an air quality monitoring module, a data analysis module, and a ventilation module. In the monitoring module, IAQ data is monitored and visualized in real time by integrating IoT sensors and the BIM database. In the data analysis, two methods were presented: 1) a method of estimating the probable area of the air pollutant sources using sensor locations and wind vectors from CFD simulation and 2) a method of excluding the areas with less probability using a sequence of triggered sensors. In the ventilation module, the system sends notifications on the occupants' mobile phones about a strategy to mitigate air pollution, based on the seat information, outdoor air quality, status of the window, and air purifier. To demonstrate the framework, a system prototype was developed, and two pilot tests and a case study were implemented in a university laboratory. The results show that the system was able to accurately estimate the probable area of the pollutant sources, and it could push notifications on the occupants' mobile phones to alert the air pollution and to suggest viable solutions depending on outdoor air quality, the statuses of the windows, and air purifiers.

Although this system worked well following the proposed framework, there are several limitations: Firstly, as an initial phase of the research, the experiment was conducted under the controlled conditions that the HVAC system was turned off, and no occupants' or objects' movements, including opening/closing windows or doors, were in the space to ensure that air flow is relatively consistent. This is because supply air flow and all the movements in the space affect indoor air flow and make it too complex to estimate the probable area using CFD simulation that cannot reflect the occupants' movement. In future works, we will further develop the framework to make it robust for the influence factors by adding wind speed and direction sensors to the sensor modules for estimating more complex wind flow in real time. And then, the robustness and reliability of the estimation will be tested for multiple scenarios. Secondly, the experiments were conducted in a small meeting room and laboratory in order to derive clear results; however, to validate its feasibility and practicality, it is necessary to test the proposed framework in a larger-scale space because the indoor air flow should be more dynamic and complex than the air flow in a small space. Thirdly, the used sensors in this study only detected TVOC gases; but, other types of air pollutants should be studied, such as particulate matter, ozone, or formaldehyde, because the behaviors of these pollutants are different. It should be tested whether the algorithm is still valid for the other gases or not. Fourthly, the performance of the proposed and the previous systems should be measured and compared to show how accurately the system can estimate the probable area of a pollutant source.

The main contribution of this study is to open a new way of monitoring IAQ and estimating the probable area of an air pollutant source based on building information and high resolution of spatio-temporal air quality information from sensors. This study shows a potential use case that a BIM model from the design and engineering phases can be reused and provide contextual data to interpret sensor data from IoT sensors in the operating & maintenance phase. The database assists the system to estimate the probable area, alert occupants, and suggest a viable solution via their mobile phones. We aim to integrate this system with a building management system or a web platform so that potential users can easily monitor IAQ and get a notification on their phones if indoor air pollution happens. The potential users include occupants, office managers, or facility managers who are responsible for managing acceptable indoor environmental quality. Currently, we are developing a BIM & IoT-based web platform adopting the proposed framework to improve accessibility to this system. Furthermore, in future works, the proposed system will be integrated with a prediction model with machine learning techniques so that the system can predict IAQ and suggest preventive actions in advance. This research can be a basis to develop air quality monitoring infrastructure in buildings or cities to understand where the air pollutant comes from and how to deal with the pollution problems for public health.

References

Klepeis, N.E., et al.: The national human activity pattern survey (NHAPS): a resource for assessing exposure to environmental pollutants. J. Eposure Sci. Environ. Epidemiol. 11(3), 231–252 (2001). https://doi.org/10.1038/sj.jea.7500165

Bahar, Y.N., Pere, C., Landrieu, J., Nicolle, C.: A thermal simulation tool for building and its interoperability through the building information modeling (BIM) platform. Buildings 3(2), 380–398 (2013). https://doi.org/10.3390/buildings3020380

Castell, N., et al.: Can commercial low-cost sensor platforms contribute to air quality monitoring and exposure estimates? Environ. Int. 99, 293–302 (2017). https://doi.org/10.1016/j.envint.2016.12.007

EPA: Building Air Quality Action Plan (1998). https://www.epa.gov/sites/default/files/2014-08/documents/baqactionplan.pdf

EPA: Building Air Quality - A Guide for Building Owners and Facility Managers (2014). https://www.epa.gov/sites/default/files/2014-08/documents/iaq.pdf

Floris, A., Porcu, S., Girau, R., Atzori, L.: An IoT-based smart building solution for indoor environment management and occupants prediction. Energies 14(10), 2959 (2021). https://doi.org/10.3390/en14102959

Golabchi, A., Akula, M., Kamat, V.: Automated building information modeling for fault detection and diagnostics in commercial HVAC systems. Facilities 34(3/4), 233–246 (2016). https://doi.org/10.1108/F-06-2014-0050

Guyot, G., Sherman, M.H., Walker, I.S.: Smart ventilation energy and indoor air quality performance in residential buildings: a review. Energy and Buildings 165, 416–430 (2018). https://doi.org/10.1016/j.enbuild.2017.12.051

Idrees, Z., Zou, Z., Zheng, L.: Edge computing based IoT architecture for low cost air pollution monitoring systems: a comprehensive system analysis, design considerations and development. Sensors 18(9), 3021 (2018). https://doi.org/10.3390/s18093021

Kang, J., Kwang-Il, H.: A comprehensive real-time indoor air-quality level indicator. Sustainability 8(9), 881 (2016). https://doi.org/10.3390/su8090881

Kota, S., Haberl, J.S., Clayton, M.J., Yan, W.: Building information modeling (BIM)-based daylighting simulation and analysis. Energ. Buildings 81, 391–403 (2014). https://doi.org/10.1016/j.enbuild.2014.06.043

Li, Y., Manoharan, S.: A performance comparison of SQL and NoSQL databases. In: Proceedings of the 2013 IEEE Pacific Rim Conference on Communications, Computers and Signal Processing (PACRIM), pp. 15–19 (2013). https://doi.org/10.1109/PACRIM.2013.6625441

Luo, L., Zhang, Y., Pearson, B., Ling, Z., Yu, H., Fu, X.: On the security and data integrity of low-cost sensor networks for air quality monitoring. Sensors **18**(12), 4451 (2018). https://doi.org/10.3390/s18124451

Mahbub, M., Mofazzal Hossain, M., Md. Shamrat Apu Gazi.: IoT-cognizant cloud-assisted energy efficient embedded system for indoor intelligent lighting, air quality monitoring, and ventilation. Internet of Things **11**, 100266 (2020). https://doi.org/10.1016/j.iot.2020.100266

Marques, G., Ferreira, C.R., Pitarma, R.: Indoor air quality assessment using a CO2 monitoring system based on internet of things. J. Med. Syst. **43**(3), 1 (2019). https://doi.org/10.1007/s10916-019-1184-x

Micolier, A., Taillandier, F., Taillandier, P., Bos, F.: Li-BIM, an agent-based approach to simulate occupant-building interaction from the building-information modelling. Eng. Appl. Artif. Intell. **82**, 44–59 (2019). https://doi.org/10.1016/j.engappai.2019.03.008

Morawska, L., et al.: Applications of low-cost sensing technologies for air quality monitoring and exposure assessment: how far have they gone? Environ. Int. **116**, 286–299 (2018). https://doi.org/10.1016/j.envint.2018.04.018

Murphy, M.: Sick Building Syndrome and the Problem of Uncertainty. Duke University Press (2006)

OurWorldData: Indoor Air Pollution (2022). https://ourworldindata.org/indoor-air-pollution

Rautmare, S., Bhalerao, D.M.: MySQL and NoSQL database comparison for IoT application. In: Proceedings of the 2016 IEEE International Conference on Advances in Computer Applications (ICACA), pp. 235–38 (2016). https://doi.org/10.1109/ICACA.2016.7887957

Saini, J., Dutta, M., Marques, G.: A comprehensive review on indoor air quality monitoring systems for enhanced public health. Sustain. Environ. Res. **30**(1), 6 (2020). https://doi.org/10.1186/s42834-020-0047-y

Senthilkumar, R., Venkatakrishnan, P., Balaji, N.: Intelligent based novel embedded system based IoT enabled air pollution monitoring system. Microprocess. Microsyst. **77**, 103172 (2020). https://doi.org/10.1016/j.micpro.2020.103172

Spandonidis, C., Tsantilas, S., Giannopoulos, F., Giordamlis, C., Zyrichidou, I., Syropoulou, P.: Design and development of a new cost-effective internet of things sensor platform for air quality measurements. J. Eng. Sci. Technol. Rev. **13**(6), 81–91 (2020). https://doi.org/10.25103/jestr.136.12

Tham, K.W.: Indoor air quality and its effects on humans—a review of challenges and developments in the last 30 years. Energ. Buildings **130**, 637–650 (2016). https://doi.org/10.1016/j.enbuild.2016.08.071

Wesseling, J., et al.: Development and implementation of a platform for public information on air quality, sensor measurements, and citizen science. Atmosphere **10**(8), 445 (2019). https://doi.org/10.3390/atmos10080445

Yi, W.Y., Lo, K.M., Mak, T., Leung, K.S., Leung, Y., Meng, M.L.: A survey of wireless sensor network based air pollution monitoring systems. Sensors **15**(12), 31392–31427 (2015). https://doi.org/10.3390/s151229859

Zauli-Sajani, S., Marchesi, S., Pironi, C., Barbieri, C., Poluzzi, V., Colacci, A.: Assessment of air quality sensor system performance after relocation. Atmos. Pollut. Res. **12**(2), 282–291 (2021). https://doi.org/10.1016/j.apr.2020.11.010

Zimmerman, N., et al.: A machine learning calibration model using random forests to improve sensor performance for lower-cost air quality monitoring. Atmos. Measur. Tech. **11**(1), 291–313 (2018). https://doi.org/10.5194/amt-11-291-2018

Coupling Co-presence in Physical and Virtual Environments Toward Hybrid Places

Davide Schaumann[1]([⊠]) [iD], Noam Duvdevani[2], Ariel Elya[2], Ido Levin[2], Tom Sofer[2], Ben Drusinsky[1], Ezra Ozery[1], Ofek Raz[1], and Tom Feldman[1]

[1] Faculty of Architecture and Town Planning, Technion – Israel Institute of Technology, Haifa, Israel
davide.schaumann@gmail.com
[2] Faculty of Computer Science, Technion – Israel Institute of Technology, Haifa, Israel

Abstract. Recent advancements in information and communication technologies (ICT) including Internet-of-Things (IoT), Extended Reality (XR), and Digital Twins (DT) enabled a massive leap toward the coupling of physical and virtual systems for working, living, learning, and playing. Current approaches mostly focus on augmenting physical environments with digital information to improve the design and operations of buildings and cities. Fewer approaches explored a deeper integration of physical and virtual environments to create places for meaningful social interactions across realities. In this paper, we discuss the concept of 'hybrid places' that couple elements of physical and virtual places to unlock new opportunities for situated social interactions. Specifically, we conceptualize a framework for hybrid places where a virtually-augmented *physical* place (the hybrid-*physical*) and a physically-augmented *virtual* place (the hybrid-*virtual*) dynamically interact to enable co-presence across realities. Following, we demonstrate a preliminary workflow where the presence of people in a physical place is detected through occupancy sensors and displayed in the coupled virtual place through VR technology; the presence of people in the virtual place is then displayed in the coupled physical place through AR technology. This paper aims at revisiting the concept of 'place' considering recent technology developments, and laying the foundation for future work that will explore the benefits and opportunities for hybrid places to promote social interactions and collaboration across physical and virtual realms.

Keywords: Co-Presence · Physical and Virtual Environments · Digital Twins · Extended Reality · Hybrid Places

1 Introduction

Recent advancements in sensing and communication technologies including the Internet of Things (IoT), Extended Reality (XR), and Digital Twins (DT) unlocked new opportunities to design 'places' where physical and virtual realms converge to create interactive experiences for working, living, learning, and playing [1]. Rather than considering the design of hybrid 'spaces', this work focuses on the concept of hybrid 'places' which

M. Turrin et al. (Eds.): CAAD Futures 2023, CCIS 1819, pp. 532–546, 2023.
https://doi.org/10.1007/978-3-031-37189-9_35

consider the 3-way interaction between a 'space', the 'people' that inhabit it, and the 'activities' they perform [2, 3]. While 'spaces' originate from material and geometric properties, 'places' arise out of how human activity takes "place" within space [3].

Physical places have been studied for centuries in a variety of disciplines, including architectural design, to analyze the reciprocal relationship between people and the space they inhabit [2–4]. Recent developments in IoT augmented the capabilities of spaces to proactively collect data and reveal space utilization patterns [5–7] effectively transforming spaces into cyber-physical systems [8]. Being bound to the laws of physics, physical environments are inherently static and do not afford co-presence, interactions, and collaboration between geographically distant people. *Virtual* places, instead, explore the creation of adaptive environments where geographically distributed people can have a shared sense of presence and interact with each other for a variety of purposes including entertainment or collaborative work [9]. Nevertheless, this approach mediates interactions through virtual reality, limiting multisensory perception and often creating a disconnect between people and the physical environment they inhabit.

As new technologies such as XR are increasingly more widespread, new opportunities arise to explore social interactions across realities that could potentially combine the qualities of physical and virtual places for co-presence and interaction. Koskela et al. [1], for example, present the concept of Hybrid Avatars to enable avatar-based interactions between VR and AR users visiting the same place both in the physical world and its virtual counterpart. Other studies analyzed human-environment interactions where digital and physical information are superimposed to create a hybrid experience [10]. These approaches can be conceptualized through the lenses of Digital Twins, an established framework to couple physical and digital systems for improved building design and operations [11–13]. However, Digital Twins mostly consider virtual environments as means to improve the performance of physical environments, rather than places for social interactions that complement human activities in the physical world.

In this paper, we introduce the concept of 'hybrid places' where a virtually-augmented *physical* place (the hybrid-*physical*) and a physically-augmented *virtual* place (the hybrid-*virtual*) dynamically interact to impact the behavior of their occupants. We demonstrate a preliminary workflow where the presence of people in a physical place is detected through occupancy sensors and displayed in the coupled virtual place through VR technology. In turn, the presence of people in the virtual place is displayed in the coupled physical place through AR technology. We test this concept in an academic library for the following reasons: (a) physical workplaces are currently being reconceptualized to promote social interactions and collaboration between geographically distributed teams, (b) an academic library setting provided the appropriate level of accessibility to conduct this preliminary study; and (c) this type of setting is consistent with prior work on hybrid spaces [14, 15].

The contributions of this paper can be summarized as follows: (a) we introduce the concept of a hybrid place; (b) we describe a preliminary workflow to couple co-presence in physical and virtual environment toward the design of hybrid places; (c) we present our prototype implementation, and (d) we discuss conceptual and implementation-related challenges together with new opportunities for future developments.

2 Literature Review

2.1 Physical Places

Built environments are designed to host a variety of social, cultural, educational, and work-related activities. While the geometric and functional attributes of built environments can be described with the concept of 'space', the social and cultural attributes of spaces, which depend on the users of the environment, contribute to the creation of a 'place' [2, 3] and 'sense of place' [16–18]. Canter [2] defined Place as the confluence of space, the people who perform them, and their conceptions. Kalay [19] defined 'place' as a setting that affords human activities (including physical, economic, and cultural), while affecting, and being affected by, social and cultural behavior. Dourish later on revisits the space-place duality to clarify that both spaces and places are social constructs [20] without contradicting his prior description of place as a higher-level ensemble including spaces, their inhabitants, and the activities taking 'place' in them.

Seminal studies in anthropology, environmental psychology, and architectural design have investigated how places actively affect and regulate people's interactions among themselves and with the built environment [21–23]. Whyte [24] described the ordinary activities of people in public urban spaces with a special focus on spaces that attract people (or not). Gehl [25] analyzed several outdoor activities and the physical conditions that affected them. Similarly, Alexander [4] focused both on the principles of physical design and the behavior patterns occurring in buildings and cities.

These approaches mostly considered spaces as static containers which impact the behavior of their occupants. Recent developments in ubiquitous computing and IT systems fostered the introduction of sensing technologies into the very fabric of built environments [5, 6]. Wearable devices have been deployed to monitor people's physiological conditions and inform environmental adaptations to improve people's well-being [26]. Ambient sensing technologies analyze the presence and activities of people and respond to them while freeing the inhabitants from wearing devices [7]. Alavi et al. [27], for example, analyzed space-use patterns in offices by combining both ambient sensing and wearables devices. Albeit highly digitally connected, physical environments are still bound to the laws of physics, limiting opportunities for interactions and collaboration across teams and organizations that are becoming progressively more geographically distributed.

2.2 Virtual Places

In the last decades, efforts have been directed to create online virtual environments that afford remote social interactions and collaboration. These efforts have been propelled by progressive technological developments in virtual reality and immersive technologies. Key to these approaches is the concept of 'presence', which resembles the sense of 'being' in a virtual place. Slater [28] defines presence as "the strong illusion of being in a place in spite of the sure knowledge that you are not there". In virtual places, users can perceive a sense of physical presence, but also a social one. Social presence (often referred to as co-presence) refers to the extent to which users can perceive (and be perceived by) other users co-located in the same virtual environment [29].

Kalay [30] describes the required features of virtual places to create meaningful social interactions including, (a) being the setting for a rich set of events, (b) involving engagement with objects and/or people, (c) providing a relative location as a context for activity, (d) creating a sense of authenticity, (e) exhibiting adaptability and appropriation to the needs of the users, (f) enabling a variety of experiences, (g) allowing choice and control over transitions in space through a narrative, and (h) being memorable. These features are impacted by the affordances of the virtual environment where interactions happen to create the desired sense of place [31].

In virtual environments, interactions occur using 'avatars', "a digital representation of a human user that facilitates interaction with other users, entities, or the environment" [32]. Avatars can interact with other human-controlled avatars or with agents (also called Non-Player Characters), computer-controlled characters that respond to the presence of avatars as well as other agents [33]. Virtual worlds are often designed for entertainment purposes although other applications related to architectural design afford experiencing places that no longer exist [34] or are yet to be built [35].

2.3 Coupling Physical and Virtual Places for Hybrid Interactions

Technological advancements in Augmented Reality (AR) enabled the creation of hybrid interactions that couple the affordances of both physical and virtual environments [14]. While VR completely immerses a user inside a synthetic environment, AR allows users to see the world with virtual information superimposed upon it. Therefore, it supplements reality, rather than completely replace it [36]. Several applications exist that explore AR applications to regulate human-building interactions in day-to-day use scenarios as well as in the design process. For example, Lee et al. [37] studied the impact of AR-enabled virtual partitions in open-plan offices. Seo et al. [10] proposed a method where users experience the same context using VR as well as AR by integrating an egocentric virtual reality-based immersion and exocentric augmented reality-based visualization. These approaches are often described as mixed or extended reality (XR) to indicate that digital information is combined with physical information.

Other approaches stress the fact that physical and virtual environments can be coupled to promote collaboration, co-learning, and situated analytics [38]. These approaches have been defined as dual reality [39], hybrid reality [40], hybrid spaces [14, 41], cross-reality [42], or fused twins [38, 43]. Among them, the concept of 'hybrid space' mostly focuses on integrating digital technology into physical spaces [44]. Underlying these approaches are principles of Digital Twinning between physical and virtual systems where physical objects, processes, or systems are represented via a continuously updated digital model that provides analytical capabilities to inform the operations of the physical entity [11]. With few exceptions [15, 45], DT as well as the approaches previously discussed mostly conceive of digital environments as means to improve the performance of physical environments, rather than as inhabitable environments that afford social interactions, while complementing – rather than replicating – physical environments. Therefore, a framework to create holistic 'places' for human social interactions in hybrid environments is yet to be fully explored.

3 Hybrid Places

In this paper, we introduce the concept of 'hybrid places' to support co-presence across physical and virtual environments. Figure 1 outlines the proposed conceptual model. Starting from the physical-virtual continuum described by Milgram [46], we first characterize the extremities of the continuum. *IoT-enabled physical places* extend the capabilities of traditional physical places since they are equipped with networked devices to sense the state of the environment including the presence and movement of the inhabitants and stream this information to the cloud for remote data access and manipulation. On the other side of the spectrum, *Multi-user virtual places* host human-controlled avatars and potentially synthetic autonomous agents to enable virtual interactions.

At the nexus of physical and virtual places, we outline two main components of hybrid places. **Virtually-augmented *physical* place**s (the hybrid-*physical*) extend the experience of physical places by superimposing virtual information about the presence and activities of people inhabiting the virtual world to enable cross-world interactions. **Physically-augmented *virtual* places** (the hybrid-*virtual*) augment the traditional capabilities of virtual environments by dynamically updating the virtual world through data collected in a physical context. The behavior of both human-controlled avatars and agents can then be affected by the behavior of physical people.

We define as **hybrid places** the sum of the aforementioned worlds together with the interactions that regulate them. This concept builds upon and extends a substantial body of work on VR, AR, and XR technologies as well as digitally twinned systems.

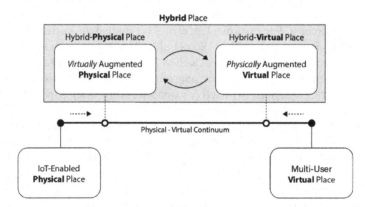

Fig. 1. Conceptual Framework for Hybrid Places

4 Case Study

The concept of a hybrid place is demonstrated in the library of the Faculty of Architecture & Town Planning at the Technion – Israel Institute of Technology. This preliminary study aimed at exploring the viability, challenges, and opportunities to create co-presence between physical and virtual inhabitants. First, low-cost custom-made occupancy sensors were developed and deployed in the library to count people entering and exiting the space and measure people's occupancy at individual workstations as in an open collaborative area. The sensed information updates in real-time a 3D virtual model of the library to reveal the spatial behavior of the physical building occupants. The digital model of the library is populated by human-controlled avatars as well as computer-controlled agents, which respond to each other's presence and movement. Finally, the movement of the human-controlled avatars is displayed in the physical library through AR technology. This study has been approved by the Ethics Committee of the Technion, and library users have been notified about the data collection process through a poster located at the entrance of the library. Figure 2 outlines the different components of the system.

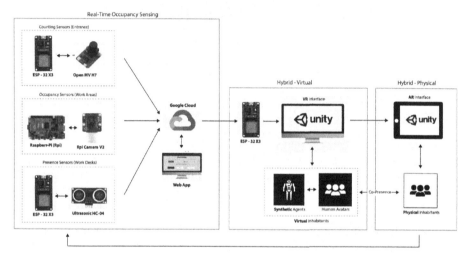

Fig. 2. Preliminary system for designing a hybrid place

4.1   Real-Time Occupancy Sensing

A variety of occupancy sensors have been developed and deployed as part of this study. The collected data is dynamically streamed to a cloud platform for managing the data collection process and streaming the relevant data to the hybrid-*virtual* place. The system does not store or broadcast any images collected by the sensors to ensure privacy compliance. The computing is done on the edge to calculate the number of people detected. The data transmitted to the cloud does not contain any sensitive information (an example of the data is provided in Table 1). Figure 3 shows a floor plan of the library with the location of the sensors, while Fig. 4 shows the custom-made sensors.

Counting Sensor. Placed on the doorway entrance, the sensor detects the direction of objects' movement to determine if someone entered/exited the library. The sensor includes an Open MV camera connected to an ESP board. On the camera's chip, a

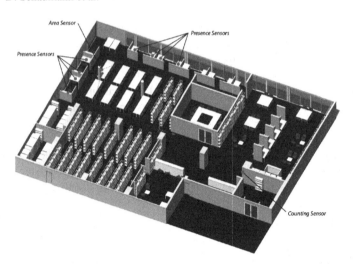

Fig. 3. Library model and sensors' location

custom-made algorithm has been installed that captures subsequent frames to determine the direction of people's movement by comparing frames to the previous ones.

Area Sensor. It calculates the number of people present in a captured image. It consists of a camera connected to a Raspberry Pi board. The sensor uses a YOLO pre-trained neural network to detect people's presence in an image [47]. The sensor performs the calculation on the edge and outputs the number of people detected for every analyzed image frame. The image detected by the sensor is not stored for privacy reasons.

Presence Sensor. It determines if a person is located at a desk by calculating the distance between the sensor and the detected object (a person in our case) via ultrasonic soundwaves. Each sensor is connected to an ESP32 microcontroller measuring and filtering data to determine people's presence. The microcontroller uses a moving average filter and outlier rejection algorithm to minimize errors and maximize accuracy.

Cloud Interface. A Google Cloud database stores the information received from the sensors. A web-based graphic interface allows users to (a) start, stop and restart the data collection system, (b) download CSV reports of the collected data within a specific time range; (c) monitor the system's live status to detect errors or faulty sensors; (d) schedule the system activation/deactivation within a specific time range; (e) map the sensor location in the space; (f) catalog the sensors using images and text descriptions. The system retains recent data (collected within 24 h) making it available in real-time, and it backs up older data using CSV files. A web interface running on an EC2 instance (AWS) is connected directly to the cloud storage and hosts the web-based user interface app. The ESP32 boards and the Raspberry Pi that are connected to the cloud can recover from Wi-Fi/firebase disconnections and power failures.

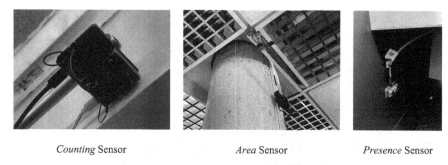

Counting Sensor   Area Sensor   Presence Sensor

Fig. 4. Custom-made sensors installed in a university library

4.2 The Hybrid-Virtual Place

Data collected in the physical world is continuously streamed to inform the behavior of people populating the hybrid-*virtual* place. A virtual model of the project setting (a library in this case) has been created in Rhino and then imported into Unity 3D, where materials and textures have been added. The sensor data is streamed to the virtual environment using an ESP 32 controller connected to the same Wi-Fi network as the sensors. The sensor data can be streamed in real-time; otherwise, historical sensor data can also be imported and visualized. The frequency of the data streaming is between 5–8 s due to the time that it takes the YOLO application on the area sensors to calculate people's presence. The data is exported in a CSV format, as shown in Table 1. In the virtual model, each sensor is associated with a specific location on the floor plan. Every time a sensor detects one or more persons, a digital representation of the detected person is visualized in the virtual model using a static humanoid agent. Figure 5 shows different timesteps of the virtual world dynamically updated through sensed data.

Table 1. Structure of the data collected using the sensors

SensorID	Type	Location	Event	Time	Day
C-01	Counting	Entrance	IN	10:43:05	20/07/22
A-01	Area	Column	2	10:43:07	20/07/22
P-01	Presence	Workstation 1	YES	10:43:07	20/07/22
P-02	Presence	Workstation 2	NO	10:43:08	20/07/22
P-03	Presence	Workstation 3	YES	10:43:08	20/07/22

The virtual place is populated with human-controlled avatars and computer-controlled agents. **Human-controlled avatars** have a humanoid shape and can wander around without colliding with obstacles. They are controlled by users using a first- or third-person camera view. **Computer-controlled agents** have a similar humanoid shape. Their behavior is scripted using a narrative-based approach [48], which directs the behavior of agents using a hierarchical planning system. In this preliminary study,

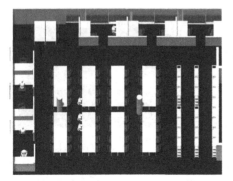

Fig. 5. Digital model of the library including white humanoid agents reflecting the real-time presence of occupants in the physical library.

agents wander through the space and visit selected targets such as the information desk, bookshelves, and study areas. They respond to the physical presence of avatars to avoid physical collision. Figure 6 shows a snapshot of the hybrid-*virtual* place.

Fig. 6. Experiencing the hybrid-*virtual* place (in 'yellow' is the human-controlled avatar in the virtual library, in 'blue' is the computer-controlled agents, in 'white' is the sensor-based agents)

4.3 The Hybrid-Physical Place

The movement of human-controlled avatars in the hybrid-*virtual* place can be perceived by the inhabitants of the hybrid-*physical* place through an AR application. The application includes a simplified digital version of the library model for (a) calibration and (b) occlusions. To calibrate the application, the user places their AR device in a predetermined location and orientation in the physical environment, then presses a button that resets the camera's position in the virtual environment. To provide feedback during calibration, the app displays transparent bounding boxes for key features in the digital model; if the boxes tightly bound their respective physical objects, then the virtual environment matches the physical (Fig. 7). In this case, our calibration features are the library tables and the floor. The library model is also used for avatar occlusion: an avatar standing behind a table or some other object will be partially or wholly hidden by that object (assuming the app is properly calibrated, and the object exists in both the physical space and the digital model). Figure 8 shows a human-controlled avatar perceivable in the hybrid-*physical* place. Figure 9 represents both sides of the hybrid place including the hybrid-virtual place (experienced using a laptop for VR), and the hybrid-physical place (experienced using a phone for AR). In future work, the proposed approach could be implemented using more immersive technologies.

Fig. 7. Calibrating the hybrid-physical visualization. In 'blue' the digital model of the tables is used as a reference to determine the position of the AR device

Fig. 8. Experiencing the Hybrid-Physical Place (in 'yellow' is the human-controlled avatar inhabiting the Hybrid-Virtual place)

Fig. 9. Comparing the visualization of the hybrid-*virtual* (using the laptop on the left for VR) and the hybrid-*physical* (using a phone on the right for AR)

4.4   System Architecture

Our implementation consists of a desktop app, in which users control avatars in the hybrid-*virtual* place, and a mobile AR app, which allows users in the hybrid-*physical* place to see the virtual avatars (and be seen by their users). One instance of the desktop app functions as a master client, meaning its version of the hybrid world is considered authoritative; other instances transmit their respective users' avatars' states to the master client, which integrates them (as well as sensor data) into its world representation. It then relays the world state to the other clients. The master client is also responsible for populating the hybrid-*virtual* place with autonomous agents. In this study, AR functionality is provided by Unity's AR Foundation framework (which utilizes Apple's ARKit on the iOS devices used in our demo). Networking is provided by the Photon Fusion library. Figure 10 provides a schematic representation of the system architecture.

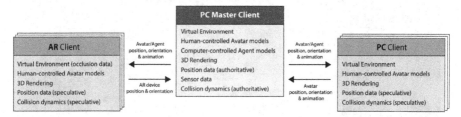

Fig. 10. System architecture and communication protocols

5   Conclusion and Discussion

This paper aims at revisiting the concept of 'place' through the lenses of recent technological advancements in extended reality. Specifically, it discusses the concept of
'hybrid places' where a virtually-augmented *physical* place (the hybrid-*physical*) and a
physically-augmented *virtual* place (the hybrid-*virtual*) dynamically interact to impact
the behavior of their occupants. We lay the conceptual foundations for hybrid places
while building upon previous conceptualizations [15, 45], and we demonstrate a preliminary system mockup to explore its technical feasibility. This approach also provides
an opportunity to revisit Digital Twin concepts to explore a deeper connection between
inhabitable and mutually interacting physical and virtual environments.

The presented study has a few limitations. First, we do not discuss possible interactions between inhabitants of the hybrid-virtual and hybrid-physical places. Only co-
presence is considered at this stage. This will be the main venue for future work together
with a technical evaluation of the proposed system and a public release of the implementation mockup. Another limitation pertains to area sensors, which can only count
people in a selected area without knowing their exact position. So, in the hybrid-virtual
place, we randomized the position of the sensor agents within the zone boundary. Further, computer-controlled agents merely respond to the presence of human-controlled
avatars. Future work will explore how to make computer-controlled agents more realistic
to enhance the experience of the hybrid-virtual place inhabitants.

In the long term, potential benefits of hybrid places include: (a) collaboration and
social interactions among members of geographically distributed organizations, which
aim to have a shared sense of place to improve experience, well-being, and productivity; (b) co-presence in social events co-hosted in physical and virtual environments to
improve the experience of the participants; (c) teleoperation of avatars controlled in the
hybrid-virtual and visualized in the hybrid-physical place to facilitate human activities
in day-to-day or emergency scenarios; (d) hybrid reconstruction of social experiences
in historical sites where behavior narrative controlled in the hybrid-virtual are displayed
in-situ in the hybrid-physical; and (e) manifestation of online social activities occurring in the hybrid-virtual into hybrid-physical places to foster the creation of hybrid
communities of physical and virtual participants.

Broader themes requiring further attention related to the design of hybrid places
(some of which have been discussed by [1, 49]) include:

- **Vacancy**. As identified by Lifton and Paradiso [39], people participating in one world
  may create a noticeable and profound absence in the other.
- **Safety**. Avatar manifestations in the hybrid-physical world may occlude danger such
  as a person or car approaching.
- **Realism** and **Self-Expression**. Research from Oh et al. [29] shows that photographic
  and anthropomorphic realism do not always yield better interactions as suggested
  by the uncanny valley theory [50]. Instead, agents' behavioral realism is a powerful
  predictor of social presence [29] especially when they demonstrate awareness of
  users' presence (e.g., mutual gaze) and provide interactivity [49].
- **Self-representation.** The visual appearance of a user's avatar may impact social
  interactions with other avatars. This phenomenon, named the Proteus effect [51], is

the result of the sense of embodiment with a virtual avatar and the tendency to adapt behaviors to match expectations implied by the avatar's body features.

- **Interaction**. Open research issues include *visibility* (who can see my avatar?), *movement* (how does my avatar move?) *interaction* (who can interact with my avatar?), *collision* (how collisions between avatars and objects should be handled?), and *control* (can a global administrator regulate avatars' behavior?).
- **Technology**. Several issues arise when designing hybrid interactions, such as (a) estimating the position and rotation of AR users with respect to the displayed virtual content, (b) streaming sensor-generated data to update a digital model in real-time, and (c) creating context awareness and real-time interactions between physical and virtual realities.
- **Ethics**. Virtual embodiments could possibly lead to negative consequences such as the preference for inhabiting virtual worlds over physical ones [49].

Despite these outstanding challenges, hybrid places hold promises to open the door for exciting research to design social interactions across physical and virtual realms.

Acknowledgments. The authors wish to thank Viky Davydov, head librarian of the Faculty of Architecture & Town Planning at the Technion – Israel Institute of Technology, for her support.

References

1. Koskela, T., et al.: Hybrid avatars: enabling co-presence in multiple realities. In: Proceedings of the 21st International Conference on Web3D Technology, pp. 69–72. Association for Computing Machinery, New York (2016)
2. Canter, D.: The Psychology of Place. The Architectural Press LTD, London (1977)
3. Harrison, S., Dourish, P.: Re-place-ing space: the roles of place and space in collaborative systems. In: Proceedings of the 1996 ACM Conference on Computer Supported Cooperative Work, pp. 67–76. ACM (1996)
4. Alexander, C.: The Timeless Way of Building. Oxford University Press, New York (1979)
5. Li, X., Lu, R., Liang, X., Shen, X., Chen, J., Lin, X.: Smart community: an internet of things application. IEEE Commun. Mag. **49**, 68–75 (2011)
6. Alavi, H.S., Verma, H., Mlynar, J., Lalanne, D.: On the temporality of adaptive built environments. In: Schnädelbach, H., Kirk, D. (eds.) People, Personal Data and the Built Environment. SSAE, pp. 13–40. Springer, Cham (2019). https://doi.org/10.1007/978-3-319-70875-1_2
7. Haque, A., Milstein, A., Fei-Fei, L.: Illuminating the dark spaces of healthcare with ambient intelligence. Nature **585**, 193–202 (2020)
8. Darwish, A., Hassanien, A.E.: Cyber physical systems design, methodology, and integration: the current status and future outlook. J. Ambient. Intell. Humaniz. Comput. **9**(5), 1541–1556 (2017). https://doi.org/10.1007/s12652-017-0575-4
9. Batty, M., Miller, H.J.: Representing and visualizing physical, virtual and hybrid information spaces. In: Janelle, D.G., Hodge, D.C. (eds.) Information, Place, and Cyberspace, pp. 133–146. Springer, Berlin (2000). https://doi.org/10.1007/978-3-662-04027-0_8
10. Seo, D.W., Kim, H., Kim, J.S., Lee, J.Y.: Hybrid reality-based user experience and evaluation of a context-aware smart home. Comput. Ind. **76**, 11–23 (2016)
11. Kaur, M.J., Mishra, V.P., Maheshwari, P.: The convergence of digital twin, IoT, and machine learning: transforming data into action. In: Farsi, M., Daneshkhah, A., Hosseinian-Far, A., Jahankhani, H. (eds.) Digital Twin Technologies and Smart Cities. IT, pp. 3–17. Springer, Cham (2020). https://doi.org/10.1007/978-3-030-18732-3_1

12. Miller, C., et al.: The Internet-of-Buildings (IoB) — digital twin convergence of wearable and IoT data with GIS/BIM. J. Phys. Conf. Ser. **2042**, 012041 (2021)
13. Alavi, H.S., et al.: Human-building interaction: sketches and grounds for a research program. Interactions **26**, 58–61 (2019)
14. Bilandzic, M.V.: The embodied hybrid space: designing social and digital interventions to facilitate connected learning in coworking spaces (2013)
15. Ylipulli, J., Pouke, M., Luusua, A., Ojala, T.: From hybrid spaces to "imagination cities": a speculative approach to virtual reality. In: The Routledge Companion to Smart Cities. Routledge (2020)
16. Norberg-Schulz, C.: Genius Loci: Towards a Phenomenology of Architecture. Rizzoli (1980)
17. Malpas, J.: Heidegger's Topology: Being, Place, World. MIT press (2008)
18. Relph, E.: Place and Placelessness. Pion (1976)
19. Kalay, Y.E.: Architecture's New Media: Principles, Theories, and Methods of Computer-Aided Design. MIT Press, Cambridge, MA (2004)
20. Dourish, P.: Re-space-ing place: "place" and "space" ten years on. In: Proceedings of the Conference on Computer Supported Cooperative Work (2006)
21. Goffman, E.: Behavior in Public Places. The Free Press, New York (1963)
22. Barker, R.G.: Ecological Psychology: Concepts and Methods for Studying the Environment of Human Behavior. Stanford University Press (1968)
23. Altman, I.: The Environment and Social Behavior: Privacy, Personal Space, Territory, Crowding. Brooks/Cole Pub. Co., Monterey (1975)
24. Whyte, W.: The Social Life of Small Urban Spaces. The Conservation Foundation, Washington, DC (1980)
25. Gehl, J.: Life Between Buildings: Using Public Space. Wiley John and Sons, New York (1987)
26. Verma, H., Alavi, H.S., Lalanne, D.: Rethinking wearables in the realm of architecture. In: Clemmensen, T., Rajamanickam, V., Dannenmann, P., Petrie, H., Winckler, M. (eds.) INTER-ACT 2017. LNCS, vol. 10774, pp. 16–23. Springer, Cham (2018). https://doi.org/10.1007/978-3-319-92081-8_2
27. Alavi, H.S., Verma, H., Mlynar, J., Lalanne, D.: The hide and seek of workspace: towards human-centric sustainable architecture. In: Proceedings of the 2018 CHI Conference on Human Factors in Computing Systems, USA (2018)
28. Slater, M.: Place illusion and plausibility can lead to realistic behaviour in immersive virtual environments. Philos. Trans. R Soc. Lond. B Biol. Sci. **364**, 3549–3557 (2009)
29. Oh, C.S., Bailenson, J.N., Welch, G.F.: A systematic review of social presence: definition, antecedents, and implications. Front. Robot. AI **5** (2018)
30. Kalay, Y., Marx, J..: Changing the metaphor: cyberspace as a place. In: Digital Design - Research and Practice, Proceedings of the 10th International Conference on Computer Aided Architectural Design Futures (Taiwan), pp. 19–28 (2003)
31. Chen, X., Kalay, Y.: Making a liveable 'place': content design in virtual environments. Int. J. Herit. Stud. **14**, 229–246 (2008)
32. Nowak, K., Fox, J.: Avatars and computer-mediated communication: a review of the definitions, uses, and effects of digital representations. Rev. Commun. Res. **6**, 30–53 (2018)
33. Wooldridge, M.: An Introduction to Multiagent Systems. John Wiley and Sons, Newyork (2009)
34. El Antably, A.H.: Experiencing the Past: The Virtual (Re)Construction of Places (2011)
35. Yan, W., Culp, C., Graf, R.: Integrating BIM and gaming for real-time interactive architectural visualization. Autom. Constr. **20**, 446–458 (2011)
36. Azuma, R.T.: A survey of augmented reality. Teleoperators Virtual Environ. **6**, 355–385 (1997). https://doi.org/10.1162/pres.1997.6.4.355

37. Lee, H., Je, S., Kim, R., Verma, H., Alavi, H., Bianchi, A.: Partitioning open-plan workspaces via augmented reality. Pers. Ubiquit. Comput. **26**, 1–16 (2019). https://doi.org/10.1007/s00 779-019-01306-0

38. Grübel, J., et al.: The hitchhiker's guide to fused twins: a review of access to digital twins in situ in smart cities. Remote Sens. **14**, 3095 (2022)

39. Lifton, J., Paradiso, J.A.: Dual reality: merging the real and virtual. In: Lehmann-Grube, F., Sablatnig, J. (eds.) FaVE 2009. LNICSSITE, vol. 33, pp. 12–28. Springer, Heidelberg (2010). https://doi.org/10.1007/978-3-642-11743-5_2

40. de Souza e Silva, A.: From cyber to hybrid. Space Cult. **9**, 261–278 (2006)

41. Ciolfi, L., Bannon, L.J.: Designing hybrid places: merging interaction design, ubiquitous technologies and geographies of the museum space. CoDesign **3**, 159–180 (2007)

42. Nguyen, B.V.D., Simeone, A.L., Vande Moere, A.: Exploring an architectural framework for human-building interaction via a semi-immersive cross-reality methodology. In: Proceedings of the 2021 ACM/IEEE International Conference on Human-Robot Interaction, pp. 252–261 (2021)

43. Grübel, J., Gath-Morad, M., Aguilar, L., Thrash, T., Sumner, R.W., Hölscher, C., Schinazi, V.: Fused twins: a cognitive approach to augmented reality media architecture. In: Media Architecture Biennale, vol. 20, pp. 215–220. Association for Computing Machinery, New York (2021)

44. Tan, F.: Production of hybrid public space: the potential of the hybrid space concept for public space production. J. Eng. Sci. Technol. (2021)

45. Ron, R., Weissenböck, R.: The aesthetics of hybrid space. In: Virtual Aesthetics in Architecture. Routledge (2021)

46. Milgram, P., Takemura, H., Utsumi, A., Kishino, F.: Augmented reality: a class of displays on the reality-virtuality continuum. In: Telemanipulator and Telepresence Technologies, vol. 2351 (1994)

47. Redmon, J., Divvala, S., Girshick, R., Farhadi, A.: You only look once: unified, real-time object detection. arXiv:1506.02640 (2016)

48. Schaumann, D., Putievsky Pilosof, N., Sopher, H., Yahav, J., Kalay, Y.E.: Simulating multi-agent narratives for pre-occupancy evaluation of architectural designs. Autom. Constr. **106**, 102896 (2019)

49. Kyrlitsias, C., Michael-Grigoriou, D.: Social interaction with agents and avatars in immersive virtual environments: a survey. Front. Virtual Reality **2**, 168 (2022)

50. Mori, M., MacDorman, K.F., Kageki, N.: The uncanny valley [from the field]. IEEE Robot. Autom. Mag. **19**, 98–100 (2012)

51. Yee, N., Bailenson, J.N., Ducheneaut, N.: The proteus effect: implications of transformed digital self-representation on online and offline behavior. Commun. Res. **36**, 285–312 (2009)

Social Signals: An Adaptive Installation for Mediating Space During COVID-19 and Beyond

Eric Duong$^{(\boxtimes)}$ ⓘ and Stefana Parascho ⓘ

Swiss Federal Institute of Technology Lausanne, Lausanne, Switzerland
{eric.duong,stefana.parascho}@epfl.ch

Abstract. During the COVID-19 pandemic, signs and barriers were used to guide people through spaces in accordance with social distancing mandates. Such measures were unable to accommodate for fluctuating traffic and complexities of human behavior – leading to inconsistent effectiveness and a diminishing of social life in once-active public spaces. Toward a more flexible and engaging means of guidance, this research brings together computer vision, human behavioral modeling, and light projection to create an installation that suggests paths to pedestrians in real-time. Through the processing of positional data, it predicts optimal trajectories for each pedestrian that accounts for the movement of others and the environment around them. The projected visuals then present these animated paths as part of a fluid and open-ended spatial interface that aims to improve pedestrian efficiency while sparking social interactions. In a month-long installation of the system in a frequently trafficked space, we observed individuals serendipitously connect through the projections. Though limitations in the scale of the installation constrained its depth, Social Signals has the potential to improve the built environment not through material interventions but rather through highlighting human interactions.

Keywords: Media Installation · Interactive Environment · Computer Vision · Pedestrian Modeling · Social Spaces

1 Introduction

The COVID-19 pandemic challenged many architectural and urban design conventions, revealing how existing spaces are not designed to handle health crises of such a scale [1]. One problem that many experienced directly was the diminishing public life of pedestrianized spaces due to restricted access during the lockdown. This shift has left urbanists and citizens alike questioning how the character of public spaces could persist in such a crisis and whether new urban design approaches and technologies could be used to reconfigure them. Distancing protocols have been responsible for much of the change in public spaces; with the urgent need for occupant guidance came the application of signage, stickers, and tape to remind pedestrians to stay safely apart from one another.

M. Turrin et al. (Eds.): CAAD Futures 2023, CCIS 1819, pp. 547–561, 2023.
https://doi.org/10.1007/978-3-031-37189-9_36

Though their simplicity and universality have proven effective in certain contexts – particularly in areas with a consistent and predictable flow of pedestrians – they have also proven unnecessarily restricting, unaesthetic, and ignorable once conditions like occupancy levels exceed expectations. As a result, such measures have become either overly controlling or inadequate in the way they fail to manage crowds.

This paper introduces Social Signals, a media installation created in response to the limitations of social distancing signage with the aim of providing a more flexible, effective, and joyful means of guiding pedestrians. To do so, it connected computational methods of sensing and analyzing spatial occupancy to a generative visualization that responded dynamically to the state of pedestrians within a space. In this way, it could account for variable occupancy levels as well as the activity of the people within the area. An overhead projector was used to display the animations at the feet of pedestrians, while a camera provided the system with video frames to be processed into spatial data. In addition to the flexibility of the visuals, adaptability extended to enabling the system to be installed with ease and in many different spaces. As a result, the hardware used was intended to be light, accessible, and not overly expensive. Not all goals were reached in this iteration as will be described in Results.

As a complete sensing, planning, and spatial interface, the contributions of the project to the fields of architecture and computation are:

- utilizing people detection, tracking, and prediction in real-time and in tandem to inform computational suggestions
- developing a visual system that communicates paths and encourages interaction
- Installing the System in a Physical Space to Evaluate Its Effectiveness Over Time

The following sections describe the methods for hardware setup and development of people detection, tracking, and visualization. Finally, the proposed system is deployed as a case-study installation at the Princeton University School of Architecture.

2 Background

2.1 Building Occupancy Analysis

Recent analyses of building occupancy have revealed performance gaps between building design and how they are actually used. This has many consequences such as wasteful energy allocation [2], underutilized spaces [3], and inefficiencies in occupant wayfinding [4]. In addition, COVID-19 has made it clear that our spaces are not designed to accommodate for the health and safety of the public in the event of a global pandemic [1]. Today, many computational techniques are utilized to capture and understand how buildings are occupied, as well as speculate on how future spaces will be used.

In post-occupancy building evaluations, digital sensors are utilized to capture occupancy information of existing spaces in an automated and objective manner. Computer vision has shown great promise in this application due to its ability to scale and collect high resolution data [5]. More specifically, modern developments with convolutional neural networks allow for detection and tracking accuracy that match traditional non-image techniques while providing more spatiotemporal detail [6]. Social distancing mandates led to the development of many such detection models that would not only identify the location of people, but also determine whether they are following guidelines [7, 8]. Though these applications are able to make accurate assessments, they have exclusively been used post-occupancy and therefore remain one step removed from the spaces they intend to change. Other research has experimented with the use of IoT devices to affect behavior, but these require specific personal devices to function [9].

Building occupancy can also be evaluated in a predictive manner, using computational simulations to model occupant behavior in a space. Since the introduction of the original social force model [10], contemporary research has found success in utilizing deep learning for more realistic predictions [11]. From convolutional neural networks (CNN) to generative adversarial networks, many of these frameworks have yielded greater accuracy in testing yet come with a significant performance cost. More recently, agent-based and deep learning crowd modeling have been used to determine how social distancing protocols will affect existing spaces [12]. Like social distancing-specific detectors, these analytical methods are applied upon collection of a complete dataset rather than as the data is collected. As such, they are not able to inform people of optimal and efficient paths in the moment they are occupying the space. Social Signals takes a different approach in utilizing observatory and speculative occupancy analyses in real-time that are immediately reflected in a human-scale visualization.

2.2 Media Art and Architecture

Media installations employ digital media technologies to transform physical space. By introducing interactivity through visuals, touch, or sound, the medium overlays new meaning onto the behavior of participants. Since Myron Krueger's first experiments in responsive environments [13], such spatial concepts have become largely possible due to advancements in sensor, computer, and display technologies. On the computation side, miniaturization of hardware has allowed for "ambient intelligences" to be ubiquitously embedded in our physical environments. As demonstrated by the internet of things and its continued expansion into urban life, these smart artifacts can participate in social situations as mediators that redefine our relationship to spaces [14].

Light projection is a particularly versatile display technology that is capable of mapping walls, floors, or even objects. On a building scale, these installations become "media façades" that shift from moment to moment. In Kissing Architecture, Lavin posits that the collision of projected digital image and physical surface has the potential to deal with uniquely urban and architectural problems, like navigation of social spaces [15]. In the context of the pandemic, only a few precedents have used projection mapping to define personal space in public [16]. However, these remain proofs-of-concept and were not informed by predictive analyses to guide participants. With these ideas in mind, Social Signals takes cues from the artistry of media installations and integrates computational methods of behavioral modeling to create a functional system that offers a depth of interactions.

3 Methods

Physical signage for social distancing has proven insufficient for spaces with occupancy demands that change throughout the day, especially in combination with the tendency of pedestrians to ignore or not notice it. An intelligent projection system provides a more flexible and engaging solution for guiding occupants in accordance with distancing protocols (Fig. 1). Towards an adaptable, unified, and non-intrusive system, the research set out to meet the following design and development criteria:

- The visuals must remain legible and clearly communicate path suggestions across variable lighting conditions with at least 24 frames per second (FPS) on average
- Trajectory predictions need to be accurate to maintain natural pedestrian flow.
- When appropriate, predictions and visuals should account for social distancing by encouraging participants to stay a safe distance apart.
- In case of deviations from behavioral predictions, the system should be capable of dynamically adjusting paths while reflecting changes smoothly in the visualization.
- To preserve privacy, collected data must minimize use of captured images.
- The system should be joyful in presentation and offer a platform for continued explorations in sparking social interaction.

Fig. 1. Social Signals is a digital projection system that detects pedestrians and shows suggestions.

The following sections will describe the individual parts of this setup – from the hardware decisions to the implementation of pedestrian detection and the visualization.

3.1 Development Framework

The installation consists of a ceiling-mounted projector and a NVIDIA Jetson Nano computing board connected to a wide-angle camera module (Fig. 2). Compared to other compact computers like the Raspberry Pi, the Jetson has much more computing power for machine learning applications and graphics due to its onboard GPU. It accomplishes this within a small form factor, enabling it be mounted more easily than a full-size PC while remaining at a fraction of the cost. Both the projector lens and camera point directly at the ground from about the same position in 3D space, such that the captured images and the projected visualizations span the same area. During setup, this alignment is calibrated by projecting the video stream onto the floor and adjusting the position of the camera until the edges of the camera view and projection area are in-line. To account for the camera field-of-view, the projector image size is adjusted via zoom options provided by the lens. With the ability to run all necessary software onboard, hardware setup is as straightforward as mounting the projector and the attached computer to a compatible ceiling with access to electricity. This computationally lightweight assembly also means that the installation is capable of running autonomously and without connection to the internet.

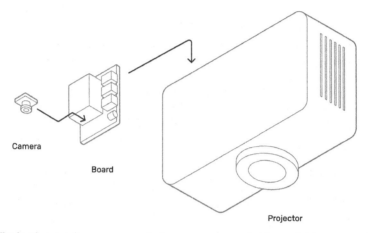

Fig. 2. The hardware scheme composed of a camera connected to a miniature computer which outputs to the projector.

The development approaches to vision, pathing, and visualization further serve the aim of easing setup difficulties and improving adaptability. In the first processing step, people detection on the camera stream allows the installation to adapt to any space – regardless of variation in the flooring or the presence of objects. Subsequent analysis in the form of path predictions and suggestions balance quality and efficiency to maintain performance. Finally, these analyses are presented as floor-projected signifiers that visually translate the input and respond to pedestrian behavior in real-time (Fig. 3).

3.2 Pedestrian Detection

Data collection begins with reading the video stream from a single 160° field of view camera module connected to the Jetson. To ensure input resolution matches the visual output, frames are captured at a resolution of 1920 × 1080 pixels. Otherwise, mapping coordinates at a lower resolution would lead to noticeably stepped movement in the projected visuals. Before pedestrian detection, OpenCV [17] is used to dewarp the fisheye distortion from the wide-angle lens. Each frame is then analyzed to locate pedestrians. Traditional computer vision techniques using shape, texture, or motion-based features

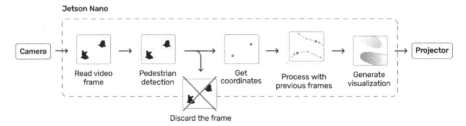

Fig. 3. Data processing is carried out on each video frame by reading from the camera, detecting pedestrians, translating to true points, matching to prior detections, generating paths, and then outputting a visualization to the projector.

are the most efficient, but have limitations in variable light conditions that are only exacerbated by the constant lighting changes of the projections around each person [18]. Deep learning techniques are far more accurate in such conditions, but come with significant trade-offs in computational efficiency and complexity [19]. Though deep learning is unable to match the performance of traditional methods, it can certainly approach real-time detection with the right hardware and network. In particular, running a Single Shot Detector (SSD) with the MobileNet CNN architecture on the dedicated GPU of the Jetson provides reliable detections with potential for further optimizations to reach steadier real-time frame rates.

Prior to using the SSD model, it was trained on a custom dataset of about 3500 pedestrian images from an overhead perspective over 70 epochs, with the resulting detector exhibiting accurate coordinates in qualitative tests on multiple people within the installation space. This trained detector running through the optimized Jetson inference library sees a framerate of just under 30 FPS.

3.3 Tracking

Following the detection stage, pedestrians in the current frame need to be matched to recent detections in past frames. This allows for detections to be continuously attributed to the same person over the time they are present in the frame, such that the resulting projected signifiers unique to each person are able to properly follow them in the space. In order to do this, the locations of people in the current frame are compared to detections in the previous frame. Detections are pair-wise matched across these two instances based on closest distance and assigned identifiers. Therefore, a detection in the current and last frame that share the shortest distance are assumed to be the same person (Fig. 4) and are assigned the same ID. If there are more detections in the current frame than the last, it can be assumed that one or more new people have entered the frame and need to be assigned unique IDs of their own. This distance must also be less than a set maximum to ensure that someone exiting the frame is not matched to another person entering on the other side of the frame at the same time or shortly after. The assignment of unique IDs across consecutive frames allows for path prediction and the management of complex states for each individual within the visualization. From this point, only the unique IDs and their Cartesian coordinates are saved for further processing in order to preserve individual privacy and storage space.

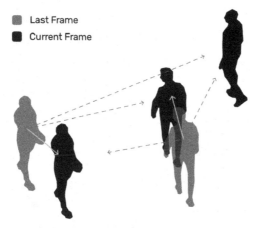

Fig. 4. Detections are matched across frames using a proximity heuristic that compares the distances between detections in the last and current frame. This allows for consistent tracking.

3.4 Path Prediction

Once pedestrians are detected, the computer suggests paths for each based on their location history. It is important that these suggestions are flexible and able to respond quickly to changes, as pedestrians will not necessarily adhere to them. Path suggestions are generated by running a short-term motion simulation of the space, where successive location data for each pedestrian informs predictions of destination point and velocity. It also predicts how pedestrians will resolve eventual conflicts in their pathing – no matter if they are stationary or moving. There are many approaches to modeling pedestrian behavior, from more traditional force models to modern neural network-based methods of prediction. Of these, Social Signals utilizes a modified social force model in conjunction with the detection data. This provides a very lightweight simulation that yields predictions which can also be used as path suggestions.

The social force model and its latest iterations abstract pedestrian behavior into point nodes in a 2D space that have their own velocities. Each pedestrian node, as well as environmental factors like walls or windows, have repulsion or attraction vectors that influence how pedestrians behave [10]. As such, repulsive forces reflect the tendency to keep a certain distance from others and boundaries of a space, while the exertion of attractive forces from the environment represents objects of interest. Though the modeling technique may not be as nuanced as current methods, it is still able to describe pedestrian behavior realistically. Furthermore, the relative simplicity of the outputs actually works toward better path suggestions within the context of the project (Fig. 5). The goal is not to accurately replicate real behavior but to provide an optimized path suggestion that is still based on a close approximation of pedestrian intentions; the fidelity of social force actually serves this need out of the box. In addition, the model allows for adjustments to force parameters that make it applicable to simulating different scenarios. With social distancing in mind, the model was tuned to ensure that repulsive forces are stronger. From this model, each pedestrian has a predicted direction and end point, as well as a confidence value that will inform the visualization.

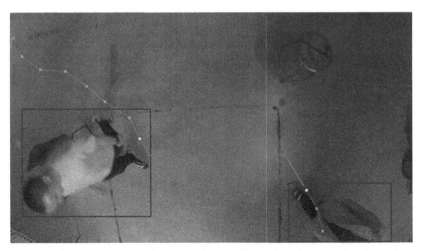

Fig. 5. The output of the path prediction model visualized. Based on past locations, the model outputs a direction and confidence of prediction.

3.5 Interaction Design

As an interactive environment for mediation of space, the system is activated only by the presence of people within the space. Upon entering, each pedestrian is shown a personal signifier that follows their movement and visually changes to reflect their state and suggested path. Previous detection and tracking steps collect participant input that affects the state of their signifier, including speed, direction, and prediction confidence. Although each iteration of the visuals is unique, they are all reflections of these same inputs. Throughout these developments, conveying individual agency and open-ended paths emerged as a central consideration. In addition, signifiers can interact with others in the space, such that they can communicate safe distances of 2 m between individuals. The way that signifiers react to the presence of others also sets the stage for unplanned social connections between strangers. Though the visual design began as a means of maintaining social distance, is has evolved toward livening shared spaces by sparking spontaneous interactions in a post-COVID-19 context.

3.6 Design Iterations

Early iterations utilized vector drawings for simplicity in generating the visualization – from line segments representing paths to dense point fields responding to presence. Clarity and openness were the primary design considerations toward creating a visual system that would be quickly understandable but capable of representing choice. This led to the idea of the "compass" – a ring of radiating line segments that modulate in length to indicate direction while remaining visually open and adaptable (Fig. 6). In practice, however, it had mixed effectiveness. Some thought that the extended lines were clear in showing direction but that the concentration of line segments made them feel too closed off from others. Others thought that the lines did not have enough presence in the space, which required them to look down at their feet in order to see their compass.

However, simply extending the lines out more did not prove to be a suitable solution. It became clear that doing so would increase the chances of collision with other compasses in the space, which could not be visually resolved due to the number and complexity of lines that would cross. This fundamental limitation also became evident when trying to incorporate interactivity between compasses – a scenario where such collisions need to be resolved. Overall, the compass and other iterations made with vector geometry were unable to keep up in this regard.

Fig. 6. The compass concept, showing interactions between more than one compass in the digital output and in space. Line colors change in response to the presence of other people, as well as the suggested direction.

Spotlight System. This led to the development of a more dynamic visual system through the use of graphical shaders [20]. Parallel processing of shaders allows for intricate pixel-by-pixel renderings of gradiented colors, as well as more complex interplay between each personal space. The ability to control gradation in an efficient way led to the development of the "spotlight", a soft directional cone of light originating from each pedestrian (Fig. 7). To reflect open-ended possibilities, the edges of the spotlight soften farther out from the center. Prediction data translates to the geometry of the cone, with more confident trajectories becoming longer. The length of the light also allows for more intricate representations of paths, such as curving to avoid an obstacle or stationary person. Directionality is much clearer in this system, with the cone being visible in peripheral vision while being able to resolve interactions much more naturally. Color plays a more important role in this system as the way to distinguish different signifiers, rather than to display state-specific information like proximity. To ensure that colors would remain unique with multiple people, the initialization of spotlights would assign a random hue that is distinguishable from the color of any existing spotlights.

Ultimately, the spotlight is more conducive to showing complex interactions between each pedestrian. Unique colors clearly mark where each pedestrian is able to move given their predicted path. Should the paths of pedestrians reach a negotiable crossing, the edges of each spotlight warp or sharpen (Fig. 8) to indicate priority. Far enough away, however, colors blend together to create unique gradients and keep the space open to movement. The ability of these edges to soften and sharpen demarcate space when necessary and facilitate interaction at safe distances. The spotlight is also able to effectively reflect the

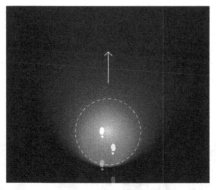

Fig. 7. An annotated spotlight in motion, with direction indicated by the expanding gradient. Personal space is shown by the dashed line annotation.

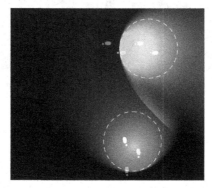

Fig. 8. An example of the spotlight responding to an approaching pedestrian. In this case, the purple spotlight is close enough to the path of the green that the edge of the spotlight sharpens in response to its presence.

state of the pedestrian it is following, by shrinking when stationary or widening with more available space.

3.7 Installation Setup

To evaluate the real-world performance of the project, the installation was set up in the Princeton University School of Architecture building during the month of December 2021 (Fig. 9). An Optoma ZU720TST was chosen to project the visuals for its short throw ratio and high brightness. The space was a highly trafficked landing in the main stairwell of the building, measuring 6.5 × 4 m. Mounted from a height of 5 m, the projector and the camera view were able to cover the entire landing. Over the course of three weeks, the installation was remotely monitored and maintained with an average of 387 pedestrian instances passing through daily. Within the space and in the areas around it, posters were put up with information and a link to the project website. Key observations and findings from this installation are described in the following section.

Fig. 9. Photo of the hardware system with projector and camera running via the attached board.

4   Results

With its integration of computational methods for detection, pathing, and visualization, Social Signals was capable of guiding pedestrians along safe paths through a space of variable occupancy. Put together, the unified process laid out in the Methods of capturing image data, detecting people, predicting trajectories, and visualizing the output ran within desired performance goals for the duration of the installation. Though it only just met the minimum threshold of 24 FPS, the resulting visualization proved responsive to movement with animations appearing smooth to match. This remained mostly consistent throughout the life of the installation, with framerates only going below 24 for 6% of the duration in which the space was occupied. These dips can be attributed to higher volumes of people; most any occupancy level at or above five would begin to degrade performance due to the increased computational load at every step of the data processing pipeline. However, the space did not often experience these volumes – with over 63% of the detection frames consisting of just one occupant and the remaining 31% for four or less. Because of this, the installation space (Fig. 10) proved to be within the performance window for long-term testing of this particular hardware configuration. It is also worth noting that with the pedestrian behavior model being the most computationally intensive part of the process, it would sometimes throttle for the rest of the system to reach the target 24 FPS. This led to situations where the direction of the signifier would update less frequently than position.

Detections ran throughout the duration of the installation while remaining accurate outside of naturally occurring occlusions from the single camera perspective. However, there was no method to evaluate the accuracy of detections outside of researchers observing the space for extended periods of time. Privacy criteria shared among Princeton University and the researchers stipulated that images of occupants were not to be stored – meaning that there was no ground truth to compare detections against. However, in the combined 4 h of direct observation over the duration of the installation there were no missed detections.

Fig. 10. Two people moving past each other through the space.

Despite path predictions relying on the relatively simple social force model, they proved to be mostly accurate. Continuous comparisons between predictions and the direction that occupants moved in following frames observed that the predicted trajectory was correct 81% of the time. This was tested with an angular tolerance equivalent to the confidence value of the given trajectory, so lower confidences meant that the measure was more forgiving. However, this is not a perfect measure of the effectiveness of the predictor. Confidence values were relatively low for the duration of most paths due to the small size of the space, leaving many of the paths with tolerances of up to 30° in either direction of the trajectory center. The space was also relatively simple with only four points of entry and exit, such that the possible paths were limited. In combination with generally low confidence values, the accuracy measure of predictions was certainly not itself perfect. In practice, these trajectories appeared to align with occupant intentions when observing the space. However, without more thorough tests it is difficult to conclude whether that was due to the characteristics of the space or whether occupants were following the projected signifiers. In situations with more than one participant, the predictor performed well in routing individuals around each other by never recommending paths that would directly collide.

The design of the visualization and relatively light computation behind it meant that the system was equipped to adapt on the fly to deviations. Visual openness as a built-in parameter simultaneously communicated uncertainty in the predictions but also afforded participants with the flexibility in choosing a path. However, other visual parameters were not as functionally successful. In particular, the reactive spotlight edges that communicated when occupants got too close were not entirely effective at maintaining social distance. This could have been due to the limited amount of space available, which obligated participants to be physically close as they moved through. It also came as guidelines were being relaxed, so participants were certainly not as intent to remain apart. What emerged instead were frequent interactions where participants would experiment and play with the reactivity of the signifiers to see how they would change based on different conditions (Fig. 11). In these cases, the projections acted as a social mediator – becoming a collective puzzle that led to many spontaneous connections in the early days of the installation. The distinct colors played a part in encouraging participants to experiment with how they would combine, while also remaining clear in communicating distinct personal spaces and directing participants along paths.

Fig. 11. The spotlight of a moving pedestrian in red reacting to the proximity of a stationary one in green.

4.1 Limitations

Space and layout limitations of the installation area meant that the adaptability of the installation could not be tested to the full extent. External light conditions were completely consistent since natural light did not reach the landing, while the space was otherwise empty. As such, the detection model and projections could be tested more rigorously in spaces with furniture and variable daylight. Furthermore, the single camera/projector setup led to occlusions in the camera view. On the visualization side, the central placement of the projector also meant that shadows cast outward would obscure projections in the direction of light. More cameras and projectors would address this need by providing additional angles of exposure, but with increased complexity.

On the hardware side, short throw projectors that are bright enough (>6000 lumens) to show the visualization in daylight conditions are not available at the consumer level.

Furthermore, despite how it is possible to run the installation on a single computer board, current boards are still not powerful enough to handle all scenarios. This was evident by the performance inconsistencies when there were more than five people within the space. At the time of this paper, projector technology and compact processing power are not yet capable of allowing for an economical, lightweight, and performative solution.

5 Conclusion

By incorporating building occupancy analysis into a media installation, Social Signals advocates for a tangibility of the same computational technologies that have become increasingly part of studies of the built environment. Computer vision techniques and pedestrian models allow the system to study a given space and provide suggestions in real-time via dynamic projections. These visual signifiers are made clear through distinct color gradients that combine uniquely and transform the nature of the space as people pass through. Their interaction with the installation informs the computation, closing a previously open loop by bringing together the processing of spatial data and behavior change in occupants. The result is a flexible system that changes how we occupy space by highlighting the interactions that define it.

5.1 Future Directions

Further research could work toward improving the adaptability of the installation, while allowing for unique environmental changes through the agency of participants. To do so, its framework could be scaled up to encompass larger and more complex spaces. Additionally, an accompanying upgrade in computational capability would allow for more granular methods of quantitative evaluation, such as detailed maps of occupancy patterns and snapshots of visualizations and paths. Such efforts to improve the rigor of experimental design would benefit the continual development of the installation.

More space also opens the door to many possibilities of path suggestions given an increase in continuous movement data for the behavior model. It would also present opportunities to consider interactions like group movement, the formation of queues or lanes, and other attractive social forces. Introducing these scenarios would improve the depth of path suggestions – opening up possibilities for new interactions, visuals, and evolutions of space. Toward this interest of making room for emergent patterns, the behavior of the installation could change over time to behave according to changes in the space and the way pedestrians occupy it over a long time scale.

In addition to exploring complexity in the behavioral model, it would be worth experimenting with how changes in the visual system could affect behavior of pedestrians and vice versa. For instance, a highly congested section could bias its suggestions to encourage pedestrians to keep moving. On the other hand, particularly social areas can transform into colorful spaces in which occupants are able to just play with the visualization. The technical and artistic potential of this research signals that this approach to interactive computation can not only be used as an advanced tool for improving the way we use buildings, but also as a means of bringing people together in serendipity.

References

1. Frumkin, H.: COVID-19, the built environment, and health. Environ. Health Perspect. **129**(7), 1–14 (2021)
2. Happle, G., Fonseca, J., Schlueter, A.: A review on occupant behavior in urban building energy models. Energ. Buildings **174**, 276–292 (2018)
3. Eldib, M., Deboeverie, F., Philips, W., Aghajan, H.: Towards more efficient use of office space. In: Proceedings of the 10th International Conference on Distributed Smart Camera, pp. 37–43 (2016)
4. Krijnen, T., Beetz, J., de Vries, B.: Airport schiphol: behavioral simulation of a design concept. In: Computation: The New Realm of Architectural Design (2009)
5. Brunetti, A., Buongiorno, D., Trotta, G.F., Bevilacqua, V.: Computer vision and deep learning techniques for pedestrian detection and tracking: a survey. Neurocomputing **300**, 17–33 (2018)
6. Vergara, L., Mauricio, A.: A computational method for quantitative post occupancy evaluation of occupants' spatial behavior in buildings. ProQuest Dissertations and Theses. Ph.D., Princeton University (2019)
7. Ahmed, I., Ahmad, M., J.J.P.C. Rodrigues, G. Jeon, S. Din: A deep learning-based social distance monitoring framework for COVID-19. Sustain. Cities Soc. **65**, 102571 (2021)
8. Yang, D., Yurtsever, E., Renganathan, V., Redmill, K.A., Özgüner, Ü.: A vision-based social distancing and critical density detection system for COVID-19. Sensors **21**(13), 4608 (2021). https://doi.org/10.3390/s21134608
9. Fazio, M., Buzachis, A., Galletta, A., Celesti, A., Villari, M.: A proximity-based indoor navigation system tackling the COVID-19 social distancing measures. In: 2020 IEEE Symposium on Computers and Communications (ISCC), pp. 1–6 (2020). https://doi.org/10.1109/ISCC50 000.2020.9219634
10. Helbing, D., Molnár, P.: Social force model for pedestrian dynamics. Phys. Rev. E **51**(5), 4282–4286 (1995). https://doi.org/10.1103/PhysRevE.51.4282
11. Zou, J., Zhao, Q., Yang, W., Wang, F.: Occupancy detection in the office by analyzing surveillance videos and its application to building energy conservation. Energ. Buildings **152**, 385–398 (2017)
12. Allison, P.R.: Social distancing with crowd simulation modeling. Computer Weekly, 1 December 2020
13. Krueger, M.W.: Responsive environments. In: Proceedings of the June 13–16, 1977, National Computer Conference. AFIPS 1977, pp. 423–33. Association for Computing Machinery, New York (1977)
14. Kaptelinin, V.: Designing mediation. In: Proceedings of the European Conference on Cognitive Ergonomics, pp. 1 4 (2015). https://doi.org/10.1145/2788412.2788413
15. Lavin, S.: Kissing Architecture. Princeton University Press (2011)
16. van der Wiel, J., Verstand, N.: Smart distancing system. Dutch Design Awards. https://www.dutchdesignawards.nl/en/gallery/smart-distancing-system/
17. OpenCV: Open Source Computer Vision Library (2022)
18. Nguyen, D.T., Li, W., Ogunbona, P.O.: Human detection from images and videos: a survey. Pattern Recogn. **51**, 148–175 (2016)
19. Benenson, R., Omran, M., Hosang, J., Schiele, B.: Ten years of pedestrian detection, what have we learned? In: Agapito, L., Bronstein, M.M., Rother, C. (eds.) ECCV 2014. LNCS, vol. 8926, pp. 613–627. Springer, Cham (2015). https://doi.org/10.1007/978-3-319-16181-5_47
20. Gonzalez Vivo, P., Lowe, J.: The Book of Shaders, 17 March 2022. https://github.com/patric iogonzalezvivo/thebookofshaders

Quantifying Occupant Behavior Uncertainty in Spatio-Temporal Visual Comfort Assessment of National Fitness Halls: A Machine Learning-Based Co-simulation Framework

Yu Li[1,2], Lingling Li[1,2], Pengyuan Shen[3(✉)], and Xue Cui[4]

[1] School of Architecture, Harbin Institute of Technology, Harbin 150001, China
[2] Key Laboratory of Cold Region Urban and Rural Human Settlement Environment Science and Technology, Ministry of Industry and Information Technology, Harbin 150001, China
[3] School of Architecture, Harbin Institute of Technology, Shenzhen 518055, China
shenpengyuan@hit.edu.cn
[4] Department of Building and Real Estate, Faculty of Construction and Environment, The Hong Kong Polytechnic University, Hong Kong, China

Abstract. Occupant behavior has been recognized as the main factor influencing visual comfort gaps between simulated and actual conditions. However, existing daylight and glare simulation methods mostly fail to deal with occupant behavior uncertainty. This study proposes a machine learning-based co-simulation framework to better quantify sports behavior uncertainty in the visual comfort assessment of national fitness halls. The framework improves upon existing approaches by formulating an artificial neural network (ANN)-based multi-objective optimization (MOO) workflow to improve overall visual comfort in large sports spaces. Additionally, a data-driven prediction model is developed to quantify the player's occupancy probability in all possible view scenes during sports. Moreover, the new term, *'probabilistic visual comfort'* is introduced to evaluate visual comfort accounting for sports behavior uncertainty. To verify the effectiveness of the proposed framework, a case study is conducted using tennis sports as the research scenario. The results indicate that the ANN-based MOO significantly improves the efficiency of searching for the final optimum solution(s) with validated accuracy. Furthermore, probabilistic visual comfort fits better than conventional metrics in evaluating occupant's actual visual comfort perception in sports scenarios. The proposed framework provides new insights into incorporating occupant behavior modeling with visual comfort assessment.

Keywords: Occupant Behavior · Visual Comfort · Uncertainty Analysis · Machine Learning · Co-simulation

1 Introduction

Mass construction demand for large sports facilities has been witnessed in China since Beijing 2008 Olympic Games, with the implementation of the 'National Fitness Plan' in recent years, there has been a new shift of constructing Chinese sports buildings

© The Author(s), under exclusive license to Springer Nature Switzerland AG 2023
M. Turrin et al. (Eds.): CAAD Futures 2023, CCIS 1819, pp. 562–576, 2023.
https://doi.org/10.1007/978-3-031-37189-9_37

from professional sports halls towards national fitness halls (NFH, a new kind of indoor multi-sports complex mainly for public fitness beyond professional matches).

Professional sports halls are deprived of daylight due to the critical lighting standard requiring the stable, sufficient and glare-free lighting during competitive matches, while for NFH, the light environment only requires the *'Lighting Class III: low level (local, small club, training, and school sports)'* [1] that can support occupant's daily fitness sports and exercises. Hence, daylight, is a potential lighting source for NFH that is greatly preferred by non-professional users. Daylighting contributes to stimulating human circadian system, improving occupant's health, well-being and productivity, as well as saving building energy consumption, while excessive sunlight is also the main cause of discomfort glare that should be avoided by proper building design strategies. To make full use of daylight for achieving player's visual comfort in NFH, it is thus crucial to understand player's uncertain behaviors within sports spaces, based on which proper building massing and fenestration systems could be designed.

Previous studies regarding visual comfort assessment of large sports spaces mainly resorted to computational simulations, in-situ measurements and human subject studies to evaluate daylight provision and glare performance in the target sports area. For instance, Shi et al. [2] conducted a subjective questionnaire and objective physiological measurements under different lighting conditions in a gymnasium to identify the interfering factors for user assessment of daylight environment, visual comfort thresholds for mass sports activities were defined. Yang et al. [3] proposed a novel simulation-based multi-objective optimization method and applied it to a top-daylighting system design case. Recently, Fan et al. [4]. Presented a multi-objective façade optimization method to balance the daylight comfort and solar radiation in gymnasiums, daylight, glare and energy performances were significantly improved in a real case study by using the proposed approach. However, to realistically investigate the impact of daylight on occupant's visual comfort, it is essential to shift the representation of daylight performance from the building space to the performance of its users [5]. Therefore, robust approaches of quantifying occupant behavior uncertainty across time and space are highly required.

Despite significant research progress of visual comfort assessment and optimization in large sports spaces, most of them are based on time-consuming daylight and glare detections in limited fixed view scenes, i.e., the specific view positions (VP) and view directions (VD), which inevitably simplified player's real sports scenarios. Moreover, existing visual comfort metrics were basically designed for spaces with relatively simple and stationary visual tasks (such as office-computer, library-reading, etc.), which cannot fully capture the complexity of player's changeable view scenes when playing sports. To this end, existing methods present great challenges and difficulties for studying visual comfort in NFH. To author's best knowledge, very few studies specifically set out to investigate player's sports behavior uncertainty in visual comfort assessment of large sports spaces. Thus, following knowledge gaps still exist: i) What kinds of visual comfort metrics can be utilized to better assess player's visual comfort in NFH? ii) How to computationally incorporate sports behavior uncertainty into visual comfort assessment? and (iii) How does players' sports behavior affect their actual visual comfort perception during sports? Hence, this study aims to propose a machine learning-based co-simulation framework that can better quantify sports behavior uncertainty in visual comfort assessment of NFH. The applicability of the proposed framework will be verified through a case study in Shenzhen, China using a parametric NFH model taking tennis single as the research scenario.

2 Methodology

2.1 Methodological Framework

The proposed framework entails 4 steps, i) Design of Experiment (DoE); ii) ANN-based MOO; iii) Sports behavior uncertainty quantification; and iv) Generation of the solution space of court layout and location, as illustrated in Fig. 1. First, Design of Experiment (DoE) is conducted to collect adequate uniformly-distributed samples for developing artificial neural network (ANN) models, after which an ANN-based multi-objective optimization (MOO) workflow is formulated to quickly improve the overall visual comfort of the whole sports space, the final optimum design solution could be determined from numerous Pareto solutions through multi-criteria decision-making (MCDM). Then, a data-driven prediction model is developed to identify player's occupancy probability in all possible view scenes (view positions and view directions) within the sports area. Afterwards, the *'probabilistic visual comfort'(VC_p)* metric is used to evaluate player's actual visual comfort by coupling player's occupancy probability with corresponding visual comfort values in each view scenes. Eventually, the solution space of court layout and location that achieve desirable visual comfort could be identified. It should be mentioned that the whole framework is enabled by a developed computational workflow using the GH-modeFRONTIER (GH-MF) platform, referred to [3]. The following sections will describe the above steps in detail.

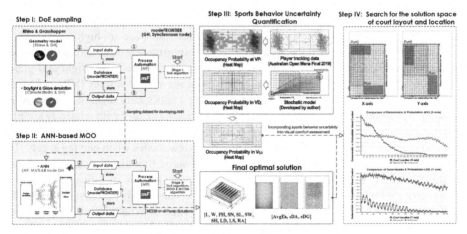

Fig. 1. Methodological framework.

2.2 Design of Experiment (DoE)

Parametric NFH Model. National Fitness hall (NFH) is a new kind of multi-sports complex named by the State General Administration of Sport (SGAS) of China, which is dedicated to mass sports and fitness activities. It differentiates traditional sports halls in its user, spatial configuration and sports activities. Since NFH users have changed from professional athletes to ordinary fitness people, most spaces are designed to support their public fitness and exercises instead of competitive matches. Besides, the large spectator area in traditional sports halls is replaced by multiple independent fitness halls that are horizontally and vertically allocated. This study develops a parametric NFH model that consists of two-story sports spaces and four-story accessory spaces, as shown in see Fig. 2. 10 design variables are determined to control the building massing, skylight and shading configurations, as listed in Table 1, the large open-plan sports space in the second story is selected as the target area in this study. Previous study indicated that roof skylight is efficient to introduce the natural light deep into large sports spaces whilst effectively avoiding discomfort glare [3]. Additionally, given that interior blinds are always down to block out glare during daily use of NFH, therefore, side windows are not considered in the developed NFH model in this study.

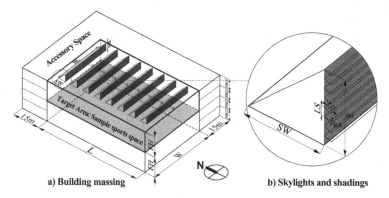

a) Building massing   b) Skylights and shadings

Fig. 2. Parametric NFH model.

DoE Sampling. This study utilizes the uniform Latin Hypercube Sampling (LHS) algorithm to generate 1000 design samples through modeFRONTIER (MF) software, the as-generated input variables are transferred to Grasshopper (GH) and ClimateStudio (CS) for generating geometry model and then calculating corresponding visual comfort values (output variables). The integrated computational platform GH-modeFRONTIER (GH-MF) could automate the parametric modelling, performance simulation and data exchange/storage, the specific process refers to [6]. Finally, 1000 generated design samples are utilized for training and validating the ANN models.

Table 1. Description of design variables of the NFH.

	Design variable	Range	Step	Unit
Building Massing	Sports Space_Length (L)	[50, 70]	1	(m)
	Sports Space_Width (W)	[30, 50]	1	(m)
	Sports Space_Floor Height (FH)	[9.0, 15.0]	0.1	(m)
Skylights	Skylights_Numer (SN)	[3, 12]	1	–
	Skylights_Length (SL)	[20, 35]	1	(m)
	Skylights_Width (SW)	[1.0, 4.0]	0.1	(m)
	Skylights_Height (SH)	[1.0, 4.0]	0.1	(m)
Shadings	Shades_Louver Depth (LD)	[0.20, 0.50]	0.01	(m)
	Shades_Louver Spacing (LS)	[0.20, 0.50]	0.01	(m)
	Shades_Rotation Angle (RA)	[−45, 45]	5	(°)

2.3   ANN-Based MOO

Objective Functions. The objective functions of visual comfort include daylight avail-ability and protection from glare, three corresponding metrics are selected, namely, Average Horizontal Illuminance (AvgE$_h$), Spatial Daylight Autonomy (sDA$_{300/50\%}$) and Spatial Discomfort Glare (sDG$_{38\%,5\%}$).

Average Horizontal Illuminance (AvgE$_h$): The average illuminance over the regularly occupied floor area over annual occupied hours (8:00am–6:00pm of 365 days, 3650 h):

$$AvgE_h = \frac{\sum_{i=1}^{n} \sum_{j=1}^{3650} A(j)}{\sum_{j=1}^{3650} t_j * \sum_{i=1}^{n} P_i} \tag{1}$$

Where: P_i is the ith sensor point (view position) in the target area (i $\in [1, n]$); $A(j)$ represents the illuminance at timesteps t_j (j $\in [1, 3650]$).

Spatial Daylight Autonomy (sDA$_{300/50\%}$): The percentage of a space that receives a minimum target illuminance of 300 lx for temporal fraction threshold of 50% of annual occupied hours [7]:

$$sDA_{300/50\%} = \frac{\sum_i S(i)}{\sum_i P_i} \in [0, 1], \quad S(i) = \begin{cases} 1 & if\ DA_i \geq \tau \\ 0 & if\ DA_i < \tau \end{cases} \quad (\tau = 50\%) \tag{2}$$

Where: $S(i)$ represents a function; DA_i is the occurrences exceeding the illuminance target at points P_i; τ is annual temporal fraction threshold.

Spatial Discomfort Glare (sDG$_{38\%,5\%}$): The fraction of view scenes in a space that report discomfort glare exceeding a temporal fraction threshold of 5% of annual occupied hours [8]:

$$sDG_{38\%,5\%} = \frac{\sum_i G(i)}{\sum_i V_i} \in [0, 1], \quad G(i) = \begin{cases} 1 & if\ DGF_i \geq \sigma \\ 0 & if\ DGF_i < \sigma \end{cases} \quad (\sigma = 5\%) \tag{3}$$

Where: $G(i)$ represents a function; DGF_i is the occurrences exceeding the DGP_{limit} (0.38) [9] at the given view scene V_i; σ is annual temporal fraction threshold, suggested as 5% in British Standard Institution (BSI) BS-EN-17037: 2018 [10].

MOO Problem Formulation. After declaring objective functions, three optimization objectives are defined, namely, max_AvgE$_h$, max_sDA, min_sDG, the MOO problem could be formulated as:

$$min: \; F(X) = \left[-f_1(X), -f_2(X), f_3(X)\right],$$

$$X = [L, W, FH, SN, SL, SW, SH, LD, LS, RA] \; and \; X \in S, \qquad (4)$$

$$subject \; to: \; f_1(X) \geq 300, f_2(X) \geq 55\%, f_3(X) \leq 50\%$$

where: $-f_1(X), -f_2(X), f_3(X)$ correspond to the objective function of max_AvgE$_h$, max_sDA, min_sDG, respectively; X denotes 10 design variables; S is the entire search space; To achieve the MOO, a common multi-objective genetic algorithm, Non-dominated Sorting Genetic Algorithm II (NSGA-II) is employed.

ANN Modelling. ANN is a self-adaptive machine learning method that can extract the non-liner relationship between inputs and outputs by different network learning mechanisms. Various ANN models have been used to predict building performances, among which the back-propagation (BP) neural network is the most frequent one and thus been used in this study. To achieve the best prediction accuracy, the independent ANN model for each output variable with different ANN structure (Hidden layers) is developed. The specific structure and algorithm of the ANN model are as follows:

Input Layer. For all three ANN models in this study, inputs are identically the 10 design variables of the parametric NFH model.

Hidden Layer. To determine the right number of hidden neurons of each ANN model, we developed an optimization program in MATLAB to execute the loop of training the ANN model with a hidden neuron number from 0 to 100, through which the MATLAB 'nftool' can automatically train the ANN model and calculate its performance, the Mean Squared Error (MSE) and R-squared (R^2) are used to evaluate the ANN performance.

Output Layer. The output of each ANN model is AvgE$_h$, sDA$_{300/50\%}$ and sDG$_{38\%,5\%}$.

Training Algorithm. The Levenberg-Marquardt (LM) learning algorithm in MATLAB toolbox (trainlm) is chosen as the ANN training algorithm in this study due to its efficiency and accuracy in generating the desirable ANN model.

As a result, 1000 samples generated from the DoE are used to develop the ANN model of each output variable adopting the selected ANN structure and algorithm, the proportion of samples for the training, validation and test dataset is 80%, 10% and 10%.

ANN-Based MOO Workflow. This study establishes an ANN-based MOO workflow (Fig. 3) to replace the traditional simulation-based MOO for saving computational cost, in which three independent ANN prediction models are employed to quickly approximate the AvgE$_h$, sDA and sDG values (using the *MATLAB* node in MF) instead of physics-based simulation. Note that those non-sensitive variables are set as constant in MOO process. To synchronize the optimization of all three objective functions, the *synchronizer* node in MF is executed that enables the parallel run of three independent MATLAB programs to simultaneously calculate three output variables.

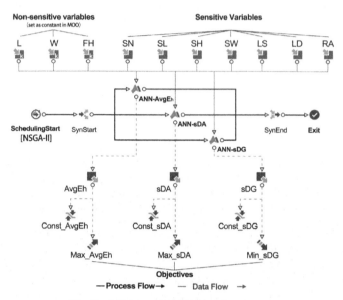

Fig. 3. ANN-based MOO workflow in MF.

2.4 Quantification of Sports Behavior Uncertainty

Fitness sports and exercises often yield critical visual tasks involving continuously changing view scenes, wherein player's real-time spatial location and next-moment movement remain great uncertainties and are difficult to forecast. As it is known that glare is highly sensitive to spatial and temporal parameters including the relative position of the viewer to glare sources, glare size and intensity in the viewer's field-of-view, it thus can be assumed that sports behavior uncertainty might directly affect player's real visual comfort perception across sports spaces. To quantify player's sports behavior uncertainty for more comprehensive visual comfort assessment in NFH, this study develops a data-driven prediction model to identify player's occupancy probability (OP) in all possible view scenes, $V_{(i,j)}$ (VP_i, VD_j) during tennis sports (tennis single as the research scenario), the prediction model consists of three parts: i) Extracting the player-tracking data from website database to calculate player's occupancy probability at all view positions (VP_i), denoted as OP_i; ii) Using a developed stochastic model to predict player's occupancy probability in all possible view directions (VD_j), denoted as OP_j; iii) Coupling OP_i and OP_j to quantify player's occupancy probability in $V_{(i,j)}$, denoted as $OP_{(i,j)}$, as illustrated in Fig. 4.

In this study, a detailed tennis dataset (.csv file) of Australian Open Men's Final 2019 was used as the fundamental player-tracking data [11]. This dataset consists of

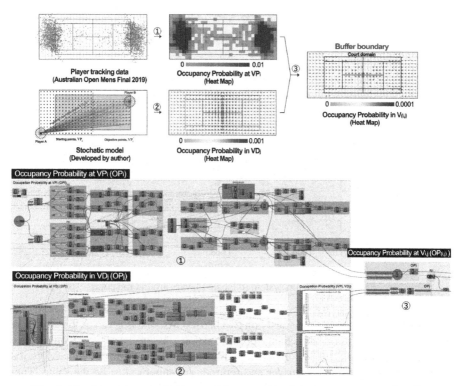

Fig. 4. The data-driven prediction model to quantify player's occupancy probability

all points in the match recorded hierarchically from events, to rallies, to actual points, among which the events data that captures the position of the striker and returner is the focus of this study. To properly locate both players around the court, we first define x, y coordinates and the coordinate origin, based on which, a set of regions for player's occupied area are then determined employing grids with the spacing of 1m. Afterwards, (x, y) positions of strikers and returners (from dataset) with respect to the court domain and buffer boundary (boundary of player's occupied area) are mapped in Grasshopper, as shown in Fig. 5. Given that daylight and glare performances are simulated based on sensor grids, to coordinate player's positions with it, we assume that those player's positions within the region i are approximately considered in the center of the region, represented as VP_i, with $i \in [0, 475]$. Therefore, player's occupancy probability at each view position (OP_i) could be calculated as Eq. (5).

$$OP_i = \frac{Occur(VP_i)}{M} \tag{5}$$

where: $Occur(VP_i)$ represents the occurrences of player's location at VP_i, and M denotes total player's locations throughout the half-court (M = 1690), both are counted from the selected player-tracking dataset.

To predict player's occupancy probability in all possible view directions (VD_j), a stochastic model for tennis single was developed. In this model, two players would

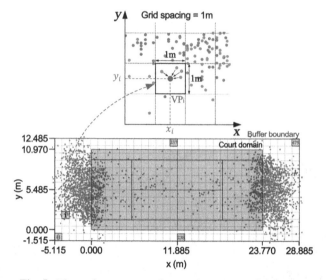

Fig. 5. The region system to locate players around the court.

participate in the sport simultaneously on two half-courts, one is given as the striker, the other as the receiver, their roles change during the point, assuming players would behave identically on different half-courts with similar locations and movements. Moreover, player's tendencies as both a striker and returner are supposed to be the same. Then, we define player's view positions on one half-court as the starting points, denoted by VP_i, all possible objective points from VP_i are the remaining view positions on another half-court, denoted by VP_i'. Let $VD_p(i)$ represent all possible view directions at VP_i, described by the connection between VP_i and VP_i'. Let VD_j represent the established view directions in glare simulation, with $j \in [0, 23]$. Thus, VD_j and VD_{j+1} denote any two adjacent VDs ($\theta = 15°$), let α represent the included angle between $VD_p(i)$ and VD_j. For any $VD_p(i)$, if $\alpha \leq {}^1/_2\,\theta$, then we approximate the $VD_p(i)$ as VD_j, otherwise, approximate $VD_p(i)$ as VD_{j+1}, as illustrated in Fig. 6. Thus, player's occupancy probability in each view direction (OP_j) could be calculated as Eq. (6).

$$OP_j = \frac{Occur(VD_j)}{N} \tag{6}$$

where: $Occur(VD_j)$ denotes the occurrences of player's view direction in VD_j, calculated by the developed stochastic model; N denotes total possible view directions at VP_i ($N = 96$). By combing the OP_i and OP_j, player's occupancy probability in each view scene ($OP_{(i,j)}$) could be calculated as Eq. (7).

$$OP_{(i,j)} = OP_i * OP_j, i \in [0, 475], j \in [0, 23] \tag{7}$$

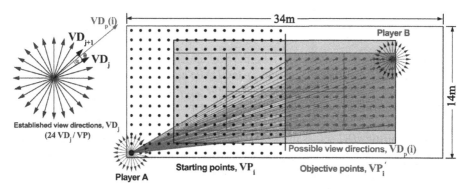

Fig. 6. The stochastic model for tennis single. Where: green and red radials denote all possible view directions in sports, and established view directions in glare simulation, respectively. This figure shows that player A's view position locates on the left half-court, and vice versa. Generation of court layout and location

2.5 Generation of Court Layout and Location

Calculation of Probabilistic Visual Comfort (VC_p).

In this study, we introduced the term, *'Probabilistic Visual Comfort (VC_p)'* to evaluate player's actual visual comfort taking account of the sports behavior uncertainty. To be more specific, the player's occupancy probability at VP_i (OP_i) is incorporated with the daylight performance $\{AvgE_h(i), DA(i)\}$ at the corresponding sensor, meanwhile, the player's occupancy probability in $V_{(i,j)}$ ($OP_{(i,j)}$) is integrated with the glare performance ($DGF_{(i,j)}$) in the corresponding view scene. Thus, the probabilistic visual comfort $\{AvgE_{hp}, sDA_p, sDG_p\}$ could be calculated as Eqs. (8–15):

$$AvgE_{hp}(i) = AvgE_h(i) * OP_i \tag{8}$$

$$AvgE_{hp} = \frac{\sum_{i=0}^{475} AvgE_{hp}(i)}{2} \tag{9}$$

$$DA_p(i) = DA(i) * OP_i \tag{10}$$

$$sDA_p = \frac{\sum_{i=0}^{475} A_p(i)}{476}, \quad A_p(i) = \begin{cases} 1 & if \ DA_p(i) \geq \tau_p \\ 0 & if \ DA_p(i) < \tau_p \end{cases} \tag{11}$$

$$\tau_p = \frac{1}{238} * \tau = 0.2101\% \quad (\tau = 50\%, \ referred \ to \ Eq.(2)) \tag{12}$$

$$DGF_{p(i,j)} = DGF_{(i,j)} * OP_{(i,j)} \tag{13}$$

$$sDG_p = \frac{\sum_{i=0}^{475} \sum_{j=0}^{23} G_{p(i,j)}}{476 * 24}, \quad G_{p(i,j)} = \begin{cases} 1 \ if \ DGF_{p(i,j)} \geq \sigma_p \\ 0 \ if \ DGF_{p(i,j)} < \sigma_p \end{cases} \tag{14}$$

$$\sigma_p = 1/(238 * 24) * \sigma = 8.7535e^{(-6)} \quad (\sigma = 5\%, \; referred \; to \; Eq. \; (3)) \quad (15)$$

Where: $AvgE_{hp}(i)$ and $AvgE_h(i)$ are the probabilistic and simulated $AvgE_h$ at each view position (VP_i), respectively; $DA_p(i)$ and $DA(i)$ are the probabilistic and simulated DA at each view position (VP_i), respectively; $DGF_{p(i,j)}$ and $DGF_{(i,j)}$ are the probabilistic and simulated DGF in each view scene ($V_{(i,j)}$), respectively; $AvgE_{hp}$ and sDA_p are the probabilistic $AvgE_h$ and probabilistic sDA across whole player's occupied area; sDG_p is the probabilistic sDG across the whole space; τ_p and σ_p are annual temporal fraction thresholds of $DA_p(i)$ and $DGF_{p(i,j)}$, respectively, which are calculated assuming that players share equal occupancy probability at each view position ($\frac{1}{238}$) and in each view direction ($\frac{1}{238*24}$).

Definition of Court Starting-Point and Layout. To locate the court domain within the whole sample sports space, we first define the court starting-point (P_0) and court layout (X/Y-axis), given the court size (24.77*10.97m) and buffer distance requirement for tennis single, player's occupied area and ranges of all possible P_0 in different court layouts could be identified. Then, we remap all possible P_0 for X/Y-axis, namely, $P_0|_{x-axis} \in [0, 275]$ and $P_0|_{y-axis} \in [0, 675]$ (see Fig. 7). Among all possible P_0, those with desirable VC_p are declared as available solutions. In this study, the GH plug-in, *colibri* is used to calculate the VC_p at all P_0 under two different court layouts.

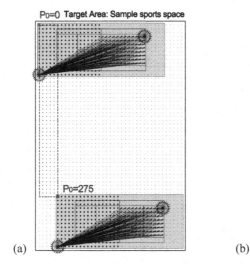
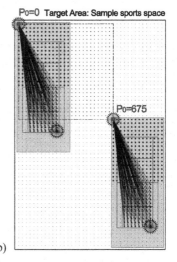

(a)   (b)

Fig. 7. Court starting-point, layout and ranges of P_0. (a) x-axis; (b) y-axis. Where: the blue domain refers to the player's occupied area, the red dashed area represents all available P_0.

3   Results and Discussion

3.1   ANN Model Training and Validation Results

Figure 8(a–c) depicts the desirable neuron number in the hidden layer of each ANN model considering the R^2 and RMSE value simultaneously. For $AvgE_h$, sDA and sDG, the respective neuron number is $24, 12, 41$. Figure 8(d–f) shows the ANN model training, validation and test results, accordingly. The MSE and R values of each ANN model indicate the good prediction accuracy.

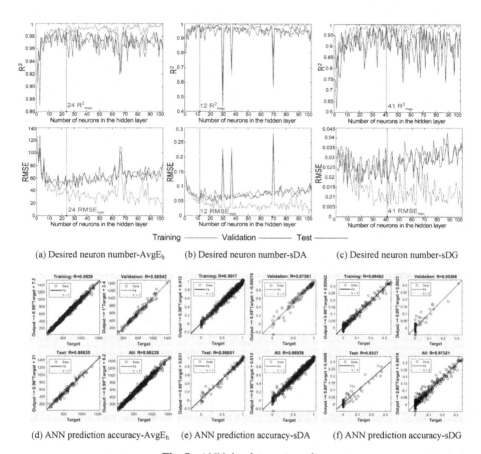

(a) Desired neuron number-$AvgE_h$ (b) Desired neuron number-sDA (c) Desired neuron number-sDG

(d) ANN prediction accuracy-$AvgE_h$ (e) ANN prediction accuracy-sDA (f) ANN prediction accuracy-sDG

Fig. 8.   ANN development result.

3.2   Pareto Solutions and Final Optimum Solution

188 Pareto solutions among all 1880 alternative designs are obtained by ANN-based MOO, as depicted in Fig. 9. By performing MCDM on all Pareto solutions using equal weights on three objectives, the best-ranked solution (ID = 1010) and the least-ranked

Pareto solution (ID = 1738) are identified. It can be known that the best one upon the worst one observes 84.27% higher visual comfort, which is determined as the final optimum solution. Since the ANN model can only predict the overall visual comfort, we then rerun the physical simulation of the final optimum solution to obtain the daylight and glare performance in each VP and VD.

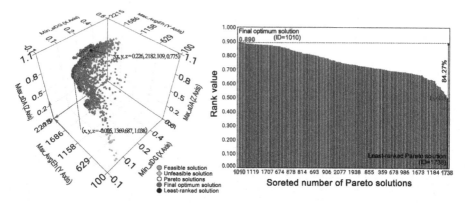

Fig. 9. Pareto solutions and final optimum solution.

3.3 Solution Space of Court Layout and Location

By incorporating player's occupancy probability into the visual comfort assessment using the developed approach, we calculated the probabilistic visual comfort (VC_p). Meanwhile, the deterministic visual comfort (VC_d) was calculated as the baseline using the equal occupancy probability in each VP and VD, as illustrated in Fig. 10(a–c).

Discrepancies between the VC_p and VC_d are observed and the degree of deviation differs from metric to metric. For $AvgE_{hp}$ and $AvgE_{hd}$, the discrepancy is generally stable under both X-axis and Y-axis except for some fluctuations at a few P_0 ($P_0|_{x-axis} \in$ [230, 275], $P_0|_{y-axis} \in$ [550, 675]. Besides, the discrepancy is not very significant under Y-axis but quite noticeable under X-axis; For sDA_p and sDA_d, the discrepancy is quite stable and considerable which is mainly due to the fact that almost all VPs achieved the acceptable DA and thus the uneven OP_i plays the key role leading to this discrepancy, as depicted in Fig. 10(b); While for sDG_p and sDG_d, there exist differences under X/Y-axis scenarios with regard to the discrepancy between sDG_p and sDG_d. Regarding the X-axis scenario, sDG_d is larger than sDG_p when $P_0|_{x-axis} \in$ [0, 79] and the discrepancy reduces with the increase of P_0, the discrepancy keeps insignificant when $P_0|_{x-axis} \in$ [79, 96], while for $P_0|_{x-axis} \in$ [96, 275], the sDG_p is larger than sDG_d and the discrepancy gradually increases until $P_0|_{x-axis} = 200$; Regarding the Y-axis scenario, considerable deviations can be observed between sDG_d and sDG_p, which shows regular fluctuations. In general, the sDG_p is more stable than sDG_d across all P_0 under both X/Y-axis.

Concerning the effect of sports behavior uncertainty on visual comfort, it can be deduced from Fig. 10(c) that uneven $OP_{(i,j)}$ can significantly affect player's actual

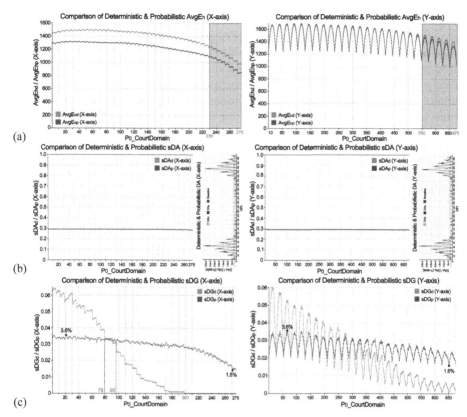

Fig. 10. Comparison of deterministic and probabilistic visual comfort. (a) $AvgE_{hd}$ vs. $AvgE_{hp}$ (X/Y-axis); (b) sDA_d vs. sDA_p (X/Y-axis); (c) sDG_d vs. sDG_p (X/Y-axis).

glare performance. Meanwhile, from Fig. 10(a), it could be concluded that uneven OP_i can slightly impact the actual daylight performance. Thus, it could be concluded that glare performance, compared with daylight performance, is more vulnerable to occupant behavior uncertainty. Another important founding is that court location (P_0) can greatly influence the overall daylight and glare performances, which could be conjectured from the changing amplitude of each line in Fig. 10(a–c). Since there is currently no exact building code restricting sDG for sports spaces, the sDG_p under X/Y-axis scenario in this study range within [1.5%, 3.6%] and [1.6%, 3.6%], indicating quite good glare performance, refers to [8]. For $AvgE_{hp}$ and sDA_p, almost all P_0 meet the requirement.

4   Conclusion

This study proposes a machine learning-based co-simulation framework that better quantifies sports behavior uncertainty in visual comfort assessment of national fitness halls. The contribution of the presented work lies in three aspects. First, the '*probabilistic visual comfort*' metric is introduced to evaluate player's actual visual comfort perception in sports spaces that outperforms conventional ones in capturing the complexity of sports behavior; Second, an ANN-based MOO workflow is formulated to improve the overall visual comfort in large sports spaces, which accelerates the traditional simulation-based MOO process and effectively saves computational cost; Last, a data-driven prediction model is developed to quantify player's occupancy probability in all possible view scenes during sports which enables more comprehensive visual comfort assessment. Case study results clarified the impact of sports behavior uncertainty on player's visual comfort assessment. Future work will investigate more kinds of sports and spaces and further validate the proposed framework at different climates.

References

1. En, B.S.I.: Light and Lighting - Sports Lighting. BSI, London, UK (2018)
2. Shi, L., et al.: Luminance parameter thresholds for user visual comfort under daylight conditions from subjective responses and physiological measurements in a gymnasium. Build. Environ. **205**, 108187 (2021)
3. Yang, D., et al.: Dynamic and interactive re-formulation of multi-objective optimization problems for conceptual architectural design exploration. Autom. Constr. **118**, 103251 (2020)
4. Fan, Z., Liu, M., Tang, S.: A multi-objective optimization design method for gymnasium facade shading ratio integrating energy load and daylight comfort. Build. Environ. **207**, 108527 (2022)
5. Danell, M., Ámundadóttir, M.L., Rockcastle, S.: Evaluating Temporal and Spatial Light Exposure Profiles for Typical Building Occupants (2020)
6. Yang, D., et al.: Multi-disciplinary and multi-objective optimization problem re-formulation in computational design exploration: a case of conceptual sports building design. Autom. Constr. **92**, 242–269 (2018)
7. IESNA, LM-83-12 IES Spatial Daylight Autonomy (sDA) and Annual Sunlight Exposure (ASE). IESNA Lighting Measurement, New York, NY, USA (2012)
8. De Luca, F., Sepúlveda, A., Varjas, T.: Multi-performance optimization of static shading devices for glare, daylight, view and energy consideration. Build. Environ. **217**, 109110 (2022)
9. Wienold, J., et al.: Cross-validation and robustness of daylight glare metrics. Light. Res. Technol. **51**(7), 983–1013 (2019)
10. En, B.S.I.: Daylight in Buildings. BSI, London, UK (2018)
11. Seidl, R.: Tennis ATP Tour Australian Open Final 2019: A tribute to Novac Djokovic and Rafael Nadal. https://www.kaggle.com/code/robseidl/australian-open-mens-final-2019-data-exploration

Building Massing and Layouts

Topological Operations: A New Method Toward Retrievable Design Operations

Yang Li$^{(\boxtimes)}$ ⓘ and Rudi Stouffs ⓘ

Department of Architecture, National University of Singapore, 4 Architecture Drive,
Singapore 117566, Singapore
`l.y@u.nus.edu`, `stouffs@nus.edu.sg`

Abstract. The process of architectural design can be complex due to the engagement of problem-solving with design thinking. However, in computer-aided approaches, there is no effective method of retrieving an understanding of the design process. This research aims to demonstrate a newly developed process-oriented workflow for design generation via retrievable design operations. This concept of retrievable design operations emphasizes design thinking as an accumulation of design strategies expressed as small operational steps. The transformation of design thinking can take place between projects and/or design alternatives through a recomposition of these steps. This paper exemplifies the method by using a proposed toolset on the Rhino-Grasshopper platform to achieve the *retrievability* of design operations. Intermediate results of this study are presented. Firstly, 3 categories of components in Grasshopper are presented that allow to define an operation-retrievable architectural massing. Secondly, the potential to amplify retrievable design operations into feasible design derivations is demonstrated via 1) transforming design models across different site conditions by retrieving the entire design operations, 2) and switching the individual typology by retrieving corresponding partial design operations.

Keywords: Topological Operations · Retrievable Design Operations · Computer-aided Architectural Design · Process-oriented Representation · Graph Representation

1 Introduction

Architectural design is a nonlinear process [1] and inherently a wicked problem [2], which requires an amplification strategy [3] to find results. Parametric systems are often used to generate design alternatives to facilitate the finding process. They allow architects and designers to rapidly develop design iterations and derivations based on predefined constraints and relations [4]. Although the parametric approach describes certain process relations at dimensional levels, the result can only exist within a 3D coordinate model that does not support *retrievability* to unpack the design process or architectural thinking. Specifically, an interpretation of the design process independent of the coordinate model cannot be readily retrieved from the parametric definitions [4] because of the way of codification.

© The Author(s), under exclusive license to Springer Nature Switzerland AG 2023
M. Turrin et al. (Eds.): CAAD Futures 2023, CCIS 1819, pp. 579–593, 2023.
https://doi.org/10.1007/978-3-031-37189-9_38

If design results or iterations and derivations can be considered as an accumulation of design strategies on an initial input geometry, then, design processes can be deconstructed as a series of design operations [5]. Di Mari and Yoo illustrated this process called operative design by using operative verbs [6] to operate on architectural forms according to design strategies. Meanwhile, if these operations are aligned with architectural ideas and strategies, the design process could be reordered according to a new design agenda toward different series of iterations and derivations. However, when a design needs to be deployed under new site conditions, the previous actions that were used to create the design model generally cannot easily be retrieved and reapplied [3] to construct the new models. This interruption makes the design amplification process difficult to update and damages its consistency. Therefore, effectively recording the design process without reference to a specific coordinate model becomes the key to resolving this translation issue, which means the process of designing a 3D model can be properly recorded in a consistent representation and be able to amplify alternatives based on retrievable design operations.

The conception of *retrievability* in design space exploration is the integration of backup, recall, and replay coined by Woodbury and Burrow [3]. *Backup* is achieved by the Grasshopper definition of each retrievable design operation. Each stage of design is recorded individually based on its thread rather than a linear stack. *Recall* represents the ability to deconstruct certain portions of design processes without affecting the relations between objects. The concept of *replay* emphasizes the projectivity and universality of codification. All three notions are embedded in the proposed toolset and workflow. In addition, the proposed system differs from the parametric method in the vehicle of controlling the form. The system of *retrievability* focuses on spatial topology and design operations as the foundation of design thinking, instead of defining flexible relations of dimensions to manipulate form.

2   Related Work

The pioneer of architectural computation, William Mitchell, defined the notion of design operation as a state change process in the design world and explained that design solutions can be searched from the combination of the set of all states [5]. Architectural design is essentially a process of finding optimal solutions from permutations and combinations. Kalay [7] and Gips [8] have defined design as a searching process. Galle [9] and Lömker [10] specified architectural design as a combinatorial problem because of constraint-based searching procedures.

Prior methods of exploration of the 3D architectural form include constructive solid geometry (CSG), shape grammars, parametrics, voxels, and non-manifold topology (NMT). CSG utilizes a binary tree to store the primitives and their corresponding operational logic. Chang and Woodbury have experimented with using CSG to generate design alternatives by interpreting the logic of undo [11]. The tree data structure has a significant impact on traversing the logical path. The system is limited to a one-to-one operation with a binary structure, which means the traversal of the combination is relatively fixed. When implementing the undo, the chain of operations is easy to break apart. A shape grammar specifies a rule-based formal language [12, 13] but also a process-oriented formal description. It is sophisticated enough to describe complex geometry such as the Chinese Ice-ray screen [14, 15]; however, the architectural applications of the Palladian grammar [16], the grammar of Queen Anne houses [17], and the Dirksen Courthouse[18] primarily translate shape rules in a two-dimensional manner such as extrusions from 2D shapes and cooperation of 2D sectional shapes. Froebel's building example has direct 3D translations [19, 20], but the design alternatives are limited. The SortalGI shape grammar interpreter is able to describe a 3D form based on an extension of descriptions from 2D shapes [21, 22]. However, while the specifications of shape rules may be easy to comprehend, the process of composing them can be intricate; for instance, the Dirksen grammar takes 78 rules [18], and the Palladian grammar takes 72 rules [16]. Parametrics is a dimension-based chain system [4], usually cooperating with Boolean operations and topology. The DigiWo plugin to Grasshopper generates residential building volumes via an automated parametric approach [23, 24]. It hard-codes three principal typologies with different urban strategies to generate different design options. The EvoMass plugin to Grasshopper integrates parametric systems with Boolean operations. Because of the codification strategy, the formal characteristics of the generation are quite limited in a squarish style. March and Steadman presented a voxel-based approach to represent the massing of the Seagram building, which utilizes matrices for performing Boolean operations [25]. The process is comparable to sculpting, where the material is removed from an initial cuboid to create the desired result. Finally, the NMT concept is implemented as a geometric library called Topologic [26]. The topological data is encapsulated in rigorous hierarchical classes such as vertex, edge, face, graph, etc. It is mainly used to query geometric elements based on the topological relations in graph class. It is typically not used independently, but rather in conjunction with parametrics.

A graph-based representation has significant effects on searching architectural solutions and representing abstract spatial characteristics. Frank Lloyd Wright experimented with the idea between formal expression and spatial connectivity through three distinct housing projects [25]. It demonstrates the value of graph representation and the correlation between formal architectural expression. The integration between graph-based representation and design operations provides an opportunity to achieve retrievable design operations.

3   Method

By defining retrievable design operations that are closely aligned with the design agenda and strategy using the proposed method, architects and designers can establish a library of design operations (Fig. 1). This library can be updated independently without direct connections to practical projects when any new thoughts emerge. It can also be developed through projects. Therefore, efforts in individual projects can be converted into commonly used design operations and strategies to support other future projects.

Specifically, the method of this research concentrates on the development of a computational design toolkit based on a rigorous and compatible codification of identifiable design operations through a newly developed graph-based representational approach to achieve retrievable design operations.

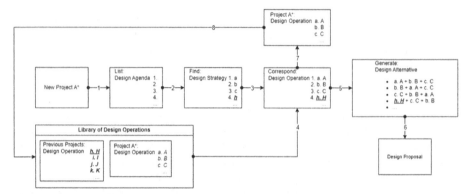

Fig. 1. Method of retrievable design operations (a previous design operation, *h. H*, is retrieved from the library of design operations according to agenda 4 and strategy *h* in the design process between steps 3 and 5. The generation of design alternatives utilizes the recomposition of retrievable design operations to create derivations between steps 5 and 6. In steps 7 and 8, the operations that are developed in the current project are collected in the library of design operations for future reference.)

3.1 Graph Algebra for Topological Operations

Instead of focusing on the infinite number of massing typologies, this research attempts to employ a finite group of architectural operations to construct different variations of architectural massing types. The method is designed to be user-oriented and interactive instead of involving complicated technical knowledge. From a technical perspective to achieve that level of simplicity, all the operations in this proposed method demand a consistent data structure to support entire operations including topological computation and visual representations.

Two major categories of Grasshopper components are considered: topological and algebra components. These two categories of components are designed to retrieve topological relations and execute graph computation. These components can be understood as a collection of foundational operations. Each component or operation can be stacked, reordered, and reused in different places in the design workflow. A combination of these components or operations can be used to describe most types of architectural massing and design processes. The interaction between components is established via the class of *algebraCells*. This class maintains the data in the same structure between the communications; *algebraCells* objects exist in both categories of components and the grid cell generator component to maintain consistency of data type.

AlgebraCells Class. The *algebraCells* class as an essential part of code exists in all the components of the algebra graph. It carries all the graph data and corresponding geometric information consistently along with the entire workflow. It helps each component to retrieve previous stage data and original geometric information to compute the current results.

The *algebraCells* class contains 5 fields:

1. *BoundaryBrep*: The current stage boundary of the shape is stored in the Brep format.

2. *FullJoinedBrep*: The current stage geometric shapes are stored in the joined Brep format.
3. *InitialBrep*: The original Brep generated by the grid cells generator. It stores all the faces with the corresponding indexes.
4. *FacesIndex*: The collection of current stage faces in relation to the *FullJoinedBrep*.
5. *adjacencyList*: The graph data of all the cells is stored as an adjacency list with consistent integers along the operations.

Below is the custom class *algebraCells* written in C# using RhinoCommon library.

```
public class algebraCells
  {public Brep BoundaryBrep {get; set;}
   public Brep FullJoinedBrep {get; set;}
   public Brep InitialBrep {get; set;}
   public List<int> FacesIndex {get; set;}
   public List<List<int>> adjacencyList;
   public algebraCells(Brep fullJoinedBrep, List<int> facesIn-
     dex, Brep boundaryBrep, Brep initialBrep)
     {FullJoinedBrep = fullJoinedBrep;
      FacesIndex = facesIndex;
      BoundaryBrep = boundaryBrep;
      InitialBrep = initialBrep;
      adjacencyList = new List<List<int>>();
      for (int i = 0; i < initialBrep.Faces.Count; i++)
        {
          adjacencyList.Add
          (initialBrep.Faces[i].AdjacentFaces().ToList());
        }
     }
  }
```

The general steps and all the corresponding components are listed in Fig. 2. The overall workflow of utilizing the graph algebra to represent architectural massing can be understood as 4 main steps.

1. Defining site boundary and grid lines as basic inputs for the grid cells generator.
2. The grid cells generator takes two inputs and divides them into the *algebraCells* and *Brep* cells that are used to visualize the physical form of the output.
3. Three different topological components are distinguished that can be used together or separately to select certain cells from the grid based on specific topological relations. These relations are associated with the building typology.
4. Using mainly algebra components to represent a slightly complex massing typology. It includes 5 distinct components that can be used in a hybrid manner.

Grid Cells Generator. The grid cells generator provides a fundamental framework for the exploration of architectural massing. It converts all the basic site information with

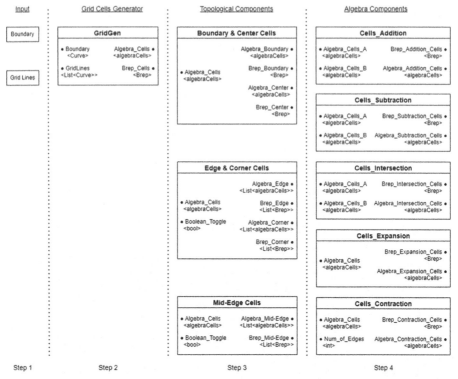

Fig. 2. Overview of the method for topological operations with 3 categories of components

divided grid cells into the customized class (*algebraCells*). The most important characteristic of this component is that the boundary curve and grid lines can be dynamic. The change of boundary shape and grid line density will automatically and instantaneously reflect on the outcomes and affect the latter operations. The grid lines can control the resolutions (Fig. 3) and patterns of massing as well as structural rhythm (Fig. 4). In Fig. 3, the first row of group *a* is the input boundary curve. The second row of group *b* is the input grid lines that are used to split the site surface created by the boundary curve. The last row of group *c* is the output of joined Brep created by the split cells. The first column *1* has a 4 × 4 grid system. The second column *2* has a 6 × 6 grid system. The third column *3* has an 8 × 8 grid system.

Topological Components. Topological components identify three types of cells based on their adjacency: 1) boundary and center cells, 2) edge and corner cells, and 3) mid-edge cells. The selection of cells is based on topological relations that are not restricted by the grid types, cell and boundary shapes, the number of cells, or any dimensions. The exact dimensions of shapes can be flexible and controlled by the grid cells generator. This flexible selection process has a wider range of adaptability to be rearranged and retrieved in the proposed operational process. The algorithm of the boundary & center cells component is built on the detection of adjacency between boundary curves and cells (Fig. 5).

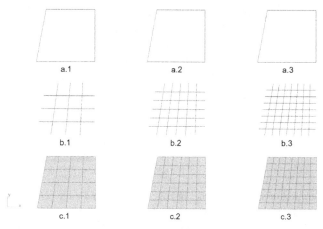

Fig. 3. The illustration of grid density and resolution

Fig. 4. The example of three different types of grid system (The grid can be evenly distributed such as *(a)*, a superimposed cubic grid *(b)*, or an irregular grid *(c)* that has different spacing between grid lines.)

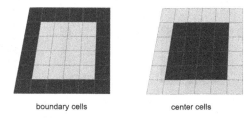

boundary cells   center cells

Fig. 5. Demonstration of selecting boundary & center cells (On the left diagram, only the cells that are adjacent to the boundary are selected. The remaining cells are defined as center cells on the right.)

The corner cells selection is based on the detection of the adjacency between boundary corner points and cells. The corner & edge cells component split the boundary into edges, and all cells are grouped and selected according to their adjacency to each edge (Fig. 6 top row). The component of the mid-edge cells is designed to create a void or public space in the middle side of the massing model, for instance, a U-shaped typology or an H-shaped typology. Technically, the mid-edge cells are selected by removing the corner cells on the basis of edge cells (Fig. 6 bottom row).

Algebra Components. The algebra components include 5 different operations: addition, subtraction, intersection, expansion, and contraction. The first 3 components are

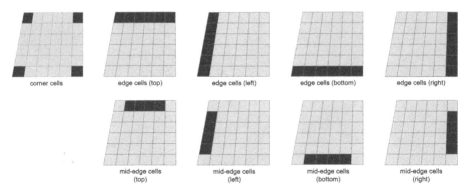

Fig. 6. Demonstration of selecting corner & edge cells and mid-edge cells (The group of edge and mid-edge cells is decided by the segment of a boundary. The demonstration uses a four-sided polygon with the selection of 4 groups' edge and mid-edge cells)

similar to the Boolean operations. The last 2 components are based on the adjacency or graph data to operate. The algebra components are the most representative operations that have the strong potential to create complex massings. It enhances the overall workflow and provides a way to interact and communicate with the topological components.

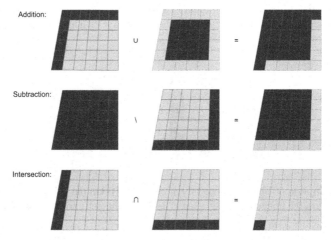

Fig. 7. Demonstration of addition, subtraction, and intersection

The addition component can be used to combine two areas together if the two are adjacent or overlap (Fig. 7 first row). The operational logic is similar to set union. This idea of subtraction can be interpreted as the architectural concept of solid and void. This subtraction operation can be employed to create open space (Fig. 7 second row). The calculation of subtraction can be understood as set difference. The intersection component calculates the shared part between two cells or algebraCells objects (Fig. 7 third row). The algorithm of this component can be understood as the calculation of set intersection.

Expansion: Contraction:

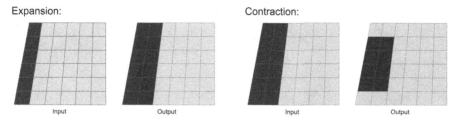

Fig. 8. Demonstration of expansion (left) and contraction (right)

Expansion and contraction are slightly complex operations in the algebra components. Expansion follows the topological order that is defined by the grid system. It expands the original area according to the adjacent condition (Fig. 8 left). Contraction is designed to shrink or cut the cells based on the topological condition (Fig. 8 right). The cells that have fewer adjacent cells will be cut first.

4 Demonstration

Two types of exploration are demonstrated. The first demonstration defines an operation-retrievable architectural massing. The second demonstration uses this massing to derive alternative design solutions.

4.1 Operation-Retrievable Massing

This demonstration uses a built project as a precedent to illustrate the capability to represent a relatively complex architectural massing and design process using the proposed method.

The project is a healthcare center and regional government office in Spain designed by BAT + ARQUITECNICA (Fig. 9) [27]. The process of dividing the plot and the logic of massing blocks are fully articulated. First, the site is divided into two parts. Secondly, the small part continues to be divided into 2 parts. One becomes a higher courtyard form; the other becomes a blockish form. Lastly, the largest part becomes a combination of an off-centered courtyard and an L-shaped form. In addition, the overall complexity of cluster typologies covers the courtyard, cuboid, and L-shape. Because of the intelligible steps and sufficient complexity, it can serve as a proper example to demonstrate the method. Figure 10 is the massing model that is generated by the graph algebra description. The description that is used to create the massing below transferred the most essential parts of the form from the precedent. Two courtyard conditions and one strip void are all represented by the proposed description.

Major Steps

- Step 1: the site is divided into grid cells by using the Grid Cell Generator.
- Step 2: the site is divided into 2 uneven parts by using the components of Edge & Corner Cell, Expansion, and Subtraction.

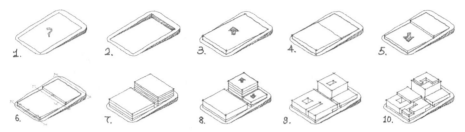

Fig. 9. Design diagram of healthcare center and regional government office in Spain redrawn from ARQUITECNICA + BAT [27]

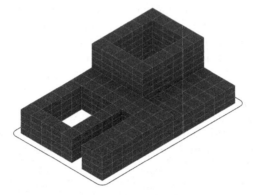

Fig. 10. Massing model of healthcare center and regional government office in Spain

- Step 3: the two parts are divided into 4 parts by using the same components as in the previous step.
- Steps 4 & 5: the red hatch areas are removed from their corresponding parts.
- Step 6: the massing is generated by the footprints regarding their different heights.

Fig. 11. Major steps demonstration

Figure 11 illustrates the above six major steps. Figure 12 elaborates on the Grasshopper definition of precedent-building massing and its outputs. This Grasshopper definition is made from the proposed components explained in the methodology section. Each important step of the component is numbered; the components and the outcome of the component are listed in the corresponding order.

Three Dimensionalization. The three-dimensionalization processes in step 6 (Fig. 11) or steps 18, 19, and 20 (Fig. 12) are based on the definitions of the number of floors

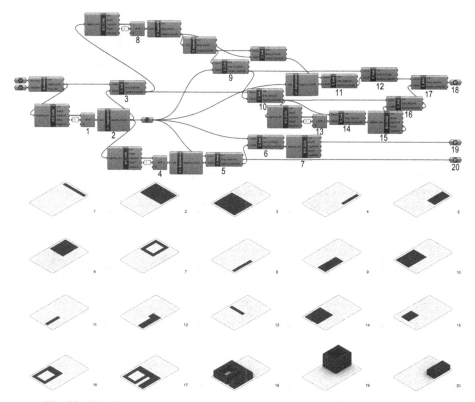

Fig. 12. The Grasshopper definition and corresponding results of the components

and floor-to-ceiling height. The algorithm of three-dimensionalization takes 2 inputs as mentioned above. The floor-to-ceiling height is a list of doubles. Each number corresponds to the floor numbers, which means it allows each floor to have a different height. The output of this component is a closed *Brep* of the entire massing.

4.2 Retrievability

Retrieving Entire Operations in Different Sites. Figure 13 illustrates the ability to transform the retrievable operations of the precedent massing to an L-shaped site (left) with an orthogonal grid system and a four-sided polygon site (right) with an irregular grid system. The original massing is situated in a rectangular site. The experiment of the transformation on the left is tested in an L-shaped site with an evenly distributed grid. The one on the right is tested in an irregular polygon site that has two edges perpendicular to each other and an irregular grid system. The proposed retrievable operations successfully transfer the massing features into other sites within their grid system.

Retrieving Typologies. As Fig. 9 illustrated, the site is divided into 4 parts with 4 different typologies (higher courtyard, off-centered courtyard, L-shape, and cuboid). Because of the *retrievability* of the proposed method, these 4 building typologies can be

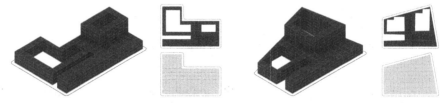

Fig. 13. Demonstration of retrieving entire design operations in different sites by using the precedent of operation-retrievable massing

switched or reordered between the divisions of the site. Figure 14 illustrates 4 different combinations of typologies by using retrievable design operations. These operations can also be stored in the library of design operations (Fig. 1) for future use.

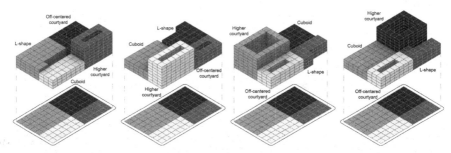

Fig. 14. Derivations generated by switching the 4 typological forms using retrievable design operations

5   Discussion

The study of form-making is very important to architectural design. Modern designers pay less attention to the logical study of formalism instead they rely on more artistic expression [28]. The process of exploring architectural solutions heavily depends on architectural formal strategies. It is difficult to create a variety of forms with a monotonous architectural formal expression. A systematic architectural formal representation is a prerequisite for solving design problems.

5.1   Design Space Exploration

The notion of design space abstractly describes searching processes of design solutions as explorations of design space [29]. Different searching approaches will highly affect the reachability of undiscovered design space. The proposed approach focuses on the interpretation of design operation, which is different from the parametric approach. If these two approaches are projected onto the diagram of the design space, it will display different areas of discovered space because of the internal mechanisms.

5.2 Design Translation

The difficulty with computational design lies in the limitations of representing design strategies as mathematical machine-driven operations [29]. However, not every design strategy can be translated into executable machine language. The specialty of retrievable design operations is that it can express the relationship between spaces and objects rigorously. The architectural strategy corresponds exactly to the executable machine operations. It helps to list combinations and derivations as solutions for comparison. As the articulation of architectural design strategies becomes more refined, the searching and comparing process will gain more opportunities to uncover larger design spaces.

5.3 Graph-Based Representation

An appropriate representation is very important in computational approaches. It bridges the gap between visual understanding and communication with machines. Computer-aided generation heavily relies on the organization of processing data. A viable representational framework requires a robust data structure and efficient algorithm. This research utilizes graph data structures and grid systems to build a comprehensive representational framework that can describe most building typologies.

The operations or components created in this framework are developed on the topological relations of the cells, which means it does not build on a coordinate-based system. The graph-based representation is an abstraction of physical forms. It can exist independently without the physical form, which offers the ability to apply operations to a completely different grid system or site conditions. This transformability or *retrievability* can be used to rapidly test design strategies or operations on different sites or reordered design strategies or operations through design processes. These operations can be also understood as the record of the design process that can retrieve and apply to different dimensional contexts (grids) via graph-based representation.

6 Conclusion

This paper demonstrates a process-oriented approach to achieving a rigorous exploration of computational formalism. This research method exhibits well-built controllability based on interpreting retrievable design operations through a graph-based massing representation.

It is critical in the way that the proposed method focuses on the topological relations to achieve retrievable design operation. It provides a new opportunity to understand and describe form and space rapidly from a topological approach rather than only relying on dimension chain systems. In addition, this graph-based massing representation will sustain the consistency of architectural design strategies and operations throughout the entire design without wasting time on the dependency of coordinate reference.

Looking from a slightly larger scope of this research, it is an essential part to construct the larger workflow of architectural design automation. The inconsistency of different computational methods prevents the application of computational design. As the graph-based representation and corresponding computational methods are fully unpacked, the design automation will be shifted into a new generation.

Acknowledgment. This research is supported by Singapore's Ministry of Education AcRF Tier 1 Grant (WBS A-0008265-01-00). Any opinions, findings and conclusions or recommendations expressed in this material are those of the authors and do not reflect the views of the Ministry of Education.

References

1. Bachman, L.R.: Two Spheres: Physical And Strategic Design In Architecture. Routledge, London; New York (2012)
2. Rittel, H.W.J., Webber, M.M.: Dilemmas in a general theory of planning. Policy Sci. **4**, 155–169 (1973). https://doi.org/10.1007/BF01405730
3. Woodbury, R.F., Burrow, A.L.: Whither design space? AIEDAM **20**, 63–82 (2006). https://doi.org/10.1017/S0890060406060057
4. Woodbury, R.: Elements of Parametric Design. Routledge, London; New York, NY (2010)
5. Mitchell, W.J.: The Logic Of Architecture: Design, Computation, And Cognition. MIT Press, Cambridge, Mass (1990)
6. Di Mari, A., Yoo, N.: Operative Design: A Catalogue Of Spatial Verbs. BIS Publishers, Amsterdam (2012)
7. Kalay, Y.E.: The computability of architectural design. In: Proceedings of the 5th Jerusalem Conference on Information Technology, 1990. Next Decade in Information Technology. pp. 372–378 (1990)
8. Gips, J.: Shape grammars and their uses: artificial perception, shape generation and computer aesthetics. Birkhäuser, Basel, Stuttgart (1975)
9. Galle, P.: An algorithm for exhaustive generation of building floor plans. Commun. ACM. **24**, 813–825 (1981). https://doi.org/10.1145/358800.358804
10. Lömker, T.M.: Designing with Machines: solving architectural layout planning problems by the use of a constraint programming language and scheduling algorithms (2006)
11. Chang, W., Woodbury, R.: Undo Reinterpreted. Presented at the ACADIA 2003: Connecting: Crossroads of Digital Discourse , Indianapolis (Indiana), USA (2003)
12. Stouffs, R.: Shape rule types and spatial search. In: Lee, J.-H. (ed.) CAAD Futures 2019. CCIS, vol. 1028, pp. 474–488. Springer, Singapore (2019). https://doi.org/10.1007/978-981-13-8410-3_33
13. Stiny, G.: Introduction to shape and shape grammars. Environ. Plann. B. Plann. Des. **7**, 343–351 (1980). https://doi.org/10.1068/b070343
14. Stouffs, R., Wieringa, M.: The generation of Chinese ice-ray lattice designs on 3D surfaces. In: Communicating Space(s). In: 24th eCAADe Conference Proceedings, Volos, September 6–9, 2006. pp. 316–319 (2006)
15. Stiny, G.: Ice-ray: a note on the generation of Chinese lattice designs. Environ. Plann. B. Plann. Des. **4**, 89–98 (1977). https://doi.org/10.1068/b040089
16. Stiny, G., Mitchell, W.J.: The palladian grammar. Environ. Plann. B. Plann. Des. **5**, 5–18 (1978). https://doi.org/10.1068/b050005
17. Flemming, U.: More than the sum of parts: the grammar of queen anne houses. Environ. Plann. B. Plann. Des. **14**, 323–350 (1987). https://doi.org/10.1068/b140323
18. Park, J., Economou, A.: The dirksen grammar: a generative description of Mies van der Rohe's courthouse design language. Nexus Netw. J. **21**(3), 591–622 (2019). https://doi.org/10.1007/s00004-019-00441-8
19. Stiny, G.: Kindergarten grammars: designing with Froebel's building gifts. Environ. Plann. B. **7**, 409–462 (1980). https://doi.org/10.1068/b070409

20. Grasl, T., Economou, A.: From topologies to shapes: parametric shape grammars implemented by graphs. Environ. Plann. B. **40**, 905–922 (2013). https://doi.org/10.1068/b38156
21. Stouffs, R.: A Multi-formalism Shape Grammar Interpreter. In: Gerber, D., Pantazis, E., Bogosian, B., Nahmad, A., Miltiadis, C. (eds.) CAAD Futures 2021. CCIS, vol. 1465, pp. 268–287. Springer, Singapore (2022). https://doi.org/10.1007/978-981-19-1280-1_17
22. Dy, B., Stouffs, R.: Combining geometries and descriptions - a shape grammar plug-in for grasshopper. Presented at the eCAADe 2018: Computing for a better tomorrow , Łódź, Poland (2018)
23. Osintseva, I.: DigiWo | DeCodingSpaces Toolbox. https://toolbox.decodingspaces.net/digiwo/
24. Osintseva, I., Koenig, R., Berst, A., Bielik, M., Schneider, S.: Automated Parametric Building Volume Generation : a Case Study for Urban Blocks (2020)
25. March, Lionel: The Geometry of Environment: An Introduction to Spatial Organization in Design. Routledge (2020). https://doi.org/10.4324/9780429343346
26. Jabi, W., Aish, R., Lannon, S., Chatzivasileiadi, A., Wardhana, N.M.: Topologic - a toolkit for spatial and topological modelling. Presented at the eCAADe 2018: Computing for a better tomorrow , Łódź, Poland (2018)
27. Healthcare Center and Regional Government Offices / BAT + ARQUITECNICA. https://www.archdaily.com/493654/healthcare-center-and-regional-government-offices-bat
28. Alexander, C.: Notes on the Synthesis of Form. Harvard University Press, Cambridge, Mass (2002)
29. Woodbury, R.F., Burrow, A.L.: Notes on the structure of design space. Int. J. Archit. Comput. **1**, 517–532 (2003). https://doi.org/10.1260/147807703773633518

A Physics-Based Constraint Solver for the Site Layout Optimization of Non-convex Buildings with Multiple Requirements

Jinmo Rhee[1]([🖂]) [ID] and Pedro Veloso[2] [ID]

[1] Carnegie Mellon University, Pittsburgh, PA 15206, USA
jinmor@andrew.cmu.edu
[2] University of Arkansas, Fayetteville, AR 72701, USA
pveloso@uark.edu

Abstract. In this paper, we address the generation of realistic apartment complex layouts. We define this layout problem as finding the optimal locations and angles of the buildings within the site boundary according to the building codes, zoning regulations, and design preferences. In contrast to manual design heuristics, which are laborious and typically produce homogeneous solutions, our method focuses on efficient design exploration of realistic and varied alternatives. To avoid exploring the large design sub-spaces with solutions, we introduce a constraint solver based on rigid-body simulation. Based on the results of performance comparison in two experiments, (a) with and (b) without the constraint solver, we address a design problem with an actual site and realistic floor area ratio, building coverage, and height restrictions. We describe four different layouts of apartment complexes generated by our method with realistic site planning patterns and considerable efficiency improvements. Discussing the potential for developing design tools based on this method, we describe ongoing research on the generation of more scenarios and suggest the future exploration of a constraint solver.

Keywords: Generative Design · Physics Simulation · Design Optimization · Site Layout

1 Introduction

In this paper, we introduce and evaluate a computational method to support the design of realistic apartment complex layouts with heterogeneous and non-convex building footprints, using constraints based on realistic regulations and custom goals to address design preferences. The proposed method is generalizable to different layout problems, but we focus on high-density apartment complexes that combine many tall residential buildings, special building types, and landscape features, in a large site with multiple restrictions defined by code.

Overall, this type of layout problem is a challenging problem with a considerable impact on the different scales and experiential dimensions of the urban environment. Such as thermal and visual comfort, human, and ecological flows, and placemaking.

© The Author(s), under exclusive license to Springer Nature Switzerland AG 2023
M. Turrin et al. (Eds.): CAAD Futures 2023, CCIS 1819, pp. 594–606, 2023.
https://doi.org/10.1007/978-3-031-37189-9_39

While the use of numerous buildings to fulfill the construction potential of a site results in a large design space, the strict constraints, such as overlapping, containment, and separation, create large portions of infeasible solutions in this space. To tackle this combination of large design space and strict spatial constraints, manual design methods tend to rely on design heuristics that are laborious and restrict the variety of building shapes, spatial configurations, and relationships. By reducing the design space exploration in favor of feasible design process, these heuristics typically result in homogeneous site configurations with negative consequences for the urban environment.

In contrast, black-box design optimization is a potential candidate to support design decisions and to accelerate the generation and evaluation of varied site layout configurations, which can lead to innovative designs. However, as black box optimization typically relies on modeling the spatial constraints with penalizations, it results in an inefficient search in vast areas of infeasible and uninformative solutions.

The proposed method in this paper contains different mechanisms to make the apartment complex layout amenable for black-box design optimization. It uses conditional rigid-body dynamics to automatically solve separation, overlap, and containment conflicts with heterogeneous and non-convex buildings. The adaptation translates infeasible to feasible solutions, enabling black-box optimization algorithms to perform an efficient search.

2 Background

Physics simulation comprehends mathematical models with variables that represent the state of the different parts of a physical system, such as mass, position, or velocity. A physics engine contains equations that define how the state variables of interacting parts change over time—i.e., its dynamics. Physics simulations have been widely adopted in spatial layout configurators over the last decade, and they can roughly be defined in two categories: soft-body dynamics and rigid-body dynamics.

Soft-body dynamics typically includes the use of particles or nodes and springs to model bodies that can be deformed over the simulation. For example, [1, 2] propose a space planning method where spaces are polygons composed of constrained nodes interconnected by springs, and topological objectives are represented by custom springs between different spaces. Recently, springs and nodes have also been used to model architectural layouts, in non-orthogonal arrangements of convex rooms [3].

In the case of rigid-body simulation, the system contains non-deformable bodies that interact under the action of external forces, joints, and constraints. Examples in space planning rely either on rigid-body dynamics libraries or in custom attraction and repulsion forces. [4] combines constrained dynamics simulation with a local search as an interaction method to modify floor plan layouts with discrete operations such as swapping, moving, grouping, or rotating rooms. The idea is that user manipulations of the layout that would break topological and geometric constraints are guided to a nearby state configuration with good performance and satisfied constraints. [5–7] use custom rigid bodies to create interactive bubble diagrams. [8] includes bubble diagrams generated by a force-directed graph algorithm as part of a computational toolkit for spatial layout configuration. [9] uses agents based on attraction, repulsion, and other operations to create a 3D bubble diagram that is converted into a voxelized layout representation that is refined with an evolutionary optimization. [10] uses neuro-evolution to evolve undirected graphs that are translated into a bubble diagram based on rigid discs and springs, and then partitioned with a Voronoi algorithm. [11] combines a multi-objective evolutionary optimization that generates urban block configurations with a repulsion mechanism that prevents overlapping.

Recently, [12] proposed a method to solve spatial layout problems by modeling the spaces as rectangular rigid bodies and the multiple requirements as a sequential application of forces. For short tests of problems of different sizes, the GPU-enabled rigid-body physics library converges faster than black box optimization algorithms using CPU. This method requires that every constraint and objective is explicitly modeled as sequential physical operation, which can be challenging for more sophisticated requirements. Furthermore, it requires reinitialization or direct manipulation of the solutions to support design exploration.

Most of the examples also rely on a high-level of spatial abstraction for the physics simulation, with bubble diagrams [5–10, 13, 14], rectangular arrangements [1–3, 5, 8, 12], and Voronoi diagrams [10]. Except for [11], the examples mentioned above emphasize spatial relationships of architectural plans, such as adjacency, circulation, and area, and not building forms or code requirements. In contrast, this research addresses the problem of generating apartment complex layouts with realistic building footprints and building codes.

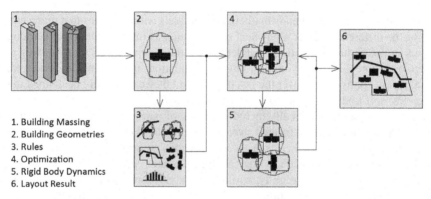

1. Building Massing
2. Building Geometries
3. Rules
4. Optimization
5. Rigid Body Dynamics
6. Layout Result

Fig. 1. A Method for Building Layout Optimization Reinforced by Physics Simulations

3 A Computational Method for Layout Optimization

In our formulation, the different elements of the apartment complex are represented as 2-D polygons or polylines. The building footprints represented as polygons with the outermost boundary of a building on the ground level, which are positioned and oriented by the optimization algorithm to define a layout configuration. The minimum distance constraints are also represented as polygons. They are defined by the union of the rectangular projections of the vertical faces of a building mass into the ground. The depth of this rectangle is defined by a multiplication of the building height and a building separation factor in building code. As a result, the multiple minimum distance polygons represent the minimum distance area that a building should secure. They are attached to their footprints, representing the smallest separation to each surface of the building envelope (Fig. 1 – step 2).

Other elements, such as site, site zones, parks, roads, and pre-existing facilities define the environment. The site boundary is represented as a polygon defining the maximum area where buildings can be located. Zone boundaries are polygons inside the site boundary that represent zoning conditions for certain types of buildings, such as different height regulations. Facilities are also polygons representing non-residential programs, such as schools or parks. Main roads across the site boundary are represented as polylines. Figure 2 shows an example of a simple site configuration with the elements mentioned above.

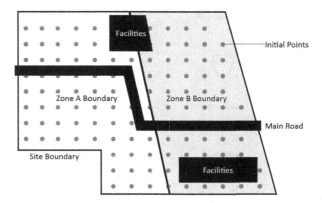

Fig. 2. Two-dimensional Representation of Site Information

Given an initial set of building footprints defined by a previous generative method, a heuristic optimization searches for layout candidates with different building positions and angles that satisfy the constraints and minimize the error according to custom functions representing design preferences. We combine this optimizer with a series of adaptation mechanisms to guide the search process into design sub-spaces with feasible layouts (Fig. 1 – steps 4 and 5). These are state configurations where all buildings are inside the site and legal zones, while avoiding overlapping with other buildings' minimum distance polygons, program polygons, site, and zone boundaries.

The following sections describe the components of our method.

3.1 A Constraint Solver Based on Rigid-Body Dynamics

Given the state of the buildings defined by an optimization algorithm, a physics engine translates the buildings, minimum distance polygons, site edges, etc. into rigid bodies. Each building footprint and its minimum distance boundaries are represented as a set of dynamic shapes that share a single rigid-body with mass proportional to its area. We decompose every non-convex shape, such as a building layout with concavities and cores, into convex parts to benefit from the quality and efficiency of geometric queries and computations of 2D rigid body physics libraries. Particularly, we tested Pymunk, which is Python library built on the top of a physics library written in C (Chipmunk2D). To address the trade-off between shape accuracy and efficiency, we developed convex hull representations with different levels of resolution for the buildings and their respective separation geometries (Fig. 3). Increasing the resolution of the decomposition increases the number of polygons. Therefore, it makes the representation more accurate but increases the computational cost of the simulation.

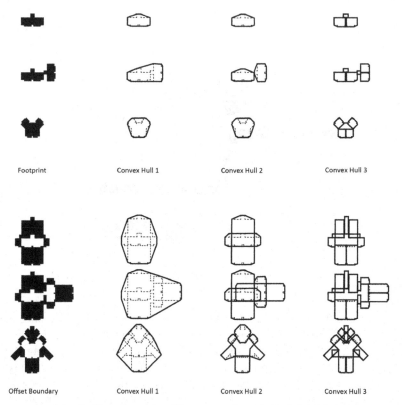

Fig. 3. Two-dimensional Representation of a Building Mass

The Convex Hull 1 of a building is the convex hull of the entire footprint. The Convex Hull 2 comprehends the convex hulls of the sets of units centered around each core. The Convex Hull 3 comprehends the convex hulls of each unit and core. This can become even more granular and sophisticated to represent every building shape accurately at the cost of slowing the computation of rigid-body dynamics. Similar resolutions can also be applied to the decomposition of the separation geometry.

Facilities are also represented as a polygonal rigid body with infinite mass so that they do not move in the simulation. In the examples used in this paper, they are convex, but they can use the same decomposition method used for the buildings otherwise.

In the case of the site boundary or of roads, every edge of the polylines or polygons is converted into rectangular rigid-body with infinite mass that do not move in the simulation. For example, six edges of the site boundary in Fig. 2 are represented as six rectangular rigid-bodies. Roads are also decomposed into their edges. This edge-based polyline representation breaks down concave polygons into line segments which can be used in rigid-body simulations.

Once all elements are converted into rigid bodies based on two-dimensional geometries, the next step is to set relationships for physics simulations that can satisfy the different constraints of the problem. Relationships are based on overlaps and conflicts. When a rigid body is allowed to be overlapped with another, the engine ignores the collision. Otherwise, the engine will apply separation forces according to their physical properties.

Figure 4 illustrates the relationships between different rigid-bodies. For instance, if a building footprint overlaps with a site boundary, it will trigger a separation until they do not touch each other. However, the minimum distance polygon can overlap with the site without triggering a separation.

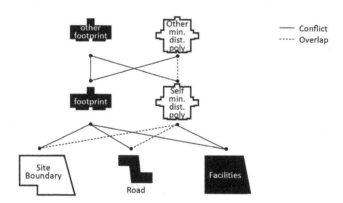

Fig. 4. Relationships Between Different Rigid Bodies

3.2 Other Adaptation Mechanisms

The proposed method also relies on the discretization of the available positions for the building. Initial points on a custom grid are created inside the site boundaries and are categorized by site zones. Therefore, when the heuristic optimization generates a numerical representation for position it can be translated into one of the points in the set of points inside the feasible zones.

The second filtering mechanism is based on angles. The user can either let the simulation define the building angles or define a range of acceptable angles. In the latter, the solutions are mapped to a set of acceptable angles and the moment of inertia of the rigid-bodies associated with the buildings is multiplied by a large factor to prevent rotation.

3.3 Optimization

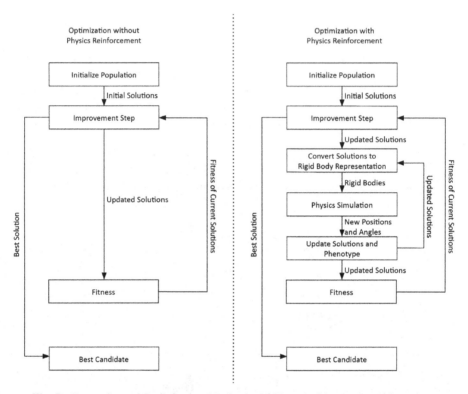

Fig. 5. Comparison of Optimization Mechanism with and without a constraint solver.

Figure 5 illustrates the optimization process using a constraint solver. Once the solutions are initialized in the population, the optimizer works to improve and update them. These solutions are then converted into rigid body representations, which are used by the physics engine to update the solution and phenotype in a simulation. This simulation runs for several iterations based on the physics parameters. Once the simulation iteration is complete, the updated solution's fitness is fed back into the improvement step, and the conversion and physics simulation steps are repeated based on the optimization parameters.

Optimization looks for the best fitting positions and angles with respect to a fitness function. For the objective function of the optimization, we used a composite fitness F based on a series of soft constraints ($c_i \in C$) and goals ($g_j \in G$) that return a value in the interval [0, 1]:

$$F(s) = \frac{1}{|C| + |G|}\left(\sum_{i=1}^{|C|} 1000c_i(s) + \sum_{j=1}^{|G|} g_j(s)\right) \tag{1}$$

The weight associated with the constraint functions works as a large penalty that dominates the goal function; when there is a violation, the fitness will reflect primarily the conflicts rather than the performance with the respect to the preferences. These are some of the functions used in this paper:

- Conflicts between buildings (c_1): returns the percentage of buildings and minimum distance polygons colliding with other building structures.
- Conflicts between buildings and a site boundary (c_2): returns the percentage of buildings that are not completely inside the site boundary.
- Zone regulations (c_3): returns the percentage of buildings that are not completely inside legal zones.
- Angle preference (g_1): returns the average fitness based the closest angle in a list of angle preferences defined by the user.
- Centrality (g_2): returns the percentage of buildings outside of the desired distance ranges from the boundary, based on association between building properties and ranges by the user.

To select a heuristic solver for the optimization with the objective function, we defined two experiments with a simple layout problem to test seven different solvers in Pygmo [15]. The problem had 20 buildings composed by using one unit plan, each building had 30 floors—3m for each floor, and simple rectangular geometry was used. A computer with the following specifications is used: 'Intel(R) Core (TM) i9-10900x @ 3.70 GHz', 256 GB memory, and nvidia RTX A6000 graphic processing units.

In the first experiment, we ran the algorithms with and without the constraint solver twenty times over 600 iterations using a function $F(s)$ that consists of conflicts between buildings (c_1), conflicts between buildings and a site boundary (c_2), and angle preference (g_1). We measured the average time, average final objective value, and number of iterations for convergence. We considered that an algorithm converged when an objective value smaller than 0.2 occurs more than 10 times. Algorithms that converge faster were considered better.

The solver consists of two steps of simulation for convex hull 1 and 2, with specific parameters of duration that affect the convergence. In the first step, the convex hull 1 both for the building footprints and the minimum distance polygons runs over 50 iterations and 0.1 as step size. This step provides a more uniform layout distribution. In the second step, convex hull 2 has 300 iterations and 0.001 as step size. Convex hull 2 for the minimum distance polygon allows buildings to occupy concave spaces adjacent to them. Convex hull 2 for the building footprints enables further refinement in the occupation of convex spaces between building units.

Without the constraint solver, PSO converged to the smallest error. SGA results in the second smaller error slightly higher than PSO (0.001), but on average it converged about 200 iterations earlier and finished the simulation 103 s faster. Considering all these factors, SGA was considered the best heuristic solver for the layout problem without physics reinforcement (Table 1).

With the constraint solver, all the algorithms converged in fewer iteration and much faster. This is an indication that gains due to the physical adaptation to feasible solutions prevails over the additional computation spent running simulations. Besides, the SGA performed better than the other algorithms again, so it was considered the best heuristic solver for the layout problem with physics reinforcement (Table 2).

Table 1. Heuristic Solver Test for a Layout Problem[a]

	ABC	DE	SADE	GACO	GWO	PSO	SGA
Time (s)	357	222	271	285	295	287	**184**
Fitness	0.120	0.139	0.119	0.132	0.116	**0.115**	0.116
Convergence (iterations)	300	590	310	350	590	280	**90**

[a]ABC (Artificial Bee Colony), DE (Differential Evolution), SADE (Self-adaptive DE), GACO (Extended Ant Colony Optimization), GWO (Grey Wolf Optimizer), PSO (Particle Swarm Optimization), SGA (Simple Genetic Algorithm), xNES (Exponential Evolution Strategies), CMAES (Covariance Matrix Adaptation Evo. Strategy).

Table 2. Heuristic Solver Test for a Layout Problem with Physics Reinforcement

	ABC	DE	SADE	GACO	GWO	PSO	SGA
Time (s)	28	16	9	18	21	13	**8**
Fitness	0.118	0.144	0.125	0.118	0.115	0.114	**0.089**
Convergence (iterations)	22	41	13	25	38	18	**4**

The second experiment was measuring the performance of the eight heuristic solvers in the same layout problem with additional components added to the objective function: zone regulation (c_3) and centrality preference (g_2). The solver uses the same steps and parameters from the first experiment. We tested the solvers five times and measured the average time for 100 iterations with 50 population and the final fitness.

Without physics simulation, CMAES was the fastest algorithm, but it did not converge to a feasible solution, and it resulted in the worst performance. SADE was the only algorithm that converged to a feasible solution (Table 3).

With physics simulation, xNES and CMAES were the faster solvers, but they did not converge to feasible solutions. Despite not converging faster in terms of time, SGA converged to the smallest error in the minimum number of iterations. All the algorithms with physics ran faster and converged in fewer iterations. Except for CMAES, all the algorithms converged to better fitness values with physics (Table 4).

Based on these two experiments, we consider that SGA is a good heuristic solver for the apartment complex layout problem. Further, these experiments demonstrate that the combination of an optimization algorithm with the proposed constraint solver performs better in all aspects under evaluation—time, fitness, and number of iterations.

Table 3. Heuristic Solver Test for a Layout Problem[1]

	ABC	DE	SADE	xNES	CMAES	GWO	PSO	SGA
Time (s)	1228	576	588	364	**321**	372	653	960
Fitness	4.686	3.865	**0.84**	4.971	5.712	5.166	4.584	3.341
Convergence (iterations)	100	100	**92**	100	100	100	100	100

Table 4. Heuristic Solver Test for a Layout Problem with Physics Reinforcement

	ABC	DE	SADE	xNES	CMAES	GWO	PSO	SGA
Time (s)	612	306	293	235	**233**	331	306	311
Fitness	0.191	0.186	0.180	4.063	5.981	4.027	0.184	**0.154**
Convergence (iterations)	40	52	44	100	100	100	45	**32**

4 Design Application

We used a design scenario to evaluate if the proposed method can produce feasible and realistic layouts, using constraints c_1, c_2, and c_3, and goal g_1 and g_2. We selected an existing site that has recently undergone construction of an apartment complex in Korea and used the same building codes. The area of the site is 161,163 m^2, the maximum floor area ratio is 270%, and the maximum building coverage is 20%. The maximum and minimum building heights are respectively 43 floors and 29 floors, with a floor height of 3 m. Based on this capacity regulation and combinations of four different unit plan types, we generated 22 or 28 buildings including 5,307 units. Then, we defined the minimum distance factor for the main facades (window-facing surfaces) and blind walls respectively as 80% and 20%. The angle preference follows this order: −45° (southern-east), 0° (south), and +45° (southern-west).

Figure 6 shows the design that has been constructed (Actual Design) and illustrates the four layouts produced by our method for this realistic design scenario under different conditions: L-shape buildings without road and facilities (A), L-shape buildings only with a road (B), L-shape buildings with a road and facilities (C), and Y-shape buildings without road and facilities (D).

The optimization for the four cases using the same parameters for SGA: cross over 0.95 and mutation 0.10. The first rigid-body simulation runs over 50 iterations and 0.1 as step size. The second step of simulation has 100 iterations and 0.05 as step size (Table 5).

All four cases converged to a small error within 500 iterations and in less than 20 min. All the four layouts resembled the existing design in that buildings were aligned to the most preferable direction following the road and site boundary. Specifically, the layout with Y-shaped buildings (Fig. 6-D) showed the regular arrangement of buildings near the perimeter of the site boundary like the actual design.

Table 5. Running time and best fitness of four different layouts on an actual site

	Case A	Case B	Case C	Case D
Time (s)	995	982	1147	1054
Fitness	0.03	0.03	0.01	0.01
Convergence (iterations)	513	488	536	514

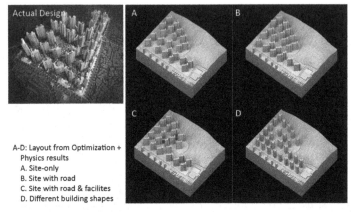

Fig. 6. Four different layouts using an actual site with realistic constraints. The figure labeled 'Actual Design' was provided by DL E&C.

5 Conclusion

In this paper we introduced a computational method to generate realistic high-density apartment complex layouts using site-relevant information, custom building geometries, constraints derived from building code, and design preferences encoded in objective functions. The proposed method searches the optimal positions and angles of buildings using optimization heuristic combined with adaptation mechanisms based on discretization and rigid-body dynamics.

One of the contributions of this paper is the design of a custom constraint solver based on rigid body dynamics. This mechanism encodes the different layout elements into rigid bodies to run a physics simulation. Particularly, it relies on decomposing the representations of buildings and minimum distance geometries into sets of convex shapes with shared rigid-bodies. Then, the simulation uses separation forces to translate illegal layouts produced by the optimization algorithm into configurations with good performance and satisfied constraints. This translation addresses realistic building separation, overlap, and containment conflicts for non-convex geometries, resulting in a more efficient design search.

Using a design scenario based on real site and constraints, we observed that the method generated realistic layouts that satisfy building codes and is informed by user preferences. In addition, the method has a large potential as tool for design exploration for apartment complexes. By simply changing a few parameters, such as the shape of the building or the functions representing design preference, designers can quickly explore many alternatives and evaluate their performance with respect to different demands. We intend to apply our method not only to design scenarios with more buildings, restrictions, and preferences but also to more innovative designs settings with unconventional building types and landscape strategies.

To preserve the efficiency of our approach we will experiment with multi-processing for the optimization algorithm and potentially for the rigid-body dynamics. For the heuristic solver, the multi-processing can compute different layout candidates or populations in different processors, reducing the total computation time. The multi-processing can also be helpful for solving early convergence to local minima in optimization. Despite the use of adaptation mechanisms, the solver can potentially get stuck in a local minimum of the fitness landscape and miss the best solutions. Communicating with large number of candidates during the optimization with multi-processing can reduce this risk of early convergence.

Finally, another alternative for future research is to approximate the constraint solver or even the whole optimization process with deep learning. This approach can benefit from the capacity of the current method to produce large datasets for supervised or even unsupervised learning. On one hand this would require addressing different issues of representation in terms of geometric learning, but on the other hand it would make the whole process more efficient, and amenable to parallelization and real-time design interactions.

References

1. Arvin, S.A., House, D.H.: Making designs come alive: using physically based modeling techniques in space layout planning. In: Computers in Building: Proceedings of the CAADfutures' 99 Conference: Proceedings of the Eighth International Conference on Computer Aided Architectural Design Futures held at Georgia Institute of Technology, Atlanta, Georgia, USA on June 7–8, 1999. p. 245. Kluwer Academic Publishers (1999)
2. Arvin, S.A., House, D.H.: Modeling architectural design objectives in physically based space planning. Autom. Constr. 11, 213–225 (2002). https://doi.org/10.1016/S0926-5805(00)000 99-6
3. Christensen, J.T.: The generation of possible space layouts. In: Thompson, E.M. (ed.) Fusion: Data Integration at its best: Proceedings of the 32nd eCAADe Conference, pp. 239–246. eCAADe, Newcastle upon Tyne (2014)
4. Harada, M., Witkin, A., Baraff, D.: Interactive physically-based manipulation of discrete/continuous models. In: Mair, S.G., Cook, R. (eds.) SIGGRAPH 1995: Proceedings of the 22nd Annual Conference on Computer Graphics and Interactive Techniques, pp. 199–208. Association for Computing Machinery, New York (1995)
5. Hao, H., Ting-Li, J.: Floating bubbles. In: Dave, B., Li, A.I., Gu, N., Park, H.-J. (eds.) New Frontiers: Proceedings of the 15th International Conference on Computer-Aided Architectural Design Researchin in Asia (CAADRIA), pp. 175–183. CAADRIA, Hong Kong (2010)
6. Veloso, P.: Exploring the bubble diagram. In: Amen, F.G. (ed.) Design in Freedom: Proceedings of the 18th SIGraDi Conference, pp. 115–119. Blucher, São Paulo (2014)
7. Bazalo, F., Moleta, T.J.: Responsive Algorithms. In: Ikeda, Y., Herr, C.M., Holzer, D., Kaijima, S., Kim, M.J., Schnabel, M.A. (eds.) Emerging Experience in Past, Present and Future of

Digital Architecture: Proceedings of the 20th International Conference on Computer-Aided Architectural Design Researchin in Asia (CAADRIA), pp. 209–218. CAADRIA, Hong Kong (2015)

8. Nourian, P., Rezvani, S., Sariyildiz, S.: A Syntactic Architectural Design Methodology: integrating real-time Space Syntax analysis in a configurative architectural design process. In: Kim, Y.O., Park, H.T., Seo, K.W. (eds.) Proceedings of the 9th International Space Syntax Symposium. Seoul, Sejong University (2013)

9. Guo, Z., Li, B.: Evolutionary approach for spatial architecture layout design enhanced by an agent-based topology finding system. Front. Architect. Res. **6**, 53–62 (2017)

10. Carta, S., St Loe, S., Turchi, T., Simon, J.: Self-organising floor plans in care homes. Sustainability. **12**, 4393 (2020). https://doi.org/10.3390/su12114393

11. Koenig, R., Standfest, M., Schmitt, G.: Evolutionary multi-criteria optimization for building layout planning: exemplary application based on the PSSA framework. In: Fusion: Proceedings of the 32nd eCAADe Conference, pp. 567–574. Newcastle upon Tyne, England, UK (2014)

12. Nordin, A.: GPU-enabled physics-based floor plan optimization based on work place analytical data. In: Newnes, L., Lattanzio, S., Moser, B.R., Stjepandić, J., Wognum, N. (eds.) Transdisciplinary Engineering for Resilience: Responding to System Disruptions: Proceedings of the 28th ISTE International Conference on Transdisciplinary Engineering, pp. 455–464. IOS PRess BV, Clifton (2021)

13. Ireland, T.: A cell inspired model of configuration. In: Combs, L., Perry, C. (eds.) Computational Ecologies: Design in the Anthropocene: Proceedings of 35th ACADIA Conference, pp. 136–147. ACADIA, New York (2015)

14. Ireland, T.: An artificial life approach to configuring architectural space. In: Martens, B., Wurzer, G., Grasi, T., Lorenz, W.E., Schaffranek, R. (eds.) Real Time: Proceedings of 33rd eCAADe Conference, pp. 581–590. eCAADe, Wien (2015)

15. Biscani, F., Izzo, D.: Pygmo (2019). https://esa.github.io/pygmo2/credits.html

Big Data, Good Data, and Residential Floor Plans

Feature Selection for Maximizing the Information Value and Minimizing Redundancy in Residential Floor Plan Data Sets

Martin Bielik[(✉)], Luyang Zhang, and Sven Schneider

Bauhaus-Universität Weimar, Marienstr. 1a, 99423 Weimar, Germany
martin.bielik@uni-weimar.de

Abstract. Despite the apparent benefits of machine learning (ML), for many practical applications, with architecture being no exception, the current bottleneck limiting its full potential is not the amount but the quality of available data.

We argue that in order to increase the quality of floor plan data sets, we need a priori approach to feature selection. In essence, to optimally allocate limited resources, we need to identify the most valuable features before collecting the data instead of the current practice of discarding irrelevant features after being acquired.

For this purpose, we evaluate 52 features describing a large variety of geometrical and contextual properties of 5558 Swiss residential floor plans and identify a set of five most valuable features. The value of each feature depends on the novel information it provides in combination with other features in the set. Consequently, the selected feature set aims to maximize the joint information value while minimizing redundancies. It is important to note that the proposed method and selected features are generalizable only for use in unsupervised ML applications but might not suit the specific needs of different supervised ML tasks.

We also discuss the possibilities of overcoming the limitation of our findings to the Swiss context. We show that the proposed feature selection method is robust enough to be fitted on small sample sizes ($N = 20$) and thus can be applied in other contexts to determine the most valuable feature before acquiring the complete data set.

Keywords: Floor plan · Feature selection · Machine learning

1 Introduction

In recent years, machine learning (ML) driven applications have become everyday companions aiding human decisions on everything from where to go for dinner up to how to structure global supply chains efficiently. Most recently, ML models able to generate text, sound and images became ubiquitous tools in creative, design oriented fields long reserved to human professionals [1]. When it comes to architecture, ML presents an opportunity to support planners in the increasingly complex challenge of designing our

© The Author(s), under exclusive license to Springer Nature Switzerland AG 2023
M. Turrin et al. (Eds.): CAAD Futures 2023, CCIS 1819, pp. 607–622, 2023.
https://doi.org/10.1007/978-3-031-37189-9_40

built environment. In this context, we observe an ever-growing number of ML applications addressing the spatial layout of residential buildings as the most widespread
building type [2]. These applications utilize floor plans as a common representation of
the spatial organization of buildings to predict certain qualities [3], identify clusters and
outliers [4], or generate new floor plans [5].

The ability of the ML model to learn how to accomplish any of the above-mentioned
tasks depends, in the first place, on the quality of the available data. Suppose the data
does not provide enough information. In that case, regardless of the computational power
or sophistication of the ML model, there is simply no way to learn how to perform the
required task. As pointed out by Saltz [6], the quality of available data is the current
bottleneck for most ML applications, with the design of residential buildings being no
exception to this trend [7].

In this respect, we focus on the quality of data sets representing residential floor plans
and elaborate on what makes high-quality floor plan training data for ML applications
and how to acquire it. In general, the data needs to encompass at the same time enough
observations and enough features to provide the ML model with a) the relevant qualities
and b) their variation.

However, It is important to realize that due to the limited resources one can invest
in data acquisition, a decision has to be made on the best trade-off between the amount
and type of features and observations to include in the data set. In other words, when
assembling the training data, the question is, which features and observations have the
biggest added value in terms of the ML model performance? In particular, do we get a
better model by adding observations or features, and most importantly, which features
are most beneficial?

1.1 Problem

In this paper, we focus on the question of feature selection, as this is the first decision
to be made before collecting any observations. Moreover, most ML algorithms tend to
require highly asymmetrical input data, meaning that the data matrix consists of much
more observations (rows) than features (columns). Consequently, adding new features
requires more effort in terms of additional data fields to be acquired than adding new
observations, and thus, each feature has to be chosen with particular care.

In general, a good data set maximizes the information content while minimizing
the number of features [8]. To achieve this, it shall consist of unique features with high
information value, or in other words, it consists of features that are not only relevant but
also independent.

For this purpose, there are multiple well-established feature selection (e.g., Filter,
Wrapper methods) and feature engineering techniques (e.g., PCA, SVD, Factor analysis)
readily available [9]. The former cut down the complexity of data by selecting the most
relevant features and discarding the redundant ones. The latter reduces the dimensionality
of data by creating a new smaller set of features while maintaining the information value
of the original data set. It is important to note that even though these approaches are
routinely adopted in the ML project pipeline to reduce the bias of the ML model, they
are usually applied after the data has already been collected. Thus these techniques do
not help to inform the fundamental decisions on how to maximize the value of the data

set. They merely remove redundant information to avoid overfitting and speed up the training of the ML model [10].

In essence, the problem is that the a-posteriori feature selection reduces the data's potential value as we invest in acquiring features just to discard them later. Additionally, we argue that this approach discourages the prioritization of the collection of features over new samples, as there is always the threat that these features will be discarded.

1.2 Research Goal and Question

We aim to support the feature selection process prior to collecting data to maximize the information value of the resulting floor plan data sets. The central question is, which floor plan features constitute a feature set with maximum information value and minimum redundancy?

1.3 Limitations

First, we limit the scope of this paper to selecting features exclusively for the purpose of unsupervised ML applications (e.g., to synthesize new or cluster existing floor plans). These ML models require training data that holistically describe the concept of the floor plan as opposed to the supervised ML applications, which are meant to predict specific labels (e.g., rental price) and thus need specific features. As a result, the optimal feature set for the supervised ML might differ from case to case, while we are able to provide a generalizable feature set for a wide range of unsupervised ML models.

Second, the generalization of the results presented in this paper is constrained by the geographic, cultural, and socio-economic scope of the data set used for the feature selection. The open-source data set [11] used in this research was acquired in Switzerland from commercial clients of Archilyse AG, specializing in the digitization and analysis of buildings. Since the information value of the floor plan features depends on their ability to holistically describe a particular set of observations, the conclusions drawn on the basis of our data set might not be generalizable to other contexts.

2 Method

To quantify the information value of different floor plan features, we evaluate a data set containing Swiss residential floor plans described by a comprehensive vector of 350 distinct features. We base our methodology on the established feature selection techniques presented in the following chapter. Since the aim is to select features based on their ability to holistically describe floor plans for the purpose of unsupervised ML, we briefly review the literature on unsupervised feature selection. Then, we continue with an in-depth description of the derived methodology and the above-mentioned Swiss floor plan data set.

2.1 Feature Selection Methods Review

In general, the feature selection methods can be classified as supervised [8], semi-supervised [12], and unsupervised [13], with the main difference being the structure of the underlying data set. The supervised methods require all data to be labeled (i.e., we need to have the dependent variable - what we want to predict), while the semi-supervised methods require some labeled data. The unsupervised methods are applied when no labels are available.

As mentioned by Tang and colleagues [10], most feature selection methods developed over the last decades have been targeting supervised classification tasks. However, due to the vast amount of unlabeled data generated in different applications such as text mining, bioinformatics, image retrieval, and social media, to name a few; the unsupervised feature selection (UFS) methods have gained significant interest in the scientific community [14].

This is mainly because the UFS methods have two advantages over the supervised feature selection. They are unbiased and perform well when prior knowledge is unavailable, and they can reduce the risk of data overfitting and thus are more generalizable to new data [15]. In line with the aim of this research, in the following, we discuss the UFS methods, which can be broadly categorized into three types - the Filter, Wrapper, and Hybrid methods.

Filter feature selection methods "apply a statistical measure to assign a scoring to each feature" [14]. In essence, this means that we evaluate the relevancy of each feature individually and select the best-ranking ones. However, as pointed out by Dy and Brodley [16], in the case of the UFS, this is a tricky problem due to the difficulty of defining feature relevance without having any goal function (i.e., there are no labels we want to predict). This difficulty is addressed by assuming that more relevant features contain more information regardless of their type or purpose. The information content of a given feature is measured via methods built around the Information Theory [17], revolving around the concept of self-information. The self-information is a probabilistic measure expressing the level of surprise of a particular outcome defined as

$$I(x) = log_b\left(\frac{1}{p(x)}\right) \tag{1}$$

where $p(x)$ is the probability of event x.

"The basic intuition behind information theory is that learning that an unlikely event has occurred is more informative than learning that a likely event has occurred" [18]. The mathematician Claude Shannon extended in his 1948 paper the concept of self-information to a measure that captures the information value of entire features [19]. It represents the feature's uncertainty (i.e., randomness) and depicts the number of bits[1] needed to store the information in a variable, as opposed to its raw data.

Thus, Shannon's entropy or Information entropy (IE) captures the "amount of information" in a variable where the higher the entropy, the higher the information value of

---

[1] The base 2 of the logarithm makes it possible to interpret the Shannon entropy as a measure of a lower bound on the number of bits [...] needed on average to encode symbols drawn from a distribution." [18].

a given feature.

$$H(X) = \sum_i^n p(x_i) log_b \left(\frac{1}{p(x_i)} \right) \tag{2}$$

Wrapper feature selection methods consider selecting a set of features as a search problem, where different combinations are prepared, evaluated, and compared to other combinations. The core idea behind the wrapper methods is that the best N features are not the N best features [20]. In other words, the features that perform well on their own (they have high information value) do not necessarily make up the best feature set. This happens if two or more relevant features are correlated - they share information. The problem can be illustrated via a Venn diagram, with each feature representing one circle (see Fig. 1). The diameter of the circle represents the information entropy of each feature (the bigger the circle, the more information it contains), and the shared information between features is represented as their intersection. The information value of the whole feature set is expressed by the joint entropy of all its features defined as:

$$H(X_1, \ldots, X_n) = -\sum_{x_1 \in X_1} \cdots \sum_{x_n \in X_n} P(x_1, \ldots, x_n) log_2 [P(x_1, \ldots, x_n)] \tag{3}$$

The key point is that the most valuable feature set is the one with the biggest area (i.e., carrying the most information) after performing the union of the whole set (i.e., accounting for the shared information). As illustrated in Fig. 2 on an exemplary data set consisting of 5 features, the two most valuable features, when considering their value in isolation, might constitute a relatively weak feature set (Fig. 2b) when compared to the optimal set composed of two features maximizing the overall amount of encoded information (Fig. 2c).

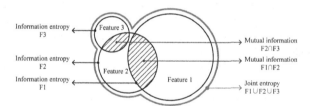

Fig. 1. Ven diagram representing the information entropy of individual variables, their mutual information, and the joint information of the feature set.

The advantage of wrapper methods is that due to their focus on the value of the whole set rather than the individual features, they tend to select better feature sets than the filter methods [14]. It is important to note that picking features based on their cross-correlations (i.e., mutual information) belongs to filter methods as it does not provide information about the value of the entire feature selection but just pairs of features[2]. However, due to

---

[2] For feature selections consisting just of two features ($N = 2$), the cross-correlations as used in filter methods provides enough information to select set maximizing the joint information entropy. For $N > 2$, wrapper search will lead to set with higher joint IE.

the combinatorial nature of the search problem, they are computationally demanding and do not scale well as the number of features grows. The use of heuristics such as stepwise forward selection or backward elimination can speed up the search. Nevertheless, the wrapper methods remain significantly slower than the filter methods.

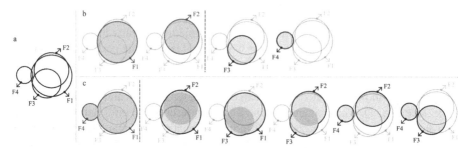

Fig. 2. Given the feature set a), we graphically show the difference between selecting b) N best features and c) best N features for N = 2. The joint IE of the feature set is larger when selecting the best N features as done in by the *Wrapper* methods.

Hybrid feature selection methods take advantage of ranking features by their individual information content as well as the joint information value of the whole feature set. In essence, the hybrid methods aim to combine the search efficiency of Filter methods and the selection quality of Wrapper methods.

Feature Selection Method. In the following, we describe two feature selection methods derived from the literature review and applied to select the most valuable floor plan features. The first method, called *Tabula rasa*, assumes that no features were acquired in advance of the selection procedure. It might be used to define the feature set a priori to the data acquisition. The second method we present is the *Context-aware* selection used to select floor plan features in addition to existing data to enrich its information value.

Both methods utilize the same hybrid feature selection procedure and differ only in the initialization step. The wrapper part of the selection aims to identify the best N features and is based on stepwise forward selection. This means that we start either with a) an empty feature in the case of the *Tabula rasa* method or b) a set of existing features in the *Context-aware* selection, and we add the feature with the highest contribution to the joint IE of the current feature selection in each step. To illustrate this, in the first step, we select the feature with the highest IE (see Fig. 3). In the subsequent steps, we search for the feature that, combined with the already selected features, maximizes the joint entropy of the data set. It is important to realize that we do not simply select the feature with the highest information entropy (represented as the diameter of the circle in Fig. 3) but the one which contributes most to the joint IE of the feature set (represented by the area of the red outline in Figs. 1, 2 and 3).

Consequently, to select the best next feature, we have to calculate the joint entropy of each available feature (i.e., the ones not selected yet) and the current feature selection. We compare how much IE each feature adds to the current selection and choose the highest (see Fig. 3 - step 2). By doing this, we not only maximize the information value of the feature selection but also minimize the redundancy. Features that do not provide new information but only share information already present in the selection will not be chosen. To speed up this computationally expensive search process[3], we pre-rank the potential candidate features based on their individual Shannon's entropy calculated in the first step of the forward selection. We test the candidates in the ranked order (i.e., we start with the feature with the highest IE) and stop if we find a feature with information gain higher than the IE of the next candidate in the ranked order.

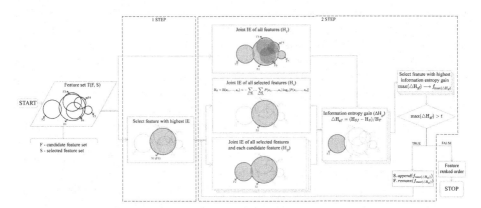

Fig. 3. Flow chart illustrating the proposed feature selection process.

The feature selection procedure results in a ranking representing the information entropy gain (further referred to as information gain) obtained by the consecutive addition of each feature to the data set. For example, if the information gain of the first two features in the ranked set accounts for 40% and 10%, they capture 50% of the joint IE of the entire feature set. By doing this, the ordered feature set can be used as a guideline to a) identify the best set of N features, maximizing the information value and minimizing redundancy, and b) identify how many and which features are necessary to account for a certain critical amount of IE (e.g., if the feature set should cover at least 95% of the IE, we need to select first N features).

Finally, to improve the stability of the selected feature set, we use an ensemble of feature selection models instead of a single model. The ensemble is trained on boot-strapped data (i.e., selecting a subset of data with replacement) which is an established method in ML (e.g., used in Decision Trees) to improve the robustness and accuracy of machine learning algorithms [21].

Data Pre-processing. As Solorio-Fernández and colleagues (2020) pointed out, the robustness toward different data types is a central issue when devising the feature selection method. They argue that mixed data (i.e., numerical and categorical) is very common and appears in many real-world problems [14]. This is also true for floor plan data as some features are numerical continuous (e.g., area), some are numerical discrete (e.g., number of rooms), and others are categorical (e.g., room type). In general, data can be

---

[3] The search requires n(n + 1)2 calculations.

transformed from continuous to discrete and from discrete to categorical feature type but not in the other direction. Thus we consider the categorical feature type as the "smallest common denominator" and turn the continuous and discrete variables into clusters of similar observations, with each cluster representing one category.

In the process of transforming all variable types to categorical, we also reduce the dimensionality of the original feature vector in order to increase the interpretability and speed of the feature selection process[4]. The data set used for the feature selection contains 350 individual features for each room in the floor plan database. As discussed more in detail in the following section, this data represents 52 higher-order features and their distribution throughout each room. This means that features such as the view of mountains with different values for different locations in one room are expanded to capture this variation. As a result, the data set does not contain one value for mountain view for each room but rather the max, min, mean, median 20th percentile, 80th percentile, and standard deviation to capture the variation within the room. Important to realize is that all these expanded features require the same data and share the computational costs (i.e., one cannot calculate the minimal view of mountains in a given room without calculating the maximum view of mountains). This means that in terms of deciding which features are most beneficial to acquire, it is more meaningful to run the feature selection on the higher-order set of 52 features as opposed to the original set of 350 features.

The type transformation and the dimension reduction are achieved via the K-means clustering of the normalized data set[5]. When it comes to the choice of K - the number of clusters for each feature, the traditional approach is to find a trade-off between the model complexity (i.e., number of classes) and the information loss. However, to compare the IE of individual variables, we need to keep the number of clusters constant[6]. In other words, all 52 higher-order features need to have the same number of clusters. We choose the global K-value by analyzing the optimal K for each feature and select the highest overall K. By doing this, we prioritize the information value of our data set over the model complexity.

2.2 Evaluation

As already discussed, a major difficulty with unsupervised ML in general and unsupervised feature selection in specific is the evaluation of the results. In other words, it is unclear how to measure the quality of the proposed feature selection method or the resulting feature set. Since, by definition, the unsupervised feature selection poses only a problem but does not provide the right answer, the common approach to evaluating the

---

[4] Computational complexity follows a quadratic function, which means that even a small increase in the set of feature candidates can dramatically increase the computation time of the feature selection method.

[5] Due to scale sensitivity of K-means all numerical features were standardized to z-score before running the k-means clustering.

[6] The number of clusters represents the number of distinct categories in each higher-order categorical feature. Since the number of categories directly affects the IE and causes bias of the IE towards features with higher number of categories, we keep the number of categories constant throughout the data set.

performance of the selected feature set is to use supervised classifiers such as kNN [22], Naive Bayes [23] or others. In a similar manner, we test the performance of the proposed unsupervised feature selection method and the resulting selection of floor plan features employing supervised learning tasks. We train three supervised classifiers - Naive Bayes, KNN, and Decision Tree to predict the room labels (e.g., kitchen, bathroom) based on the unsupervised feature selection. We compare the performance of the classifiers when trained on a) all features, b) the features sub-set identified by the proposed unsupervised feature selection method, and c) the features sub-set identified by the state-of-the-art supervised feature selection method mRMR.

In addition to the performance of the proposed feature selection method, we test its sensitivity toward the sample size. In other words, we ask how much data does the proposed feature selection method need to identify a reliable set of floor plan features?

3 Case Study

For the purpose of feature selection, we use a large commercially acquired open-source data set on 43 466 unique residential floor plans located in Switzerland [11]. The floor plan data was acquired by the digitalization of existing floor plans into semantically labeled vector data. The unique quality of this floor plan data-set compared to other available alternatives is that it comprises geometrical, visual, configurational, and contextual features for a) each floor plan, b) individual rooms in the floor plan, and c) a fine-grained analysis grid of 25×25 cm[7]. Each of these three spatial representations embody a trade-off between the granularity and complexity with the room level representation being the preferred compromise used in the scope of this research (Fig. 4).

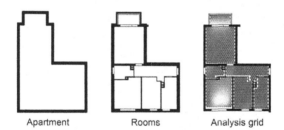

Apartment   Rooms   Analysis grid

Fig. 4. Three levels of spatial aggregation. Source: Archilyse AG

Overall, the floor plan features can be categorized into 52 high-level (see Table 1) and 350 low-level features. The high-level features stand for different geometrical, visual, and configurational properties of each room. In contrast, the low-level features capture the variation of these properties inside the room (e.g., daylight, outside view). In essence,

---

[7] For details on data acquisition and methods applied to derive the floor plan features please see to the online reference by the data-set provider Archilyse AG: https://zenodo.org/record/707 0952#.YymPDtJBy-Z.

the low-level features described the average, minimum, maximum, median, standard deviation, 20th percentile, and 80th percentile of selected high-level features[8].

It is important to note that due to the fully automated process of floor plan acquisition based on highly diverse data, in some cases, the data set might contain incomplete observations. Therefore, we randomly selected and manually checked 7 000 floor plans (containing 53 356 rooms), out of which 1 442 floor plans were discarded due to missing or erroneous data. The feature selection was conducted on a reduced data set of 5 558 residential floor plans (42 498 rooms).

Table 1. Overview of the 52 higher-order features by category.

Layout	Room element	Connectivity	View		Noise	Sun light
shape Area basics / geometry	**layout_connects_to_bathroom** True if the area connects to a bathroom.	**connectivity_balcony** Shortest distance to the balcony. (m)	**view_buildings** Visible buildings. (Steradian)	**view_pedestrians** Visible pedestrians. (Steradian)	**window_noise_traffic** Car traffic noise received on the area's windows from daytime and night-time (dB(A))	**spring_equinox_morning** Solar radiation received in the Spring Equinox morning [06:00-12:00). (klx)
door Area doors	**layout_connects_to_private_outdoor** True if the area connects to an outside area that is private to the apartment.	**connectivity_bathroom** Shortest distance to the bathroom. (m)	**view_greenery** Visible greenery. (Steradian)	**view_primary_streets** Visible primary_streets. (Steradian)	**window_noise_train** Noise received on the area's windows from daytime [06:00-22:00] and night-time [22:00-06:00] train traffic. (dB(A))	**spring_equinox_noon** Solar radiation received in the Spring Equinox noon [12:00-16:00). (klx)
window Area windows.	**layout_has_bathtub** True if the area has a bathtub.	**connectivity_betweenness_centrality** The Betweenness-Centrality value.	**view_ground** Visible ground. (Steradian)	**view_railway_tracks** The amount of visible railway_tracks. (Steradian)		**spring_equinox_evening** Solar radiation received in the Spring Equinox evening [16:00-20:00). (klx)
room The area's share to the apartment's room count.	**layout_has_entrance_door** True if the area is directly leading to an exit of the apartment.	**connectivity_closeness_centrality** The Closeness-Centrality value.	**view_highways** Visible highways. (Steradian)	**view_secondary_streets** Visible secondary_streets. (Steradian)		**summer_solstice_morning** Solar radiation received in the Summer Solstice morning [06:00-12:00). (klx)
	layout_has_shower True if the area has a shower.	**connectivity_eigen** The eigen centrality value.	**view_isovist** Visible isovist. (Steradian)	**view_site** Visible site. (Steradian)		**summer_solstice_noon** Solar radiation received in the Summer Solstice noon [12:00-16:00). (klx)
	layout_has_sink True if the area has a sink.	**connectivity_entrance_door** Shortest distance to the entrance door. (m)	**view_mountains_class_2** Visible mountains of UN mountain class 2. (Steradian)	**view_sky** Visible sky. (Steradian)		**summer_solstice_evening** Solar radiation received in the Summer Solstice evening [16:00-20:00]. (klx)
	layout_has_stairs True if the area has stairs.	**connectivity_kitchen** Shortest distance to the kitchen. (m)	**view_mountains_class_3** Visible mountains of UN mountain class 3. (Steradian)	**view_tertiary_streets** Visible tertiary_streets. (Steradian)		**winter_solstice_morning** Solar radiation received in the Winter Solstice morning [06:00-12:00). (klx)
	layout_has_toilet True if the area has a toilet.	**connectivity_living_dining** Shortest distance to the living-dining room. (m)	**view_mountains_class_4** Visible mountains of UN mountain class 4. (Steradian)	**view_water** Visible water. (Steradian)		**winter_solstice_noon** Solar radiation received in the Winter Solstice noon [12:00-16:00). (klx)
	layout_is_navigable True if the area is navigable by a wheelchair.	**connectivity_loggia** Shortest distance to the loggia. (m)	**view_mountains_class_5** Visible mountains of UN mountain class 5. (Steradian)			**winter_solstice_evening** Solar radiation received in the Winter Solstice evening [16:00-20:00). (klx)
		connectivity_room Shortest distance to the bedroom. (m)	**view_mountains_class_6** Visible mountains of mountain class 6. (Steradian)			

---

[8] Some high-level room features such as area do not vary inside the room and thus are not accompanied by any low-level features.

4   Results

First, we identified the optimal number of clusters (K) for transforming the 350 lower-order multi-type features (e.g., continuous, discrete, categorical) to 52 categorical higher-order features. We first determine the optimal K-value for each higher-order feature (see Fig. 5) via a) the Elbow method [24] and b) the Silhouette method [25] and choose the highest. As a result, we end up with optimal different K for each higher-order feature going from three (room count feature) to nine clusters (connectivity room feature). Finally, we choose the highest optimal K = 9 to transform the 350 lower-order mixed-type features into 52 higher-order categorical features via K-means clustering. The higher-order feature set serve as the basis for the feature selection presented in this section.

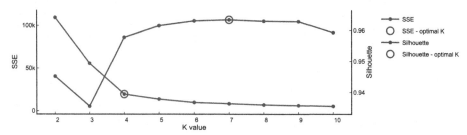

Fig. 5. Determining the optimal K - number of clusters for the floor plan feature *connectivity to logia*. The Blue circle highlights the optimal number of four clusters identified by the Elbow method (via the sum of squared errors - SSE). The red circle shows the optimal number of clusters identified by the Silhouette method. (Color figure online)

4.1   Measuring the Information Value of Individual Features

To quantify the importance of all 52 higher-order features when describing residential floor plans, we calculate each feature's information gain by the *Tabula rasa* feature selection procedure. We start with an empty feature set and stepwise search for the most valuable feature and add it to the selection.

When conducted on the complete bootstrapped[9] data set (containing 5558 apartments and 42498 rooms), we identify a set of five most valuable features describing the spatial relationships, the traffic noise, and the floor plan shape (see Fig. 6). The highest ranked feature with information gain of 24% (i.e., it captures 24% joint IE of the whole data set) was the closeness centrality describing the connectivity between all rooms in each floor plan. The second highest ranked feature was the traffic noise, with an information gain of 20%, followed by shape and connectivity to living/bedroom adding another 17% each. When also considering the next most valuable feature - connectivity to balcony with information gain of 9%, we conclude that these first five most valuable features describe

---

[9] We select with replacement 1024 observations and repeat the process 20 times. These parameters were empirically calibrated to provide best accuracy by keeping the feature selection procedure under 30 min.

87% of the joint IE of the entire data set. All subsequent features provide marginal information gain below 5%[10].

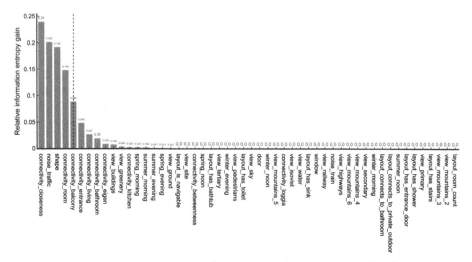

Fig. 6. *Tabula rasa* feature selection on the entire floor plan data set. Features are ranked by their relative information gain.

4.2 Evaluation

To evaluate the performance of the proposed unsupervised feature selection method, we use the selected features for supervised ML tasks and compare the results with a) a complete feature set and b) floor plan features obtained by a state-of-the-art supervised mRMR algorithm [10]. The supervised ML task consists of training three different ML models, a) Gaussian Naive Bias, b) k Nearest Neighbors, and c) Decision Tree to predict the correct type of each room in the data set based on the provided set of features.

The results presented in Table 2 show that, as expected, the full data set of 52 higher-order features provides enough information to correctly classify the room labels in 92% to 99% of cases based on the type of the ML algorithm. We consider this as evidence that the process of data transformation and dimension reduction used to derive the 52 higher-order features is achieving its goals while keeping the information value.

However, when it comes to the performance of the reduced feature set selected by the mRMR method, we found a significant drop in the classification accuracy ranging from 18% to 56% based on the classifier. The mRMR method selected the five most important features capturing 1) connectivity to the balcony, 2) navigability of layout, 3) presence of toilet in a given room, 4) presence of staircase in a given room, and 5) presence of sink in a given room.

---

[10] The threshold 5% is arbitrarily chosen and depends on the purpose of the feature selection. Nevertheless, we use this threshold for comparison purposes through out this paper.

Table 2. The classification accuracy based on the a) full feature set, b) supervised mRMR feature selection, and c) unsupervised Tabula rasa feature selection proposed in this paper.

Feature selection method	Number of higher-order features	Selected higher-order feature	NaiveBias accuracy	kNN (k = 9) accuracy	DecissionTree accuracy
Full data set	52	-	0.918	0.985	0.997
mRMR	5	layout_has_toilet layout_is_navigable layout_has_stairs layout_has_sink layout_has_shower	0.183	0.556	0.556
Tabula rasa	5	connectivity_closeness noise_traffic shape connectivity_room connectivity_balcony	0.606	0.856	0.881

On the contrary, the proposed unsupervised Tabula rasa feature selection method identified five features capturing 1) closeness centrality, 2) traffic noise, 3) shape of individual rooms, 4) connectivity to living/bedroom, and 5) connectivity to a balcony which provided enough information for the classification to achieve an accuracy as high as 88%.

It is somewhat surprising to see the supervised feature selection method outperforming the unsupervised method when evaluated on the supervised ML task for which the former was explicitly trained. As discussed by Brown and colleagues [26], the mRMR algorithm uses a series of approximations which might lead to bias. It was originally developed for to select the most relevant genes out of tens of thousands of options [27], thus the computation speed was of central importance. In contrast, our feature selection algorithm is designed for much smaller feature sets and therefore does not rely on approximations but can search the complete collection of all possible feature combinations[11]. In other words, the single most important goal of our feature selection was maximizing the information value, while in the case of the mRMR more emphasis was put on its computational efficiency.

Overall, these results confirm the ability of the proposed feature selection method to reduce the complexity of the floor plan data set while maintaining its information value. We were able to identify just 10% of the most valuable features providing accuracy comparable to the full data set.

---

[11] For comparison, the *Tabula rasa* search takes 30 min, while the mRMR feature selection takes approximately 3 s on the same data set.

5 Conclusions and Discussion

This paper introduced and tested an unsupervised feature selection method that allows us to rank floor plan features by their information value. Based on the Swiss residential data set, we found that the full feature set ($N = 52$) can be reduced to just the five most relevant features while maintaining 87% of the information value. Among these five features are three features describing the spatial relationships and configuration of the floor plan (closeness centrality, connectivity to living/bedroom, and connectivity to balcony), traffic noise, and layout shape. On the one hand, the high information value of the configurational features reflects the current tendency in floor plan generation with traditional image-based training data being extended by topological graphs [5, 28]. On the other hand, what came as surprising is the high value of the rather unusual traffic noise feature. Since the traffic noise calculation depends on third-party traffic noise data sets and contextual data (e.g., buildings blocking the noise), it is rather demanding to acquire and might not be done without prior evidence about its value.

Another unexpected result of the feature selection was the low information value of the 18 view-related features (e.g., view to mountains, view to buildings) and the nine features capturing the sunlight at different daytimes and seasons. It is important to realize that this does not mean that sunlight and view are not relevant floor plan qualities, but rather that in the context of Swiss residential buildings and other features in our data set, their information value is negligible. This might be a result of the strict building regulations or other factors guaranteeing that all residential apartments are provided with enough sunlight and certain view standards. Nevertheless, since acquiring these features requires additional contextual information and simulation, making it relatively expensive to obtain, it might be advisable to invest in other features or more data samples instead.

We tested the performance of the proposed unsupervised feature selection method and the resulting feature selection by performing a supervised ML task. We found that the five selected features were able to predict room labels 88% correctly. This was an 11% accuracy reduction compared to the full feature set (i.e., 52 features), and most importantly, we outperformed the state-of-the-art supervised feature selection method. Thus, we conclude that the proposed unsupervised feature selection method might also be applied beyond the scope of unsupervised ML application.

Another key finding was the high resilience of the proposed feature selection method regarding small sample sizes. We showed that with as few as 20 data samples, we were able to derive reliable estimates of the floor plan features information value. As a result, one can apply this method in other than the Swiss context and still use it to maximize the information value of the collected feature set. One can simply collect a low number of samples and run the feature selection to inform the decision about which features to collect for the full data set.

We conclude that the presented floor plan feature selection method as well as the ranked feature set provide orientation and guidance when acquiring data for floor plan related unsupervised ML applications. It allows us to differentiate between different floor plan features based on their information value and invest limited resources to prioritize quality before quantity.

Funding. This research was conducted in scope of the Neufert 4.0 project funded by the Federal Ministry for Housing, Urban Development and Building.

References

1. Cetinic, E., She, J.: Understanding and creating art with AI: review and outlook. ACM Trans. Multimedia Comput. Commun. Appl. **18**, 66:1–66:22 (2022)
2. Tostevin, P.: The total value of global real estate (2021)
3. Narahara, T., Yamasaki, T.: Subjective functionality and comfort prediction for apartment floor plans and its application to intuitive searches (2022)
4. Guo, X., Peng, Y.: Floor plan classification based on transfer learning. In: 2018 IEEE 4th International Conference on Computer and Communications (ICCC), pp. 1720–1724 (2018)
5. Nauata, N., Chang, K.-H., Cheng, C.-Y., Mori, G., Furukawa, Y.: House-GAN: relational generative adversarial networks for graph-constrained house layout generation. In: Vedaldi, A., Bischof, H., Brox, T., Frahm, J.-M. (eds.) ECCV 2020. LNCS, vol. 12346, pp. 162–177. Springer, Cham (2020). https://doi.org/10.1007/978-3-030-58452-8_10
6. Saltz, J.S.: CRISP-DM for data science: strengths, weaknesses and potential next steps. In: 2021 IEEE International Conference on Big Data (Big Data), pp. 2337–2344 (2021)
7. Standfest, M.: Reducing bias for evidence-based decision making in design. In: Gengnagel, C., Baverel, O., Betti, G., Popescu, M., Thomsen, M.R., Wurm, J. (eds.) Towards Radical Regeneration, pp. 122–132. Springer, Cham (2022)
8. Tang, J., Alelyani, S., Liu, H.: Feature selection for classification: a review. In: Data Classification. Chapman and Hall/CRC (2014)
9. Guyon, I., Elisseeff, A.: An introduction to variable and feature selection. J. Mach. Learn. Res. **3**, 1157–1182 (2003)
10. Zhao, Z., Anand, R., Wang, M.: Maximum relevance and minimum redundancy feature selection methods for a marketing machine learning platform (2019)
11. Standfest, M., et al.: Swiss dwellings: a large dataset of apartment models including aggregated geolocation-based simulation results covering viewshed, natural light, traffic noise, centrality and geometric analysis (2022). https://zenodo.org/record/7070952
12. Sheikhpour, R., Sarram, M.A., Gharaghani, S., Chahooki, M.A.Z.: A survey on semi-supervised feature selection methods. Pattern Recogn. **64**, 141–158 (2017)
13. Alelyani, S.: On feature selection stability: a data perspective. Arizona State University (2013)
14. Solorio-Fernández, S., Carrasco-Ochoa, J.A., Martínez-Trinidad, J.F.: A review of unsupervised feature selection methods. Artif. Intell. Rev. **53**(2), 907–948 (2019). https://doi.org/10.1007/s10462-019-09682-y
15. Devakumari, D., Thangavel, K.: Unsupervised adaptive floating search feature selection based on contribution entropy. In: 2010 International Conference on Communication and Computational Intelligence (INCOCCI), pp. 623–627 (2010)
16. Dy, J.G., Brodley, C.E.: Feature selection for unsupervised learning. J. Mach. Learn. Res. **5**, 845–889 (2004)
17. Cover, T.M., Thomas, J.A.: Elements of Information Theory. Wiley (2012)
18. Goodfellow, I., Bengio, Y., Courville, A.: Deep Learning. MIT Press (2016)
19. Shannon, C.E.: A mathematical theory of communication. Bell Syst. Tech. J. **27**, 379 (1948)
20. Cover, T.M.: The best two independent measurements are not the two best. IEEE Trans. Syst. Man Cybern. **SMC-4**, 116–117 (1974)
21. Efron, B.: Computers and the theory of statistics: thinking the unthinkable. SIAM Rev. **21**, 460–480 (1979)
22. Fix, E., Hodges, J.L.: Discriminatory analysis. nonparametric discrimination: consistency properties. Int. Stat. Rev./Revue Internationale de Statistique **57**, 238–247 (1989)

23. Maron, M.E.: Automatic indexing: an experimental inquiry. J. ACM **8**, 404–417 (1961)
24. Satopaa, V., Albrecht, J., Irwin, D., Raghavan, B.: Finding a "kneedle" in a haystack: detecting knee points in system behavior. In: 2011 31st International Conference on Distributed Computing Systems Workshops, pp. 166–171 (2011)
25. Rousseeuw, P.J.: Silhouettes: a graphical aid to the interpretation and validation of cluster analysis. J. Comput. Appl. Math. **20**, 53–65 (1987)
26. Brown, G., Pocock, A., Zhao, M.-J., Luján, M.: Conditional likelihood maximisation: a unifying framework for information theoretic feature selection. J. Mach. Learn. Res. **13**, 27–66 (2012)
27. Ding, C., Peng, H.: Minimum redundancy feature selection from microarray gene expression data. J. Bioinform. Comput. Biol. **3**, 185–205 (2005)
28. Hu, R., Huang, Z., Tang, Y., Van Kaick, O., Zhang, H., Huang, H.: Graph2Plan: learning floorplan generation from layout graphs. ACM Trans. Graph. **39** (2020)

An Integrated and Interactive 3D Space Planning Framework Considering Collective Multi-criteria Optimization Based on Multi-agent Constraints Solving

Jiaqi Wang$^{(\boxtimes)}$ (iD) and Wanzhu Jiang$^{(\boxtimes)}$ (iD)

South China University of Technology, Guangzhou, China
{ucbq121,ucbqwj0}@ucl.ac.uk

Abstract. As the core content of architectural design in the early stage, space planning involves the overall arrangement process, which has rarely eliminated the inefficient manual mode with repeated trial and error. The partiality, specificity and lack of user interaction make the current automated space planning methods disjointed from the diverse practical problems and imprison their creativity and possibility. This paper aims to coordinate the relationship between the architectural space planning process and the algorithm model operation procedure, incorporating various constraints and goals to establish an integrated and interactive algorithm framework. The model frames the whole process by four components: Interface, Agent, Solver and Generator based on the multi-agent system, expressing it as the morphogenesis of spatial agents with bottom-up intelligence in a voxelized environment with top-down influence, realizing the rapid generation of suitable planning solutions. Compared with conventional approaches, this framework concentrates on extensibility, evolvability, and partnership with roles like architects, allowing humans to become participators and maintainers of the model. This research produces a Grasshopper plug-in and performs experiments based on an activity center project under real conditions. This intelligent design tool has been verified to orchestrate complex design constraints and effectively integrate them into the planning process, assisting project deliberation and decision-making, and facilitating real-time interaction and detailed development.

Keywords: Space Planning · Integrated Framework · Complex Constraints Solving · Multi-Agent system · Multi-objective Optimization · Generative Design

1 Introduction

Space planning is the core of the early stages in urban and architectural design, organizing space elements in an appropriate way to creatively meet some criteria and/or optimize some objectives [1]. It ranges in scales from cities, blocks, parks, buildings to rooms, and relates to concepts including Spatial Allocation, Layout Generation, etc. As a crucial link in the design process, space planning realizes the leap from the site and user data

© The Author(s), under exclusive license to Springer Nature Switzerland AG 2023
M. Turrin et al. (Eds.): CAAD Futures 2023, CCIS 1819, pp. 623–638, 2023.
https://doi.org/10.1007/978-3-031-37189-9_41

collection to architectural space expression, which often involves multi-dimensional complexities such as multi-element analyzing, multi-constraint solving, multi-objective balancing, and multi-preference selecting. So far, the solution to space planning problems still relies on the inference and experience of architects' tacit knowledge, requiring constant trials and errors and consuming much time and energy. Limited by human computational capability, it can rarely complete the accurate quantitative coordination of all aspects and parameters. When the design conditions increase, decrease or change, this traditional manual method cannot adaptively perform the analogy and deduction of the results.

The automation of space planning has always been an essential issue that computational design attempts to solve. In the nearly 50 years of CAAD research, the exploration of space planning problems can be roughly divided into several directions: Rule-oriented methods based on generative algorithm models (shape grammar, cellular automata), relationship-constrained methods based on spatial configuration analysis, performance-optimized methods based on physical environment simulations, and data-driven methods based on machine learning techniques. It is true that the approaches mentioned above can offer practical, understandable, and roughly reliable solutions according to a specific set of parameters, but the limitations are also quite obvious. They usually deal with space planning problems in a simplified or abstract way, only considering a single generation goal or inference path [2], discarding a lot of effective information and precision, resulting in biased outcomes. Besides, due to the fixed formula of the computational framework, especially generative algorithm models, the input conditions and output results are specified, which makes it challenging to adapt to complex and diverse practical problems. The emphasis on the longitudinal results excludes the role of the user (architect) in the algorithm, such as machine learning, making the model a kind of black box. Under complete mechanical automation, the lack of interactive behavior restricts the creativity of users and imprisons the possibility of solutions. The absence of User-friendly "Integrated" computer program which considers "multi-criteria optimization" that be capable of considering all architectural problems is sensed [3].

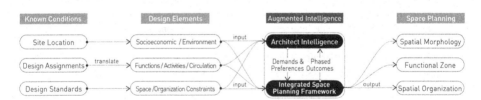

Fig. 1. The logic structure of the integrated and interactive 3D Space Planning Algorithm.

Therefore, it is a significant study to coordinate the corresponding relationship between the architectural space planning process and the algorithm model operation procedure, incorporating site social and environmental conditions, spatial function and form requirements, building codes and personal preferences, to establish an integrated and interactive algorithm framework. Compared with the conventional approach and the automated space planning that entirely relies on algorithmic generation, this framework concentrates on the extensibility, evolvability, and partnership with roles like architects (see Fig. 1) of the computational model, allowing humans to become participators and maintainers of the model. This research produces a Grasshopper plug-in and performs experiments based on an activity center project under real conditions. This intelligent design tool has been verified to orchestrate complex design constraints and effectively integrate them into the planning process, assisting project deliberation and decision-making, and facilitating real-time interaction and detailed development.

2   Literature Review

It has been argued that planning problems are "wicked problems" [4], eschewing formulations, involving contradictions, keeping transformations, and difficult to be accurately identified or rationally defined. Similarly, space planning is the coordination, making, and mediation of space and its property. It focuses on the dynamic and complex spatial system formed by the integrity of elements, constraints, objectives, and preferences, rather than linear logical reasoning, which is the principal contradiction with computational methods.

2.1   Space Planning with the Multi-agent System

Multi-Agent System (MAS) is widely employed in disciplines involving complex phenomena, which conforms to the need to establish indeterministic patterns in planning problems. It organizes multiple agents in an environment for distributed computing, modeling the space planning process as the agents' local behavior to solve the complex problem in a bottom-up manner. Each programmable agent representing a space has personalized attributes and achieves local and global goals through communication, competition, and cooperation. In theory, the attributes of a single agent, the interaction modes between multiple agents, and the general status of the agent group are editable at each level, which allows the system to accommodate all hierarchy and amount of information. Thus, the agent-based system for architectural spaces can be used as the underlying computational framework, adequately combined with other approaches such as spatial configuration analysis, physical environment simulation, multi-objective optimization, and reinforcement learning technologies to melt a method and process with high integrity and good interactivity.

According to the correspondence between agents and spaces, the current multi-agent space planning methods can be roughly divided into three types: 1) agents act as movable space units, 2) agents move to divide space ranges, and 3) agents expand to occupy space ranges [5]. There are differences among the three in terms of space expression and information integration.

The first type sets the agents as predefined patterns (residential buildings in a community [6]) or space units (rooms in a multi-story house [7]) and manipulates the repulsion or attraction between agents in accordance with the organization information to calculate their positions in the planning environment. Since the agents' intrinsic attribute settings precede the generation process and rarely change with the computation, this model has a high degree of controllability and target completion. However, the organization method based on mechanics is oversimplified with weak carrying capacity of information, difficult to describe intricate and diverse constraints.

The second type sets the agents as the anchor points to divide the planning area, and the specific partition logic determines the shape of each space range. Danil Nagy proposes an office plan generation model based on neighborhood seed partition and establishes a complete space planning process combined with multi-objective optimization [8]. A similar method is also used for the hospital layout division [9]. This model has featured a fixed contour of the planning range and clear boundaries of the divided spaces. However, this also leads to the deficient type and the convergent performance of agents. Regarding information capacity, a stable planning range means clear site simulation outcomes, convenient for guiding the agents' behaviors.

While the third type relies on the agent's morphogenic behavior to obtain space territories. This model generally adopts a voxelized environment, updating the state of agents by occupying or releasing nodes, simultaneously articulating their individual attributes and organizational tendency to generate a comprehensive spatial layout. Based on the principle of stigmergy, AnnaLisa Meyboom's study enables the agent to interact with the environment and other agents by emitting color pheromones (R, G, B) to control the agents' positions and domains [10]. Pedro Veloso utilizes multi-agent reinforcement learning to train the occupancy experience of the agents to achieve single-story residential plan optimization [11]. The advantage of this model lies in the plentiful information hierarchies. By encoding pheromone diffusion or occupation behavior, the complex constraints of the individual, clusters and entirety of the agents can be logically defined. Moreover, the voxelized planning environment is also conducive to the storage of context data. Marshall Prado adopts multi-agent morphogenesis to respond to the inclusive results of various physical environment simulations, implementing the evaluation and visualization of 3D site information [12]. The complexity of the space forms control is an important issue that affects the planning results. The outputs shown in the current research are still abstract, which is difficult to have a direct connection with the actual regular architectural spaces.

To sum up, in the three types of models from one to three, the information capacity of the agent and the environment increases sequentially, with a higher abstraction and a lower controllability of space expression.

2.2 Space Planning in an Integrated Framework

According to the approach analysis in Sect. 2.1, in order to ensure the flexibility of the space units and the inclusiveness of complex conditions, this study adopts the third model, applying and expanding the principle of stigmergy to complete the space planning by multi-agent morphogenesis. Compared with the studies that partially focus on spatial relationships [10] or environmental conditions [12], this framework aims to set up an integrated interface, incorporating relevant information in the calculation and expressing it as the agents' behaviors. The controllability and rationality of spaces can be improved by adjustable modules that restrict forms and negotiations. The users' customizability is also given attention with the interactable constraint weight values. Most critically, this research defines a comprehensive agent behavior control strategy with quantitative indicators, realizing multi-objective optimization driven by complex constraints in multi-agent systems.

3 Methodology

3.1 The Operational Process of the Algorithm

By sorting out the space planning process, we summarize it into four steps: identifying design requirements, clarifying spatial and environmental conditions, arranging space units in the environment, and modeling intentional outcomes (see Fig. 2). We apply the multi-agent system as the basis for the algorithmic construction of space planning. A voxelized 3D grid represents unstructured and heterogeneous site conditions to form a top-down planning environment, while a set of programmable agents act as independent spaces or functional units to realize bottom-up space generation. Environmental information and design requirements are collected, translated, and stored as attributes of both types of agents. In the planning process, spatial agents gradually outline their boundaries by continuously occupying grid nodes in the environment. The space generation and organization are embodied as agents' expansion and negotiation behaviors. Then, the agent planning results, after reaching steady states, are extracted to generate vectorized models for detailed design. Therefore, the abstract space planning process is reified as the algorithm flow of "relevant data input, spatial and environmental agent encoding, multi-agent operation, model generation and visualization."

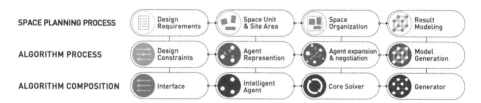

Fig. 2. Correspondence between space planning processes and algorithm operation steps.

3.2 The System Composition of the Algorithm

Corresponding to the four-step flow, this 3D space planning framework is mainly composed of four components: Interface, Agent, Solver, and Generator (see Fig. 2).

Interface. The interface serves as a data importer, gathering pertinent data and converting it into planning elements or constraints. Through the integration, quantification, and organization of various information, it guides the subsequent establishment of planning environments and the operation of spatial agents. There are mainly two types of interfaces: environment-oriented and space-oriented. Each category contains basic information and feature information, which respectively describe necessary design parameters and configurable design requirements.

As for the environmental interface, its basic information includes the scope of space planning, the scale and shape of the planning unit, etc., which frames the underlying voxelized environment, while the feature information accommodates the physical and social conditions of the site, which is divided into masking data and mapping data according to different condition types. Among them, the masking data demarcates the unchangeable range that needs to be avoided in the planning, such as terrains, obstacles, or core tubes, imported as data matrices or mesh volumes. The mapping data is related to the context conditions, such as sunlight, ventilation, and noise in terms of physical performance, and landscape, traffic, and economy with regard to social benefits. These data generally appear in the form of areas with gradients or points with distances, imported in the form of data matrices and superimposed by different channels. The conversion ratio (T_E) of different types of data (E) can be set according to empirical research or local requirements and unified into the demand degree of the same measurement ($E' = E * T_E$). Based on the above classification, it is theoretically possible to cover all quantifiable environmental conditions required for space planning.

The spatial interface corresponds to each spatial agent in a distributed manner. It is oriented towards the refined creative participation of users, encouraging different types of space customization through different parameter settings. The basic information describes the initial location, target pheromone and expected area ratio; the feature information indicates the spatial properties including the target spatial form and adjacency relationship, etc.

Agent. Intelligent agents are the essential components in this framework, which is the digitalization of site and building, including global environmental agents and local spatial agents, whose attribute settings conform to the information in the interface (see Fig. 3).

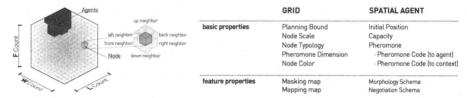

Fig. 3. Environmental agent and spatial agent.

Environmental Agent. Environmental agent builds the space planning context, represented as a 3D grid formed by arrays of discrete nodes. Each node contains basic properties such as position, size, shape, and color. Among them, the color illustrates the nodes' occupation status by the agents with the initial value [0, 0, 0]. In addition, the two types of feature information are collected into each node and converted into different hierarchies of information maps or called pheromones, formatting the feature properties of the environment to affect the spatial agents' behaviors.

Spatial Agents. Spatial agents represent independent spaces or functional areas according to different planning precision, expressed as shaped node clusters in the planning environment. During planning, an agent initialized at a specific location establishes environmental awareness by releasing and reading pheromone codes to adjacent nodes, decides to occupy or release nodes based on the calculation results from the solver, and gradually creates an internal closed geometry (see Fig. 4). The phenomenon of an agent assimilating the node's pheromone when occupying it is called paint. The process of the agent releasing pheromone to the surrounding nodes is called diffusion, and the diffusion rate shrinks as the distance from the agent increases. In each global calculation, every node will be diffused with its neighboring agents to achieve mutual perception and communication.

$$\text{Diffuse}(Pa, Pb, t) = Pa \cdot (1 - t) + Pb \cdot t, 0 \leq t \leq 1$$

(Pa, Pb are two pheromone codes, t is diffusion rate)   (1)

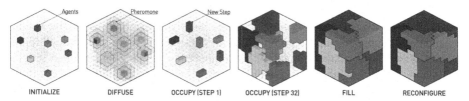

Fig. 4. Multi-agent space planning process.

The basic properties of spatial agents include initial Position, Capacity, and Pheromone. Capacity indicates the upper limit of the node number that the agent can accommodate, consistent with the volume of the space. When the number of nodes occupied by the agent reaches capacity, the agent will keep updating the territory by releasing the inner points and occupying new ones.

Based on the principle of stigmergy, the pheromone is the medium through which the agent interacts with neighbors and the environment, marking its position in the computing system as a distinct label. Specifically, the pheromone is demonstrated by a code sequence, where each position is a numerical value between 0 and 1 (see Fig. 5): The first three bits (adjustable) participate in the interaction with other agents and conduct diffusion, encoding the space functional properties and visualized as the color [R, G, B]. Subsequent digits are responsible for interacting with environmental conditions and do not engage in diffusion. They correspond to the information layer stored in nodes, controlling the agent's reaction to different environmental data. These two types of pheromones shape the main planning behaviors of the agent.

Besides, the feature properties of spatial agents are connected to the feature information in the spatial interface, representing further constraints on specific spaces, which are depicted as the superposition of different schemas. Schemas describe the agents' unique properties, currently including the morphology schema that limits the specific shape (sphere, cuboid, irregular shape, etc.) and the negotiation schema that influences the organizational relationship (close to a particular space, tending to a functional group, etc.). In contrast with the indirect effect of the pheromone on agents' behaviors, the schema strengthens the description and regulation of agents' internal state, solving the

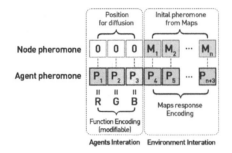

Fig. 5. Agent pheromone and node pheromone.

problem of individual space confusion in the traditional clustering algorithms. The two improve the agents' coding and enrich the generative behaviors.

Solver. As the core of the algorithm operation, the solver performs progressive scoring calculations for all input data, accomplishing the constraint-solving process of all spatial agents. Based on various attributes of environmental and spatial agents and specific calculation rules, the solver will comprehensively evaluate each node adjacent to the currently occupied range and absorb the one closest to the demand in the next step, driving the spatial agent to run inside the environment with discrete steps. Abstract properties are transformed into concrete motions and converted into point cloud models to realize multi-objective space generation and configuration.

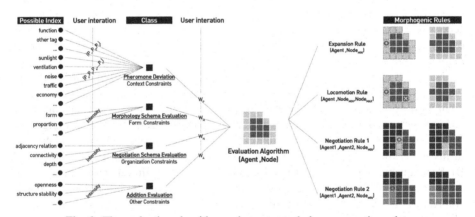

Fig. 6. The evaluation algorithm and agent morphology generation rules.

The solver is centered on an evaluation algorithm for adjacent nodes and is further extended to a set of rules manipulating the agents' morphogenic behaviors, such as expansion, locomotion, and negotiation (see Fig. 6). The evaluation algorithm takes the agent and its real-time adjacent nodes (N_n) as the calculation target and solves the complex constraints of all inputs, including pheromone and schema, to obtain the following step position (N_n^m). Precisely, the standard deviation $\sigma = \mathrm{Dev}(P_A, P_N)$ of the pheromone code (P_A) (or target pheromone) of the spatial agent and the adjacent node pheromone code (P_N) in the 3D von Neumann Neighborhood [13] is calculated sequentially. The node with the smallest value is the closest to the overall demand of the agent. Then, the agent's feature properties are added to the screening of adjacent nodes. For instance, morphological schema (MS_A) evaluates whether the node conforms to the space's geometry; negotiation schema (NS_A) judges whether the node adjoins other attracting agents. In particular, to promote user participation and creativity, each scoring parameter is set with an adjustable weight (pheromone weight and schema weight) to meet the user's need for further modification of the priority level of these constraints.

$$P_i(p_0, p_1, p_2 \ldots p_n) \tag{2}$$

$$Dev(P_N, P_A) = \sqrt[2]{(Pn_0 - Pa_0)^2 * w_0 + (Pn_1 - Pa_1)^2 * w_1 + \cdots + (Pn_n - Pa_n)^2 * w_n} \tag{3}$$

$$N_{choose} = N_n^m \rightarrow \min_{0 \leq m < n} (Dev(P_n^m, P_A) * w_p + Eva(N_n^m, MS_A) * w_m$$
$$+ Eva(N_n^m, NS_A) * w_n + Add(N_n^m) * w_{add}) \tag{4}$$

Based on the above evaluation algorithm, the expansion rule of the agent makes it constantly occupy the node with the smallest value. The locomotion rule acts when the agent reaches the capacity, helping it first release the node with the largest value in the current set before continuing to expand. The negotiation rules are triggered in two situations: 1) When two agents are about to occupy the same node, the solver will compare the deviations on the node of them and the agent with larger value yields. 2) When the adjacent nodes of an agent are all occupied, the solver will calculate the deviations with respect to their owner agents and selects the one with the largest value for the current agent to absorb.

Additionally, the planning settings in the solver manage the collective behaviors of the multi-agents. Among them, the occupancy rate demonstrates the volume proportion that all agents can hold in the environment, which is proportional to the floor area ratio. The diffusion rate and deposit rate represent the contagion ability of the agent pheromone and the rigorous level of the schema constraints, respectively.

$$GFA = OR * N * \frac{NodeScale^3}{FloorHeight}$$

(OR = Occupancy Rate, N = node count of grid, GFA = Gross Floor Area)   (5)

Generator. The generator is responsible for the generation, interpretation and visualization of spatial models, providing a variety of methods for refining point cloud and outputting planning results. The vectorized point cloud model shows great flexibility and efficiency in post-processing. We have now developed several ways of generating voxelized models, 3D blobs, 2D floors, etc. Among them, voxelized models can be replaced with different types of prefabricated modules at a later stage to produce modular assemblies quickly.

4 Activity Center Space Planning Experiment

4.1 Algorithm Model Settings

Based on the above algorithm research, we develop a Grasshopper plug-in called space planner with C# programming language on the Rhino + Grasshopper platform and apply it to an activity center planning experiment. Spatial clusters in this architecture have free-form layouts and rich functional types, conforming to various form requirements and environmental limitations, which is suitable for algorithmic space planning tests. A complete workflow adapted to architecture design will be established to verify the utility of the algorithm.

This experiment is carried out on a university campus in southern China. The site has a public square on the east and an open waterscape on the west. The building line is a trapezoid with an area of 3378 m²; the expected building area is about 5000 m², and the building density is less than 50%. The model configuration is as follows, which serves as the input data for space planning.

Environmental Agent Settings. The node size is set to 2 m in proportion to the planning fineness and storey height. The initial 3D grid (x, y, z = 22, 39, 8) with 6864 nodes is obtained after simply dividing the site range. Regarding the feature information: 1) The masking map concretizes the design specification and form manipulation, including a 2 m setback within the site scope, a specially shaped atrium and a tunnel connecting it and the east square. The points inside the masking maps are removed to form a computable grid with 3778 nodes. 2) Mapping maps set the other four necessary environmental information (sunlight, noise, view, and circulation layer). The data collection methods include model simulation and field investigation. The sunlight data are extracted by Honeybee plug-in [14] and input as data matrices. The main landscape and noise sources are determined by the sampling points in the site survey. These attractors can form fields and exert influence on each node. The entrances are determined according to the circulation relationships, assigning values on the nodes by distance. Finally, all the data can be normalized, mapped and superimposed on the grid to obtain a visual gradient map (see Fig. 7).

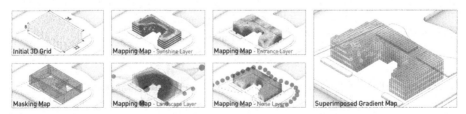

Fig. 7. Computable grid environment and environmental information mapping.

Spatial Agent Settings. According to the functional partition and spatial requirements in the design assignment, this experiment has 14 agents in 7 categories (see Table 1). Among them, 1) Each agent's capacity is the proportion of the gross floor area. 2)

The specific needs of the building define the agents' pheromone codes: P1–P3 concern functional properties, respectively indicating zoning, publicity (whether free access) and activity (whether releasing noise). P4–P7 reflects the matching degrees to various environmental conditions. For example, the learning area has higher requirements for sunlight, showing a larger P4 value; the leisure area places more emphasis on landscape, holding the highest P6 value. 3) Regarding the feature information, six shapes are designed as morphology schemas for different functions (see Fig. 8) with the intensity of 1.0. Since there is no fixed spatial topological relationship, the negotiation schema does not work temporarily, enhancing the flexibility of the planning process. 4) The initial positions of the agents affect the 3D spatial configuration. Before formal planning, the agents are allowed to wander in the grid as a single voxel to search for the optimal position with the most suitable environmental resources. Remarkably, the staircase agents do not participate in subsequent calculations, and their initial positions are fixed by genetic algorithms based on entrance locations and service ranges.

Table 1. Design specification and detail spatial agents settings.

| Design Objective | | | | Spatial Agent Settings | | | | | | | | | | |
Functional Zoning	Room Name	Room Quantity	Area (m²)	Agent (No.)	Agent Quantity	Capacity (%)	Functio n Code (P1)	Publicit y Code (P2)	Activity Code (P3)	Sunlight Code (P4)	Noise Code (P5)	View Code (P6)	Flow Code (P7)	Morpho logy Schema
Common Area (01)	Entrance Hall	1	400	0101	1	0.08	0.5	0.9	0.6	0.5	0.5	0.3	1	7
	Event Hall	1	400	0102	1	0.08	0.5	0.8	0.5	0.5	0.5	0.6	0.5	9
Gathering Area (02)	Lecture Hall	1	400	0201	1	0.08	0.9	0.7	0.1	0.2	0.4	0.1	0.7	1
	Exhibition Hall	3	200	0202	1	0.12	0.9	0.6	0.2	0.2	0.3	0.2	0.5	5
Sports Area (03)	PingPong Room	1	200	0301	1	0.08	0.3	0.5	0.8	0.5	0.4	0.6	0.4	1
	Billiard Room	1	200											
	Gym	1	200	0302	1	0.08	0.3	0.4	0.7	0.5	0.4	0.6	0.4	6
	Dance Room	1	200											
Study Area(04)	Reading Room	5	50	0401	1	0.1	0.7	0.4	0.3	0.8	0.1	0.7	0.3	2
	Study Room	5	50											
	Seminar Room	5	100	0402	1	0.1	0.7	0.3	0.4	0.8	0.2	0.8	0.4	8
Leisure Area (05)	Tea Room	1	100	0501	1	0.02	0.7	0.7	0.3	0.7	0.5	0.9	0.8	3
	Cafe	1	100	0502	1	0.02	0.7	0.7	0.5	0.7	0.5	0.9	0.8	5
	Shop	6	50	0503	1	0.06	0.3	0.7	0.5	0.2	0.5	0.5	0.9	4
Office Area (06)	Office	5	40	0601	1	0.08	0.1	0.1	0.3	0.6	0.2	0.6	0.4	2
	Conference	2	100											
Other(07)	Toilet	/	250	/	/	/	/	/	/	/	/	/	/	/
	Circulation	2	250	/	/	/	/	/	/	/	/	/	/	/
Total			5000		12	1								

Solver Settings. 1) The total occupancy rate of the planning environment can be calculated as 0.66 by formula 5. All pheromone weights remain 1. 2) The global schema is shown in Fig. 9 with the intensity as 0.2, calibrating the collective planning goal of the multi-agents.

Fig. 8. Agents initial position optimization and the morphology schemas.

4.2 Space Planning Control Experiments

Three experiments were conducted to test the model's capability to generate reasonable spatial organizations and to observe the influence of parameter changes on the results, verifying the interactivity and adaptability of the model (see Fig. 9).

In Experiment 1, the entrance hall and exhibition were located near the main entry, connecting all zones. The leisure area and office were on the ground floor. The sports area was piled up in the south wing, responding to the sunlight and view needs; while the study area was gathered in the north wing, occupying the part close to the atrium. The lecture hall was independently distributed in the northeast corner, near the secondary entrance and driveway, forming a courtyard with the study area.

In Experiment 2, the area proportions of the agents were adjusted. The seminar room (0402) was reduced by 300 m²; the exhibition hall (0202) and event hall was increased by 200 m² and 100 m². In this case, the overall spatial relationship basically remained stable, and the added spaces of 0202 overlapped on lecture hall (0201), enhancing the sense of volume of the gathering area.

In Experiment 3, the area ratio was consistent with Experiment 1, and the planning environment was endowed with a global schema as the overall scope. Some adaptive changes occurred: The study and the sports area no longer tended to the boundary but moved inward according to the shape, and the adjacency of zones was tighter. The office was squeezed to the south corner; the shop was pushed to the atrium.

5   Results and Discussions

The experiments were conducted on a laptop with a 3.30 GHz Intel Core i7-11370H processor, 16 GB RAM, and RTX3070 laptop GPU. It averagely took 17.7 s to reach 100 steps. After 295 steps with 167 s, the model reached a relatively balanced state.

The visual assessment of the models and floor plans (see Fig. 10) in experiments shows that the spaces are in reasonable location organizations and adjacency relationships, with feasible forms to meet functional requirements. Although some are still fragmented corners, available floor plans can be produced with simple processing. Some illuminating spatial relationships are generated: to cope with the excessive volume in the north, the algorithm facilitates a courtyard in the first two experiments and creates bottom overhead public spaces in all outcomes; several half-story spaces symbolizing the outer corridors also emerge. Some problems also exist: due to the agents' 3D operation mode, some spaces may cross floors (e.g. 0503_shop) or are too small on a particular floor when divided vertically.

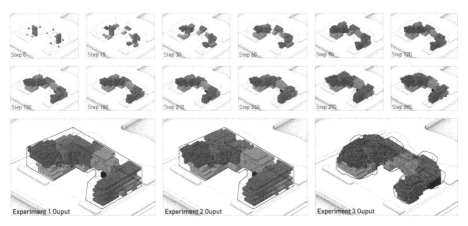

Fig. 9. Space planning process of Experiment 1 and planning results of three experiments.

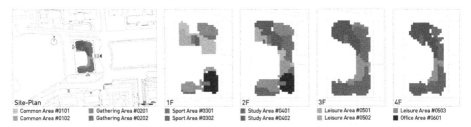

Fig. 10. Schematic site plan and floor plans of the generated result in Experiment 3.

The quantitative analysis of the data after planning demonstrates that the gross floor areas in all experiments are around 5000 m² and all space units meet the area requirements. In the three experiments, the total deviation value (see Fig. 11) of each agent is mainly distributed between 1.2 and 1.6, and their average value, i.e. the deviation for the generated proposals, is around 1.35. Combined with its upper limit of 4.732, the fitness is 71.5%. While for the three constituent parts of total deviation, their values are basically distributed around 0.4. Taking the data of Experiment 1 as an example, the environment-oriented fitness (Base Pheromone) is $(1–0.48/2) = 76\%$, the spatial function-oriented fitness (Pheromone) is $(1–0.36/1.73) = 79.1\%$, and the spatial form-oriented fitness (morphology schema) is $(1–0.5/1) = 50\%$. It means that the agents are more responsive to the first two and less responsive to morphological constraints, since the spatial adjacency may affect the formation of a geometric shape.

$$\text{Fitness} = 1 - \frac{\text{Deviation}}{\text{MaxDeviation}} \qquad (6)$$

The accurate analysis of each agent's deviation corresponding to each input data in the three experiments illustrates Fitness in the following radar chart (see Fig. 12). The satisfaction degrees of the agents for the constraints are relatively good. Among the total 108 parameters of 12 agents, 84.3% can reach 70% fitness or more. The substandard parameters are mainly for morphology schema. Based on the comparison of the three

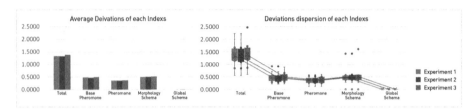

Fig. 11. The chart of the agents' average deviation values and values distribution.

experiments, it can be seen that the parameter adjustment has little effect on each index's deviation, indicating the stability of the model. In addition, the fitness values of the agents in the sports (03) and study (04) areas are the best; while the morphology schema deviation of the shop (0503) is greater than 1, the reason may be the extrusion from the adjacent agents in the negotiation.

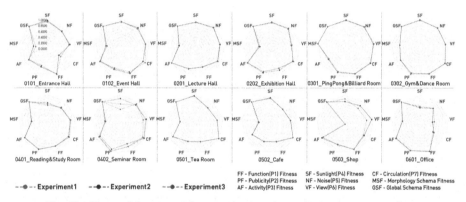

Fig. 12. Charts of the agents' fitness values corresponding to input constraints.

To sum up, this algorithm produces reasonable space planning results, effectively responding to constraints and objectives. Furthermore, after obtaining the planning model, it has the potential to participate in the subsequent design process. By combining the existing content generation algorithm (see Fig. 13), it can generate a complete architectural proposal automatically with a simple components library.

Fig. 13. Space components library and generated proposal.

6 Conclusion

This paper proposes an integrated and interactive 3D space planning algorithm framework based on multi-agent constraint solving to incorporate various conditions. In the experiment, this algorithm model is validated to produce reasonable 3D space layouts and satisfy the real-time interaction needs of architects, exhibiting the capacity to deal with complicated criteria and adapt to changing situations.

The research we present is situated within a larger body of work related to space planning tools and is still in the early stages of establishing the framework. Therefore, how to model the whole space planning processes, gradually form a structured and scalable framework, and systematically sort out and solve various constraints and objectives is the concentration and contribution of this paper. As an intelligent space design tool, this algorithm completes the leap from abstract data to concrete spaces, improves the quality and efficiency of architectural design, and provides convenience for extended development and re-creation. As a promising start, this research also has several challenges: 1) The interface types and compositions must be further developed, improving the design description and constraint coordination to promote efficiency and accessibility. 2) The precise assessment of space configurations needs to be added. 3) Due to the calculation efficiency, the environmental data cannot be updated in real-time, and ANN can be introduced for prediction.

References

1. Liggett, R.S.: Automated facilities layout: past, present and future. Autom. Constr. 9(2), 197–215 (2000)
2. Singh, V., Gu, N.: Towards an integrated generative design framework. Des. Stud. 33(2), 185–207 (2012)
3. Vatandoost, M., Ekhlassi, A., Yazdanfar, S.A.: Computer-aided architectural design: classification and application of optimization algorithms. J. Inf. Comput. Sci. 10, 18–43 (2020)
4. Rittel, H.W., Webber, M.M.: Dilemmas in a general theory of planning. Policy Sci. 4(2), 155–169 (1973)
5. Veloso, P., Rhee, J., Krishnamurti, R.: Multi-agent space planning. In: International Conference on Computer Aided Architectural Design Futures, vol. 18, p. 24 (2008)
6. Li, B., Qian, J.: Exploration of generative method based on MAS for architectural design: high FAR. New Arch. 3, 99–103 (2011)
7. Guo, Z., Li, B.: Evolutionary approach for spatial architecture layout design enhanced by an agent-based topology finding system. Front. Arch. Res. 6(1), 53–62 (2017)
8. Boon, C., Griffin, C.T., Papaefthimiou, N., Ross, J., Storey, K.: Evolutionary parametric analysis for optimizing spatial adjacencies. Arch. Res. 312 (2015)
9. Nagy, D., et al.: Project discover: an application of generative design for architectural space planning. In: Proceedings of the Symposium on Simulation for Architecture and Urban Design (2017)
10. Meyboom, A., Reeves, D.: Stigmergic space. In: Proceedings of the 33rd Annual Conference of the ACADIA, Cambridge, pp. 200–206 (2013)
11. Veloso, P., Krishnamurti, R.: An academy of spatial agents: generating spatial configurations with deep reinforcement learning. eCAADe: Anthropologic - Architecture and Fabrication in the cognitive age (2020)

12. Prado, M.: Morphogenic spatial analysis: a novel approach for visualizing volumetric urban conditions and generating analytical morphology. Tech. Arch. Des. **3**(1), 65–75 (2019)
13. MathWorld. https://mathworld.wolfram.com/vonNeumannNeighborhood. Accessed 11 Mar 2023
14. Food4Rhino. https://www.food4rhino.com/en/app/ladybug-tools. Accessed 11 Mar 2023

Demand vs Design – Comparing Design Proposals to "New Work" – Based Spatial Requirements

Andreas Mack[1,2(✉)] ⓘ, Ekaterina Fuchkina[1] ⓘ, and Sven Schneider[1] ⓘ

[1] Bauhaus-University Weimar, Belvederer Allee 1, 99423 Weimar, Germany
andreasmack@gmx.net, sven.schneider@uni-weimar.de
[2] Nething Generalplaner GmbH, Wegenerstraße 7, 89231 Neu-Ulm, Germany

Abstract. The implementation of multi-space concepts in office buildings is a challenge for architects in early design phases. Activity-based work programmes require a user-specific demand analysis of the company based on employee profiles in order to be able to determine zone relationships, areas and other requirements. The multitude of factors results in complex demand programmes, which in turn, lead to difficulties in translating them into a design.

The main objective of this paper is to reveal these difficulties for an essential parameter for efficient work, which is the interaction between employees within the company. This interaction is highly dependent on the spatial relation between the different zones in the building. A digital, grid-based model is used for quantifying zone relations in office building designs. The required zone connections are translated to the analysis model. For analysing zone connectivity, the parameters path length (metric shortest paths) and path complexity (visually shortest paths) are used.

In this paper, we investigate two design projects for multi-space office buildings with 9300 m² and 3600 m². The results of the floor plan analysis are evaluated and compared to the demand requirements.

Conspicuous deviations of the design from the demand are then discussed through expert interviews with the designing architects. The paper presents new findings on analysing path relationships in terms of the demand programme. It identifies factors that challenge the fulfilment of demanded zone connections, summarises the results and points out improvements to floor plan evaluation as well as optimisation potentials in the demand planning process.

Keywords: Floor plan analysis · Demand analysis · New work · Office building

1 Introduction

The idea of "new work" concepts is currently strongly influencing the planning of office buildings [4]. Multispace offices are the main architectural feature of these office structures. In multispace concepts, different activities are performed as activity-based working in corresponding zones of the office. Studies found that multispace offices can increase well-being, motivation and performance of employees due to their versatile spaces [4].

© The Author(s), under exclusive license to Springer Nature Switzerland AG 2023
M. Turrin et al. (Eds.): CAAD Futures 2023, CCIS 1819, pp. 639–654, 2023.
https://doi.org/10.1007/978-3-031-37189-9_42

The Covid-19 pandemic is also affecting the future of offices. While there are numerous different future scenarios for the share of remote work in different industries, most studies see a hybrid model of remote work and face-to-face interaction as the future model in most industries. Thus, offices will continue to play a major role in the future, with a focus on communication [3, 6]. The need for communication and flexibility supports the trend towards multi-functional spaces.

For architects, planning these multispace concepts in early design phases is challenging. For their planning, a user-specific demand analysis of the company is required. The analysis of the employees is first translated into employee profiles. The combination of these profiles with expert experience results in zone areas, zone connections and other parameters, which are subsequently formulated in a demand programme [15]. The variety of influencing factors relevant for successful office planning results in difficulties when translating these into a design proposal. In this paper, these difficulties will be analysed by considering the parameter of zone connections. As Hillier & Penn and Penn, Desyllas & Vaughan found out, zone connections and resulting path relationships in the office building are an essential factor for social interaction in the building [2, 9]. This communication in turn strengthens the potential for team success [16].

The importance of the connections between zones is determined in functional diagrams which serve as a basis for the architect's design. These connections are usually described as: strong connections, medium connections and weak connections. The stronger a connection is targeted, the shorter is the desired length of this connection. However, no normalized target values exist for these connections. This leads to problems in design evaluation (comparing the demanded performance with the actual performance of the design proposal). Also, there are hardly any scientific approaches for the detailed study of design proposals in terms of fulfilling demanded zone connections.

Space Syntax formulates three parameters for measuring the distance between two points in a building: Topological distance – the minimum number of direction changes on a path between two points, angular distance – the angular change from one space to another and the metric distance – the Euclidean distance in metres from one space to another [20]. In this paper we focus on metric distance (path length) and topological distance (path complexity).

Allen defined the influence of metric distance between workplaces in the Allen Curve and showed that shorter distances between employees and face-to-face meetings have a positive influence on communication [1]. Various studies such as Waber, Magnolfi and Lindsey and McElhaney found that Allen's findings are still relevant in times of digitalization [8, 18]. Other research, such as Sailer & Penn and Sailer & McCulloh, supports this hypothesis by showing increased measured interaction between office workers when workstations are in increased proximity [13, 14]. There is also a lot of research that changes in direction also have a big impact on interaction. Sadalla & Magel showed that the number of direction changes influences the perceived distance [11]. The higher the number of direction changes, the longer we perceive a path to be [12]. Peponis et al. and Wineman et al. found that spatial proximity, measured by directional changes in the building, leads to better collaboration between colleagues in the knowledge sector [10, 19].

In this paper, two case studies of office buildings are used to investigate the quality of fulfilling the demand programme. The zone connections are quantified through automated analyses and then compared to the demand. Then the results are evaluated and compared both statistically and qualitatively. Finally, conclusions are drawn for the conceptualization of new work spatial demand programmes.

2 Aim of the Study

The overarching goal of this paper is to gain new insights for demand planning by investigating how well demands for "new work" office buildings can be fulfilled in a design proposal. This is achieved by analyzing the feasibility of user-specific demand plannings through a target-performance comparison. The focus of this paper lies on the parameter of zone connections in the building, resulting in two sub goals:

1. Development of an analysis model that allows the use of existing graph analysis in terms of the demand programme and evaluating deviations, frequencies and correlations between parameters.
2. To gain new insights into the characteristics of zone connections and the challenges that their fulfillment represents. The following detailed research questions arise:

- Q 2a) Do path length and path complexity differ from each other?
- Q 2b) What are the characteristics of strong, medium and weak connections in spatial configurations?
- Q 2c) Are there deviations between demanded zone connections and the actual connections in the floor plan? And, subsequently, if so,
- Q 2d) which factors lead to these deviations?

3 Methodology

In order to answer the research questions 2a)–d), an analysis model was developed and applied to a case study. In the following both are described.

3.1 Calculating Zone Connections

Figure 1 (left) shows a simple example for a functional diagram. The example includes four zones, which are linked by connections of varying importance. These connections are divided into three types (strong, medium, weak). The more important a connection is, the smaller the targeted distance between these zones should be. In addition, there are zones which have no connection to each other. To enable an automated analysis of the zone connections, the connections shown in the functional diagram are transferred to a connectivity matrix using numerical values for different connection strengths.

Figure 1 (right) shows the four placed zones, their connections, the shortest paths, and the analysis grid used to create a visibility graph. In a visibility graph, all grid points that are intervisible (no walls obstructing the view) are specified as connected [17]. A fast multilevel implementation of the Floyd-Warshall algorithm [7] is used to compute both path length and path complexity values based on obtained adjacencies. Path length is a distance measure for metrically shortest paths (see green lines in Fig. 1b). It is calculated as the sum of the Euclidean distances of the connected lines needed to get from one zone to another. Path complexity is a distance measure for visually shortest paths (see blue lines in Fig. 1 right), indicating the least possible number of directional changes required to get from one zone to another. Therefore, complexity is calculated as the number of lines in a path minus 1. For example, if two points are mutually visible, the complexity is equal to the smallest possible value 0 (1 "connecting line" − 1 = 0 direction changes). Although there are many works that evaluate paths based on metric length and direction changes [5, 17], in this work the complexity value is not calculated on the geometry of the metric shortest path, but a second visual shortest path with the same origin and destination is used.

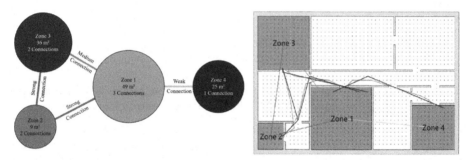

Fig. 1. Scheme of example demand with zones and their connection strengths (left) and its translation into the analysis model with calculated metrically (colored in blue) and visually (colored in green) shortest paths (right). (Color figure online)

3.2 Case Studies

The first case study (CS 1) is the design of a six-story office building with a floor space of 9.300 sqm. The programme comprises 123 zones. The second case study (CS 2) is an office building with 3.600 sqm consisting of 72 zones distributed on four floors. Case study 1 has a conventional development via two cores. An inner courtyard leads to a ring-shaped horizontal development. In this case, sightlines are often interrupted by walls or enclosed zones, and numerous enclosed rooms are found. Case study 2 has a vertical and horizontal development around an inner courtyard. The overall configuration of the floor plan is more open than case study 1, there are few enclosed areas. Due to the intended "new work" concepts, both buildings were planned on the basis of a user-specific demand analysis. The planners used functional diagrams, which graphically represent the relationships between the zones resulting from the user analysis, as a basis for the zone connections. The dataset includes 1.122 (case study 1)/1.182 (case study 2) relevant zone connections. The connections defined in the demand planning are divided into strong (306/272), medium (360/634) and weak (456/276) connections, as described in 3.1. These functional diagrams were translated into a connectivity matrix. The floor plans of the design proposal were redrawn in a simplified format in Rhinoceros 3D. The zones from the programme are located as points with an ID in the floor plan. The origin-destination connections between the zones are then automatically generated based on the connectivity matrix. Figure 2 shows these connections and thus the complexity of the requirements of demand planning.

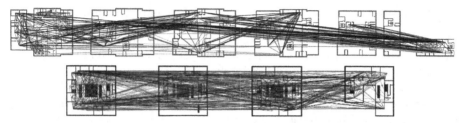

Fig. 2. Origin-destination-pairs of zone connections - CS1: 1122 zone connections (above) and CS 2: 1182 zone connections (bottom) (strong – purple, medium – blue, weak – black). (Color figure online)

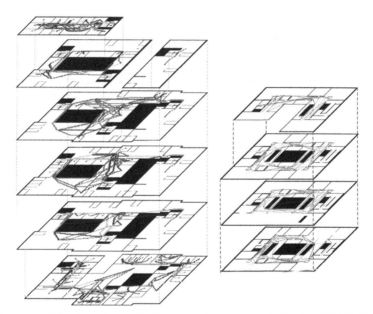

Fig. 3. Isometry of the floor plans with shortest paths (metric and visual) – CS 1 (left) and CS 2 (right).

In the next step, the metrically shortest path and the visually shortest path are calculated for each origin-destination pair. The path length and the number of direction changes are used for the analysis under "4. Results". Figure 3 shows the calculated paths of the two case studies in isometry.

4 Results

In the following, the calculated paths of both case studies are analyzed statistically for answering Q2a) to Q2d) (see Sect. 2).

4.1 Does Path Length and Path Complexity Differ from Each Other?

This question aims at testing whether in the forthcoming investigations, path length and the path complexity need to be considered separately. Therefore, for all 1122 connections of case study 1 (Fig. 4 left) and all 1182 connections of case study 2 (Fig. 4 right) the relationship of the metric distances and direction changes is visualized in a scatter plot. Thus, peculiarities and outliers as well as a general tendency of the relationship between factors can be read. The correlation coefficient is used in the following observations to quantify the direction and strength of the relationship between two factors and to allow correlations to be compared. The results show that the parameters "path length" and "path complexity" are strongly correlated with $r = 0.88$ (CS 1) and $r = 0.97$ (CS 2). Thus, in both case studies, longer paths also have a higher number of directional changes and are related to more complexity. Nevertheless, while in case study 2 there only minor deviations from the trend line, the residuals in some cases of case study 1 are quite large: For example, a connection with a length of 40 m can have both three and 10 changes of direction, respectively a connection with 6 changes of direction has a possible range of 10–60 m. In case study 2, this range is much smaller, with a maximum of 3–6 directional changes at 40 m and a maximum of 30 m variance per direction change. Due to these discrepancies, the following analyses for both parameters are examined separately.

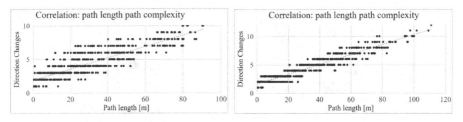

Fig. 4. Scatter Plot of path length and path complexity – CS 1 (left) and CS 2 (right).

4.2 What Are the Characteristics of Strong, Medium and Weak Connections in Spatial Configurations?

To examine the characteristics of strong, medium, and weak connections, two approaches were taken. In the first approach, thresholds were identified by partitioning the data set into three parts according to the frequency of the different connection types. Figure 5 shows the distribution of path lengths (left) and path complexity (right) of the different connection strengths and the two case studies. Two threshold values are indicated in each case: the lowest values up to the number of strong connections (CS 1: 306, CS 2: 272), middle values up to the number of medium connections (CS1: 360, CS 2: 634) and the highest values up to the number of weak connections (CS 1: 456, CS 2: 276). This results in the following thresholds. CS 1: 11 m and 28 m (metric)/3 and 5 (direction changes). CS 2: 11 m and 53 m (metric)/1 and 6 (direction changes). These mark three target ranges for strong (left), medium (middle) and weak connections (right). The higher threshold between medium and weak connections in case study 2 compared to case study 1 is notable. This results from the higher number of connections of medium strength.

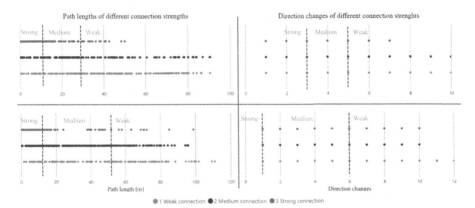

Fig. 5. Metric distance (left) and direction changes (right) of different connection strengths – Case study 1 (above) and case study 2 (bottom).

Table 1 shows, that all connection types have a large number of path lengths and direction changes outside their respective range in both case studies. Path lengths are all at least 30% out of their respective ranges. The weak connections of case study 2 have the largest deviation with 57%. The path complexity generally shows smaller deviations. Here, most of the deviations are around 20%. The medium connections in case study 1 and the weak connections in case study 2 with 59% each represent large deviations.

Table 1. Percentage of connections outside the respective range

	Connection type	Total number of connections	Percentage of connection outside the respective range	
			Metric	Direction changes
Case study 1	strong	306	43%	20%
	medium	360	49%	59%
	weak	456	36%	25%
Case study 2	strong	272	37%	16%
	medium	634	33%	23%
	weak	276	57%	59%

It is noticeable, that the strong and medium connections in case study 2 are rather in their respective range than in case study 1. Nevertheless, the mean deviation of the metric distance (43%/42%) and the direction changes (35%/33%) of both case studies are very similar. This means there is a wide spread of values and a significant overlap of the different connection strengths.

In the second approach, a more detailed view on the spectrum and the overlaps of path lengths and path complexity is taken. For this purpose, a box-plot diagram is used to evaluate in which range the data is located. Quartiles and outliers are used to check the tendency of compliance with the defined demand and to detect deviations. Thereby for each connection type, the median and the thresholds for the four quartiles are visualized (Fig. 6).

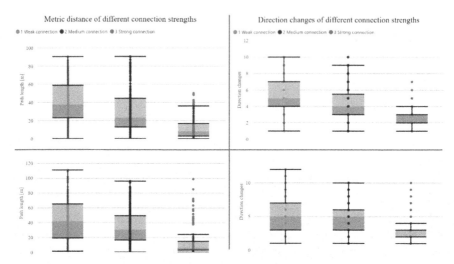

Fig. 6. Relation of different connection strengths – metric distance (left) and direction changes (right) for case study 1 (above) and case study 2 (bottom).

When looking at the median and the central quartiles (2nd and 3rd), it can be seen that both path length and complexity increases the weaker the demanded connection is. The more important and thus stronger the connection specified in the demand, the shorter the measured metric distance and the fewer directional changes are in both case studies. The 2nd and 3rd quartiles are used to describe the major ranges for the respective connection type. (Table 2).

Table 2. Characteristics of the connections – 2nd and 3rd quartile thresholds.

		Path length [m]	Path complexity [direction changes]
Case study 1	Strong connection	3,2–17	2–3
	Medium connection	13–44,1	2–5
	Weak connection	25,6–60,6	3–8
Case study 2	Strong connection	3,3–15,0	0–1
	Medium connection	16,7–49,6	1–6
	Weak connection	19,4–65,4	1–7

Both case studies show similar patterns. Also, it is remarkable that although case study 1 has 2.6 times the floor area, it shows the same patterns in metric distances. This shows the more open floor plan with fewer obstructions and larger sightlines/straight paths as described in Sect. 3.2. Although with increasing weakness of the connection, a general trend of increasing path length and complexity can be observed, there is no hard distinction between the three connection types. All connection types show path lengths and path complexities that are in the range of the other two connection types. No significant overlaps are to be found between the 2nd and 3rd quartiles of strong and weak connections. When looking at the spread of path length and complexity, strong connections show a much smaller range than the medium and weak connections. Comparing the two case studies, it is noticeable that despite the similarity of the metric connections, there is a clear difference in the changes of direction. These have a wider range in the second case study for the weak and medium connections and start in a lower range. This means that despite the metric distance remaining the same, fewer changes of direction are required here to cover the distance. This is also due to the more open configuration of the floor plan, the clear structure (resulting in straight lines of sight) and less enclosed spaces. Also, the central development without cores located in the peripheral areas provides fewer changes of direction.

4.3 Comparison of Demand and Design: Are There Deviations in Zone Connections

Figure 5 showed a large spread of path lengths and direction changes across all connection strengths in both case studies. The box-plots in Fig. 6 showed in which range the path lengths and direction changes are located. The weaker a connection strength gets an increase of the medians and 2nd and 3rd quartile could be observed. This trend reflects the demand. Allen defined a curve in 1977 showing how communication frequency decreases with increasing metric distance [1]. This curve can be divided into two significant ranges: <16 m (strong curve slope, communication frequency >5%), >16 m (low curve slope, communication frequency <5%). Transferring this curve to the results in Table 2, it can be seen that the strong connections in both case studies are in the strong curve slope region and provide high communication. No thresholds are available for comparison for the medium and weak connections, or for the measurement of connection length via path complexity. However, Table 1 showed that many connections are not in their respective ranges. In order to measure these deviations from the demand, we first need to define, what a deviation means: The longest paths and the paths with the most changes in direction, are the connections with the greatest deviations from demand. The shortest paths and paths with the least complexity are not considered deviations because short and less complex connections have no negative effect on the communication frequency. The top quartiles in the box-plots describe the 25% of the longest or most complex paths. These top quartiles show a clear overlap with the other connection strengths and thus deviate strongly from demand. In the following, the connections in these top quartiles are described as "deviations".

4.4   Which Factors Lead to Deviations?

In the following, three parameters are examined for their influence on the deviations from the demand to better understand how other design parameters might influence the fulfilment of required zone connections.

The Effect of the Number of Demanded Zone Connections

Figure 7 shows scatterplots depicting the relationship between the number of demanded connections of a zone and the number of its deviations. It is remarkable that in the second case study, despite the smaller area and smaller number of zones, a similar number of relevant connections were defined. For both, the metric distance and the directional changes a strong positive correlation (CS1: $r = 0.79$ metric, $r = 0.75$ direction changes/CS2: $r = 0.84$ metric, $r = 0.73$ direction changes) is found. Thus, the more a zone should be connected to other zones, the more difficult it is to ensure these connections. For case study 1, both scatterplots show a similar pattern in this regard. Particularly large deviations occur in the range between 20 and 30 total connections per zone. No zone with more than 20 connections has a number of deviations smaller than 6. Case study 2 shows a similar result, the largest deviations are in the range between 25 and 35

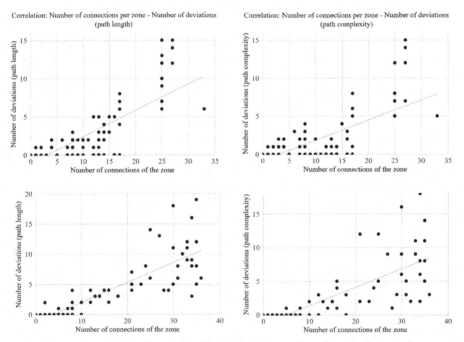

Fig. 7. Scatterplots showing the relationship between the number of required connections – and the number of deviations in path length (left)/path complexity (right) of case study 1 (above) and case study 2 (bottom).

required connections. No zone with more than 20 connections has fewer deviations than 4. Here, despite a larger number of connections, fewer deviations could be achieved in some cases. It is noticeable that the number of deviations in path complexity is low in case study 2 despite the high number of connections, which can be attributed to the open plan configuration of the floorplan.

The Effect of the Demanded Zone Area

In Fig. 8 the relationship between the zone area and the number of deviating connections (length and complexity) is depicted. The results of the two case studies differ strongly. While case study 1 shows no significant correlation for both metric ($r = 0.04$) and direction change deviations ($r = 0.04$), case study 2 shows a medium correlation for metric ($r = 0.46$) and direction change deviations ($r = 0.38$).

For CS 1, most of the deviations occur at small zones (5–36 sqm). Only two zones with an area larger than 80 sqm have a high number of deviations. These are two working zones (200 sqm and 219 sqm) with 13 and 6 (metric)/12 (direction changes) deviations each.

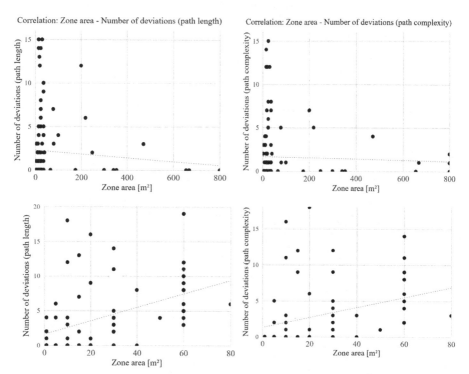

Fig. 8. Correlation of the zone area and the number of deviations metric (left)/the number of direction changes (right) to the zone area – Case study 1 (above) and case study 2 (bottom).

CS 2 shows a correlation between zone size and their number of deviations. Thus, the larger a zone is, the higher the number of its deviations tends to be. Metric and directional deviations show a similar picture. Despite the positive correlation, even small zones between 10 and 30 sqm have a large number of deviations as in CS 1.

Zone Function

Lastly, we checked, if the type of use of a zone does affect the deviations. Therefore, we clustered the 123 (CS1) and 72 (CS2) zones into 19 uses. Figure 9 shows the average number of deviations for each function (both path length and path complexity).

The following functions have a number of deviations, which is strongly above the average number of deviations in both case studies: management office, individual office, working zones. Particularly noticeable is the "management office" zone, which has the

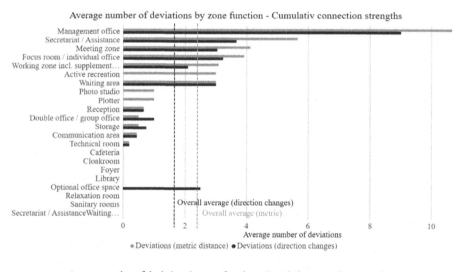

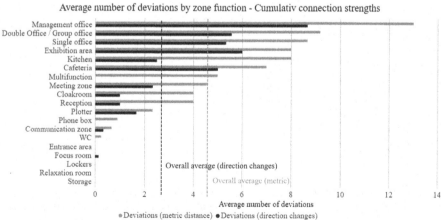

Fig. 9. Average number of deviations by zone function (case study 1 – above, case study 2 – bottom)

largest number of average deviations for both path characteristics and both case studies. One reason for this is that management zones, according to the functional diagram, have connections to almost all other departments. It can also be seen that all work zones are above the average deviation number, except for the Double office/group office zone in Case Study 1 while these show high deviations in case study 2. The two case studies also differ in other functions: Secretariat/Assistance differs greatly in case study 1, this function does not exist in case study 2. Meeting zones are ranked worse in case study 1, in case study 2 they are average. It is noticeable that in case study 2, the average metric deviations are significantly higher than those of the directional changes, while in case study 1, the directional changes have a larger average deviation.

5 Discussion of the Results with the Architect

Finally, the statistical results were presented and discussed in expert interviews. For this purpose, deviations were evaluated with two project-managing architects per case study. All architects interviewed have many years of professional experience in leadership positions in project management. Qualitative assessment identifies difficulties in the planning process and discusses the translation of demand planning. The following is the qualitative assessment of the results.

5.1 Number of Zone Connections

The interview responses support the results of the analysis that too many targeted connections cannot be met in demand. The architects stated that there are too many connections overall in the functional diagram, but especially too many medium and weak connections, so that the focus in the design process was on the strong connections. This can be also seen in the results, as strong connections have a lower spread and a lower overlap with other connection strengths. The architects describe that they interpreted the connection strengths as metric distance, suggesting that directional changes had no direct planning influence. This interpretation relates to both case studies.

5.2 Zone Area

In the analysis of the first case study, no correlation between the parameters "zone area" and the number of deviations of a zone could be found. Small zones (<36 sqm) have a larger number of deviations overall. However, there were two large zones that had high deviations. This result is supported by the architect's statement that large zones are placed first due to the limited area per floor. Nevertheless, the large zones can be located less flexibly due to their area. The two strongly deviating large zones are related working areas, which had to be distributed over two floors due to their large area and the limited floor space. Smaller zones are placed in the next step and are flexible due to their area, but they have to be adapted to the large zones, which makes their localization less flexible. The correlation in case study 2 can be justified by the fact that the large zones with the most deviations are working zones. These have a large number of connections. Thus, the correlation is not due to area, but due to use, as described in Sect. 5.3.

5.3 Zone Function

From the interviews, the following findings emerge for the different zone functions:

1. The deviations of the management offices are mainly due to the numerous connections of the management to other departments. Here, too many connections were specified in the demand, which could not be met in the design.
2. In both case studies, the offices and workspaces have a large number of variations. These zones effect a large number of employees and the majority of paths have a work zone as a start or finish point. For the architect, deviations in these zones are more critical than in small zones where only a few employees are affected. For CS 1, a division into smaller zone areas with fewer connections would make sense in order to better arrange the zones internally as well. Logically, zones would be prioritized by the number of employees in each zone. The connections concern human interaction, so the effect of too long connections is more relevant for frequently used zones.
3. There are different sizes of meeting rooms. In the design of CS 1, smaller meeting rooms were placed close to their departments. Large meeting rooms deviate more often because they were placed centrally in the public area of the building. Large meeting rooms are used with external partners, necessitating a location in the public realm. Smaller meeting rooms are also used for short informal communication meetings, so their near-department location is more important. In CS 2, the meeting zones were placed decentrally. These are used as meeting and project rooms. Due to the decentralized location and without the requirement to place them in the public area of the building, more connections could be maintained in CS 2.
4. One department of CS 1 (including the two deviating work zones in Sect. 4.4) shows many and partly large deviations in many associated zones. The reason for this is that the department had to be split into two floors due to its large area (541 sqm) and the constraints in building geometry. This meant that many important relationships could not be maintained.

6 Conclusion and Outlook

In this paper we analysed the reasons for difficulties in translating different types of zone connections from the demand programme in two "new work" office building designs. We used a grid-based analysis model and evaluated the results both statistically and qualitatively. The analysis model provides a data set from which insights can be gained about the translation of a demand into a design, correlations between different requirements and connection characteristics. The analysis of this data set allows to get project-specific insights into the prioritization of parameters, functions as well as the emergence of deviations. The qualitative evaluation by the architect provided insights into the planning process and the difficulties in implementing the formulated demand.

The main findings of the paper relate to the relation of the number of connections and the number of deviations. Both the metric distance and the number of directional changes show that the greater the number of connections of the given zone, the higher the number of deviations. As an inference to demand planning, this means that fewer connections should be specified in order to be able to meet them. The interviews also revealed two main reasons for the difficulties: large zone areas and too many demanded connections. For optimising demand planning a more fine-grained division of the zones with more detailed, but fewer connections could be beneficial. Additionally, demand planning would require consideration of whether different meeting room sizes should have different levels of connectivity to their respective departments and whether strong connectivity is necessary for large meeting rooms. In the future, we will conduct further case studies to determine whether they share similar characteristics and correlations, and analyse, if a zone prioritization in the demand could solve the problem of too many required connections. Further, the analysis model should be further developed for other demand parameters, such as the frequentation of paths.

References

1. Allen, T.J.: Managing the Flow of Technology: Technology Transfer and the Dissemination of Technological Information Within the R&D Organization. MIT Press, Cambridge (1977)
2. Hillier, B., Penn, A.: Visible colleges: structure and randomness in the place of discovery. Sci. Context **4**(1), 23–50 (1991)
3. Hodgson, C., et al.: COVID-19 – Global Real Estate Implications Paper II, JLL Research (2020)
4. Jurecic, M., Rief, S., Stolze, D., Bauer, W. (Hg.): Office analytics - success factors for designing a work type-based working environment. Fraunhofer IRB-Verlag (2018)
5. Lee, J.K., Eastman, C.M., Lee, J., Kannala, M., Jeong, Y.S.: Computing walking distances within buildings using the universal circulation network. Environ. Plann. B. Plann. Des. **37**(4), 628–645 (2010)
6. Lund, S., et al.: The future of work after COVID-19. McKinsey Global Institute (2021)
7. Lund, B., Smith, J.W.: A Multi-Stage CUDA Kernel for Floyd-Warshall (2010). CoRR, abs/1001.4108
8. McElhaney, B.: Advancing Collaboration Between Joint Terrorism Task Forces and Fusion Centers. Doctoral dissertation, University of Southern California. ProQuest LLC (2019)
9. Penn, A., Desyllas, J., Vaughan, L.: The space of innovation: interaction and communication in the work environment. Environ. Plan. **26**(2), 193–218 (1999)
10. Peponis, J., et al.: Designing space to support knowledge work. Environ. Behav. **39**(6), 815–840 (2007)
11. Sadalla, E.K., Magel, S.G.: The perception of traversed distance. Environ. Behav. **12**(1), 65–79 (1980)
12. Sadalla, E.K., Burroughs, W.J., Staplin, L.J.: Reference points in spatial cognition. J. Exp. Psychol. Hum. Learn. Memory **6**(5), 516–528 (1980)
13. Sailer, K., McCulloh, I.: Social networks and spatial configuration—how office layouts drive social interaction. Soc. Netw. **34**(1), 47–58 (2012)
14. Sailer, K., Penn, A.: Spatiality and transpatiality in workplace environments. In: 7th International Space Syntax Symposium, Stockholm, Sweden (2009)
15. Spath, D., Kern, P., Bauer, W. (Hg.): Office 21 - push for the future - better performance in innovative working environments. Fraunhofer-IAO (2003)
16. Stryker, J., Santoro, M.: Facilitating face-to-face communication in high-tech teams. Res. Technol. Manag. **55**(1), 51–56 (2012)
17. Turner, A., Doxa, M., O'Sullivan, D., Penn, A.: From isovists to visibility graphs. a methodology for the analysis of architectural space. Environ. Plan. **28**(1), 103–121 (2001)
18. Waber, B., Magnolfi, J., Lindsay, G.: Workspaces that move people. Harv. Bus. Rev. **92**(10), 68–77 (2014)

19. Wineman, J., Hwang, Y., Kabo, F., Owen-Smith, J., Davis, G.F.: Spatial layout, social structure, and innovation in organizations. Environ. Plan. B Plan. Des. **41**(6), 1100–1112 (2014)
20. Analysis of spatial relations Space Syntax (2023). https://www.spacesyntax.online/overview-2/analysis-of-spatial-relations/. Accessed 01 Mar 2023

The Impact of Spatial Layout on Orientation and Wayfinding in Public Housing Estates Using Isovist Polygons and Shape Matching Algorithms

Jiahua Dong$^{(\boxtimes)}$ ⓘ and Jeroen van Ameijde ⓘ

Chinese University of Hong Kong, Shatin, N.T., Hong Kong
{Jiahuadong,Jeroen.vanameijde}@cuhk.edu.hk

Abstract. Hong Kong progressively constructs high-density public housing estates to accommodate the increasing population. Standardised apartment construction systems often result in homogeneous environments, so increased efforts have been made to diversify housing estate spaces through varied tower arrangements and public space layouts. There is a lack of research, however, on the impact of these variations on the visual perception of the in-between spaces, and on the potential for orientation and wayfinding. Computational analysis of visibility characteristics along pedestrian circulation pathways can be conducted, to better understand the spatial and configurational properties of public housing estates in relation to people's visual orientation and sense of place. In this paper, we explore a new method for quantifying the uniqueness of spaces within high-density urban environments using visibility analysis. Our approach adopts recently emerged computer vision techniques and shape matching algorithms to conduct large scale and fine-grained spatial analysis of isovist polygon outlines, as isovists' geometric properties have been linked to social behaviour in urban environments in neurophysiological research and behavioural studies. Using two case study estates in Hong Kong, we demonstrate how our methodology can quantify aspects of spatial perception of public open spaces in high-density urban environments, giving detailed insights into the estates' repetitive or unique spatial characteristics and their potential impact on residents' sense of place and community identity.

Keywords: Orientation and Wayfinding · Urban Morphology Analysis · Isovist Analysis · Shape Matching Algorithm · Public Housing · Hong Kong

1 Introduction

Walkability in urban neighbourhoods has grown in importance as a measure to evaluate the quality of urban life and build towards more liveable and healthy cities [6]. Since Jane Jacobs introduced walkability as a concept in urban studies, it has been widely used to evaluate the qualities of the built environment, considering network distributions, visual qualities along roads and other indicators [19]. In the study of walkability, indexes

M. Turrin et al. (Eds.): CAAD Futures 2023, CCIS 1819, pp. 655–669, 2023.
https://doi.org/10.1007/978-3-031-37189-9_43

focusing on functionality, efficiency and enjoyment are fundamental [15]. The perception of walkability is not only dependent on the characteristics of the pathways, but on the spatial qualities of the built environment in general.

Hong Kong has constructed large scale public housing estates since the late 1950s, resulting in a programme that houses around 3.2 million people, accounting for 45% of the population [14]. As the land available for housing development in Hong Kong is limited, public housing estates are planned as high-density, high-rise estates contained multiple tower blocks in compact arrangements. Open views in their public spaces are often limited, recreation areas are compressed, and pedestrian networks can be constrained and crowded. The visual qualities of public housing estate spaces are impacted by the repetitive surroundings as a result of low-cost standardised construction. As Hong Kong continues to plan large amounts of new public housing in several strategic New Development Areas, there is an important opportunity to not only address housing shortages, but also build towards more inclusive and vibrant communities by expanding on the premise of non-standard construction. The complexity and diversity of pedestrian networks can play a significant role in shaping vibrant urban spaces, which are important stimulants for casual neighbouring, socialising, and community engagement [19]. In order to improve walkability in public housing estates, the Hong Kong Government's Transport Department has suggested that the study of orientation and wayfinding efficiency in pedestrian networks is essential [15]. This also connects to the wider aspiration of making Hong Kong a walkable city.

Wayfinding is an important visuospatial behaviour, connected to walking as part of daily activities. Introduced by Lynch [23], the term 'wayfinding' describes the cognitive response of locating oneself through environmental information in direction-driven movements through walking. Orientation precedes wayfinding and refers to the cognitive transformation of visual information into abstractions, to navigate complex environments [22]. Lynch also argued that the continuity of certain spatial properties in the built environment can enhance human perception of complex urban areas [23]. As a spatial understanding of environments is built up through visual information experienced through movement, orientation and wayfinding are important mechanisms to correlate behavioural patterns with the morphological aspect of urban space [5, 21].

Researchers have demonstrated that visibility analysis can serve as a reliable method to predict human behaviour in open spaces, as orientation, wayfinding, and human perception are influenced by the geometry and intervisibility of the visual field [5, 10, 12, 30, 34]. Isovist analysis has been widely used for visibility studies, defining an isovist as "the set of all points visible from a single vantage point in space in relation to an environment" [1]. Isovist polygons can effectively reflect the spatial layout of environments and can be used to describe various geometric properties of a viewing area, including its size, perimeter length, and occlusivity. The geometric properties have been associated with findings in neurophysiological research and behavioural studies, linking boundary conditions to social behaviour in urban environments [27, 31]. Empirical studies have found that wayfinding ability is constrained when people are exposed to a homogeneous built environment [8, 25]. To better understand how homogeneity is created and experienced in high-density public housing estates, we can conduct visibility analysis to quantify and visualise isovist polygon shapes across various public open spaces. This knowledge can inform suggestions to improve wayfinding ability in future estate planning projects, which can improve walkability, socialising and community forming in Hong Kong's future public housing environments.

2 Related Works

Wayfinding is widely discussed in the field of environmental psychology, as part of spatial cognition. Hegarty et al. found that spatial layout is the foundation of spatial understanding [12], when assessing wayfinding experiments at different scales. Van Nes and Yamu used visual graph analysis to represent intervisibility connections in modernist neighbourhoods [34], to illustrate the relationship between wayfinding and conditions of vision boundaries. As behavioural analysis of wayfinding processes is constrained in real world practice, an increasing range of studies is employing experiments in virtual environments. Researchers have shown these innovative methods to be efficient in studying wayfinding behaviour and pedestrian movements, and yield results that closely match real-world testing outcomes [9, 24]. Virtual Reality (VR) has emerged progressively since 2007 as a keyword related to wayfinding experiments [7]. In particular, Meneghetti et al. used VR to test visuospatial thinking, through recollection tasks experiment [24]. Virtual environments have been used to test healthcare facilities and improve wayfinding abilities in homogeneous indoor environments [20].

The similarity of features has been explored in psychology research since the contrast or similarity between spatial properties are analysed through a 'feature-matching process' as part of human spatial perception [33]. Detecting the similarity of shapes is an important part of the cognitive processes of shape retrieval, recognition and classification, alignment and registration, approximation, and simplification in computer graphics [35]. The most versatile shape matching algorithms are based on Hu Moments, a two-dimensional moment invariant for geometric patterns [16]. The Hu Moments are utilised to characterise an object's shape, size, and orientation in an image. The computation of Hu Moments involves the derivation of normalized central moments from an object's image. These moments capture the distribution of intensity values around the object's centroid [18]. A shape matching similarity algorithm can calculate the similarity of two shapes and operationalise a search for corresponding objects in a wider context. The algorithm utilises a representation of shape contours as sets of points, which are subsequently transformed into symbol and number strings through alignment and matching techniques [2]. They process images based on a density probability function which ignores scale and rotation [11]. Recently made available OpenCV shape matching algorithms can be used for efficient calculation of the similarity value between two polygons [2]. As computer vision technologies have become increasingly advanced, shape matching has been adopted in various fields, for instance, to recognise objects in automated production environments or to recognise road signs in driver assistance systems [17, 29]. The algorithms have been shown to produce efficient and accurate results in identifying similar elements within complex and variable environments.

Despite the widespread use of shape matching algorithms in science and industry, there have not yet been any studies to employ shape similarity analysis and computer vision towards urban morphology analysis. Related literature shows that morphological aspects of the environment and wayfinding are strongly related, however there is a gap in quantifying specific urban morphology parameters in relation to spatial behaviour, specifically orientation and wayfinding. Hence, our hypothesis is formulated as follows: The shape of isovist polygons derived from visibility analysis affects human orientation and wayfinding ability. Isovist polygons with homogeneous shapes increase the cognitive load required for spatial orientation and decision-making, resulting in a confusion perception. Conversely, irregular isovist polygons derived from distinct spatial layout can enhance wayfinding ability by providing memorable environment. Our study aims to develop an innovative method which uses shape matching algorithms to quantitatively analyse the similarity between isovist polygons, to quantify the homogeneity or uniqueness of environmental conditions in high-density environments. By using visibility analysis, it explores how people's orientation and wayfinding ability is influenced by the geometric properties of Hong Kong's high-density public housing estates. The study aimed at identifying positive examples of urban configurations and connectivity nodes which contribute to people's sense of place, safety and psychological comfort, which can lead to improved socialising, community interaction, health and well-being.

3 Methodology

Our study has focused on two Hong Kong public housing estates as case study sites. Both having been constructed recently under the 'site-specific' planning approach which incorporates environmental analysis and micro-climate studies [26]. The analysis of the estate layouts was done in four stages. First, the circulation routes in the estates were redrawn as axial-line maps, referring to space syntax theory to identify decision points at the intersections of circulations lines [13], which were used as vantage points for visibility analysis. Second, the isovist polygons obtained at these decision points were analysed with the OpenCV shape matching algorithm, to calculate the similarity values within this dataset and saved as a matrix for each estate. Based on these similarity values, we derived a 'dominance' and 'uniqueness' rate indicating which spatial layout occurs seldomly and which shape occurs frequently throughout the estate plan. To visualise the results, we plotted VGA Graphs to show the similarity values results on the master plan of each separate estate. Additionally, we compiled charts of the quantitative outcomes to evaluate the variation in data and compare the two case study sites. Third, we conducted separate virtual reality (VR) based experiments with volunteers, documenting their orientation and wayfinding ability in digital models of the two housing estate public spaces. Finally, outcomes from the previous processes were combined and compared, to assess the correlation between both assessment methods and verify the accuracy of the computational analysis method in prediction human perception.

3.1 Decision Points and Isovist Analysis

The detailed steps to analyse the estates' pedestrian networks, involved converting digital map information retrieved from external sources into axial-line maps. Within these axial-line maps, intersection points were identified as a split or intersection of circulation routes, which were labelled as decision points (see Fig. 1). These points were used as the vantage points for isovist analysis, which was conducted using commonly used plugins and components in Rhino 3D and Grasshopper. Every decision point was labelled by the 'point list' component.

The average visual acuity of a human is typically considered to be 20/20, which means that a person can see objects clearly at a distance of 6 m. This measurement is based on the Snellen eye chart, which is a standard tool for measuring visual acuity [3]. However, people's ability to sense the place extends beyond the limitations of visual acuity. Nonetheless, as the distance increases, their sense of place may decrease [28]. 50 m is a generally used buffer on spatial perception analysis, relating to the average distance at which people can clearly distinguish objects and other people [4, 32]. Hence, for the isovist analysis, we set a visual length limitation of 50 m. The isovist polygons obtained at each intersection were labelled similarly to the decision points.

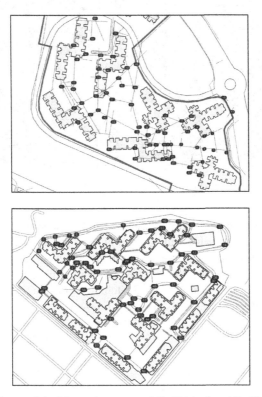

Fig. 1. Axial-lines and decision points maps of On Tai (up) and So Uk Estate (below)

3.2 Shape Matching and VGA Graph

After obtaining the isovist polygon shapes (see Fig. 2), the shape matching algorithm was employed using Python OpenCV, to calculate the similarity values between each isovist polygon shape and all other shapes in the entire estate. Lower values meant a

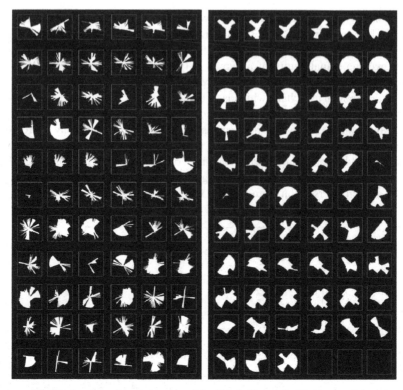

Fig. 2. Isovist polygon datasets of On Tai (left) and So Uk Estate (right)

high similarity of the polygon shapes, larger value the opposite. If two polygon shapes were exactly the same, the value returned would be 0. The similarity values used as the morphological similarity of the spatial layout in the context of isovist polygon shapes in housing estates. For each estate, the outcome is a similarity matrix $S = (s_{ij})$, $[s_{ij}]$. The code used to employ the OpenCV shape matching algorithm, return similarity values and write these values into an excel sheet is as follows:

```
import cv2 as cv
import numpy as np
from glob import glob
import pandas as pd
from pandas import DataFrame

path = glob('.\\image\\*')
p_num = len(path)
result_list = [ ]
for x in path:
    tag_list = [ ]
    for y in path:
        img1 = cv.imread(x,0)
        ret, thresh = cv.threshold(img1, 127, 255, 0)
        img2 = cv.imread(y,0)
        ret, thresh2 = cv.threshold(img2, 127, 255, 0)
        contours,hierarchy = cv.findContours(thresh,2,1)
        cnt1 = contours[0]
        contours,hierarchy = cv.findContours(thresh2,2,1)
        cnt2 = contours[0]
        ret = cv.matchShapes(cnt1,cnt2,1,0.0)
        # ret = np.float64(ret)
        tag_list.append(ret)
    result_list.append(tag_list)

result = np.asarray(result_list)
print(result)

data: DataFrame = pd.DataFrame(result)
writer = pd.ExcelWriter('isovist.xlsx')
data.to_excel(writer, 'sheet_1', float_format='%.5f', header=False, index=False)
writer.save()
writer.close()
```

After exporting the data to Excel, it was possible to evaluate the maximum, minimum, median, first quartile, last quartile and average values of each decision point and its comparison group. For a more descriptive analysis of location across each estate, we derived a 'dominance' or 'uniqueness' rate, which indicates if the shape occurs frequently or seldomly across the estate. Evaluating these outcomes could lead us to discuss whether and where the spatial layouts of the case study sites contain relatively repetitive or unique places. To identify similarities that are not an exact match but fall within a defined range of limited variation, we set a 'tolerance threshold', within which 'almost similar' were counted as similar ones. In the experiments presented in this paper, we used a tolerance threshold of 25% of the maximum value within the comparison group determined through pretesting of the algorithm and analysing the similarity values. There are two statements worth noting: Firstly, to gauge the uniqueness or dominance rate of a shape within a given environment, the average similarity value is not used directly, but rather a percentile-based measure is employed to provide a more equitable indicator of its spatial condition relative to others. Secondly, the tolerance threshold can be adjusted to suit specific contexts. The calculation method for dominance rate represents as follows:

In the 'A' estate, the similarity comparison matrix represents as $A = (a_{ij})$, $[a_{ij}]$. The elements represent the similarity value of isovist polygon shape i and isovist polygon shape j. The tolerance threshold value is denoted by a_0, representing the 25% quantile of all elements in the matrix A. Since a small a_{ij} means the isovist polygon shapes i is relatively similar to isovist polygon shape j. For each i-th row in the matrix A, the amount of elements that are less than a_0 is identified as the amount of similar shapes for the i-th isovist polygon shape, denoted by N_i. N provides the amount of all isovist polygon shapes obtained from decision points in 'A' estate. Therefore, in notation form the calculation for every shape's dominance rate is expressed as:

$$D_i = \frac{N_i}{N} \tag{1}$$

As a final step, a VGA graph representing the dominance rate results was generated on the axial-line map, to visualise the dominance or uniqueness of the place found at the various locations across the estates.

3.3   VR-Based Wayfinding Experiments

To conduct virtual reality-based wayfinding tests with volunteers, we used high-resolution digital photogrammetry models as well as 3D spatial massing models of the two case study estates obtained from the Hong Kong Planning Department model resources website, based on aerial photography conducted in 2017 and 2018. Virtual reality environments were composed using Blender Version 3.1, which allowed the experiments to be conducted in first person perspective. A total of five participants joined our experiment on a voluntary basis, after verifying that their medical history and vision were suitable to take part in extended virtual reality experiences. The participants were university students, who had not visited the case study site before.

The experiments were conducted in three stages, which included familiarisation, location recognition tasks within the VR environments and a brief interview. As part of the familiarisation, a predefined route through the estate spaces was shown. To conduct the location recognition task, participants were shown a virtual reality location and asked to mark their location on a separate map of the housing estate. The decision points were classified according to three categories, which were 'dominant place', 'intermediate place' and 'unique place'. The time that participants needed to point out each location, and how many locations were correctly marked was documented. In addition to these behavioural data, questionnaire results were obtained through a brief interview. Participants were asked to provide feedback on the experiment with the aims of obtaining qualitative information to expand and evaluate our findings through the following questions:

1. *Which method did you use for orienting yourself?*
2. *Are any parts of the environment that are easily recognisable or potentially confusing? Please point them out.*

The overall results of the experiments contain the behaviour performance accuracy scores and questionnaire responses for all participants, indicating the wayfinding difficulty of a series of selected locations across the two high-density estates.

4   Results

From the OpenCV shape matching algorithm returned results, we obtained different similarity values for all the decision points in the two estates written in Excel. Based on 66 and 63 decision points, a total of 4,356 and 3,969 results were obtained for On Tai and So Uk Estate, respectively. The code is executed within an Anaconda environment, utilising a Python 3.7 interpreter. The script's runtime is recorded as 56.39 and 50.15 s, respectively, which demonstrates the high efficiency of the algorithm in computing multiple similarity values simultaneously. To visualize the overall similarity values, we created a chart with the vertical axis representing the similarity values and the horizontal axis representing the labels of isovist polygon shapes (see Fig. 3). The folding lines in the graph indicate the maximum, third quartile, average, first quartile and minimum values for each isovist polygon shapes, from top to bottom. The graphs indicate that the overall similarity values from So Uk are lower than On Tai Estate, which implies that the spatial layout of So Uk Estate is more regular than On Tai Estate. Most of the results are staying at a low level, which means that both public housing estates contain a strong similarity in morphology. Also, comparing the maximum values identified in both estates, we can conclude that each estate contains only a few locations which have a high degree of uniqueness compared to the rest of the estate spaces. On Tai has a higher number of unique places (three major outliers, or five unique places depending on the threshold) than So Uk Estate (only one major outlier, or three medium to highly unique places). The result corresponds to our reading of the estate plans from the perspective of an urban designer, which shows that So Uk has many parallel spaces due to the fixed orientation and alignment of the tower blocks. On Tai, in contrast, has part of its layout containing more varied tower arrangements.

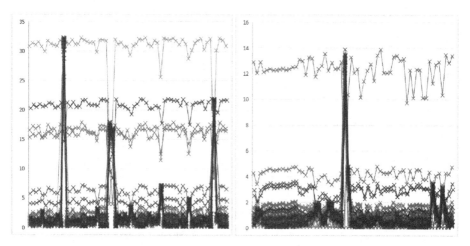

Fig. 3. Line chart of similarity values of On Tai (left) and So Uk Estate (right).

After obtaining all the similarity values for each estate, the dominance rate calculation was conducted. Firstly, our attention is drawn towards the dominance rate of the On Tai Estate, which is reflected by the blue line in Fig. 4. Inspection of Fig. 4 up reveals that Shape Label 6, 13, 24, 28, 29, 35, 41, 45, 54, 62, 63 exhibit a markedly low dominance rate, indicating the presence of distinct spatial layouts within the associated locations. Conversely, for instance Shape Label 7, 8, 9, 14, 27, 37, 49, 50, 53, 55, 56 display a comparatively high dominance rate, signifying the homogeneous morphology around the whole estate. Figure 4 shows the dominance rate comparison graphs which evaluate the differences between the two case study estates. Both graphs show that So Uk Estate's dominance values are significantly higher than On Tai, which indicates that On Tai's spatial layout has more diversity.

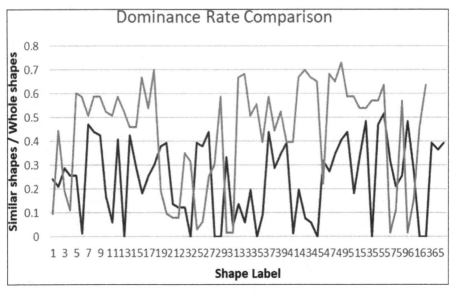

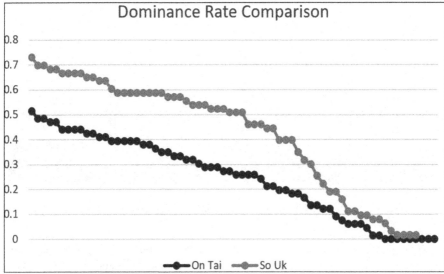

Fig. 4. Dominance rate comparison graphs according to label index (moving spatially across the estate) and in relation to sorted values (high-low)

After the descriptive analysis of the similarity values, we used VGA graphs to visualise the dominance rate the two case study estate spaces (see Fig. 5). Warmer (red)

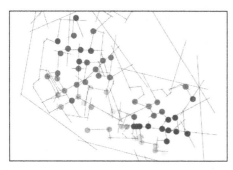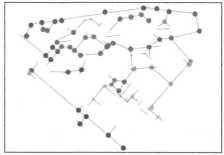

Fig. 5. VGA Graph plots of dominance and uniqueness shapes across On Tai (left) and So Uk Estate (right)

colours indicate that a location's isovist polygon occurs frequently in the whole estate; colder (blue) colours indicates that the isovist shape occurs seldomly. From the visualisations, we can see that the northern part of On Tai Estate's has more dominant polygon shapes, while the southern part's isovist polygon have uniqueness values. Reflecting on the tower block arrangement in the estate's master plan (see Fig. 1 up), we can see the spatial layout is more varied in the south but more regular in the north. In So Uk Estate, the northern part has more dominant spatial characteristics as well. Interestingly, although the southern part's building typologies are very similar to each other, the isovist polygon shapes have more variation here, as there are some small differences in the spatial layout which create diversity.

When evaluating the VR-based wayfinding experiments and feedback from our volunteer navigators, it becomes clear that the average 'thinking time' is highly varied across the different location categories of 'dominant place', 'intermediate place' and 'unique place'. The duration of cognitive processing is ranked by locality as follows: dominant places took the longest to identify, intermediate places took a middle position, and unique places took the shortest amount of time. Second, we found that the participant accuracy during location recognition tasks for dominant places is significantly lower than for unique places. Specifically, accuracy in identifying dominant places was found to be 10%, while accuracy in identifying unique places was 90%. Intermediate places fell in between with an accuracy rate of 60%. By comparison, the overall accuracy of location recognition tasks in On Tai Estate was 66.67%, with 10 places correctly identified, while the accuracy in So Uk Estate was 46.67%, with only 7 places correctly identified, which indicates that the wayfinding ability is generally better in On Tai than So Uk Estate. Notably, there were minimal accuracy differences observed in identifying dominant and unique places across On Tai and So Uk Estate. However, while all participants correctly identified intermediate places in On Tai, only one participant correctly identified intermediate places in So Uk Estate. These results suggest that the intermediate places in On Tai are more 'unique' than those in So Uk Estate.

Questionnaire responses were collected orally following the location recognition tasks. When asked about the navigation method they used, the participants all acknowledged the significant impact of spatial layout diversity on orienting. As one participant noted:

"When I navigate around the city, I often lose the signal between huge towers. So, to solve this problem, I simply look around and find a matching top view from Google Maps to figure out where I am. It's kind of like what we did in this experiment - I found the plot on the map with the same shape of the 'surroundings' to determine my 'current' location."

When asked about the easy recognised and confusing places, the participants all identified the dominant places as confusing places but unique places as relatively recognisable places. As one participant noted:

"It's so hard to orient in So Uk. The spatial layout is too regular there. This place (refer to dominant place) occurs a lot in the whole estate. On Tai is pretty cool with irregular public spaces."

5 Discussion

The results obtained during the testing of our methodology indicate the relative accuracy of our computational method in analysing repetitive or unique spatial characteristics in high-density urban neighbourhoods. In previous literature related to this topic, scholars have shown how monotonous urban layouts are associated with negative psychological effects including confusion, disorientation, impatience and stress during orientation and wayfinding. While being a relative small-scale experiment aimed at delivering a proof-of-concept application, our workflow reflects the feasibility of adopting computer vision methodologies towards the analysis of morphological aspects of public housing estates. From the results shown above, we can conclude that the OpenCV shape matching algorithm is efficient in finding similarity conditions within spatial layouts, which can effectively predict how people might experience difficulties with orientation and wayfinding. Through comparison with navigation experiments by volunteers, our hypothesis that dominant places have a negatively impact on people's sense of place is partially proven. Similarly, we found that unique places contribute positively to the cognitive processes of interpreting and negotiation complex urban layouts.

Based on the questionnaire responses associated with our wayfinding experiments, we found that people who are not familiar with the built environment rely heavily on the geometric properties of the public open spaces for successful wayfinding. Rather than using intuitive memorialisation of spatial features, people are forced to study, rationalise and memorize spaces to a higher level of technical detail, similar to the difficulties and abstract reasoning processes when people navigate through reading and interpreting maps. From these findings, it can be suggested that for high-density housing estates such as those built in Hong Kong, additional circulation pathways and unique identifiable public space design elements can be added offer spatial and visual diversity. As these would contribute to people's orientation and wayfinding ability, positive impacts on the usage intensity and socialising activities along estate circulation path could be expected, improving the overall walkability of public housing estates and their capacity to promote community making.

The study presented here has several limitations, which can be addressed in our future development of the methods and the scope of their testing. We acknowledge that isovist analysis only represent the 2D geometric environment of the public housing estates, which does not fully represent the complex visual qualities of Hong Kong's multi-layered public space networks. Improved and three-dimensional spatial analysis methods have recently emerged and could potentially be incorporated in our workflow. In addition, the methods of testing wayfinding in VR-based environments can be expanded by letting participants experience more comprehensive routes, to build up more complex cognitive maps of the various navigational possibilities.

Future research directions based on the methodology explored in this study could incorporate a range of other emerging research techniques, in particular to take advantage of the capacity of current computational infrastructure to process large datasets, making data-driven analysis tasks scalable and able to process high resolution of sampling data. Current research around Machine Learning tools for Street View Imagery can enable this method to be used in places where 3D models are not available. In addition to analysing urban morphology, it could be feasible to consider more detailed aspects of the urban spatial experience and perception, for instance in relation to facade materials, transparency and signage, public space greening and landscape features. The computational analysis and VR-based testing can also be employed as a community engagement tool to enable more user-centric or participatory design approaches.

6 Conclusions

This paper has presented a new methodology for quantifying aspects of spatial perception in high-density public housing environments, focusing in particular on the impact of repetitive environments on orientation and wayfinding. The study has shown the effectiveness of using the OpenCV shape matching algorithm as a tool to calculate isovist polygon similarity values, to analyse the degrees and distribution of spatial variation in high-density housing estates. From the shape matching algorithm's results, we derived dominance or uniqueness values that can be used to visualise the similarity in spatial layout conditions across housing estate spaces, which influence people's ability to navigate confidently through a series of decision points within the pedestrian networks. A virtual reality-based wayfinding experiment conducted in parallel has produced outcomes that support our findings. Our study found that the homogeneous environments and repetitive spatial configurations, resulting from standardised design approaches in building design and estate planning, can affect the human perception of public open spaces and reduce people's ability for orientation and wayfinding. From the analysis of our case study estates, we found that irregular and diverse tower block arrangements that produce uniquely shaped public in-between spaces help create more memorable public space environments and atmospheres, which can contribute to the sense of place, place attachment and community identity. Our methodology can help identify spatial arrangements that improve wayfinding, socialising and neighbouring in housing estate shared environments. Through this capacity, the approach can contribute to improved planning approaches aimed at increasing community engagement, openness, and interaction with surrounding urban areas, to promote the social integration and mobility of public housing residents in the context of liveable and inclusive cities.

References

1. Benedikt, M.L.: To take hold of space: isovists and isovist fields. Environ. Plann. B. Plann. Des. **6**(1), 47–65 (1979). https://doi.org/10.1068/b060047
2. Bradski, G., Kaehler, A.: Learning OpenCV: Computer Vision with the OpenCV Library. O'Reilly Media, Inc. (2008)
3. Colenbrander, A.: The historical evolution of visual acuity measurement. Vis. Impair. Res. **10**(2–3), 57–66 (2008)
4. Curtis, J.W., Shiau, E., Lowery, B., Sloane, D., Hennigan, K., Curtis, A.: The prospects and problems of integrating sketch maps with geographic information systems to understand environmental perception: a case study of mapping youth fear in Los Angeles gang neighborhoods. Environ. Plann. B. Plann. Des. **41**(2), 251–271 (2014)
5. Darken, R.P., Peterson, B.: Spatial orientation, wayfinding, and representation. In: Handbook of Virtual Environments, pp. 533–558. CRC Press (2002)
6. De Certeau, M., trans. by Rendall, S.: The Practice of Everyday Life. University of California Press, Berkeley (1984)
7. Deng, L., Romainoor, N.H.: A bibliometric analysis of published literature on healthcare facilities' wayfinding research from 1974 to 2020. Heliyon, e10723 (2022)
8. Dogu, U., Erkip, F.: Spatial factors affecting wayfinding and orientation: a case study in a shopping mall. Environ. Behav. **32**(6), 731–755 (2000)
9. Dong, W., et al.: Wayfinding behavior and spatial knowledge acquisition: are they the same in virtual reality and in real-world environments? Ann. Am. Assoc. Geogr. **112**(1), 226–246 (2022)
10. Emo, B., Dalton, R.C.: Wayfinding and spatial configuration: pedestrian behaviour at street corners. In: Kim, Y.O., Park, H.T., Seo, K.W. (eds.) Proceedings of the Ninth International Space Syntax Symposium, Seoul, South Korea (2013)
11. Flusser, J.: On the independence of rotation moment invariants. Pattern Recogn. **33**(9), 1405–1410 (2000)
12. Hegarty, M., Montello, D.R., Richardson, A.E., Ishikawa, T., Lovelace, K.: Spatial abilities at different scales: Individual differences in aptitude-test performance and spatial-layout learning. Intelligence **34**(2), 151–176 (2006). https://doi.org/10.1016/j.intell.2005.09.005
13. Hillier, B., Hanson, J.: The Social Logic of Space. Cambridge University Press, Cambridge (1984)
14. Hong Kong Housing Bureau. Housing in Figures (2022)
15. Hong Kong Transport Department. Walk in HK (2022)
16. Hu, M.K.: Visual pattern recognition by moment invariants. IRE Trans. Inf. Theory **8**(2), 179–187 (1962)
17. Huang, C., Chen, D., Tang, X.: Implementation of workpiece recognition and location based on opencv. In: 2015 8th International Symposium on Computational Intelligence and Design (ISCID), vol. 2, pp. 228–232. IEEE (2015)
18. Huang, Z., Leng, J.: Analysis of Hu's moment invariants on image scaling and rotation. In: 2010 2nd International Conference on Computer Engineering and Technology, vol. 7, p. V7-476. IEEE (2010)
19. Jacobs, J.: Death and Life of Great American Cities. Random House, New York (1961)
20. Kalantari, S., et al.: Evaluating the impacts of color, graphics, and architectural features on wayfinding in healthcare settings using EEG data and virtual response testing. J. Environ. Psychol. **79**, 101744 (2022)
21. Lidwell, W., Holden, K., Butler, J.: Universal Principles of Design (2003)
22. Lohman, D.F.: Spatial abilities as traits, processes, and knowledge. In: Advances in the Psychology of Human Intelligence, pp. 181–248. Psychology Press (2014)

23. Lynch, K.: The Image of the City. MIT Press (1964)
24. Meneghetti, C., Miola, L., Toffalini, E., Pastore, M., Pazzaglia, F.: Learning from navigation, and tasks assessing its accuracy: the role of visuospatial abilities and wayfinding inclinations. J. Environ. Psychol. **75**, 101614 (2021)
25. Montello, D.R., Sas, C.: Human factors of wayfinding in navigation (2006)
26. Ng, E.: Urban renewal and environmental design. J. Hong Kong Inst. Plan. **25**(1), 26–35 (2010)
27. Peponis, J., Wineman, J., Rashid, M., Kim, S.H., Bafna, S.: On the description of shape and spatial configuration inside buildings: convex partitions and their local properties. Environ. Plann. B. Plann. Des. **24**(5), 761–781 (1997)
28. Rijnks, R.H., Strijker, D.: Spatial effects on the image and identity of a rural area. J. Environ. Psychol. **36**, 103–111 (2013)
29. Russell, M., Fischaber, S.: OpenCV based road sign recognition on Zynq. In: 2013 11th IEEE International Conference on Industrial Informatics (INDIN), pp. 596–601. IEEE (2013)
30. Schwering, A., Krukar, J., Li, R., Anacta, V., Fuest, S.: Wayfinding through orientation. Spat. Cogn. Comput. **17**(4), 273–303 (2017). https://doi.org/10.1080/13875868.2017.1322597
31. Stamps, A.E., III.: Isovists, enclosure, and permeability theory. Environ. Plann. B. Plann. Des. **32**(5), 735–762 (2005)
32. Su, N., Li, W., Qiu, W.: Measuring the associations between eye-level urban design quality and on-street crime density around New York subway entrances. Habitat Int. **131**, 102728 (2023)
33. Tversky, A.: Features of similarity. Psychol. Rev. **84**(4), 327–352 (1977). https://doi.org/10.1037/0033-295X.84.4.327
34. Van Nes, A., Yamu, C.: Orientation and wayfinding: measuring visibility. In: Introduction to Space Syntax in Urban Studies (2021).https://doi.org/10.1007/978-3-030-59140-3_3
35. Veltkamp, R.C.: Shape matching: similarity measures and algorithms. In: Proceedings International Conference on Shape Modeling and Applications, pp. 188–197. IEEE (2001)

Author Index

M. Turrin et al. (Eds.): CAAD Futures 2023, CCIS 1819, pp. 671–672, 2023.
https://doi.org/10.1007/978-3-031-37189-9